The Dance Language and Orientation of Bees

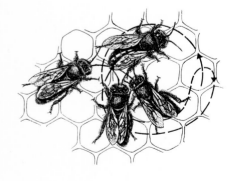

The Dance Language
and Orientation of Bees

Karl von Frisch / University of Munich

Translated by Leigh E. Chadwick

The Belknap Press of Harvard University Press
Cambridge, Massachusetts 1967

© Copyright 1967 by the President and Fellows of Harvard College / All rights reserved / Distributed in Great Britain by Oxford University Press, London / Library of Congress Catalog Card Number 67-17321 / Printed in the United States of America

Translated from *Tanzsprache und Orientierung der Bienen* (Springer-Verlag, Berlin, Heidelberg, New York, 1965).

This book was set in 10 pt. Times Roman directly on film by the Intertype Fototronic machine. Composition was by Graphic Services, Inc.

The book was printed by photo-offset by Murray Printing Co. and bound by The Colonial Press.

Book design by David Ford with page make-up by Graphic Services, Inc.

Preface

For more than 50 years the bees, in our laboratory and during vacation times on the Wolfgangsee, have been the favorite animals for my scientific work. Their color vision, their smelling and tasting and the relations of their senses to the world of flowers, their "language" and their capacities for orientation—this was the wonderland of puzzles that drew me ever onward. Little by little much that is new has been revealed from the treasure chest of their being. In the course of the years these findings have mounted ever higher, and have become more and more insistent in their demands for contemplation and review.

Thus I have had the intention of writing this book for a long time. But great gaps in our knowledge have been as distracting as the large white areas used to be for a person looking at ancient maps. The desire to present as complete a picture as possible constantly summoned one from the writing desk to experimentation, in order to gain a deeper insight—and each new insight raises new questions. The end does not come automatically. Ultimately one simply has to call a halt.

Many readers may wonder whether it is proper to call the communication system of insects a "language." The use that is made here of this word must not be misunderstood, as though what bees inform one another of were to be regarded as the equivalent of human speech. In its wealth of concepts and its articulate mode of expression the language of man stands on quite a different plane.

But G. Révész is unwilling to speak of a language among animals at all. In his view, doing so would require the demonstration that, with the object of communicating with their congeners, they use a system of sounds or gestures in such a way that "each sound or complex of sounds, every motion or complex of movements, carries a definite meaning." But, as we shall see, this requirement exactly fits the dances of the bees. Through their precise and highly differentiated sign language the bees are unique in the whole animal kingdom in regard to their ability to communicate. The use here of the term "language" is intended to give emphasis to what is extraordinary in the realm of animal behavior (Révész 1953; v. Frisch 1953; see also Lotz 1950, Kainz 1961).

Devoting to the bees' *orientation* a considerable part of a book about their "language" may also call for some justification. Bees seeking food may well fly out more than 10 km from the hive. For a creature 13 mm long that is a respectable distance. And yet at the end of such a flight the bees are able to find their way home and then back again to the identical flower. Beyond this they can give their hivemates so precise a description of the location of the feeding place that the latter, without having known the route previously, can steer correctly to the goal. Thus, bees have exceptional ability for orientation and at the same time the gift of transmitting to one another the data essential for finding the proper course. Hence, their "language" is intermeshed intimately with the means by which they find their way, and study of their ability to communicate has led of necessity to the problem of their orientation.

When nearly 45 years ago I reviewed the results of 3 years' work (v. Frisch 1923), I thought I knew the significance of the bees' dances. Following an interruption of 20 years the shortage of food during the Second World War became the occasion of taking the theme up again. We wanted to be able to tell the bees in their own "language" that they should gather food more intensively, and where they should fly. Then I became aware that previously I had missed the most interesting aspects. As the depth of knowledge grew, the yet unsolved problems also expanded. Many co-workers came to join in a common undertaking. What we found out during two decades of work is scattered through numerous publications, some of them difficult of access. A summary revealing the guiding thread of these investigations seems to me timely. It will make evident what has been clarified hitherto, and will point out gaps that future research may be able to close. The informed reader will also find many items not previously published.

I have endeavored to write in a manner understandable to all, and hope that even in our day, when the pressure is toward greater and greater specialization and one knows fewer and fewer leisure hours, the bee will be able to catch the interest of wider circles. She has been able to do so for millenia.

I am conscious of many shortcomings. I could hope for nothing better than that they should impel others to carry the work forward.

I owe thanks on many hands—first to all my students who by their doctoral dissertations have contributed one stone after another to the construction of this edifice; in temporal succession they are Werner Jacobs, Gustav A. Rösch, Fritz Rauschmayer, Herbert Baumgärtner, Ingeborg Beling, Berta Vogel, Elisabeth Opfinger, Oskar Wahl, Ruth Lotmar, Maria Hörmann, Elisabeth Kleber, Jesus del Portillo, Ilse Körner, Hans Engländer, Hildtraut Steinhoff, Martin Lindauer, Therese zu Oettingen-Spielberg, Hedwig Stein, Herbert Heran, Egon Palitschek, Josef Bretschko, Irmgard Kappel, Herta Knaffl, Wolfgang Schick, Wolfgang Steche, Therese Lex, Karl Stockhammer, Rudolf Boch, Karl Daumer, Ahmad-H. Kaschef, Wilhelm Fischer, Rudolf Jander, A. Ruth Bisetzky, Elmar Meder, Peter Görner, Lore Becker, E. Maria Schweiger, Una F. Jacobs-Jessen, Friedrich Otto, Ahmad Schah sei Djalal, Hannelore Dirschedl, Alexandra v. Aufsess, Johannes Schmid, Dieter Bräuninger. Many of them have continued their concern with bees beyond their doctoral programs. Ruth Beutler, whose thesis was in quite a different area, soon turned to the bees and has advanced our knowledge in matters of biological chemistry and flower biology. Max Renner joined our group, Harald Esch too came in, and Martin Lindauer became my extremely close colleague. His students (J. O. Nedel, H. Kiechle, H. Markl, W. Rathmayer, H. Martin, U. Maschwitz, V. Neese, B. Schricker) have contributed new findings, and thus a hopeful third generation continues research in the same direction. They will not run out of material for study.

Many students, technicians, and other people have been of valuable and indispensable assistance in the experiments. Ing. Richard Woksch frequently helped with technical requirements. Prof. Hans Benndorf (Graz), Prof. K. O. Kiepenheuer (Freiburg im Breisgau), and Prof. W. Rollwagen (Munich) gave me helpful advice and support in regard to questions in physics. With respect to matters of the beekeeper's art I was aided by G. Bamberger (Munich) and M. Ellmauer (St. Wolfgang), whose receptivity toward scientific problems and whose great practical experience worked to our advantage. With gratitude I recall also how through the years Prof. K. v. Goebel, Prof. O. Renner, and Prof. L. Brauner, as directors of the Munich Botanical Gardens, showed great understanding in giving us expensive plants and how many experiments they put up with in the garden precincts. Chief Forester F. Promberger allowed us to set up hives in the Forstenrieder Park near Munich and to drive cars on the forest roads. In this way he made a valuable experimental territory available to us. Also we are indebted to the Bavarian Department of the Interior and its State Patrol.

I thank Dr. H. Fernandez-Moran, Dr. T. H. Goldsmith, and Dr. J. J. Wolken for kindly providing some electron micrographs for this book.

I am indebted to my former students Prof. M. Lindauer and Dr. R. Jander for looking over the manuscript and for many valuable suggestions.

The German Society for the Encouragement of Scientific Research (*Deutsche Forschungsgemeinschaft*) has repeatedly supported the studies with grants as well as fellowships for co-workers. Finally, my especially warm thanks are due to the Rockefeller Foundation, which for so many years has generously removed the financial obstacles to a free, unworried planning of the experimental program.

K. v. Frisch

Munich, March 1965

Contents

PART ONE: THE DANCES OF BEES

I / Historical   3

II / Methods in General   7

    1. The observation hive   7
    2. Heatable observation hives   12
    3. Rooms for bees   13
    4. Marking the bees with numbers   14
    5. We set up an artificial feeding place   17
    6. Automatic recording of visits to the observation place   20
    7. Cleaning the equipment; scents as sources of error   22
    8. How bees are put to work or brought home   23
    9. Measurement of the tempo and direction of dancing   23
   10. Selection of the bees   27

III / The Round Dance as a Means of Communication When Nectar Sources Are Nearby   28

    A. The objective is known to the bees informed   28

    1. The scouts   28
    2. The round dance   29
    3. Contact without dancing can also be effective with other members of the same group   30
    4. Diffuse information given when bees are fed at sugar-water dishes   30
    5. Distinct information given when bees feed at flowers   31
    6. Odors as a means of communication   33
    7. The pollen collectors   34
    8. The dance floor   36
    9. Grouping according to odor   37
       SUMMARY   42

B. The objective is not known to the bees informed   43

10. Recruitment of additional helpers   43
11. New forces are aroused only when needed   44
12. How do the recruits find their objective?   46
13. The odor of the food sources as a guide to the newcomers   46
14. Experiments with flowers   47
15. The adhesiveness of odors to the bee's body   49
16. Absence of communication about colors and shapes   50
17. The role of the scent organ   50

SUMMARY   55

IV / The Tail-Wagging Dance as a Means of Communication When Food Sources Are Distant   57

1. Description of the tail-wagging dance   57
2. The transition from the round dance to the tail-wagging dance   61
3. Comparison of nectar and pollen collectors   62

SUMMARY   63

A. The indication of distance   64

4. The tempo of the dance   64
5. The influence of internal factors on the dance tempo   70
6. The influence of external factors on the tempo of dancing   75

   a. Temperature   75
   b. The wind   79
   c. Gradient of the flight path   81
   d. Pharmaceutical agents   82

7. How accurately can newcomers follow the distance indications? Stepwise experiments (Stufenversuche)   84
8. What part of the tail-wagging dance is the signal that defines the distance?   96

   a. The components of the tail-wagging dance and their correlation with distance   97
   b. Experiments to vary components of the dance independently   104
   c. Comparison of the precision of searching and the accuracy of distance indication   106

9. How does the dancer gauge the distance?   109

   a. Bees do not signal the absolute distance to the goal   109
   b. The indication of distance is not based on the duration of the flight   113
   c. The expenditure of energy as a measure of the distance   114

10. The significance of the outward and the homeward flight in the indication of distance   116
11. The shape of the curve for distance   121

SUMMARY   126

B. The indication of direction   129

12. First hints of the mode of indicating the direction of the goal   129
13. The indication of direction on a horizontal surface   131
14. The indication of direction on the surface of a vertical comb   137
15. Dances on an oblique comb surface   146
16. Individual differences in the indication of direction, and the influence of age   149
17. Comparison of the effects of round dances and tail-wagging dances   149
18. How precisely is the indication of direction followed by the newcomers? Experiments in a fan-shaped pattern   156
19. Dances when the sun is in the zenith   162
20. No indication of direction upward or downward   163
21. The significance of the outbound and homebound flights for the indication of direction   169
22. Detour experiments   173

    a. First observations and preliminary experiments   173
    b. Experiments on the Schafberg   174
    c. Experiments with Italian and Indian bees   178
    d. The biological aspect   182
    e. Detour experiments with bees on foot   183

23. The indication of direction in a crosswind   186
24. "Misdirection"   196

    a. Light-dependent "misdirection"   197
    b. "Misdirection" due to the force of gravity ("residual misdirection")   204

25. The role of the scent organ and floral odors with distant sources of food   222
26. We look for a feeding station from directions supplied by the bees   227

    SUMMARY   230

V / Dependence of the Dances on the Profitability of Foraging Activity   236

1. Factors determining the release and liveliness of the dances   236

    a. The sweetness of the sugar solution   236
    b. The purity of the sweet taste   238
    c. Ease of obtaining the solution   238
    d. Viscosity   239
    e. Load   239
    f. Nearness of the food source   240
    g. Floral fragrance   241
    h. Form of the food container   242
    i. Uniform flow from the source of food   243
    j. General status of nourishment in the colony   243
    k. Improvement of the food   244
    l. Time of day   244
    m. Weather   245

2. Regulation of supply and demand on the flower market   246
3. The clocks of bees and of flowers   253
   SUMMARY   255

VI / Guidance by Scent   257

1. Historical aspects   257
2. Methods   258
3. Results   261
4. Verification—but no useful application   262
   SUMMARY   264

VII / Application of the Dances to Other Objectives   265

1. Water   265
2. Bee glue (propolis)   268
3. Dwellings   269
   SUMMARY   276

VIII / Other Dance Forms   278

1. Jostling run, spasmodic dance, and sickle dance   278
2. The buzzing run   279
3. Grooming dance (shaking dance)   280
4. Jerking dance (D-VAV)   281
5. Trembling dance   282
   SUMMARY   283

IX / Danceless Communication by Means of Sounds and Scents   285

1. Sounds   285
2. Odors   288
   SUMMARY   291

X / Variants of the "Language of the Bees"   293

1. Racial differences ("dialects")   293
2. Differences among species; the Indian bees   300
3. From primitive to successful messenger service with the stingless bees (Meliponini)   306
4. A brief glance at other social insects   313
   SUMMARY   317

XI / Phylogeny and Symbolism of the "Language of the Bees"   321

SUMMARY   327

PART TWO: THE ORIENTATION OF BEES
ON THE WAY TO THE GOAL

XII / Orientation on Long-Distance Flights   331

A. Landmarks   331
B. The sun as a compass   333

1. Displacement experiments   334
2. Competition between the celestial compass and landmarks   339
3. The contribution of the time sense to orientation, and knowledge of the sun's course   347
4. Perception of the sun through a cloud cover   366

   SUMMARY   377

C. Orientation by polarized light   380

5. The polarized light of the sky   381
6. Demonstration of orientation by polarized light   382
7. The connection between the polarization pattern and the position of the sun. Experiments in the shadow of a mountain   392
8. The use of artificial polarization patterns when the sky is cloud covered   395
9. The relative significance of the sun and polarization of the sky   398
10. What color range is effective in the perception of polarization?   401
11. What degree of polarization is needed for orientation?   403
12. On the function of the bees' ocelli   404
13. Spontaneous orientation relative to the plane of vibration of polarized light   407
14. Is perception of polarization direct or indirect?   409

    a. Partial elimination of the eyes   410
    b. Alteration of the reflected pattern   412
    c. Indirect and direct orientation in other animals   414

15. The analyzer for polarized light   415

    a. Is the analyzer in the dioptric system?   415
    b. The radial analyzer in the insects' ommatidia   416

16. Structure of the visual rods and perception of polarized light in other groups of animals   429

    SUMMARY   434

D. A glance at other animals   438

17. Orientation to the plane of vibration of polarized light   439
18. The celestial compass   444

    a. Arthropods   444
    b. Vertebrates   451

19. Orientation to a magnetic field   459

    SUMMARY   463

XIII / Orientation When Near the Goal 465

 A. The orientation flights 465
 B. Optical orientation nearby 471

  1. The bees' color sense 471
  2. Form vision 478
  3. Vision in bees and the appearance of flowers 481

 C. Orientation nearby by means of the sense of smell and taste 491

  4. Olfactory discrimination in bees 492
  5. The location of the sense of smell 495
  6. The capacity to localize by smelling 499
  7. The bees' olfactory acuity 502
  8. The biological significance of floral odor 507
  9. The sense of taste 511
  SUMMARY 519

Retrospect 524

References 527

Index 557

Part One / The Dances of Bees

I / Historical

In his book about the life and nature of bees (*Das Leben und Wesen der Bienen*) H. v. Buttel-Reepen (1915:190ff) found it necessary to begin the section about their abilities in communication with the confession that here one could speak only of "possibilities and probabilities." That the members of a hive are able to communicate in some way or other has been known for a long time. Every beekeeper is aware that after a short interval numerous visitors appear on any honeycomb that is discovered anywhere by a single bee. Also in a large beekeeping establishment one may occasionally see that during many hours thousands of forager bees may land with heavily laden pollen baskets at a single hive, evidently because early in the day an individual scout from this hive had discovered a productive source of pollen and had spread the news among her comrades (v. Buttel-Reepen 1915:171). Bees of adjacent hives may simultaneously be bringing in pollen of a different color from other flowers that were found by *their* scouts.[1] How the information is passed on was unknown. V. Buttel-Reepen thought communication by sounds the most important, because the practiced ear could very readily distinguish the tone of a swarming hive from the "stinging tone" of irritated bees, from the "hunger tone" of a starving hive, from the cozy buzzing that follows a good period of food collection, or from the wailing sound characteristic of queenlessness. That these sounds are heard by the members of the hive had already been doubted by August Forel (1910:309–310). The relatively loud "piping" and "quacking," by means of which a young queen seems to communicate with her rivals that are about to emerge in the hive, could indeed be regarded as serious evidence in favor of the perception of sounds. A. Hansson (1945) has shown that this is mediated by the sense for perception of vibration. There is response only when there is contact with the substratum and only at a short distance. Tones of different pitch are not distinguished. Free-flying bees never react to sounds. V. Buttel-Reepen's assumption that bees possess a differentiated audible language thus has not been substantiated, even though recent discoveries indicate that perception of vibration is more important than it long seemed.

[1] In the collection of nectar, too, different populations from the same large establishment sometimes select quite different kinds of flowers (Maurizio 1962).

F. W. Sladen (1902), who in a little-noticed article called attention for the first time to the use of odor as a signal, recognized the significance of the olfactory sense in communication. Bees fanning with abdomen raised in front of the hive entrance ("sterzeln") and thereby attracting comrades are a well-known sight to the beekeeper. Sladen noticed that when doing this the bees protrude near the tip of the abdomen an integumental fold that releases a scent perceptible to us also, and that it is not their "joyful buzzing" but rather this odor, dispersed by fanning the wings, that is the lure. Later, without knowing of Sladen's work, I came to the same conclusion.

The stimulus for me to work with these problems was provided by experiments on color vision and on the sense of smell of the bees. When I wished to start an experiment I would set out on a table in the open a sheet of cardboard with some honey on it. As a rule this was found after a few hours by one bee. Then their number would grow swiftly to dozens, or even to hundreds. A further phenomenon too spoke emphatically in favor of communication among the hivemates. I trained bees to collect from a dish filled with sugar water. When the feeding dish had been emptied I paused in order to limit the number of collectors. At first large numbers of bees swarmed about the empty dish, but gradually they dispersed and after about 20 min they visited only sporadically. But if one of these now found the dish refilled, the others reappeared in swift succession after her return home.

I was curious to know how the news was spread about there at home, and built an observation hive with glass windows. It is described in the next chapter.[2] In the spring of 1919 I set it up in the old Klosterhof of the Zoological Institute in Munich, erected a feeding station beside it, and marked the forager bees with a spot of red pigment on the thorax. After a pause in the offering of food, they would sit among the others on the comb near the hive entrance. The next scout found the dish refilled. It was a fascinating spectacle when after her return home she performed a round dance, in which the red-spotted bees sitting nearby showed lively interest. They tripped along after the dancer, and then left the hive to hasten to the feeding station. Soon it became apparent that the circular running is a dance of invitation, which not only recalls the former collecting group to action but also recruits new members to strengthen the working party (v. Frisch 1920).

With pollen collectors that were returning home with filled pollen baskets from natural sources of provisions, I saw another form of dance, the tail-wagging dance, and fell into the error of thinking that the round dance was performed when sugar water or nectar was collected and the tail-wagging dance after pollen collecting (v. Frisch 1923). Henkel (1938) refuted this. Under natural conditions he observed with

[2] I am indebted to J. B. S. Haldane for pointing out that Pliny mentions an observation hive with windows of horn, in which the emergence of bees could be observed (*Natural History*, XI, 16, 49). About 1740 Réaumur described observation hives of his own construction, quite similar to ours (see Büdel 1957). Such hives also served blind Franz Huber (1814, 1856) in the fine discoveries he was enabled to make through the eyes of his gifted assistant Franz Burnen. Since then many beekeepers and amateurs have certainly constructed something similar. The mating boxes customarily used in rearing queens are also useful in many observations.

nectar collectors tail-wagging dances that did not differ in form from those of pollen collectors, and explained the round dances of the sugar-water collectors as due merely to the unnatural abundance of food at my artificial feeding stations. On the basis of new experiments I at first held to my conception (1942). Today we know that Henkel was right when he described tail-wagging dances performed by collectors of nectar, but I was right too in describing their performance of round dances. I was wrong when I regarded the round dances as dependent on the gathering of nectar, and he was wrong in ascribing them to the abundance of food. The clarification came when I gave my co-worker Ruth Beutler a piece of bad advice. She was running a feeding station with the odor of thyme 500 m away from a beehive and wanted to have bees gather quickly around a sugar-water dish at a place nearer the hive. I advised her to feed them well at the 500-m station and also to put out a sugar-water dish with thyme fragrance at the desired place near the hive. The hivemates would be stimulated by the round dances of bees harvesting from the distant point to search first nearby around the hive and would necessarily find the new feeding dish quickly. There was no success. Did the distance of the feeding place influence the manner of dancing?

Experiments directed to this point showed in fact that round dances were performed with sources of food nearby, tail-wagging dances with more distant ones, by collectors of nectar just as by collectors of pollen, and that the tail-wagging dances announced also the direction and distance of the goal. The mistake had come from my setting up the artificial feeding station with its sugar water in the immediate neighborhood of the hive, in order to keep both feeding station and honeycomb in view, whereas the pollen collectors were coming from natural, more distant sources. Under these conditions there were only round dances among the bees collecting sugar water, and only tail-wagging dances among the pollen collectors. In addition, the pollen collectors kept on dancing after their pollen baskets had been emptied. It was natural to interpret the tail-wagging dances of nectar collectors arriving from natural, distantly located flowers as the pollen dances of bees that had already scraped off their loads. But now one miracle after another was unveiled. Probably bad advice has rarely been so nobly rewarded.

There remains to be mentioned the fact that the bee dances have been discovered repeatedly. W. Park observed round dances and tail-wagging dances before we knew of one another. In these dances he recognized a method of communication (1923a, b). He also reported, probably as the first, the same dances in water-collecting bees. Like Henkel he could find no difference in the way of dancing between bees collecting pollen and those collecting nectar. Instead of assuming that I was mistaken, he was so kind as to express the suspicion that the bees in Germany must dance differently (1923c).

Neither Park nor I knew that 100 years earlier Unhoch (1823:115) had already given an excellent description of the round dance:

> To many it will seem ridiculous or even incredible if I mention that bees too, when the hive is otherwise in good condition, indulge

in certain pleasures and jollity, and that at times they even set about a certain dance after their fashion . . . Without warning, an individual bee will force its way suddenly in among 3 or 4 motionless ones, bend its head toward the surface, spread its wings, and shiver its upraised abdomen a little while, the adjacent bees do the same, bend the head toward the surface, finally they twist and turn together in something more than a semicircle now to the right then to the left five or six times, and execute what is a genuine round dance. All at once the dance mistress leaves them, joins some other sedentary bees on another spot, and again does the same as the first time and the adjacent bees dance after her. The dance mistress often repeats her dance four or five times at different places. I have frequently demonstrated this to several friends of mine, who were greatly astonished and laughed heartily at it . . . What this dance really means I cannot yet comprehend; whether it perhaps expresses some mood of joy and merriment among them or whether it occurs for some other as yet unknown reason the future will have to teach, and with this ballet of the bees I now conclude my first part.

Apparently it was unknown to Unhoch—as it was to me—that 35 years previously pastor Ernst Spitzner (1788:102) had not only observed this "ballet of the bees," but in essence had interpreted it correctly:

When a bee has come upon a good supply of honey anywhere, on her return home she makes this known in a peculiar way to the others. Full of joy she twirls in circles about those in the hive, from above downward and from below upward, so that they shall surely notice the smell of honey on her; for many of them soon follow when she goes out once again. I observed this in the glassed hive when I put some honey not far away on the grass and brought only two of the bees to it. In a few minutes, since these had made it known to the others in the manner described, they came in great numbers to the place.[3]

There is even a reference, to which J. B. S. Haldane called my attention, in Aristotle:

On each trip the bee does not fly from a flower of one kind to a flower of another, but flies from one violet, say, to another violet, and never meddles with another flower until it has got back to the hive; on reaching the hive they throw off their load, and each bee on her return is followed by three or four companions. What it is that they gather is hard to see, and how they do it has not yet been observed.[4]

None of the observers named pursued his discovery further. Thus they remained in ignorance of what will now fill the pages of this book.

---

[3] After my first communication on the subject (1920), Manger (1920) brought to my attention the reports of Unhoch and Spitzner.

[4] Aristotle, *Historia Animalium*, IX, 40, Becker 624b; modified from the translation by D. W. Thompson in *The Works of Aristotle* (Oxford English edition; Clarendon, Oxford, 1910). [In the German version of the present book the citation is Aristotle, *Tierkunde* (ed. P. Gohlke, Paderborn, 1949), IX, 624b, p. 423.]

II / Methods in General

1. THE OBSERVATION HIVE

During the first years of experimentation I used a *large* observation hive, with 6 adjacent combs (each 22 × 38 cm), that can hold a bee colony of normal strength; and a *small* observation hive for two combs of the above size or four half combs. I described and figured both in 1923 (pp. 12-19). The large hive is particularly impressive for demonstrations. It is most appropriately used in a fixed location. But often in our experiments moving the hive was necessary, and the little hive is better suited for this. There was no suggestion of abnormal behavior resulting from the smaller populations, which after all ran to several thousand individuals. According to Gontarski (1949:311), in a small population about 500 members are enough for developing the normal social organization. Geschke, with small populations of 500-1000 bees, found the same division of labor as in normal colonies (cf. Hoffman 1961).

Based on many years of experience, our observation hive today appears as shown photographically in Fig. 1 or in the sketch of Fig. 2. The important feature is that, in contrast with customary beehives, the combs do not conceal one another; they are placed so that one may see the whole extent of each through glass windows on either side of the hive. This accounts for the flat shape of the hive. Of course this is unfavorable for maintaining the proper temperature. Hence the hive, when not under observation, is protected by wooden covers insulated on the inner surface with cotton wool or similar padded material (Fig. 2b). During cool weather and especially at night, a thickly padded outer covering was very useful (Fig. 2a); this was drawn over the hive like a tea cozy. Double windows serve further to prevent overcooling. The outer pane, the frame of which is hinged to the framework of the hive, can be swung open during warm weather like a door. The inner pane is held fast by wooden pins or fasteners.

The hive entrance is on one of the narrow sides. Arriving bees are forced by means of an adjustable wedge (Fig. 2c) to run to the right or to the left as may be desired. When observations are made from one side continuously, as is customary, the wedge is set so that all incoming bees have to turn to that side and none can be overlooked. In order

FIGURE 1. (*a*) The small observation hive (closed) in a demountable roofed stand, showing the flight funnel in front of the hive entrance. (*b*) The protective covering has been removed and the two combs, one above the other, are visible. The hinged cover with its outer glass window is open; it can be taken off its hinges. Photograph by Dr. Renner.

even more surely to keep in view all bees and their dances, care must be taken in setting the hive up that the combs fully fill their wooden frames. Any gaps that exist—especially in the lower part of the combs—must be carefully closed off with bits of wax, or otherwise many homecoming bees will shift over to the other side, and all of them can no longer be watched by a single observer. As a matter of fact, the inhabitants soon construct new passages to the other side of the comb, but these are mostly at a higher level where they do not interfere.

The size of the wooden form (Figs. 1*b*, 2*a*) into which the two comb frames are fitted should be proportioned to the natural comb dimensions of the colonies from which the hive is populated. Since the forms do not always match precisely the measurements thus dictated, in constructing the hive one leaves room for a little play and then makes the frames tight, if they are loose, by pushing in wooden wedges. The distance between opposite inner surfaces of glass is 4.6–4.8 cm. *This distance is important.* If the interval is less the bees cannot move freely between comb and window; with a greater separation they are able to sit on top of one another or will build with wax on the pane, both of which interfere with observation.

Outside the entrance I usually place a funnel-shaped passage (Fig. 1) that has sides and top of glass. Bees arriving and departing customarily run some distance in it on foot, and this facilitates observation of marked individuals.

Beneath the hive two wooden supports are placed (Fig. 2*a*), to prevent the hive from tipping over. These can be screwed to the baseboard, and turned in under the hive when it is moved. A handle on top (Fig. 1*b*) makes carrying easier. During transportation sufficient ventilation must of course be supplied: the entrance and the food opening on top, which otherwise is closed off with a cork (visible in Fig. 1*b* between the handle and the hand), are barred with screening. When it is necessary to supply the colony with food, one may insert into the food opening a bottle with a glass outlet tube that dips into a metal or plastic food trough (2 in Fig. 3*a*). When nectar is in good supply, the honey cells of the observation hive are relatively soon filled, because of the limited space. The result is a decline of the foraging activity and the mood for dancing decreases. Then the only recourse is to remove a full comb and replace it with an empty one.

In general the bees soon get accustomed to light, even of long duration. But if it happens that they run about rapidly and disorientedly on the inner pane of glass, they can be calmed quickly by darkening them temporarily (but in this connection see p. 27). Direct sunlight has a disturbing effect after a little while and is therefore to be avoided.

The observation hives are not suitable for overwintering. In the late fall we either put the bees together with other colonies or else transfer them to a hive with a normal arrangement of combs.

Quite often in our experiments we wanted to study the behavior of the bees on a comb in a horizontal position or more or less inclined. For this purpose, an observation hive that can be tilted is used. This hive is borne at some distance from the baseboard by two short brass

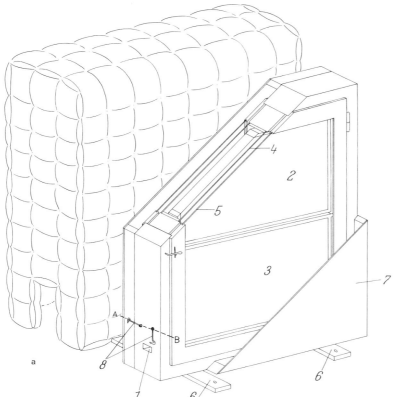
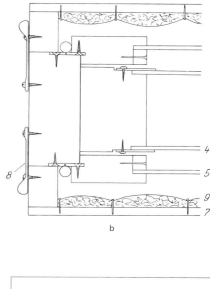
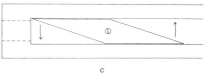

FIGURE 2. (*a*) Observation hive (oblique cutaway), with protective hood in background: 1, flight entrance; 2 and 3, upper and lower combs (only the wooden frames are shown); 4 and 5, inner and outer glass windows, the latter in a hinged frame that can be opened and taken off its hinges; 6, wooden supports (capable of being rotated); 7, protective wooden cover (only one corner shown, its padding not included); 8, hook for fastening the protective cover. (*b*) Section corresponding to line *AB* in Fig. 2(*a*): 4 and 5, inner and outer glass windows; 8, hook for fastening the protective cover 7; 9, padding of the protective cover. (*c*) A movable wooden wedge inside the hive entrance guides incoming bees to the desired side of the comb.

tubes (diameter 4 cm) that are fastened to the front and rear narrow ends of the hive and fit into suitable holes in two wooden blocks that are screwed fast to the baseboard (Fig. 3). The hive entrance lies inside the front brass tube, and the hive is held in position by a lateral iron support one end of which is screwed to the baseboard, the other end being fastened with a wing nut to the framework of the hive. After this nut has been loosened the hive can be set horizontally or at any desired angle and held in place by tightening the screw (Figs. 3*b* and 4). A pointer attached to the hive and a scale on the supporting block permit precise measurement of the angle.

The bees' dances are not only disturbed by direct sunlight but distorted by the sight of blue sky (pp. 197ff). Therefore the sky must be screened off during observations. This can be done with a canvas tent. Sunny spots must be shaded by placing pieces of canvas or the like so

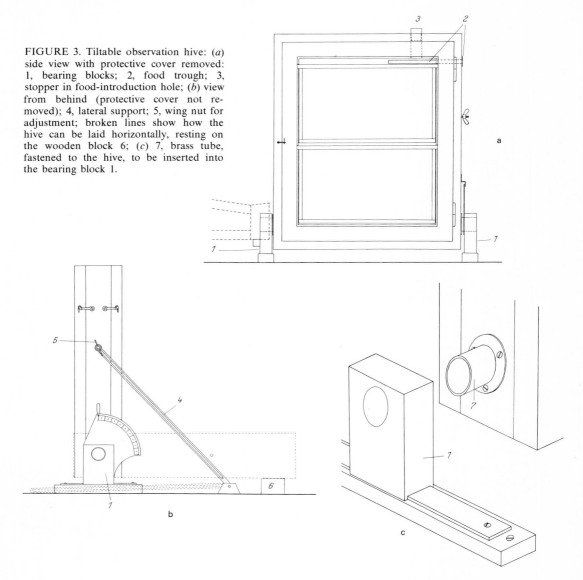

FIGURE 3. Tiltable observation hive: (*a*) side view with protective cover removed: 1, bearing blocks; 2, food trough; 3, stopper in food-introduction hole; (*b*) view from behind (protective cover not removed); 4, lateral support; 5, wing nut for adjustment; broken lines show how the hive can be laid horizontally, resting on the wooden block 6; (*c*) 7, brass tube, fastened to the hive, to be inserted into the bearing block 1.

as to afford and maintain even, diffuse light within. This is somewhat troublesome. Consequently, as a rule we use a hut made of sheets of soft wallboard (which can be cut easily to proper size) or of hard wallboard (which has the advantage of weathering better). The sheets are pushed into grooves at the side of the observation hut. On *one* side closure is provided by a shed—collapsible for the sake of easier transportability—that is covered by a horizontal panel and allows the observer sufficient room (Fig. 5). Light coming from beneath the baseboard and under the gable above ensures an adequate diffuse illumination, which at need can easily be made very uniform by shading with cloths. I have also used artificial red or white light within, but have preferred diffuse daylight. At times other than those of observation, the

FIGURE 4. The tiltable hive in an oblique position, with the protective cover removed from the upper side.

observer's shed is replaced by a fourth lateral sheet and a covering panel above; these provide additional temperature protection for the hive.

This observation hut too has undergone many changes and improvements. Under some conditions it has been necessary to allow the bees a free view of the entire sky. For this reason, I use an observation stand to the four legs of which only the baseboard is firmly attached; it thus makes a solid table, while the posts for the protective roof are attached laterally to the table legs and may be easily removed together with the roof (Fig. 6).

Another variation of the observation hut has also been useful, but is somewhat more difficult to construct and to transport (Fig. 7). It consists of three parts. The base is as high as the table on which the observation hive stands (Fig. 7a). The middle part, which fits into grooves in the base, is made of sheets of hard wallboard, like the base, and it is closed above by means of a roof that extends beyond its walls. The interior affords room enough for the observer, who can climb in and out through a side door in the middle portion (Fig. 7b). On the other side, opposite the observer's place, shutters and curtains are provided in the base and middle portions; thus a diffuse illumination can be supplied within when the shutters are open. Inside there is room enough to allow the hive to be tilted. If one wishes to give a free view of the sky one need only remove the two upper portions.

During observations in the hut, light-colored clothes or a white shirt are distracting. A black apron, reaching to the neck, does not reflect light and therefore allows a clearer view, and any distortion of the bees' behavior is avoided.

FIGURE 5. Observation hive in a hard-wallboard hut: F, flight funnel; the observer's hut is in the foreground. Photograph by Dr. Renner.

12 / The Dances of Bees

2. Heatable Observation Hives

A bee colony is able to regulate the internal temperature of the hive with astonishing precision. Maintenance of warmth during cool weather is rendered more difficult for them in the flattened observation hive, but can be facilitated by using a heatable observation hive. Such a hive was essential also for experiments in which the hive temperature was to be altered at will. This may be accomplished in many ways. One may select a simple or more complicated system according to the precision desired. Two models found useful are described below.

Figure 8 shows an arrangement designed by M. Lindauer. The observation hive stands in a protective box with sliding glass windows. The space surrounding the hive is heated by an infrared radiator U and temperature is held constant with a thermostat T.

More precise temperature regulation is permitted by an electrically heated observation hive (Fig. 9). The bees are guided from the entrance to the side of the comb facing the observer, where they perform their dances on the lower part of the comb. The cells are removed from the reverse side of this comb in order to increase the speed of heating. The heating element (chrome nickel wire) runs in pyrex-glass tubes J. A resistance regulates the amount of heating. The bees are kept away from the heater by means of wire netting. Cooling on warm days was achieved by means of an air stream from the laboratory compressed-air system; this air was fed downward through the comb passages from porous bubblers of the sort used for aerating aquaria. Mercury thermometers were built into the lower part of the comb for temperature measurements, and thermoelements for registering temperature changes of short duration. Details are to be found in Bräuninger (1964).

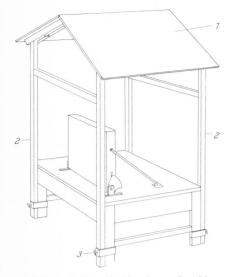

FIGURE 6. The tiltable observation hive in the observation stand, the lateral and upper panels of which have been removed: 1, roof, removable in two pieces; 2, roof supports; after the screw fasteners 3 on the table legs are loosened, the roof supports can be lifted out so that the view of the sky from the hive is unobstructed in all directions.

FIGURE 7. Three-piece observation hut of hard board: (*a*) table with observation hive, in the lower part of the hut, (*b*) the middle part of the hut and the roof placed on the base; the door is opened for the observer to enter; the shutters are closed.

3. Rooms for Bees

Some experiments require that the observation hive be set up in a closed room. At one time I used the spacious Winterhalle of the Munich Botanical Garden (Fig. 10), which stood empty during the warm season, with good results. After the hive was set up, the older bees did indeed try to get out through the skylights and were lost, but the younger ones became properly oriented during their preliminary flights and grew accustomed to the unusual circumstances.

Even in the open countryside one can install the observation hive in a screened cage. A framework of wooden strips, about 3 × 4 m and 2 m high and covered with wire screen, is satisfactory for many purposes.

The autumnal temperature decline puts an often unwelcome end to experiments with bees. But in the bee rooms designed and tested by M. Renner (1955, 1957) one can experiment with them even in winter. Besides, conditions can be kept quite constant there. One of Renner's observation colonies lived for 5 years in excellent condition in a closed space only 7.2 m long, 2.35 m wide, and 2.70 m high, produced brood uninterruptedly summer and winter, and even reared new queens. The key to such success lies in ample light and warmth and in sensible feeding.

The walls of the bee room (Fig. 11) are whitewashed, and are painted with black figures for the better orientation of the bees. An illumination of 500–1000 lux at 1 m above the floor should be provided by fluorescent tubes on the ceiling; their light can be diffused by means of a layer of transparent paper stretched beneath them, but this is not essential. The room temperature must be held at about 28°C, if neces-

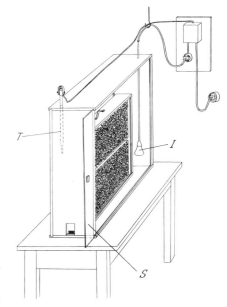

FIGURE 8. Heatable observation hive: S, sliding window of the outer box; I, infrared bulb; T, thermometer. Designed by Lindauer.

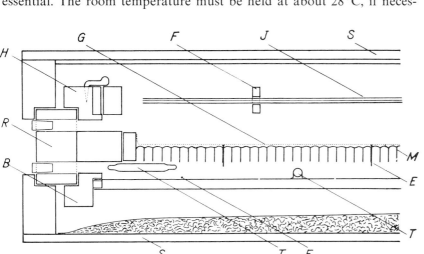

FIGURE 9. Heatable observation hive, equipped for temperature measurement; horizontal section in region of the lower comb: B, hive framework with a glass pane on the observation side; E, thermoelements; F, fastening for heating tube; G, wire screening; H, heater framework; J, pyrex-glass tubes with heating wire; M, comb for temperature measurements; R, framework of observation hive; S, protective cover; T, mercury thermometer. After Bräuninger 1964.

14 / The Dances of Bees

FIGURE 10. The Winterhalle of the Munich Botanical Garden.

sary by the addition of automatically regulated heaters. The shallow, screened water pan h holds the relative humidity at 60–80 percent. The bees of the observation hive d collect sugar water at the feeding table k; they find water in the drinking trough i on the floor and pollen (obtained from pollen baskets by "pollen traps" and later dried and pulverized) in the pan g. So that the pollen will not be blown away by the wind created by the flying bees it is proffered in a pan 20 cm deep with slanting walls. The need for it is very great. If a shortage occurs, a pollen comb is suspended in the hive (further details in Renner 1955, 1957).

In regard to demountable bee rooms see pp. 356f.

4. Marking the Bees with Numbers

Just as indispensable as the observation hive was a procedure for labeling the bees individually. The method of numbering them with colored lacquers dissolved in alcohol that I described in 1923 has proved very useful ever since. Occasional tests of other paint gave no improvement.

The numbers are defined by means of 5 different colors and their position on the bee's body. A white, red, blue, yellow, or green spot on the anterior thoracic margin means the number 1, 2, 3, 4, or 5, respectively; the same colors near the posterior thoracic margin signify the numbers 6, 7, 8, 9, or 0. Double figures are given by the appropriate combinations of two spots on the thorax, and hundreds are indicated on the abdomen. In this way the bees may be numbered from 1 to 599 (Fig. 12), and by means of small variations up to several thousand. Repeatedly my colleagues have numbered an observation colony completely for special purposes. With some practice one can read these color spots as quickly and surely as one could written numbers; they are also recognizable during flight even before the bees settle, if they

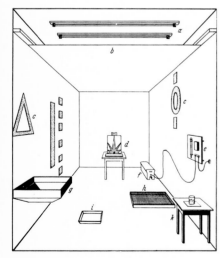

FIGURE 11. A bee room: a, lights; b, transparent paper; c, wall figures for bees' orientation; d, beehive; e, thermometer and relay; f, electric heater; g, pollen pan; h, screened water pan for regulating humidity; i, drinking trough; k, feeding table. After Renner 1957.

Methods in General / 15

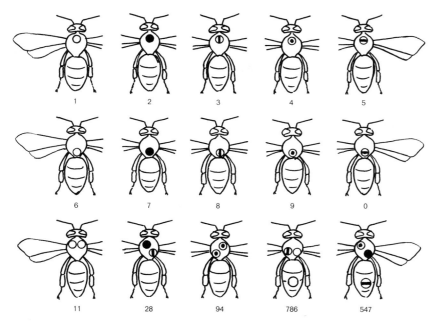

FIGURE 12. Numbering of bees by means of five different colors; ○, white; ●, red; ⦶, blue; ⊙, yellow; ⊖, green.

are not too dark; they may be made lighter by mixing them with white. If fluorescent colors are used they are also legible in the dark. When properly applied they last for several weeks.

The correct preparation and application of the colors is important. Powdered artists' pigment is ground into a paste with a solution of shellac in alcohol ("white shellac" dissolved to a high concentration in absolute undenatured alcohol). The paste must not be too viscous or it will not penetrate into the hairy cover on the bee's body and the blob of pigment will dry superficially and fall off easily. But otherwise it will usually stick fast for weeks. When a spot is lost its color often is still recognizable in bits that remain, and it can be renewed.

The shellac colors have the advantage of drying rapidly. This is a disadvantage in storing colors already prepared for application. These have to be closed up very tightly. Two successful arrangements for keeping them are shown in Fig. 13. The pigments are very conveniently ground together with the shellac solution in small embryological culture dishes, and are protected from drying by cover slips laid on the greased rims (Fig. 13a). The fine brushes used for marking the bees are stuck through holes bored in the cork stoppers of test tubes and dip into alcohol. The colors dry out quickly in the culture dishes unless the edges are kept clean and well greased. To avoid this trouble, the prepared colors may be kept in rubber-stoppered vials (Fig. 13b).

The bees are most easily marked while they are taking up concentrated sugar solution from a watch glass. They are then so intent on feeding that they tolerate the treatment calmly (Fig. 14). If specimens that are collecting from flowers are to be marked, a special pair of

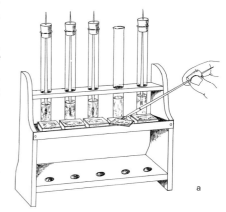

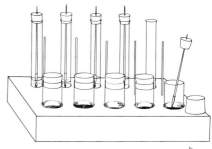

FIGURE 13. Equipment for numbering bees; explanation in text.

FIGURE 14. (*a*) A group of foragers is numbered at the food dish. (*b*) A bee is given number 16. Photographs by M. Renner.

tongs for seizing and holding them is useful (Fig. 15). The spots of color can be applied through the meshes of the net. When the tongs are opened carefully the bees usually return to their work at once.

5. WE SET UP AN ARTIFICIAL FEEDING PLACE

The natural feeding places for bees are flowers. For experimental purposes one can make bees collect under controlled conditions on flowers (see p. 31). Considerably simpler and also better for most purposes is an "artificial flower," in the form of a watch glass filled with sugar water. If necessary the color of the flower can be simulated by a sheet of colored paper beneath the dish and a floral scent can be added.

When the feeding place is set up it is important to train only bees of the observation hive to collect. When bees of another hive join in, they come as a rule from normal populations that outnumber the observation colony and soon drive it from the feeding place. Hence it is wise to begin early in the morning, before other bees are spying about. As a start a few drops of sugar solution are pipetted into the hive entrance; they are quickly found even before flight begins. By means of a line of drops the bees gradually are enticed all the way to the end of the flight funnel. Now a feeding table with a watch glass filled with sugar solution is placed very close to the funnel, so that it is at the same level with the bottom of it, and a bridge (preferably of blue or yellow cardboard) is laid from the funnel to the table; across this the bees are led by means of additional drops of sugar water to the watch glass. Then one can remove the bridge and slide the table farther away, at first only a few centimeters at a time. These first steps must be carried out with great care lest the connection be broken. Much time is lost when a wholly new start has to be made. Very gradually one can proceed by longer steps, as soon as a few of the bees have returned for a second visit to the watch glass. If the goal lies at an appreciable distance—we have gotten as far away as 12 km—then the table should finally be advanced about 100 m at a time. It is as though the bees had grasped the movement of the feeding place, for at times they are awaiting one at the next station. If during the journey one passes conspicuous objects (a road, a hedge, a tree, or the like), the bees may easily "get stuck." One must not tarry too long at such places, for otherwise the bees become accustomed to flying to these landmarks and are hard to lead farther. As a general rule they come along much more easily if the feeding place is identified by a scent, for instance if the watch glass is placed on filter paper with a few drops of an essential oil (oil of peppermint, oil of lavender, or the like). The whole procedure can be shortened by adding a scent at the beginning: if a few milliliters of a scented sugar solution are offered within the hive and a watch glass with the same scented sugar solution is placed in front of the flight funnel, the bees usually fly to the dish quickly (for the proper way of adding the odor see page 20). If the addition of a scent is to be avoided, a conspicuously colored paper will serve as a good optical cue for the feeding place.

FIGURE 15. Special tongs for seizing bees. The square end frames are 5 × 5 cm and are pierced with fine holes for sewing the netting on. The ratchets on the handles make it possible to hold a bee once it has been caught.

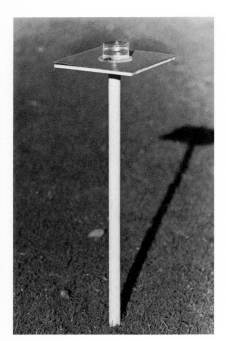

FIGURE 16. Feeding table.

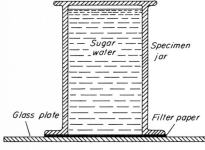

FIGURE 17. Pneumatic feeding vessel. After Baumgärtner 1928.

Success is threatened from two directions: in spring the natural honeyflow is so good that, even with concentrated sugar solution to which honey has been added, it is hard to get bees to come to the food dish. They prefer the fields with their blossoming flowers. Often all we could do was to put off the beginning of experimentation for several weeks. But in late summer, after the natural honeyflow has ceased, the colonies are so eager for anything sweet that strangers from other hives may become a pest and, unless great care is taken, will have driven off the experimental bees before one notices. Then there is only one thing to do: to mark the bees at the dish individually at the beginning of the training journey and to make certain that visitors to the dish originate only from the observation hive. Any numbered bee that has not appeared at the hive is killed on its first return to the feeding place. A more elegant method of control, which has been of great help in especially difficult instances, is to use a completely numbered observation colony (cf. Knaffl 1953:138, 139). Lack of a number gives a stranger away at once. Populating the observation hive with Italian bees is less time consuming. These are already marked naturally, so to speak, by their yellow body pattern. Using them is sensible, of course, only when no other Italian bees are lodged in the neighborhood.

As a feeding table we use a square wooden slab (about 20 × 20 cm), fastened to a sharpened pole that is pushed into the ground (Fig. 16). On hard ground we use a three-legged table.

As a feeding dish an open watch glass may be used. In 1919 (p. 16, Figs. E and F) I described a feeding vessel that refilled automatically; it was simplified by Baumgärtner (1928:59): the wide, ground rim of a specimen jar 8–9 cm high and 4–5 cm wide, filled with sugar water, rests on a plate of glass (Fig. 17). Between the glass plate and the rim of the jar is placed a disk of filter paper from which a sector 4 mm wide at the circumference is cut out. The bees can sit all around the rim imbibing sugar water. The fluid sucked out is replaced by air that enters the jar through the slit in the paper. One can also forgo the use of filter paper and instead put a straw or a splinter of wood at one spot between the plate and the rim; the bees drink from the resulting crack, which of course has to be kept narrow enough to prevent the sugar water from running out by itself. Of late I have preferred to use a grooved arrangement designed by Renner (1959:457–458). This is modeled after Baumgärtner's pneumatic feeding device, except that the sugar-water jar stands on a plate of glass or plexiglass into which radial slots, about 8 mm long and 1 mm deep, have been etched (Fig. 18). The bees drink from these little channels. These last-mentioned vessels have several advantages: they have to be refilled less often than an open watch glass, the concentration of the sugar solution is not changed by evaporation, the bees cannot soil their wings with sugar solution, and also their zeal for collecting and their eagerness to dance are increased. Every cursory observation gives this impression. Experiments carefully directed to the question have shown that on each flight bees imbibe more sugar water from narrow slits or from glass capillaries, which are more like the natural nectaries of flowers, than from watch

FIGURE 18. Feeding vessel on grooved plate. After Renner.

glasses, and that they then dance longer in the hive and nevertheless return sooner to the feeding place, because they do not loiter after dancing (Kappel 1953).

Sometimes one wishes to feed the bees so sparingly that although they keep on collecting they dance little or not at all. An arrangement for such meager feeding is shown in Fig. 19. An open glass dish is filled with layers of filter paper, which are moistened with sugar water by means of a syringe inserted through a hole in the side of the dish. Supplying the proper dose of course demands constant attention. Figure 20 shows another arrangement, which works automatically but does not permit instantaneous regulation of the amount of food.

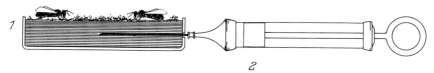

FIGURE 19. Device for "meager feeding": 1, glass dish with filter paper, somewhat frayed by the bees in feeding; 2, syringe with sugar water.

At the beginning of the experiment and during movement of the feeding table the sugar water should be a 2M solution[1] (near saturation) of cane sugar (C.P. sucrose or commercial sugar). Thus one succeeds most quickly. Later as a rule the concentration must be decreased because otherwise the number of bees at the feeding place becomes excessive. The proper degree of dilution is determined mainly by the quality of the natural food. When the latter is poor, the foraging bees will usually fetch in newcomers when fed with no more than a

---

[1] A 2M sugar solution = 68.4 g cane sugar in 100 ml water. For volume percent and weight percent see v. Frisch 1934:12.

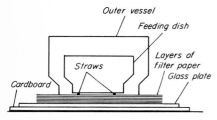

FIGURE 20. Automatic arrangement for "meager feeding." A feeding dish of the type used by Baumgärtner stands on several layers of filter paper. The feeding dish is covered by a second glass vessel at the edge of which the bees drink. After Boch 1956.

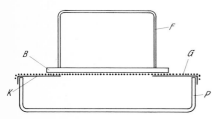

FIGURE 21. Feeding dish to which scent can be added: P, petri dish for the odorant; K, cardboard ring; G, wire netting; F, feeding dish; B, glass plate.

$\frac{1}{8}$M sugar solution; when the natural flow is rich one may occasionally have to hold to the 2M solution in order to keep a crowd. As soon as the group of numbered bees has reached the desired size, each additional newcomer is seized with forceps and killed (best put into a cork-stoppered vial of alcohol). If in the interests of the experiment concentrated food must be dispensed for hours at a time, this may mean sacrificing several hundred bees; but other methods of keeping the crowd of foragers small have disadvantages.

For marking the feeding place by means of an odor there are countless products of the perfume industry (see v. Frisch 1919:17ff, 37ff). As a rule we use essential oils (oil of lavender, oil of peppermint, orange-blossom oil, and the like).[2] The simplest procedure is to place the dish on a sheet of glass or cardboard covered with filter paper and drop a little essential oil on the paper with a fine pipette. The oil evaporates little by little and must be renewed correspondingly. If considerable variations in the intensity of the scent are objectionable, the following tested arrangement is to be recommended (Fig. 21): a petri dish (8-cm diameter) is covered with a ring of cardboard the central opening of which is somewhat smaller than the circular bottom of the feeding dish that is placed over it. Between the cardboard and the feeding dish is a fine wire gauze. The bottom of the petri dish is covered with an essential oil. The odorous substance can evaporate sufficiently through the gap afforded by the gauze between the cardboard and the feeding dish.

Instead of placing the oil beside the food the scent can be added to the sugar solution—but sparingly, for otherwise the food is made too repellent for the bees. One drop of an ethereal oil per liter of sugar solution is about right. (Further details in Kaschef 1957.)

If the bees have grown accustomed, even only briefly, to an odorless feeding place, great caution must be used in shifting to scented feeding, for otherwise the assemblage is driven off. Thus, for instance, one first puts only a single tiny drop of the scent beside the food dish. One sees that initially the bees settle at the other, unscented side. When additional scent is added gradually they grow used to it.

After distances have become greater a field telephone between observation hive and feeding place simplifies matters and saves time. Lately we have used portable radiophones, which free us from dependence on the telephone cable.

6. Automatic Recording of Visits to the Observation Place

When there is no experimental reason to catch and remove the bees that appear at an observation place or to list them individually, but only to record the number and time of their visits, a photoelectric recording device designed and tested by Renner (1959:454) serves well. One central automatic recorder replaces a lot of observers.

In Renner's studies the feeding box (and during an experiment each observation box) stood in a plastic bucket (Fig. 22b) that was let into

[2] Our sources were Schimmel and Co., Miltitz bei Leipzig, and Schmoller and Bompard, Grasse, South France.

FIGURE 22. Apparatus for the automatic recording of the frequency of bees' visits. (*a*) The apparatus sunk into the ground and ready for experimenting. Through the opening the bees can see the front face of the box, the hole in which is surrounded by a floral pattern. (*b*) Box in the plastic container with cover removed. (*c*) The cubical aluminum box cover is removed; the inner box with the feeding dish is visible. The projection at the front edge serves for photoelectric recording and contains a doorway *T* leading to the food. When the cover is in place the doorway is continuous with the opening in the flower model. After Renner 1959.

the ground and covered with a round sheet of hard wallboard 50 cm in diameter. Consequently the whole arrangement was inconspicuous to the eye. The bees flew in and out through a hole in the center of the cover (Fig. 22a). A relay, transistors and resistors, an electric lamp, and a 6-volt storage battery are housed in a sheet-iron box 11.5 × 13 cm × 7 cm high in the bucket (Fig. 22c). Wires could be connected to terminals for banana plugs in one side. The plastic (pertinax) cover of the box, attached by screws, bears at one edge a pertinax projection 6 mm thick containing an opening 7 × 7 mm through which only one bee can pass at a time. A photocell built into the top of this little doorway is illuminated by the lamp inside the metal box through a red filter (red light is invisible to bees) placed in the lid of the box directly under the doorway. Interruption of the light beam operates the relay. Beyond the doorway, in the middle of the pertinax sheet, is the feeding dish. Over it is a cubical cover of aluminum sheeting (Fig. 22b; cover removed and lying beside the bucket in Fig. 22c). A hole in the front wall of the cover coincides with the opening of the doorway with the photocell and is surrounded by a colored flower design. For the center of the latter an ultraviolet-absorbing, yellow pigment (fast yellow 51 BN[3]) was used; the outer parts of the flower were blue (cobalt blue 660[3]). This model simulates a widespread floral pattern (see pp. 483ff), and surprisingly quickly induces the bees to enter properly.

7. Cleaning the Equipment; Scents as Sources of Error

When feeding dishes and other pieces of equipment are reused, one must realize that the bees leave behind persistent traces of odors that are attractive to their comrades. If Julian Françon (1938, 1939) had noted this source of error he would not have adorned his affectionate descriptions of the "sagacity of the bees" with such fantastic statements. From his experiments he drew the conclusion that the discoverer of a hidden feeding place could tell her comrades at home whether they should enter through a trap door or through a chimney, whether from above or from below (Françon 1939:151ff), and that their language has words for the different colors (p. 180). Actually the newcomers doubtless smelled the places where their predecessor had been.

Where remaining traces of scent might interfere in experiments we work with new, unused materials so far as possible. That is so especially with paper items, wire netting, and so forth; wooden tables can be freed of remnants of scents by thorough washing followed by exposure to sunlight for several days. Forceps and other metallic objects can be heated to redness. Glassware should be soaked in chromic sulfate, rinsed for at least 12 hr in running water, given a final rinse with distilled water, and dried where possible in sunshine. If traces of fatty substances remain on the objects they should first be placed in xylene and then rinsed with warm detergent solution (cf. Fischer 1957).

The odor of shellac clings to bees recently marked with alcoholic lacquers. Therefore the newcomers aroused by them go hunting for the

---

[3]Source: Siegle and Co., Stuttgart-Feuerbach. The paints were mixed with *Oleton*.

odor of shellac—unless some stronger odor has been given to the feeding place—just as under normal circumstances they would seek the scent of the kind of flower that had been visited (see p. 33), and hence when they reach the feeding place they are more interested in the stand with the paint vials than in the feeding dish. Also the bees may in time become familiar with the individual odor of a regular experimenter at a feeding place. Then during a pause in feeding the bees may find the experimenter even if he is 100 or 200 m away and well hidden, so that he is visited stubbornly by "his" numbered bees.

These complications must be kept constantly in mind if one wishes to avoid being deceived in experiments with bees.

8. How Bees Are Put to Work or Brought Home

How often I wished to begin an experiment early, and yet the bees did not come to the feeding place! Either it was still too cool for them, or too dusky, or the numbered group had not been fed for a couple of days and were taking their time. There is an almost unfailing means of avoiding such wasted hours: with a pipette one puts a few milliliters of sugar water into the hive entrance or, if this is unoccupied, into the hive through the feeding hole above. The solution is quickly consumed and disseminated. This stimulus is enough to cause one or another of the bees to appear at the feeding place and, if she finds the table set, to give the alarm to the rest. If the feeding place was marked with a scent, addition of the same scent to the sugar water squirted into the hive hastens the effect.

A more difficult problem is gathering the bees together into the hive after their work is done and while it is still daylight. During many experiments in which the hive was moved to different places this was often very desirable. The experiment had perhaps been completed early in the afternoon and the observation hive would have to be brought back from a distant location to its home base. Then one would either have to wait until darkness fell or be faced with another trip out in the evening. This loss of time could be avoided in the following way. The hive entrance was closed with a screen. As soon as a good number of returning bees had gathered there, the entrance was opened briefly to let them in. In order to avoid a simultaneous escape of bees from the hive, the cork stopper of the feeding aperture at the top was removed and replaced by a screen closure; also the wooden covers of the observation hive were lifted at the back in front of the opening of the hive entrance (Fig. 23). Bees ready to fly out then tried to go upward and backward in the hive toward the light, and did not get out through the entrance. In this way, even in fine weather and when the bees are flying actively, nearly the entire colony can be assembled in the observation hive within half an hour.

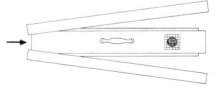

FIGURE 23. Observation hive seen from above. The arrow points to the entrance. The protective covers are spread apart at the rear. See text.

9. Measurement of the Tempo and Direction of Dancing

In order to compare the tempo of dancing under different conditions we determined with the help of a stopwatch the number of circuits made in 15 sec. The stopwatch is started at the moment when the dancer begins her straight tail-wagging run. Naturally the interval of $\frac{1}{4}$ min would not necessarily coincide with the completion of a circuit. Therefore we measure the time required for a certain number of rounds and calculate the result for 15 sec. The total duration of repeated dances between two food-collecting flights, if it is of interest, can be measured easily with an integrating timer.

The individual elements and phases of the dances can be analyzed with motion pictures (using increased frequency, about 60 frames/sec), if simultaneously one photographs a clock face with a large second hand and scale divisions of 0.1 sec. For such photographs the observation hive was suspended in a metal frame so that it could be raised or lowered, moved forward or back, without the disturbance of shaking. In this way a bee starting to dance can be brought quickly into the field of view on which the camera is focused. A bellows is attached to the hive opening so that the flight entrance can be kept at one spot during the movement of the hive; thus there is no interference with traffic coming and going (Fig. 24). It proved necessary to put a thick base of foam rubber beneath the suspending framework of the hive, since otherwise the dances were repressed by the transmission of vibra-

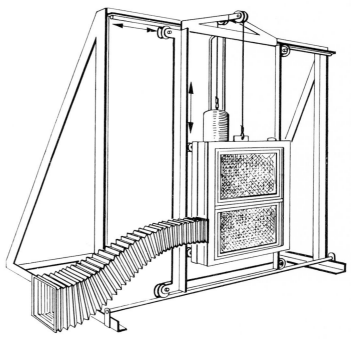

FIGURE 24. Movable suspension of the observation hive, for photographic studies. The weight of the hive is balanced by a hollow counterweight filled with shot, at the other end of the cord. The rollers on the hive and on the inner frames run on rails and allow a rapid, vibration-free movement, during which the hive entrance at the wide end of the photographic bellows remains in one position. After v. Frisch and Jander 1957.

tions through the hive floor when the camera was running (v. Frisch and Jander 1957:240).

In recent years methods have been found for having the dances recorded photographically or electrically by the bees themselves. Spots of fluorescent paint placed on the abdomen shine out so strongly in the light of a mercury-vapor lamp that the entire course of the dances can be recorded photographically in the hive. By employing phosphorescent pigments previously exposed to light the pictures can be taken in complete darkness (Hoffmann, Köhler, and Wittekindt 1956; F. Köhler 1959). New information was discovered through Esch's (1956) method of registering the bees' movements electrically with the help of a tiny magnet attached to the abdomen (see p. 58). The technical details presently were improved considerably and are described by Esch (1961a:1).

During my initial observations I merely estimated the direction of the dances. To facilitate matters I drew with a glass-marking pencil on the glass window of the observation hive in front of the dancing area a grid with lines vertical, horizontal, and inclined 30° and 60° to left and right. Since 1947 I have used the protractor shown in Fig. 25. It is held against the glass window in front of the dancer in such a way that the diameter L of the semicircle parallels the direction of dancing. The plumb bob allows one to read off on the scale the angle of dancing. The direction of dancing can be determined more easily and more exactly with a device designed by Palitschek, which we have used since 1950. It has been described by Palitschek (1952). Figure 26 shows a somewhat improved form of it. In a square plate of plexiglass, 24 × 24 cm, is cut a circular hole 19 cm in diameter, into which is set a rotating plexiglass disk with a peripheral scale. Parallel lines are scratched on this disk at 2-cm intervals. This protractor is held against the observation window in such a way that it rests against the wooden window frame below or at the side. In its initial position the lines on the rotating disk must be vertical. Above the disk, in the fixed, square plexiglass frame, is scratched an index line that corresponds precisely with the center line of the disk in its normal position. In taking a measurement, the rotating disk is turned until its lines parallel the direction of dancing. Then the angle of dancing can be read on the scale from the index line M (Fig. 26).

This protractor can also be used when the observation hive is horizontal. However, it was not good for determining the direction of dancing beneath a set of polarizers. Then the direction of dancing was better determined by means of a grid, for which I stretched white threads in front of the comb (Fig. 27). They were fastened to the inner margin of the outer glass window with sticky wax.

In some experiments with the horizontal hive it was necessary to avoid refraction of light by the various transparent panes. The glass window over the dancing area was arranged to open as a sash; the rotary plexiglass inset was removed from the protractor (Fig. 26) and its lines were replaced with thin elastic bands fastened parallel to one another on a rotary circular frame.

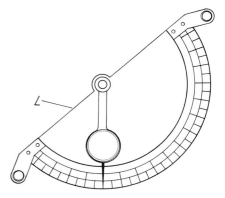

FIGURE 25. Protractor for measuring the angle of dancing: the diameter L of the semicircle is set parallel to the direction of dancing.

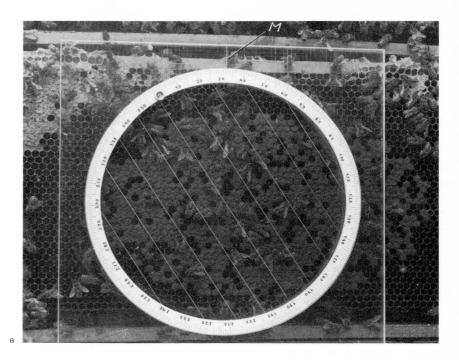

FIGURE 26. Protractor of lucite, for measuring the angle of dancing, as used by Palitschek; a rotating plate with parallel lines and a scale is set into a fixed frame: (*a*) photograph, by M. Renner; (*b*) schematic drawing, seen from above and in section; *M*, fixed line for measurements.

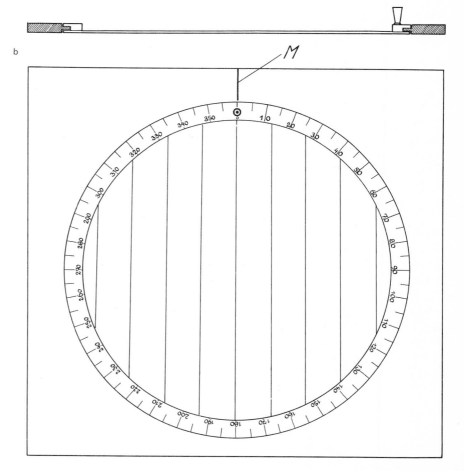

10. Selection of the Bees

A colony of bees may be calm or excitable, peaceful or eager to sting. For the observation hive calm and peaceful inhabitants are preferable, quite apart from the inconveniences associated with a desire to sting. In the great majority of experiments I used bees of the Carniolan race (*Apis mellifera carnica* Pollm.), whose quiet nature is a distinctive racial characteristic. The choice of especially peaceful and at the same time diligent stocks is a matter of experience.

In individual cases the decision of course depends on the experimental objective. It has already been mentioned that Italian bees (*A. mellifera ligustica* Spin.) are distinguished easily by their yellow body pattern from the principal races that are widespread among us, and that therefore they are favorable for many purposes. That they indicate to their comrades the direction of the goal at much shorter distances than do the Carniolan bees can be noted as a further peculiarity that is advantageous under some circumstances.

In studying racial differences in the "language of the bees" we have drawn on all available races, but have found no reason to prefer any other to the two aforementioned. Hence, unless it is expressly stated otherwise, the experiments described here refer to Carniolan bees. However, they frequently contained a trace of the Italian or German races (*A. m. ligustica* Spin. or *A. m. mellifera* L.), as is inevitable in our area unless one continually imports new bees from rearing establishments. Where we used purebred swarms this also is noted especially.

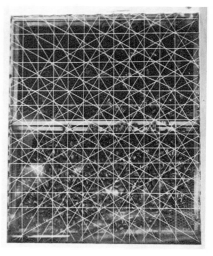

FIGURE 27. Grid of threads fastened in front of the glass window of the hive for estimating the angle of dancing. The threads are stretched vertically, horizontally, and at angles of 30° and 60° to right and to left.

III / The Round Dance as a Means of Communication when Nectar Sources Are Nearby

We set up an artificial feeding place. There, one or two dozen bees, numbered with spots of paint and hence known to us individually, gather sugar water from a watch glass. After a rather long interruption in our offering food, only occasionally will a bee make a visit to the dish. But then, when one of them finds the dish full and resumes her foraging, her marked companions reappear in rapid sequence. As the presentation of food continues, new unmarked bees join in. The site was known to the first ones, not to the others. We shall consider the two cases separately.

A. THE OBJECTIVE IS KNOWN TO THE BEES INFORMED

1. The Scouts

After an interruption of about $\frac{1}{2}$ hr in feeding one finds almost all the numbered forager bees in the observation hive, most of them near the entrance, where they are sitting motionless among their hivemates or crawling slowly about a little. Now and then one of them becomes active, running as if after a sudden decision to the entrance and appearing at the feeding place, where she inspects the empty dish hastily or even thoroughly, often for minutes at a time.

There are no specific scouting bees. Any one of the foragers may set out on her way. When artificial feeding has been stopped, they at first do this frequently and then at increasing intervals. But they display great individual differences in their eagerness; many of them come relatively often and others stay at home altogether after a few vain visits (v. Frisch 1923:26).

If a scout finds the dish empty, she crawls slowly up into the comb after her return flight and comes to rest near her fellow foragers. But if she finds sugar water in the watch glass, she drinks and fills her honey stomach, flies home, runs upward swiftly and with excitement, and immediately hunts out recipients for her burden. As a rule there is no dearth of such. Soon she halts and lets a drop of honey water from the honey stomach appear at her mouth; one to three bees, each with proboscis outstretched, take this fluid up amid mutual drumming with the antennae (Fig. 28). Mostly these are younger bees, whose task it is to distribute among their fellows any food brought in or to store it in cells. Often enough a numbered companion also gets some. This redistribution of sugar water may be repeated at various spots on the comb and requires about ½–¾ min altogether.

2. The Round Dance

Now the round dance begins. With swift, tripping steps the forager bee runs in a circle, of such small diameter that for the most part only a single cell lies within it. She runs about over the six adjacent cells, suddenly reversing direction and then turning again to her original course, and so on (Fig. 29). Between two reversals there are often one or two complete circles, but frequently only three-quarters or half of a circle. The dance may come to an end after one or two reversals, but 20 and more reversals may succeed one another; correspondingly, at times the dance lasts scarcely a second and at others often goes on for minutes. During dances of long duration the center of movement may shift gradually over the breadth of several cells. After the round dance has ended, food often is distributed again at this or some other place on the comb and the dance is then resumed; this performance may even be repeated thrice or (rarely) oftener (v. Frisch 1923:32f). The dance ends as unexpectedly as it began, and after a short period of cleaning and "refueling"[1] the bee rushes hastily to the hive entrance and takes off on the next foraging flight.

There is never any dancing on an empty or sparsely occupied comb, but only in the midst of a throng of bees. Hence the dancer in her circling comes at once into contact with other bees that, if in an appropriate mood, trip excitedly after her, holding their antennae against her abdomen. An amusing sight is the way a dancer draws along with her a train of two or three, often as many as five or six, followers who imitate her circlings and changes of direction (Fig. 29). After the dance has ended, they turn away. When the train includes numbered bees, comrades from the feeding place, as a rule these at once begin preparations for flying out. They clean themselves, load up with honey, hasten independently to the hive entrance, and fly to the feeding place. The round dance was the sign that there is something to be fetched. When the dance is repeated after each foraging flight (as when circumstances are favorable), and when the aroused comrades too dance after each productive flight, things in the hive get more and more lively and soon

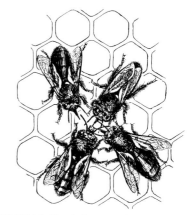

FIGURE 28. A forager (lower left) who has returned home and is giving nectar to three other bees.

FIGURE 29. The round dance. The dancer is followed by three bees who trip along after her and receive the information.

---

[1]The honey stomach serves as a fuel tank. After each foraging flight its contents are distributed, and then the honey supply necessary for the next flight is taken from other bees, in just the amount required in relation to the distance of the goal (R. Beutler 1950:104). In experiments with artificial feeding places, Istomina-Tsvetkova (1960) confirmed that when foraging bees fly out the content of the honey stomach varies according to the distance to be flown. For a distance of 5 m it was 0.782 mg; for 500 m, 1.610 mg; for 1000 m, 2.200 mg; and for 1500 m, 4.130 mg.

most of the members of the old crew come into contact with a dancer and appear at the feeding place.

Here is an example of the effectiveness of this method of communication. On the morning of 4 August 1919, of 12 bees that had been trained to a feeding station for 3 days, during one hour's observation there came to the empty watch glass only a single scout. At 9:29 bee No. 2 found the dish filled and at 9:32 danced on the comb, whereupon Nos. 10, 12, and 16 followed her. During the preceding hour No. 12 had not left the hive, while Nos. 10 and 16 each had taken a single look at the feeding place. Now all three hastened to the entrance and were at the dish within 2 min. The other marked bees, which had not come in contact with the dancer, appeared there only after additional dances had begun.

In summary it is evident from the protocols of that time that, after exclusion of doubtful cases, of 174 bees that had contact with a dancing bee from their group, 155 (89 percent) came to the dish within 5 min, after on the average they had not been there during more than $\frac{3}{4}$ hr previously (v. Frisch 1923:27–28, 35–38).

3. Contact without Dancing Can also be Effective with other Members of the Same Group

The group members waiting in the hive can also be induced by returning foragers to visit the feeding place anew even without a dance. An opportunity of seeing such behavior is not rare, since the readiness to dance varies according to the temperament of the bees and external conditions. Of 121 bees noted in the protocols, which came in contact with a successful returning but not dancing bee of their own group, 47 (39 percent) appeared at the watch glass within 5 min. When contact was made with a dancing bee it was 89 percent. Thus contact with a nondancing bee is less effective. Whether in these instances too there is an active summons on the part of the forager has not been clarified. One sees a mutual antennal drumming and often also the reception of a food sample. Perhaps the antennal drumming can be interpreted as a summons. But one can also assume that the waiting bees, because of the excited approach of the forager (see pp. 278f on the "jostling run") to whom they often turn from a distance of several millimeters, and by means of the gift of food, simply have their attention drawn to the fact that once again there is something to be fetched, and that hence they fly out to the place that is known to them.

4. Diffuse Information Given when Bees Are Fed at Sugar-Water Dishes

Under normal conditions there are in a hive several simultaneously active groups of foragers visiting different kinds of flowers. If at times the bees of one such group find no food at the accustomed flowers they do not immediately transfer to another source. In fact, bees are among the most constant of insects in their devotion to a single kind of flower.

Then for the most part one sees them sitting inactive in the hive. We ask whether such resting bees may perhaps become activated by a dancer from a different group or whether the fellow laborers know one another personally and distinguish a dancer who is not one of their own.

In order to decide this I allowed two groups of numbered bees to collect sugar water from two different stations near the hive. For easier differentiation, one group was marked with white on the abdomen, the other with yellow. After a rather long interruption in feeding at both places, sugar water was reoffered to one of the groups. The bees that came home from the refilled dish aroused *both* groups with their round dances. The members that belonged to the fed group flew to the filled dish, the others to *their* feeding place, the only one they knew, and there they examined in vain the empty feeding dish.

Example of an experiment. On 26 June 1920, 17 white-marked and 16 yellow-marked bees were visiting the two stations. After an interruption in feeding from 11:55 to 12:30 both groups were sitting still, scattered at random among one another, on the lower part of the comb in the hive. During the last quarter hour 3 scouts had appeared at each of the stations. At 12:30 the "yellow" dish was refilled with sugar water; the "white" dish remained empty. At 12:37 the first successful scout flew home. In the next 25 min 13 of the 16 bees from the yellow group came to their dish, and 14 of the 17 white ones to theirs. After a persistent search the latter had to return to the hive unsuccessful; 6 of them came a second time. Observation in the hive proved correspondingly that both yellow and white bees responded equally to the dances of the yellow one and that after such a meeting they rushed excitedly to the hive entrance. The reciprocal experiment, and its repetition 15 times, on other hives as well, confirmed the result (v. Frisch 1923:48–52).

Such behavior does not seem very astute. Whether it obtains under natural conditions needed to be investigated. Bees are accustomed to collect nectar from flowers and not sugar water from watch glasses.

5. Distinct Information Given when Bees Feed at Flowers

In order better to arrange for controlled visits to flowers I took the observation hive into the Winterhalle of the Munich Botanical Garden (Fig. 10); the hall, empty during the warm season, was a room about 30 m long, with windows and skylights. Near the hive I set up nectar-rich flower clusters of the locust *Robinia viscosa* Vent. The flowers were soon discovered. I let 17 numbered bees gather from them. On the next day fresh flower clusters were set out and (after half an hour) were found by bee No. 5; after her return home she carried out a typical round dance, in the course of which she made contact with six numbered bees. Four of them appeared at the flowers within 2 min, a fifth shortly thereafter, and the sixth after a further contact with a dancer. From this and additional experiments it is clear that bees that are collecting nectar from natural flowers are given information by successful scouts after a temporary cessation of the flow of nectar, via the same dance and with the same effectiveness as in the case of sugar-water collectors.

While no experiments with flowers were in progress, about 100–200 unmarked bees were fed with sugar water at another nearby station. It was striking that this group was not aroused by the bees that were gathering from flowers, and that on the other hand when sugar water was fed the "locust bees" stayed at home and did not join the unmarked dancers even when they came in contact with them. The two groups did not arouse one another. Only the comrades from the same group were called into action.

We approach natural conditions a step closer in allowing two numbered groups of bees to collect nectar from two different kinds of flowers. One group in the Winterhalle was visiting locust clusters, the other, at a station 2 or 3 m away, flowers of the summer linden (*Tilia platyphyllos* Scop.), rich in nectar. After an interruption in feeding at both places, on one occasion the locust flowers, on another the linden flowers, were again set out and the dances and their effects were observed. In other experiments I used instead of linden a milkweed, *Asclepias curassavica* L. The outcome in both series of experiments was the same, and was striking: thirty-five times I saw a bee coming from flowers make contact during her round dance with another numbered bee. Of the 35 bees whose contact with a dancer was seen, 21 had collected nectar from the same species as the dancer and 14 from the different flower. Not one of these 14 bees from the other group was influenced by the dance to fly out, while of the 21 bees of the same group 18 appeared at the feeding station within 5 min.

By good fortune I had among my numbered bees two shrewd individuals, Nos. 4 and 8, who—as happens exceptionally—were not very constant in their attachment to a particular flower species and who were quick to take advantage of any opportunity offered. Then it became apparent, with all clarity, that one and the same bee would mobilize the "linden bees" with her dance when she came from linden flowers, and the group of "locust bees" when she came from locust clusters. The experiments with milkweed and locust flowers taught exactly the same lesson (v. Frisch 1923:55–61). Whether the bee belongs to the same collecting group does not matter, but what kind of flower she comes from is decisive.

Simple observation of what happened in the hive pointed to how that is possible: foragers that came in from sugar-water dishes attracted the attention of their idle group comrades only on direct contact or when extremely close to them. But if they came from flowers, often it was to be seen that a group comrade would suddenly press inconsiderately through the throng of nonparticipants from as much as 2–3 cm away, run to the forager bee, and interestedly examine her with the antennae, where the olfactory receptors are located, while the collector was still sitting quietly and giving up her burden. The comrade in the hive seemed to have perceived the flower perfume that may have still clung to the body of the forager.

6. ODORS AS A MEANS OF COMMUNICATION

In order to answer the question whether the differing outcome of this last experiment is to be ascribed to the odor of the flowers, we return to our earlier experimental setup, and in the open country erect near the hive two artificial feeding stations at which two groups of numbered bees collect sugar water from watch glasses. But this time we add to each feeding place a distinct, characteristic perfume, by setting the glass on a sheet of cardboard covered with filter paper (Fig. 30) and putting around it at one station oil of citronella, say, and at the other oil of peppermint. When now after an interruption in feeding the peppermint dish is refilled, the dances of the peppermint bees quickly arouse their group comrades, while the citronella bees show no interest. Just the opposite occurs when the citronella dish is refilled. As in the experiments with natural flowers, the bees returning home often attract the attention of bees that have previously collected at the dish with the same odor at a distance of as much as 2–3 cm. This is never the case when they are fed on an odorless filter paper. Experiments like this were done several times with various scents, various colonies, in the open field and in the Winterhalle (v. Frisch 1923:62–72). The results leave no doubt that the specific odor clinging to the body is the signal by means of which the dancer summons her particular group comrades.

C. R. Ribbands (1954) fed two groups of marked bees at two feeding stations that were supplied with different odors (benzyl acetate and methyl benzoate), on two or three successive afternoons. Because of their being trained to this time of day, in the morning very few scouts came to the empty feeding dishes. But if he pipetted one of the two aromatic materials into the hive the number of visitors from the corresponding group about doubled on the average. According to Ribbands,

FIGURE 30. Numbered bees at the feeding place. The watch glass stands on a sheet of cardboard covered with filter paper. Around it are drops of an essential oil as a scent.

then, the inactive foragers can be induced to inspect their feeding place merely by perception of the accustomed odor, without any dancing and without the return of their foraging comrades.[2]

7. THE POLLEN COLLECTORS

In addition to sugar-rich nectar the bees also collect protein-rich pollen. As is well known, they carry it home in the "pollen baskets" on their hind legs. Occasionally they bring home both nectar and pollen at once, but as a rule each is collected by a different group of foragers.

The flowers of some species of plants produce no nectar and supply food only via the superfluity of their pollen. Experiments with such "pollen flowers" in the Winterhalle of the Munich Botanical Garden showed us that bees that have loaded their pollen baskets give the alarm in the same way as the nectar collectors, and that in this case too the comrades who have collected from the same plant species respond to the round dances and set forth to work.

Two numbered groups of bees were collecting pollen from poppy flowers (*Papaver nudicaule* L.) or from unblown roses (*Rosa moschata* hybr.), which were readily visible at neighboring stations.[3] The stems of the flowers dipped into screened vases. After feeding interruptions, one or the other kind of flower was set up anew. In the course of these experiments, on 25 occasions I saw a *dancing* pollen bee make contact with an idle group companion. Of these, 24 (96 percent) appeared at their feeding place within 5 min; on 20 occasions I saw that a bee returning home with laden pollen baskets met a group companion *without dancing*. At these meetings too the interest of the awaiting bee was aroused, but only 6 of the 20 of them (30 percent) came to the flowers within 5 min. Two more followed during the seventh and eighth minute; 12 stayed at home. The bees that were foreign to this group showed no interest.

Where there is a rich supply of nectar or sugar water,[4] the honey stomach is soon filled. Collection of pollen takes longer. No doubt it is connected with this difference that pollen collectors sometimes interrupt their activity soon after it has begun, fly home, dance vigorously, and then return to the collecting spot with their pollen baskets barely filled in order to take more load. I first saw this with an especially industrious bee that found roses again after an interruption. Although more than ample pollen was at hand, after 3 min she flew home with very lightly laden pollen baskets, danced 4 min on various parts of the comb as though possessed, and thereby induced many of her group companions to revisit the roses. Then with her incompletely filled pollen baskets she rushed out again and flew to the flowers, where in another 13 min

---

[2] This was in fact true only of the experiments with benzyl acetate. If the odor of methyl benzoate was injected into the hive, according to Ribbands' data (for a total of 3 experiments), 41 bees came to the methyl benzoate station and 50, hence even a few more, to the benzyl acetate station.

[3] Since the pollen is collected speedily, the flowers must be replaced by fresh ones often. For further details see v. Frisch 1923:75–81.

[4] Dancing occurs only when ample food is available (see p. 44).

she brought her pollen balls to a majestic size. Now she returned home for the second time and finally disburdened herself. This shows impressively that the dance is not a by-product of collecting activity, as has often been suspected, but has the function, within the framework of the social organization, of bringing a message to the comrades. I saw this same bee behave similarly on a second occasion and yet another pollen collector do the same (regarding corresponding behavior in propolis collectors see p. 269).

It was to be anticipated that the specific excitation of the group companions by the pollen collectors is to be attributed to their odor. Favoring this view was the observation that the comrades (as among the nectar collectors) often turned toward the homecoming bee from a distance of 2 cm even before her dance had begun. Once I saw that a bee with pollen baskets filled from poppies returned to the hive and, creeping upward, passed near a comrade who had been sitting quite still for $\frac{3}{4}$ hr. Thereupon the latter, as though startled out of her reverie, leapt suddenly across a nonparticipating bee and with her antennae excitedly examined the balls of pollen on the collector's legs.

That their attention is concentrated on the pollen baskets brings up the question whether with pollen bearers, as with collectors of nectar, the odor of flowers clinging to the body or here the odor of the transported pollen is the significant signal. As to whether pollen has any odor and how its smell may be related to that of the floral parts, botanical texts and handbooks gave no definite information. In sample tests of six different floral species I found (1923:88, 89) that the pollen had a definite odor, specifically distinct in the different flowers, and for the most part more intense than and differing from that of the petals. Alexandra v. Aufsess repeated (1960) and extended these tests, with the result that, among 17 species of flowers visited by insects, in 13 cases the pollen smelled stronger and qualitatively different than the rest of the flower, in 2 species the pollen odor was not more intense but was of a different quality, and in 2 species it was of the same nature but stronger. Training tests with two samples (*Polyantha*-rose hybr. ht. Wildfire and *Oenothera fruticosa* L.) demonstrated for bees too the qualitative difference between the pollen fragrance and the smell of the rest of the flower.

Under these circumstances it is understandable that the odor of flowers that clings to the body is overshadowed by the smell of the accompanying pollen balls brought home and that it is the latter that is decisive in communication with the group companions. The following experiments with roses (*Rosa moschata* hybr.) and Canterbury bells (*Campanula medium* L.) gave definite proof.

In the Winterhalle of the Munich Botanical Garden one group of numbered bees were collecting pollen on roses and 8 m away another group were collecting for Canterbury bells. In freshly opened Canterbury bells the stamens border the pistil on all sides; later they curl back and the pollen mostly remains stuck to the pistil (Fig. 31*a*). After an interruption in feeding at both stations I put out roses whose stamens had been removed and replaced by *Campanula* pistils with adherent

FIGURE 31. (*a*) Flower of Canterbury bell (*Campanula medium* L.), with part of the petals removed to show internal structure; most of the pollen from the recurving stamens has remained stuck to the pistil. (*b*) Canterbury-bell flower with the pollen-bearing parts replaced by rose stamens. (*c*) Rose (*Rosa moschata* hybr.). (*d*) Rose with its own stamens replaced by two Canterbury-bell pistils with adherent pollen.

pollen (Fig. 31*d*). A scout came and collected Canterbury-bell pollen from roses at the rose station. With her dancing she promptly dispatched six of seven *Campanula* bees to the *Campanula* station, where there was nothing to be obtained. But the rose bees remained on the comb. In the reciprocal experiment, after an interruption in feeding, in place of the Canterbury bells, flowers were put out whose pistils and filaments had been removed and replaced by the receptacle of a rose together with the stamens that arise from it (Fig. 31*b*). One bee came and made four collecting flights one after the other. At the *Campanula* station she gathered rose pollen from *Campanula*. Not a single one of the six other *Campanula* collectors took any interest in her dances, not one came to the station to lend her assistance, but all the bees of the rose group that met her in the hive flew out—to the empty rose station.

The pollen baskets are unloaded by the collector into cells that serve for storage of pollen. Often she repeats her dance after the pollen has been scraped off. Even thereafter its odor no doubt clings clearly enough to the hind legs, and likely to the whole body.

8. THE DANCE FLOOR

The means of communication described can be successful only when the resting bees stay on a limited area of the comb in the observation hive and when the dances take place in the same region. This prerequisite is fulfilled. In the experiments described hitherto an area of about 100 cm² on the lower comb of the observation hive, near the entrance, served as a dance floor. Here the bees wait and here they dance. However, this region is not precisely defined. When the hive is lightly populated or during cold nights, the bees withdraw to the upper comb. Forager bees returning home then may well give up part of their burden to individual recipients below, but they dance only above, in the midst of a crowd, until the latter, during continued collecting activity, gradually move downward. So the dance floor moves downward with the throng. Later it will be reported that the position of the dance floor is influenced also by the distance of the food source from the hive (p. 138). The important thing is that the forager bees are to be found at a given time in that region of the comb where successful scouts and workers perform their dances. In this way a rapid arousal of the entire group is ensured.

The arrangement of combs in the observation hive does not correspond to natural relations. Hence the question is justified, whether in a normal hive too there is a dance floor, and where it may be found. In order to test this, I marked at a feeding station bees from a hive arranged for handling from above. After a rather long interruption in feeding the hive was opened and the combs were removed serially. The marked bees were sitting almost exclusively on both sides of the third comb counting from the hive entrance. The hive had nine combs, arranged in order parallel to the anterior hive wall ("warm construction"). The combs may also be placed in the hive parallel to the side walls ("cold construction"). According to R. Götz (unpublished), here one finds the dancers thickest near the flight entrance, on the combs that abut directly against the entrance. Thus in normal bee colonies too the forager groups stay in a rather limited region (in this connection see also p. 138).

9. GROUPING ACCORDING TO ODOR

The speed with which a message is grasped by an interested group is favored yet more by a peculiarity of behavior, one might almost speak of a class consciousness, of the collecting assemblage; my attention was first drawn to this in September 1952. I had been providing numbered bees with lavender-scented food. In the afternoon, after several hours' interruption, the insects were sitting close together on the comb, where with their colorful spots they were a startling spectacle, a motley island in the midst of their unmarked hivemates. Later the group scattered somewhat, but soon they gathered together anew. Was it the familiar fragrance, still clinging to their bodies from the feeding place, that brought them together?

I have never seen this phenomenon so conspicuous since that time. But it left me with the desire to pursue the problem. It was only in the summer of 1961 that I finally got around to doing so.

I set up two feeding stations, each 80 m from the hive; 15 numbered bees were collecting at each. One group, at the northern station (34 m north of the other), was marked by an additional white spot on the abdomen, the other group, at the southern station, by a yellow spot. Between 16 and 19 August both groups regularly were fed simultaneously, with intermittent breaks, with sugar water. No scent was added. Twelve times during the pauses I cautiously opened the hive for inspection and noted down in a prepared diagram all numbers in view; this took about 5 min. For this purpose the comb had been divided by means of threads stretched in front of the glass window into squares 4 cm on a side. The bees of both groups, white and yellow, were sitting scattered at random among one another on the "dance floor" (Fig. 32).

At noon on 19 August the feeding place of the white group (the northern station) was scented with oil of lavender, that of the yellow group (the southern station) with orange-blossom oil (at each a few droplets were placed on the filter paper around the feeding dish). Feeding in the presence of a scent was continued, with intermittent breaks, from 19 to 21 August, and eleven times during the feeding pauses the distribution of the forager bees was checked. There was no mistaking the fact that bees belonging to the same group maintained closer contact on the dance floor with one another than with the other group (Fig. 33).

Now I numbered a new group of bees at both places, this time with a blue abdominal spot (northern group) and a red one (southern group). From 21 to 24 August both groups were fed with sugar water in the absence of an odor. On five inspections during the feeding pauses their distribution was random. From 25 to 27 August oil of anise was added for the blue group (northern station) and oil of peppermint for the red group (southern station). Thirteen inspections, the last on the morning

38 / The Dances of Bees

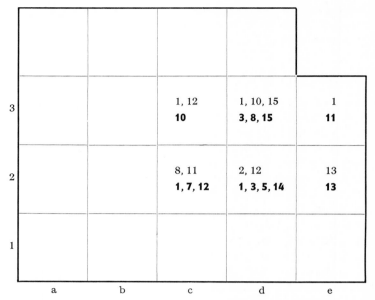

FIGURE 32. Distribution on the "dance floor" of bees of two groups, fed with unscented sugar water, as observed during pauses between feedings, 3:45 P.M., 18 August 1961. The comb was divided into 4-cm squares. The numbers in the squares are the numbers of the bees in each group, distinguished by different type. The position of each number shows which square the bees were found in, but not their exact positions in the square.

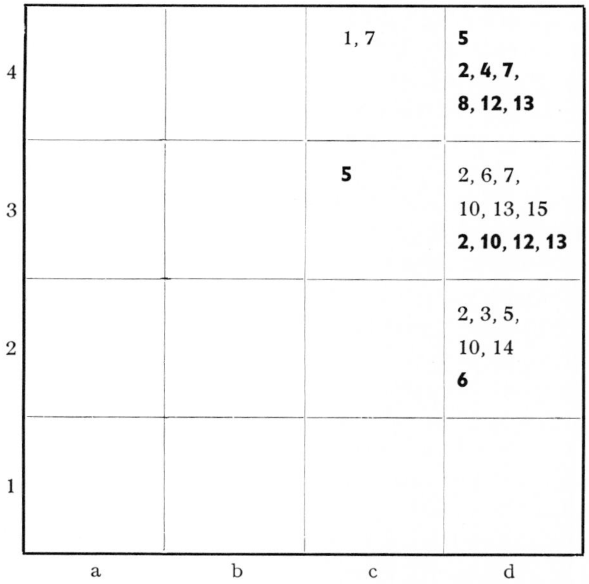

FIGURE 33. Distribution on the "dance floor" of bees of two groups fed with differently scented sugar water, as observed during pauses between feedings, 12:30 P.M., 20 August 1961. (See Fig. 32.)

of 28 August, confirmed the closer association of group companions when a scent was present with the food.

Our data show only which numbered individuals were found together in a common square, without attention to the precise places occupied. The actual spectacle was much more convincing than these records: surprisingly often the group comrades were sitting close together within a single field or on either side of the boundary between adjoining fields, so to speak shoulder to shoulder. When they were fed in the absence of odor there was nothing of the sort to be seen. After a rather long interruption in feeding and especially toward evening the foragers like to retire to undisturbed rest, and creep into empty cells. Only the tip of the abdomen remains visible, and with it the colored spots. Then not infrequently one may see two blue or two red companions asleep in neighboring cells. Neighbors with dissimilar colors are met with much less often.

In our further consideration of the problem we shall ignore this impressive behavior and confine ourselves to determining how many of the bees of one or the other group are found in the same square on inspection. This is listed in Table 1. Only those squares are included in which numbered bees were found. The figures show how many bees of one group (ordinary type) and of the other (italic type) were sitting in the same area. Dr. Rudolf Jander was kind enough to analyze this material statistically. With reference to the results he gives me the following information:

> In order to test whether the representatives of both groups of bees are distributed in the squares according to chance or whether they are grouped because of some other reason, all the records have to be summarized in a few numbers. All the squares in which only a single bee was found have to be excluded, since with a single bee one cannot judge whether it was attracted to one group or the other. For each experimental case, the number of squares occupied by more than one bee is distributed to one of three classes, according to whether $3/3-0/3$, $2/3-1/3$, or $1/3-0/3$ of the bees found in it belonged to one group or the other (Table 2). If, for example, a square contained one bee from the southern and one from the northern station, this square is assigned to the middle class; if a square contained two southern bees and a single northern one, then a value of $1/2$ is added to both classes I and II.

The result of this division into classes is shown in the hatched bars of Fig. 34. With reference to the question at issue, this figure is to be interpreted as follows: The more that bees of the same color grouped themselves together, the greater the total of squares in classes I and III, and the more bees of differing colors were grouped together, the larger is class II. An initial cursory comparison indicates indeed that after odors were added more bees of the same color became associated than after unscented feeding, yet for a highly significant conclusion it must be taken into account not only that the total number of squares in the two comparisons varied but

TABLE 1. Distribution on the comb of the bees from two collecting groups, at all inspections during feeding interruptions. The numbers show how many individuals of the two groups, distinguished by ordinary and italic type, were found in each 4-cm-square field. Only those fields are listed, seriatim, in which there were numbered bees.

Experiment	Date (August 1961) and time				Distribution of marked bees
	End of last feeding		Inspection		
1a. Odorless	16	4:12 P.M.	17	7:30 A.M.	1,*1*–1,*1*–2,*1*–1,*1*–1,0–0,*1*
				8:00 A.M.	1,*0*–0,*1*–1,0–1,2–3,*1*–1,*1*–1,0–0,2–0,*1*–0,*1*
				10:45 A.M.	0,*1*–1,0–3,3–0,2–0,*1*–1,2–0,*1*
	18	10:40 A.M.	18	11:10 A.M.	1,0–1,0–0,*1*–0,2–2,*1*–2,*1*–1,*1*–1,2
				11:50 A.M.	1,0–1,*1*–3,4–2,0–0,*1*
		3:00 P.M.		3:45 P.M.	2,*1*–2,3–3,3–3,2–4,*1*–1,*1*
				4:05 P.M.	0,*1*–0,2–1,*1*–1,*1*–1,0
				4:40 P.M.	0,*1*–1,0–1,3–0,2–3,3–3,*1*–1,0–1,0
			19	7:45 A.M.	2,3–0,*1*–0,*1*–1,3–1,*1*–1,*1*–1,0–1,0
				8:30 A.M.	0,3–2,2–1,0–1,0–1,3–1,*1*–1,*1*
	19	11:00 A.M.		11:45 A.M.	0,*1*–2,4–2,2–3,*1*–2,*1*–0,2–2,0
				12:45 P.M.	1,0–1,*1*–0,*1*–1,0–1,2–1,0–0,*1*–0,*1*–1,0–3,*1*–0,*1*
1b. Oil of lavender, *orange-blossom oil*	19	4:15 P.M.	19	4:45 P.M.	0,2–1,0–1,*1*–1,3–7,3–2,0
				5:15 P.M.	0,*1*–0,*1*–1,0–1,0–3,*1*–2,0–1,0–3,0–1,0–1,0–3,0
				6:00 P.M.	0,*1*–0,*1*–3,*1*–2,0–3,0–1,*3*
			20	8:00 A.M.	0,*1*–1,0–8,4–2,4–2,*1*–1,*1*
	20	11:40 A.M.		12:20 P.M.	2,0–0,*1*–1,6–6,4–5,*1*
				1:10 P.M.	1,3–0,2–0,*1*–5,*1*–1,2
				2:00 P.M.	1,4–1,0–6,2
			21	8:35 A.M.	0,*1*–4,0–1,3–1,0–1,2
	21	11:15 A.M.		12:00 M.	0,*1*–2,0–5,0–1,4
				12:45 P.M.	1,0–0,*1*–4,0–2,5–0,*1*
				1:45 P.M.	4,0–1,3–1,*1*–3,4–3,0
2a. Odorless	23	12:00 M.	23	1:00 P.M.	0,*1*–2,4–4,2–0,*1*–0,*1*–2,3–1,2
		5:00 P.M.		5:30 P.M.	0,*1*–1,*1*–1,2–2,3–1,0–1,0
			24	8:00 A.M.	1,0–3,4–0,2–3,4
	24	10:30 A.M.		11:30 A.M.	1,0–1,0–0,2–2,2–5,2–3,3–1,2
		4:50 P.M.		5:20 P.M.	3,*1*–1,2–2,3–2,0
2b. Oil of anise, *oil of peppermint*	25	11:00 A.M.	25	11:45 A.M.	0,2–1,*1*–1,2–4,3–0,*1*–2,*1*
		4:00 P.M.		4:45 P.M.	2,0–0,4–1,6–2,0–4,0
				3:30 P.M.	0,*1*–1,2–1,4–4,0–3,*1*–3,2
			26	7:45 A.M.	0,*1*–2,0–1,2–0,*1*–1,4–3,0–1,0–0,*1*–1,0–1,0
	26	9:25 A.M.		10:45 A.M.	0,2–5,4–3,4–1,0–1,0
		12:30 P.M.		2:00 P.M.	1,0–2,0–1,3–3,0–4,*1*–0,3–1,0–4,0
		4:00 P.M.		5:00 P.M.	1,0–3,0–7,5–0,*1*–0,3–2,0
				6:00 P.M.	1,3–0,*1*–3,*1*–1,0–1,0–2,0
	26	4:00 P.M.	27	7:15 A.M.	0,*1*–7,*1*–0,3–1,0
				8:00 A.M.	2,0–4,2–4,3–1,4
	27	10:00 A.M.		11:00 A.M.	0,*1*–3,4–3,3–0,*1*–2,0
				3:00 P.M.	0,*1*–0,3–0,3–1,*1*–1,0–1,0
				5:15 P.M.	0,*1*–0,*1*–2,*1*–1,0–2,0–0,*1*–0,*1*–1,0
			28	7:15 A.M.	2,3–1,4–2,0–1,*1*–2,0–0,*1*

TABLE 2. Classification of the squares according to the fractional distribution of bees from the two feeding places.

Classes	I	II	III
Bees from the southern station	$3/3 - 2/3$	$2/3 - 1/3$	$1/3 - 0/3$
Bees from the northern station	$0/3 - 1/3$	$1/3 - 2/3$	$2/3 - 3/3$

also that there were variations in the frequency distribution of the bees and in the ratio of the northern to the southern bees. In order to circumvent this difficulty, for each of the four tests the distribution of the squares among the three classes was calculated for comparison with the assumption that the bees from the north and south stations were associated entirely according to chance.

Calculation of the random distribution begins with the expansion of binomials of the form $(p + q)^n$. Here p and q for each comb area taken into account are the relative frequencies with which all southern and northern bees occur, in all squares with more than one

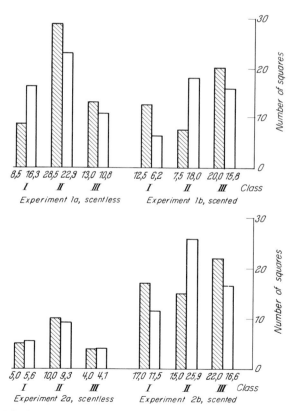

FIGURE 34. Distribution of square areas on the comb into three classes corresponding to the relative positions of bees from north and south feeding places. The hatched columns represent the observed frequencies, the white ones those to be expected according to a chance distribution. Exact values of the ordinates are given below the columns. See text. For classes I, II, III see Table 2.

marked bee. For n are substituted all actually observed numbers (2–12) of marked bees per square. The expanded binomials yield the frequencies of squares to be expected by chance for all n-values of marked bees and for all possible relations of southern and northern bees within a square. These theoretical relative frequencies of the squares are each multiplied by the actually observed frequencies of all squares with the corresponding n-values. The theoretical frequencies of the squares thus calculated for each number n of bees actually present, by distributing them over the three classes aforementioned and summing them, yield finally the chance distribution sought. These chance distributions are depicted in Fig. 34 by white bars.

In a final step the goodness of fit between theoretical and actual frequencies is now tested with the help of the X^2 method. The observed results after odorless feeding show no significant difference from a chance distribution. But after scented feeding this difference is highly significant ($p < 0.01$) in the first experiment (orange-lavender) and significant ($0.02 > p > 0.01$) in the second (peppermint-anise). Since in these two cases the frequency of the squares in classes I and III is significantly greater than would be expected from a chance distribution, it follows that bees from the same feeding station, with a similar odor, tend to associate.

As a result we record that bees whose source of food is marked by a specific odor associate during interruptions in feeding more closely with one another than with other collecting groups. Arousal of the whole hive by successful scouts is thereby speeded.

More will be said later (p. 49) about the persistence of the flower fragrance on the body.

Summary: The Arousal by Round Dances for a Goal That Is Known

1. Worker bees that are collecting from a source of food (for instance, from a watch glass of sugar water) cease making their regular flights when the source runs dry, and sit still in the hive. But they continue to scout the feeding place, often at first and at increasing intervals later. All bees of the group make scouting flights, but with differing individual zeal. After a successful scouting flight the bee returns home with a full honey stomach, makes a lively and excited run up the comb, and feeds other bees with the sugar solution.

2. By means of a round dance she arouses the interest of the surrounding bees, who trip along after her (Fig. 29). The round dance means that food is in good supply. If group comrades are among the bee's following, they are induced by the dance to fly out again to the feeding place. After their return home they dance too. In this way the entire working group can be aroused in a short time.

3. A returning food-distributing bee can also, without dancing, cause the group comrades with whom she comes in contact to revisit the feeding place; but that occurred only after about 40 percent of the

recorded meetings, whereas contact plus dancing was effective in about 90 percent of the cases.

4. If two working groups are collecting at different sugar-water dishes, and if one of the dishes is refilled after an interruption in feeding, then bees of the second group also are aroused by the dances of the successful group; they fly out to their accustomed feeding place and there keep examining the (empty) container. The bees in a given group are not known personally to one another.

5. Bees that are collecting on natural flowers definitely recognize one another in the hive. Only group companions respond to the dances. Bees that have been collecting from another kind of flower display no interest in the dance, even on direct contact. Those that have been collecting from the same kind of flower often give attention to one of their fellows from a distance of as much as 2–3 cm and hasten to her.

6. The sign for recognition is the flower fragrance that clings to the body. Hence artificial feeding stations that are given olfactory identity with ethereal oils cause, like natural flowers, a specific arousal of the fellows of a group.

7. With pollen collectors the species-specific odor of the pollen serves the group companions for mutual recognition. When the stamens of roses and Canterbury bells are interchanged, the dancers arouse bees of the wrong group (Fig. 31).

8. During an interruption in feeding the forager bees stay in a limited region of the comb (the "dance floor") where the successful returning scouts dance. By this means information is spread rapidly among the members of a group.

9. Additionally, during a rest period bees whose source of food is distinguished by a specific odor maintain closer contact with one another than with other collecting groups. In this way too the arousal of the group is expedited after successful scouting flights.

B. THE OBJECTIVE IS NOT KNOWN TO THE BEES INFORMED

10. RECRUITMENT OF ADDITIONAL HELPERS

A bee who after an interruption in feeding finds her sugar-water dish refilled and dances on her return home, arouses the (numbered) group companions she comes together with; besides these one sees in her train of followers, and for the most part in the majority, unmarked bees that have not yet been to the feeding place. If the watch glass has just been set out for the first time and then discovered by a bee from the observation hive, naturally she has as dance followers only individuals who are not acquainted with the place that has been found. In either situation, after a few minutes unmarked bees, newcomers, appear at the feeding dish and begin to participate in the collecting activity. They have been induced to do so by the dance. For they come only when there has been dancing.[5] And the more numerous and more lively

---

[5] In contrast to those bees that already are familiar with a feeding place and that in part are alarmed even without any dance (see p. 30).

the dances, the more of them that come. But since dances take place only when foraging is worth while, and since they are the more vivacious and long-lasting the better the source of food, here there is manifested a mechanism of the greatest biological significance, both for the colony and for the pollination of the flowers.

11. New Forces Are Aroused Only When Needed

It is a simple matter to convince oneself of these relations, as follows: We let a group of numbered bees collect food at a watch glass and take care that the sugar water does not become exhausted. They find "good hunting," and most of them dance after their return home. Newcomers appear in rising numbers and if we did not catch and remove them the crowd of foragers might have grown two- or threefold within half an hour. Now we substitute "bad hunting." In such a glass dish as is shown in Fig. 19, the filter paper is at first moistened with sugar water to such an extent that the bees, accustomed to better things, will not be driven off and stay away. But the less sugar solution they can extract from the filter paper henceforth, the shorter and less lively their dances on the comb, and finally they do not dance at all any more, but—if the amount of food is regulated properly—they will continue to gather food for hours. Whereas previously they would fill their honey stomachs from the watch glass in 1–2 min, now they tarry drinking at the feeding place for 10–15 min and longer, while their proboscises rake the paper into tatters (Fig. 19). Finally they fly home with an incompletely filled honey stomach, dispense their burden, and without a round dance set forth on the next collecting flight. Under such conditions there is no arousal of newcomers. The number of foragers does not increase.

This occurrence is paralleled completely during the visiting of natural flowers. Observation shows that flowers rich in nectar release dancing and that when the amount of nectar is small collection indeed continues as long as something is to be had, but the foraging flights demand more time and the dances are omitted. Only when it is profitable will recruiting dances increase the number of foragers.

In this connection the value of the food source is a complex affair. Later on, details will be given as to how many material and psychological factors are concerned in it (pp. 236ff). Here, as an elementary factor in addition to the quantity of nectar, we shall mention only its sweetness. A source of food can become just as unprofitable through reduction in the sugar content as through a decrease in the amount of solution. Since the sugar concentration can be regulated more easily and precisely than the volume of food taken up by the bees, with solutions of differing sweetness one can convince oneself especially simply and forcefully of the close connection between profitability, dancing, and arousal.

Examples: On 13 August 1927 12 numbered bees received $\frac{1}{4}$M cane-sugar solution from a watch glass beside the observation hive for 2 hr. During this time four newcomers appeared at the feeding place. When

I raised the concentration to 2M, 64 newcomers came to the dish in an equal time (v. Frisch 1934:101). As a matter of course, all newcomers to the feeding dish were caught and removed. The number of collectors remained constant.

On 24 August 1962 a group of four numbered bees were fed with sugar water 15 m from the hive between 10:00 A.M. and 1:25 P.M.; during this time the concentration was increased after the first 30 min and then again every 35 min, in six steps altogether, from $\frac{1}{16}$M to 2M (Table 3). With only four collectors active it was possible for the two observers to watch every return home, and if there were dancing to determine its total duration with an integral timer. After each increase of the concentration there was a 5-min pause in the observations. At the beginning of the experiment the temperature was 22.5°C and it had risen only 3 C deg by the end. Table 3 shows clearly the relation between sugar concentration, readiness to dance, duration of the individual dances, and effectiveness in arousal. That the number of newcomers did not become very large even with good feeding was the consequence of the small number of collectors.

Many additional observations corroborate these interrelations.

TABLE 3. Experiment of 24 August 1962 on connection between quality of food, willingness to dance, and effectiveness of arousal. The times are those when the various sugar concentrations were set out. Observations began 5 min after each change in concentration.

| Time | Concentration of sugar solution (M) | Collectors | | | Average duration of dancing (sec/bee) | Number of newcomers at feeding place |
| | | Number returning home | Dancing | | | |
			No.	Percent		
10:00–10:30	$\frac{3}{16}$	29	0	0		0
10:30–11:05	$\frac{1}{4}$	39	3	7.7	0.25	0
11:05–11:40	$\frac{3}{8}$	49	26	53	10.5	3
11:40–12:15	$\frac{1}{2}$	52	38	73	17.6	10
12:15–12:50	1	55	51	92.7	14.9	15
12:50–1:25	2	48	48	100	23.8	18

If one offers 2M cane-sugar solution, one can see long-lasting and vivacious dances. Their duration is measurable with a stopwatch. The liveliness is hard to express in words, but is most impressive for the observer. Only at its extreme can the liveliness of the dance be easily described through the vibratory movements of the abdomen that are performed intermittently during the circling—without emphasis on a particular direction, such as occurs in the tail-wagging dance (pp. 129ff).[6] When the concentration is reduced stepwise the dances become shorter and less lively; possibly one may see now and then with $\frac{1}{4}$M (8.5 per-

---

[6]Wittekindt (1960) describes this as "dances that deviate from the round-dance form previously described." I quote from my article of 1946, p. 12: "When the dances are extremely lively abdominal wagging may occur, but in the course of the circular run." Such "tail wagging during circling" has been recorded frequently in my protocols since 1945. In it I do not see a divergent form of the round dance, but only an expression of increased excitement (cf. p. 238, note 5).

cent) sugar solution the preliminaries of a round dance, but in general there are no dances at this threshold concentration although food collection continues even down to a dilution of about $\frac{1}{8}$M. The figures cited above have no general validity. For the threshold concentration, and along with it the liveliness of dancing at the different concentrations, is dependent on various conditions, among them the state of the natural food supply (v. Frisch 1934:100; Lindauer 1948). The value of food is a relative concept, and not only for humans. Regarded from the biological point of view, the problem is one of bringing working strength to bear at the place where food collection is most profitable at the time.

12. How Do the Recruits Find Their Objective?

The widespread impression that the newcomer flies out from the hive to the feeding place in the company of the discoverer does not hold. Even on the comb she hastens independently to the hive entrance, often she reaches the food dish first, and under favorable conditions it may be observed directly that the dancers fly unaccompanied—the newcomers appear nonetheless (v. Frisch 1923:101-105). But the location of the goal is unknown to them initially. Following the round dances they swarm out in all directions and examine the surroundings of the hive. One can satisfy oneself of this easily by feeding a few numbered bees about 10 m from the hive with a mixture of honey and sugar water[7] in a watch glass and setting out similar feeding stations in the meadow in various directions up to a distance of 70 m. When after about 10 min the first newcomers join the foragers, they appear in all directions at the other watch glasses (v. Frisch 1923:105-109). They do not know *where* to look. But almost always they know *what* to look for.

13. The Odor of the Food Source as a Guide to the Newcomers

Just as in the alerting of her original group comrades the odor clinging to a dancer indicates the kind of flower from which she has been gathering (p. 33), so too the others who follow her perceive on the dancer's body the specific fragrance that is peculiar to nearly every species of flower; they note this while they are still on the comb, even for newly discovered sources of food. In the course of examining the surroundings for this odor they inevitably arrive at the flowers to which the summons referred.

This is demonstrated most simply with essential oils. If one allows numbered bees to collect from a sugar-water vessel the filter paper beneath which has been supplied with a few drops of oil of peppermint, and sets out in the grass near the hive some ten dishes scented with ten different essential oils, among them a single one with oil of

FIGURE 35. A group of bees is collecting from drops of sugar water on cyclamen.

---

[7] In this instance a little honey is added to the sugar water so that the watch glasses set out in various directions will be found more easily by the newcomers.

peppermint, this one is immediately frequented zealously by newcomers while the others are visited but little or not at all. If oil of basil is added at the feeding place, the visits of the newcomers are directed specifically to the basil-scented dish, even if it is put out in quite a different direction and at a different distance from the hive than the peppermint-scented dish was previously. For this it is not necessary that the foraging bees come into actual contact with the applied drops of essential oil. In the experiment with oil of basil, for instance, a brief stay in the vapor emitted by a few drops placed 10 cm away from the feeding vessel was sufficient (v. Frisch 1923:109–116).

14. Experiments with Flowers

The subject gains interest when we shift to natural flowers. An experiment with wild cyclamen (*Cyclamen europaeum* L.) and garden phlox will clarify the arrangement.

At an experiment table a few numbered bees were fed from a sprig of cyclamen the blossoms of which had been supplied with drops of sugar water in order to yield a rich supply of food (Fig. 35). A few meters to one side I set out in the grass a dish with 200 cyclamen flowers, one with 400 open phlox flowers, and yet another (not visible in Fig. 36) with 200 cyclamen flowers. During ½ hr of feeding not a single bee came to the phlox flowers, although a total of 127 newcomers flew about the cyclamen dishes, and in part examined the flowers persistently but found no nectar there. At the feeding place the cyclamen flowers were now replaced little by little by phlox flowers bearing drops of sugar water (Fig. 37). These were visited by the same bees that hitherto had collected from the cyclamen. A half hour later a 30-min count at the flower dishes showed a total of 2 visits to the pair of cyclamen dishes and 112 to the phlox dish.

In a later similar experiment, during the artificial feeding of numbered bees at phlox flowers, I noticed that unmarked bees came in increasing numbers to the phlox plants in the surrounding flower beds and poked about with interest in the blossoms, without getting anything because the deeply concealed nectar of these "butterfly flowers" is not accessible to the bee's proboscis. Previously I had never seen bees at phlox. They had been sent there by the dancers.

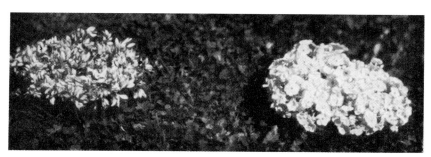

FIGURE 36. By means of their dances the cyclamen foragers dispatch newcomers that visit cyclamen flowers but take no interest in phlox.

FIGURE 37. The same bees feeding on phlox flowers.

The outcome of such experiments disclosed in two respects truly astounding capabilities of the bees:

The first relates to their ability later in the outdoors to find, amid a sea of different kinds of flowers, the fragrance initially perceived in the hive.

I set up my observation hive in the Munich Botanical Garden, at the east end of the systematic section where, in Renner's words,[8] "thousands of plants, shrubs, vines, arranged according to their relationships, are crowded into narrower quarters." At that time 700 species of them were blooming at once. In the midst of this variegated carpet a little bed with an everlasting (*Helichrysum lanatum* Harv.) formed a tiny ornament (Fig. 38). According to Knuth (vol. 2, 1, p. 604) its fragrant flowers are not visited by bees. I too was unable to find any bees on it until (on 14 July 1921) I set up a feeding place 16 m from the hive (at *F* in Fig. 38) and fed ten numbered bees from the observation hive at a sugar-water dish that was surrounded by a wreath of cut *Helichrysum lanatum* flowers, so that while drinking the bees had to stand on the flowers. During the following hour the *Helichrysum* flower bed (72 m distant from the observation hive) was visited by 13 unmarked honeybees; part of them inspected the flowers only casually; others examined them at length yet found nothing. Further experiments showed that the flights were directed specifically to this species of flower (v. Frisch 1923:121–125).

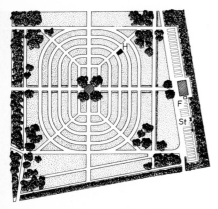

FIGURE 38. The systematic section of the Munich Botanical Garden. *St,* observation hive; *F,* feeding place; *H,* little flower bed with *Helichrysum lanatum,* 72 m distant from the observation hive.

A second surprise came during a series of experiments in which I wanted to include odorless flowers. Actually it became clear to me in the course of this work that such species are a great rarity among insect-pollinated plants. And so at first I was able to ferret out only flowers with a very weak odor. In an experimental arrangement that in principle was comparable to the previously described cyclamen-phlox tests, gentian flowers (*Gentiana asclepiadea* L.) were used in combination with tobacco flowers (*Nicotiana tabacum* L.), thistles (*Cirsium oleraceum* Scop.) with crowfoot (*Ranunculus acer* L.), crowfoot with scarlet runner (*Phaseolus multiflorus* Lmk. var. *coccineus* L.). With crowfoot and scarlet runner the delicate floral odor was so weak that it could be perceived by the human nose only when several flowers were put together in a little bouquet. In spite of this the result continued positive in all cases; the newcomers aroused flew in large numbers to the kind of flowers from which the numbered collectors came (v. Frisch 1923:127–131).

The final yield of extensive search was two entomophilous plants on whose flowers even an acute human nose could detect no trace of fragrance: the iris *Tritonia cocosmaeflora* Voss., a popular decorative plant with showy blossoms, and the wild grape (*Parthenocissus quinquefolia* Planch.) with its inconspicuous flowers that nevertheless are heavily visited by bees.[9] In addition I included the odorless flowers of a wind-pollinated plant (the grass *Holcus lanatus* L.). These experiments with odorless flowers gave negative results (v. Frisch 1923:131–133).

---

[8] O. Renner, *Karl von Goebel, der Mann und das Werk* [Karl von Goebel, the man and his work]. Ber. d. D. Bot. Ges. 68, 147–162, 1955 (p. 152).

[9] In reference to the search for odorless flowers see p. 510.

Even after the bees had been fed on them for several hours, inflorescences of the same species set out in the surroundings awakened no interest among the newcomers that we could see swarming in the vicinity. Thus the flower fragrance was the actual means of communication, and one was forced to accept the idea that flower fragrance, even when extraordinarily weak, must cling to the bee's body sufficiently to be recognizable in the hive after her return. One might suspect also that a flower fragrance that is very weak for us could have seemed manyfold more intense to the bees. But this notion was refuted by many experiments (v. Frisch 1919; Fischer 1957).

In the Winterhalle of the Munich Botanical Garden I was able to convince myself how energetically pollen collectors too with their dances dispatch newcomers to the species of flower they have been visiting. A group of numbered bees was filling its pollen baskets at a dish of wild roses. Presently, thereupon, unmarked bees swarmed out in all directions and quickly found the little patches of roses that were awaiting visitation at various spots in the room. Let us recall here the reversed alerting of the original collecting groups in the experiments with roses and bellflowers whose stamens we had interchanged (p. 35), and we will not doubt that with pollen collectors too newcomers are sent to the proper flowers by means of the species-specific odor of the pollen. That this holds true with free-flying bees despite the large number of simultaneously blooming species was manifest in Lindauer's experiments (1952:329). The young bees in a colony had been numbered and were under continuous observation. Of 91 individuals that returned home successful after their first flight out, 79 brought back the same kind of pollen as the dancer who had stimulated them to go forth.

15. The Adhesiveness of Odors to the Bee's Body

Since even an extraordinarily weak scent is enough to inform the hivemates of the species of flower, odorous substances must cling especially tenaciously to the bee's body. Hildtraut Steinhoff (1948) tested this in two ways: samples of odors were tested with human subjects as well as with bees that previously had been trained to the scent in question.

Either whole bees, killed by CO_2, or only their abdomens were exposed for a set time to an odorous material in a petri dish; the ability of human subjects to recognize the substance was then determined at intervals, while similarly treated glass, china, brass, cotton wool, filter paper, and wooden objects of like form and size served for comparison. It developed that the bee's body had a greatly superior ability to retain the odorants in recognizable form. On living bees (in little cages of wire mesh) the odors could be detected about twice as long as on dead ones, perhaps because they had penetrated the body via the tracheal system and were gradually exhaled. Of greater interest was the testing of the scented objects by the olfactory organs of the bees themselves. For this purpose bees were trained to a particular odor out of doors and then given the task of finding, among four bee abdomens

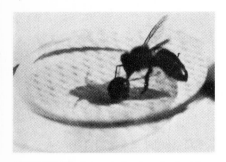

FIGURE 39. A bee trained to linalool has flown to the amputated abdomen of a dead bee that was scented with linalool, and is drumming it with her antennae. After Steinhoff 1948.

presented in a little glass vessel, the one that had been scented. Figure 39 shows how definitely a bee trained to linalool evinces by drumming with the antennae her interest in the correct object. No attention was paid to untreated abdomens.

It was evident that for the olfactory organ of the bee, too, odors are best retained by the bee's body and that natural flower fragrances, for instance, remain recognizable for $\frac{1}{2}$ to more than 1 hr after application. Since only dead bees were used for these experiments, and since on testing with the human nose the retention time was doubled if the scented bees were alive, still longer times are to be assumed for natural conditions. Inasmuch as the time of persistence was about halved by defatting the bee's body, one suspects that the wax in which the bee's integument is so rich participates in the very tenacious retention of the odorants (cf. Richards 1953; Locke 1961).

16. Absence of Communication about Colors and Shapes

When the bees continued always to direct their comrades to the correct goal even with almost odorless flowers, doubt arose temporarily whether odor alone really is the signal or whether possibly they could be giving information in the hive about the appearance of the flowers. As was to be expected, this suspicion was not confirmed. When the numbered bees were fed on a blue background, similar blue surfaces set out in the vicinity received as little attention as yellow ones. I had reported on such experiments in 1923 (pp. 117, 118), and in 1950 I once again convinced myself that this is so. In addition it was confirmed by C. R. Ribbands (1953). Even when the different colors are combined with different floral shapes the experiment does not turn out otherwise. And when I fed numbered bees on artificial yellow "gentians" (Fig. 40) to which a trace of oil of peppermint had been added, yellow "gentians" exposed nearby minus the odor went unnoticed, whereas blue "asters" with the odor of peppermint were swarmed upon by unmarked bees and were visited persistently. The odor was the deciding factor.

FIGURE 40. Artificial flowers made of colored paper: *left,* gentian shape; *right,* aster shape.

17. The Role of the Scent Organ

In many species of insects the females have dermal glands that by means of their odorous secretion attract the males. In the honeybee worker, free of any sexual meaning, such a scent organ has been turned to the service of the social organization. The organ lies on the dorsal side of the abdomen, at the anterior margin of the last visible tergite, and consists of several hundred unicellular glands; their ducts open into the pouchlike invaginated membrane between the last two abdominal segments, and the secretion collects here (W. Jacobs 1925). This pouch can be everted, whereupon the light-colored intersegmental membrane is exposed as a sickle-shaped pad (Figs. 41, 42); then the secretion evaporates and diffuses its balmy odor. Boch and Shearer (1962) identified chromatographically and spectroscopically one of its constituents

FIGURE 41. A forager at the sugar-water dish everts her scent organ. Photograph by M. Renner.

as geraniol. Its production first starts in bees about 10 days old, and increases strongly when the foraging flights begin—hence just at the time when the scent organ takes on biological importance (Boch and Shearer 1963). According to Free (1962), bees are indeed attracted by synthetic geraniol but not nearly so strongly as by the product of the scent organ, so that other components of the secretion must be presumed active. Lately Boch and Shearer (1964) were able to identify in it the isomer nerol in addition to geraniol. Mixtures of geraniol with synthetic *cis*- and *trans*-geranic acid are only slightly less attractive to bees than the natural odorant.

The queen of *Apis mellifera* lacks this organ. However, in other abdominal segments she has groups of glandular cells that are absent in the workers (Fig. 43); these cells are particularly numerous on the second, third, and fourth visible abdominal segments. They occur at the posterior margin of the segments, whereas the scent organ of the worker lies at the anterior margin of its segment. In the queen the glandular cells are similar to those of the worker. Their ducts lie dorsally, near the segmental boundary. They produce a specific aromatic odor that is strikingly noticeable on queens at the time of their nuptial flight. Apparently it—in addition to the secretion of the mandibular glands—acts as a sexual attractant; cf. p. 289 (Renner and Baumann 1964).

The significance of the scent organ of the worker as a gland of attraction was first realized by Sladen (1902). In many circumstances, for instance when a swarm moves to a new dwelling, bees with erected abdomen ("sterzelnde" Bienen) are seen in front of the hive entrance, showing a characteristic behavior: headed toward the hive, they elevate the abdomen, evert the scent glands, and fan the wings, thereby wafting

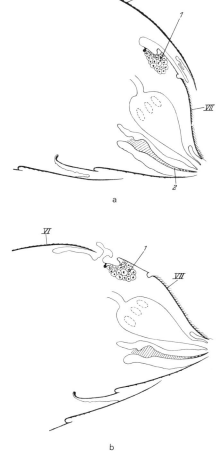

FIGURE 42. Sections through the abdomen of a worker bee, schematic: (*a*) scent organ in rest position; (*b*) scent organ everted, abdomen lifted at an angle; VI and VII, sclerotized integument (tergites) of the 6th and 7th abdominal segments; 1, scent gland; 2, sting. Drawings by W. Jacobs.

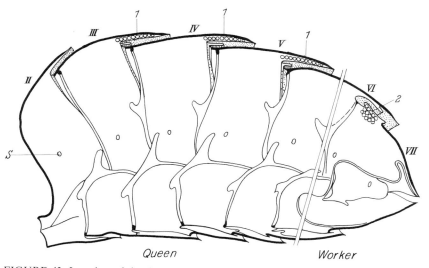

FIGURE 43. Location of the dorsal scent-gland cells on the bee's abdomen: *left*, in the queen; *right*, in the worker. II–VII, tergites (dorsal plates of the abdominal segments). Tergite I is not visible externally. 1, glandular complexes in intermediate dorsal segments of the queen; 2, scent organ in the worker. Drawing by M. Renner.

FIGURE 44. Bees sit in front of the hive entrance, evert their scent organs, and by fanning their wings waft the odor toward incoming bees (*sterzeln*). Photograph by E. Schuhmacher.

FIGURE 45. Experiment to demonstrate the importance of odor in identifying a feeding place: 1, observation hive; 2, fence palings; *a* and *b*, feeding stations, each 10 m from the hive; see text.

the air with the odor backward and showing their comrades the way (Fig. 44). That they also make use of the scent organ in facilitating the location of the feeding place by the newcomers they arouse, I saw in experiments of the following kind (v. Frisch 1923:146ff, 158ff). Of two numbered groups from the observation hive 1 (Fig. 45), one is fed copiously with sugar water at *a*, the other is fed meagerly (as in Fig. 19) at *b*. The bees coming from *a* dance in the hive, those from *b* do not dance. Since the newcomers alerted by group *a* hunt everywhere round about, while neither of the two stations is marked by a specific odor, one should expect equal numbers of newcomers at both stations. Actually, during the same time interval about ten times as many unmarked bees came to the plentiful food source as to the meager source. If now feeding was made copious at *b* and meager at *a*, the influx of newcomers changed accordingly.

This phenomenon caused some headscratching before the simple explanation was found. Many of the foragers and often practically all of them[10] that have danced in the hive and return to the rich source of food hover with everted scent organ over the watch glass for some time before they settle; in doing this they impregnate the surroundings with the odor of their glandular secretion, so that it is apparent even to our nostrils. After they have settled, they usually keep the organ everted for some time (without fanning the wings); in the further course of drinking it is pulled in. In general they make use of the scent organ only when they have danced in the hive, and the more vigorously they have attempted to recruit, as a rule the more abundantly the feeding place is perfumed. On the other hand the meagerly fed bees do not dance and when they return to the feeding place they do not evert their scent organ.

[10] The scent organ is used with individually differing frequency.

That it is really the odor that so strongly attracts to the rich source of food newcomers searching in the vicinity can be demonstrated. By means of a fine paintbrush one covers with alcohol solution of shellac the scent organ of all bees from station a (Fig. 45), for instance. The shellac dries rapidly and sticks the tergites together, so that the organ can no longer be everted. Now we offer sugar water copiously at both stations. The bees from a dance just as intensively as the other group, they hover about their feeding place just as persistently in the endeavor to perfume it, but they do not succeed in doing so. And now, in spite of the rich feeding, the group with sealed scent organs is joined by just as few newcomers as came previously to the meagerly fed group that had not used their scent glands. Table 4 summarizes the results of six experiments, in which on three different days the scent organ was sealed alternately for the collecting group at a or at b (v. Frisch 1923:163–166).

The well-fed bees not only evert the scent organ but also arrive at the feeding dish with a clearly higher flight tone than the meagerly fed insects. At first there was doubt as to which is the effective signal. But since well-fed bees fly in with as high a flight tone after the scent organ has been sealed up as they did before, and in spite of this attract just as few newcomers as do meagerly fed insects, the flight tone is without significance as a means of attraction (v. Frisch 1923:166–170). This agrees with the investigations of Hansson (1945), according to whom bees in flight do not perceive tones (sounds transmitted through the air); vibrations transmitted through the substrate can be perceived, but different frequencies cannot be distinguished.

Under natural conditions, when flowers are being visited, the everting of the scent organ is less easily observed. For instance, I watched for it in vain in the Munich Botanical Garden when, at the end of June, locust (*Robinia viscosa* Ventenat) was in full bloom and was frequented heavily by bees. But there was no nectar to be seen in the flowers. With so many visits the sugary sap exuded was licked up at once, and the foragers found "meager feeding." One rarely sees the early morning

TABLE 4. Seven numbered bees frequented each of two feeding stations, a and b (Fig. 45). Alternately at the two stations, the scent organ of the food-gathering bees was sealed. About ten times as many newcomers joined the normal bees.

Date (August 1921)	Time	Normal bees		Bees with sealed scent organs	
		Feeding station	Number of newcomers	Feeding station	Number of newcomers
6	8:40–9:40	a	30	b	5
	11:10–12:10	b	12	a	1
7	7:05–8:05	b	28	a	2
	9:15–10:15	a	23	b	0
9	7:30–8:30	a	17	b	1
	9:30–10:30	b	13	a	3
	Total		123		12

scouts or the first discoverers that come to the full calyces. When in the Winterhalle I set out cut locust branches, in the flowers of which nectar had again accumulated, these were hovered over with everted scent organs by the bees that had danced in the hive and were returning to the flowers, just as with the watch glass of sugar water. Even after they had settled on the flowers the organ remained everted at first. Bees that were collecting pollen from roses (*Rosa moschata* Herm.) behaved similarly. Not until the food supply grew slight was the perfuming of the place omitted. On Ceylon C. G. Butler (1954:141) frequently noticed the eversion of the scent organ during visits to flowers by *Apis indica* F., a close relative of our honeybee, and by the dwarf honeybee, *Apis florea* F.

Thus the scent organs of the worker bees enable them when returning to the known rich food source to direct newcomers searching close by to the right place.

Since according to repeated observations the distance covered by this signal extends several meters, the odor must be more intense for the olfactory sense of bees than it is for us. This has been confirmed experimentally. According to threshold determinations in training tests (v. Frisch 1919), methyl heptenone smells approximately as strong to bees as to ourselves. When a foraging group was fed in the presence of this odor, and a second food dish with methyl heptenone, which no bees were visiting, was set up at another spot, the newcomers came in an overwhelming majority to the place where the scent glands provided an additional attractant (v. Frisch 1923:171, 172), although the artificial odorant had a much stronger smell for us. W. Fischer (1957) checked the matter with the odorous material isolated from the organ and likewise found a severalfold superiority in the olfactory sensitivity of the bee compared with man. With the exception of the smell of wax, this is not the case with any of the other artificial perfumes and flower odors tested hitherto (v. Frisch 1919; Fischer 1957).

The question whether the odor from the scent organ is identical in different colonies or whether it is attractive specifically to the members of a single hive led to differences of opinion. The second viewpoint was advocated by Rösch and me (1926), by Kaltofen (1951), and by Kalmus and Ribbands (1952). R. Wojtusiak (1934), who pursued further in our laboratory the experiments begun with Rösch, obtained contradictory results. Renner (1960) first made decisive improvements in technique and clarified the situation. He carefully collected the secretion of everted scent glands from living bees by dabbing it up with little bits of filter paper, and found this isolated glandular product just as attractive as were living bees who were using their scent organ. The secretion was as attractive to other colonies as to the fellow members. Since it was imaginable nevertheless that the odor from another colony indeed differed even though it was interpreted as a token of a desirable supply of food, Renner trained bees to the pure attractant, with the definite result that the secretions of different colonies, even if they belonged to distinct races, did not differ in respect to their olfactory qualities (tested for *Apis mellifera carnica, ligustica,* and *nigra.*).

Renner was able to show also how the contrary opinion had arisen. The hive odor of a colony of bees, as was demonstrated for floral scents (p. 49), clings tenaciously to the bodies of the inhabitants, and these hive odors are specifically distinct. Among other factors they are influenced by the flowers favored in the collection of food by the colony in question. The affinity of the bees to a given colony, or the opposite, can be recognized through this hive odor on the body, and this was why the earlier experiments, in which entire bees as well as the odor from the glands were concerned, gave such discordant results.

Even those bees that, by fanning with erected abdomen and everted scent gland on the approach board of a heavily populated hive, point the way to the correct entrance, are not able to indicate the proper home by means of an odor peculiar to that specific colony. This may surprise many people. And it must be kept in mind nevertheless that in fanning before the entrance they waft the odor-laden atmosphere of their hive toward those arriving and thus favor recognition of the maternal dwelling. But on the other hand Renner correctly points out that the original and natural homes of the bees, hollow trees, do not stand in rows adjacent to one another like beehives. With habitations less compactly arranged, a hive-specific action of the attractant is not necessary for distinguishing the home.

Summary: Recruitment of New Bees by Round Dances

1. A bee that first discovers a source of food, by means of her round dance recruits helpers to whom the goal is at first unknown: "newcomers." Also, when an already familiar source of food begins to flow again there are in the train following the dancers in addition to their group comrades (identified by their numbers) newcomers that join the original group. They come only when there is dancing, the more of them the more numerous and lively the dances.

2. The profitability of the food source determines the number and liveliness of the dances. In the collection of nectar this is determined mainly by the quantity and sweetness of the sugar solution. When the food supply is copious nearly all the foragers dance vigorously and at length each time they come home. If the quantity or the sugar content of the food diminishes, the dances grow less lively and shorter. The influx of newcomers decreases. With a meager amount of sugar water or—more easily arranged—at a certain limiting concentration, the dances cease. Although foraging persists, no newcomers are recruited for the now relatively unprofitable supply. If matters become yet worse, the original foragers terminate their activity. If the natural food sources are good, the limiting concentration is higher than during times of scarcity. These relations are of the greatest biological significance for the strength of the colonies.

3. Newcomers alerted by the round dances search the vicinity in all directions.

4. If a feeding place is marked with an ethereal oil, the newcomers take note of the odor in following after the dancer. When they fly forth they search for this specific odor.

5. They behave in precisely the same manner in frequenting natural flowers. The specific odor of the different kinds of flowers enables the newcomers to find with certainty, amid the lush sea of flowers out of doors, the blossoms of that species of plant from which the dancer has collected successfully.

6. Even flowers with a faint odor are found in this way. That is possible because the odors cling exceptionally well to the bee's body—better than to all other objects tested in comparison.

7. Olfactory cues of two kinds facilitate the finding of the flowers by the newcomers: they not only learn the specific fragrance of the flowers frequented from the odor that clings to the dancer's body, but their location outdoors also is brought to their attention by the particular body odor of the foragers. When the latter have danced in the hive and return to the flowers (or to the sugar-water dish), during their approach and after resettling they evert their scent organ (Fig. 41 and p. 50). Its odor is readily perceptible to us, but is unusually intense for bees. It can guide the newcomers to the right place from a distance of several meters. If the food supply is so sparse that there is no dancing, the scent organ also is not everted during the approach to the flowers.

The smell of the scent organ is identical for different colonies and—so far as has been tested—for different races of bees too; but the hive odor of different swarms differs.

IV / The Tail-Wagging Dance as a Means of Communication when Food Sources Are Distant

When the goal is 100 m or farther from the hive the round dance is replaced by the tail-wagging dance (Fig. 46). Like the round dance, it announces in the first place the existence of a profitable source of food, secondly the kind of flower frequented (by means of the floral odor clinging to the dancer), and thirdly the potential productivity of the food source (by means of the liveliness of the dances); and here too the scent organ is set into action on the return to the feeding place and the newcomers' locating this correctly is thereby facilitated. But beyond this the tail-wagging dance makes known the distance and the compass direction to the goal, the first by the tempo of the dance, the latter through the direction of the straight segment of the run in the dance pattern. This description of the location enables the newcomers to fly rapidly and with certainty to the indicated flowers, even when these are kilometers away—an accomplishment on the part of the bees that is without parallel elsewhere in the entire animal kingdom.

1. Description of the Tail-Wagging Dance

In the typical tail-wagging dance the bee runs straight ahead for a short distance, returns in a semicircle to the starting point, again runs through the straight stretch, describes a semicircle in the opposite direction, and so on in regular alternation. The straight part of the run is given particular emphasis by a vigorous wagging of the body. This results from rapid rhythmic sidewise deflections of the whole body that are greatest at the tip of the abdomen and least at the head. The axis about which the sidewise oscillation is to be envisaged lies close before the bee's head and perpendicular to the substrate (Fig. 46). The movement to and fro is repeated 13–15 times a second.[1] An additional emphasis, discovered much later, is given to the tail-wagging movement by the fact that the dancer, simultaneously with her tail-wagging

FIGURE 46. The tail-wagging dance. Four followers are receiving the message.

[1] Hoffmann, Köhler, and Wittekindt (1956) record about 15 cycles/sec, likewise Esch (1956, 1956a). Steche (1957) gives 14–15.5 cycles/sec, v. Frisch and Jander (1957) 13 cycles/sec. It is possible that the rate of waggling varies somewhat with different swarms. There is general agreement that it is not influenced by the distance to the goal.

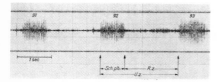

FIGURE 47. Sample recording of sound production during three cycles of a tail-wagging dance. Feeding station 300 m distant. *Sch. ph.*, tail-wagging phase; *R.z.*, duration of return run; *U.z.*, duration of total circuit. After Esch 1964.

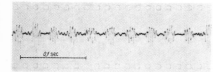

FIGURE 48. Pattern of vibratory movements during the tail-wagging run, recorded acoustically; from a dance for a feeding station 4000 m distant. After Esch 1961a.

FIGURE 49. Structure of a vibratory episode, recorded acoustically at a higher recording speed. After Esch 1961a.

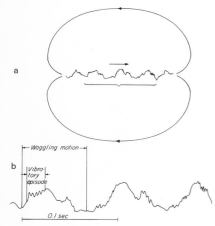

FIGURE 50. Electromagnetic recording of the tail-wagging and vibratory movements: (*a*) pattern of the tail-wagging dance, with vibratory movements during the waggling segment; (*b*) three tail-wagging movements, enclosed by the brace in (*a*), with superimposed vibratory movements. Feeding station 150 m distant. After Esch 1961a.

and during this portion of the dance pattern only, emits a buzzing sound that is readily heard if one inserts a plastic tube in the ear and holds the other end at a distance of 1 cm above the bee's thorax.[2] By means of a microphone the sound can be recorded. Figure 47 shows an example. Four circuits of a dance are reproduced. The production of sound takes place each time during the tail-wagging phase (*Sch. ph.*) of the dance.

A more detailed analysis is afforded by Fig. 48, in which a segment of a single wagging phase is recorded at higher speed. One sees that the oscillations arise as the result of repeated short vibratory episodes (Fig. 49). Each individual episode lasts only a fraction of a second (15 msec) and is separated from the next one by an equally short pause. On the average there are about 30 episodes/sec. It is this frequency that our ear perceives as sound. The frequency of oscillation within the short individual episodes themselves is about 250 cycles/sec and thus corresponds to the frequency of wingbeat. Since the source of the sound lies in the thorax and the latter contains the flight musculature, evidently we are dealing here with short buzzing noises, movements denoting an intention to fly, without the development of an actual wing stroke.[3]

In the acoustic recording of Fig. 48 the form of the sound vibrations stands out very clearly. After fastening a tiny magnet to the dancer's back, Esch was able also to record the vibrations electromagnetically on tape, together with recordings of the waggling movements of the abdomen (Fig. 50). Now one sees that the short vibratory episodes are superimposed independently on the waggling movements and are not confined to certain parts of them. Yet the duration of sound production corresponds precisely to the duration of the waggling run, which is therefore accentuated not only by the waggling movement but at the same time by the production of sound (Esch 1961, 1961a; Wenner 1962).

At a low sugar concentration, which releases only a feeble dancing activity, "silent dances" often are to be observed. Sound production is omitted. Esch (1963) determined that in 6000 cases of silent waggling runs no newcomer came to the feeding place (at a distance of 150 m). On the other hand it happened also that on rare occasions vibratory episodes occurred throughout the entire dance, even during the return phase. Such dances too were without success in recruiting. From these observations Esch concludes that the emphasis of sound during the waggling phase is indispensable for successful alerting.[4] This conception is attractive, but it is not yet established. The inefficacy of weak inceptions of dancing can be attributed to their feebleness as well as to their silence—as is indeed the case with round dances, which are silent in general (see p. 238, note 5). Emission of sound during the return phase also occurs only exceptionally, so that any connection with success in recruiting is hard to demonstrate.

[2] According to Esch (1961a), the sound intensity 1 cm dorsal to the thorax is 70–80 phon.
[3] In the same manner, according to Simpson (1964), the queen produces her tooting and quacking sounds (see p. 286).
[4] Sound alone—the noise given by a loudspeaker—does not result in any arousal (Esch 1963, comment during discussion).

During the tail-wagging run the wings are spread slightly and to differing extents, often scarcely noticeably.

When the distance to the feeding place is not more than 300 m the straight portion of the tail-wagging dance extends over only one or two cells of the honeycomb, but grows larger with increasing distance to the goal, since the tempo of dancing slows down and the duration of wagging increases (cf. pp. 97ff). For a distance of 500 m the tail-wagging run covers two or three cells, for 1000 m three cells, for 2000 m three or four cells, for 3500 m four cells, for 4500 m four or five cells.[5] These distances of run are not adhered to very exactly, however, and may vary by more than a cell width for a single distance. The amplitude of waggling may change even within a given run. Often the abdomen is swung strongly and regularly to and fro, often only weakly or even scarcely noticeably. In general the waggling motion is limited to the linear portion of the run, but exceptionally it may begin somewhat earlier, during the incurve, or with a few wagglers may be prolonged beyond the straight run into the outcurve. It may start sharply and energetically, or it may be augmented gradually.

The semicircles over which the dancer returns approximately to the starting point after the completion of her tail-wagging run are as a rule traversed alternately to right and left, but exceptionally two right turns or two left turns may succeed one another. In about 1000 tail-wagging dances recorded with this in mind (by motion pictures), this occurred in about 1.5 percent of the cases, without there being any recognizable relation between the frequency of such a happening and the distance (200 to 4500 m) to the feeding place. It is worth noting that at times the dancer does not run any closed semicircles at all, but resumes the straight waggling run after following an S-shaped curve. When the goal is at a great distance this may be seen quite often. Counts made on photographs showed S-curves only after 3–4 percent of the waggling runs for food places as much as 1000 m distant ($n = 544$), after 14 percent for 2000 m ($n = 43$), and 33.3 percent for 4500 m ($n = 48$). In the course of such behavior the length and shape of the return run vary appreciably. Figure 51 shows some examples from films.

Just as with the round dance, the tail-wagging dancer is trailed after by other bees that, with extended antennae, seek to keep in touch with her and acquire information. In Fig. 46, for the sake of clarity only the dancer and her train of followers are shown. In actuality the event takes place in the midst of the massed throng, while the dancing group in its energetic rounds pushes aside the uninterested portion of the other hive members and thus makes itself sufficient open space. Figure 52 (copy from a photograph, feeding place 2000 m distant) gives some conception of this situation. In this case the train of eight following bees was larger than usual. But exceptionally I have seen ten.

Like the round dance, the tail-wagging dance of long duration is interrupted one or more times; during the pauses the dancer distributes to the followers food from the contents of her honey stomach (cf. Fig. 28). Esch (1964) in listening to the dancers with a microphonic probe,

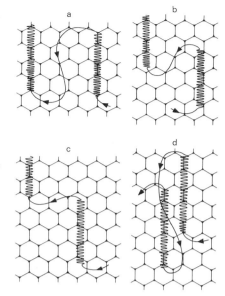

FIGURE 51. Four photographically recorded examples of tail-wagging dances with S-curves during the return run. Feeding station 4500 m distant.

FIGURE 52. Tail-wagging dance with a large number (eight) of followers. In this sketch are shown also the nonparticipating bees in the vicinity. The movements of the dancing group clear a little free space on the thickly populated comb. The positions of the bees are taken from a photograph. Feeding station 2000 m distant.

---

[5] According to films and direct observations.

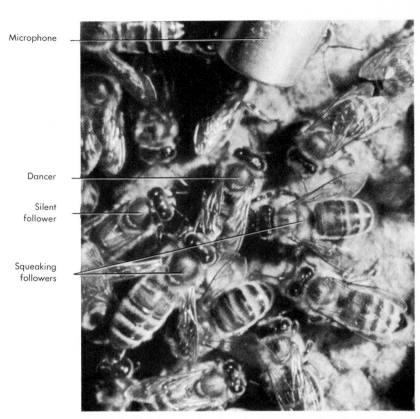

FIGURE 53. Photograph of a dance scene: microphone above, a dancer in the middle. A squeaking follower approaches from the side at the right; a second squeaking bee is climbing over other bees behind the dancer. During squeaking the wings are held spread apart and upward, like a wedge. After Esch 1964. Photo H. Miltenburger.

FIGURE 54. Oscillogram of the tail-wagging run of a dancer, and a squeak of a follower: *Sch.ph.*, tail-wagging phase; *P.l.,* squeak. After Esch 1964.

FIGURE 55. Excerpt from an oscillogram of a squeak; the time scale different from that of Fig. 54. After Esch 1964.

discovered that they may be brought to a halt by a squeak from the followers. In doing this a follower will suddenly stop, spreading its wings to a V (Fig. 53); simultaneously there is heard a squeaking tone with a duration of 0.1–0.2 sec. The frequency lies between 300 and 400 cycles/sec. An oscillogram of a waggling phase (of a dancer), together with the associated "squeak" (of a follower), may be seen in Fig. 54; an excerpt on a different time scale in Fig. 55. The dancer and the surrounding bees are "paralyzed" by the squeak. For the most part the dancer is approached immediately thereafter for food by the bee that utters the squeak. Hence it is probable that the squeaking may be regarded as a demand for delivery of a food sample.

When the squeaking, recorded on a tape, is piped into the hive over a loud speaker, all the bees are seen to become motionless for the duration of the sound. But since this occurs only when the sound is of very high intensity or is transmitted directly to the substrate, perception in this case too seems to be conveyed by immediate contact.

According to Esch's observations, almost every follower emits squeaks in the autumn, but only an occasional one does so in the spring and in the summer. The phenomena deserve thorough study.

2. The Transition from the Round Dance to the Tail-Wagging Dance

If the feeding place is 10 m from the hive, only round dances are seen (Fig. 29). From about 25 to 100 m is the region of transition from round dance to tail-wagging dance. The shift is gradual and may take place in the following way (Fig. 56, upper row of diagrams).

The course traversed, which hitherto has encompassed the diameter of approximately one comb cell, is extended in alternation to one of two neighboring cells; the single circle becomes two, the path a figure eight. Presently the pattern assumes a definite position with reference to the direction of flight, and at the place where the circles cross a brief waggling ensues. As the distance to the feeding place increases, the waggling grows livelier and there become evident *two* straight portions of the run, two divergent waggling segments, the bisector of the angle between which indicates the direction to the goal. At greater distances of the goal the divergence of the waggling segments decreases until the two portions of the run are superimposed, leading to the typical picture of Fig. 46.

I have seen this form of transition as the rule among Carniolan bees. With other races, for example *Apis mellifera mellifera* or *A. m. ligustica*, the shift occurs as depicted in the lower row of Fig. 56. In place of the ∞-course there occur sickle-shaped runs, open arcs, resembling a compressed ∞. Before the turning points, at the ends of the sickles, here too there develops a waggling, at first merely suggested, that with increasing distance becomes a true waggling segment.

I have seen both transitional forms side by side in a single swarm. Probably this is connected in part with the fact that truly racially pure stocks are rarely available. But it must be emphasized also that the illustrations in Fig. 56 are to be looked on as schemata, as types that are recognizable over and over in the dances but that are varied in a multitude of ways. Mostly the dancer does not remain precisely on a single spot. As she shifts across the comb the course traversed is somewhat deformed. Even in a single location the pattern may be repeated exactly or with less regularity, one might say in a slovenly way. In this respect clearly there are individual differences.

Within the limits given above the distance at which an indication of the direction becomes recognizable and at which a definite tail-wagging dance is performed is variable in general and also differs with individ-

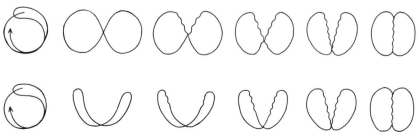

FIGURE 56. Transition from round dance to tail-wagging dance, schematic: *upper*, via ∞-form; *lower*, via a sickle-shaped dance.

uals. As a rule when the distance of the feeding place is 25 m, a few short, definitely directed waggling runs are to be noted. At 50 m, such runs may be seen with about half the dancers, and at 100 m the tail-wagging dance is dominant; even there scattered round dances may still occur from time to time. Some degree of divergence of the waggling runs may be preserved at distances up to several hundred meters.

3. Comparison of Nectar and Pollen Collectors

In the historical review I have mentioned already that at first (in 1923 and even as late as 1942) I took the tail-wagging dance for the form typical of pollen collectors. In order simultaneously to keep in view what was taking place both on the comb and at the feeding place, I set up the latter beside the observation hive; and since with even this simple arrangement the bees afforded one surprise after another it did not occur to me initially to establish the feeding station at a distance. Hence, with my collectors of sugar water and nectar I saw only round dances, but with the pollen collectors, who were coming from natural food sources at a great distance, there were the conspicuous tail-wagging dances. Surely on occasion there must have been also tail-wagging dances by nectar collectors coming from natural food sources at a distance, but since the pollen collectors frequently repeat their dances several times after they have stripped the loads from their legs one could not judge immediately whether a dancer had been collecting nectar or pollen. When finally (1946) the dependence of the form of dance on the distance became clear, I was so thoroughly entangled in my error that a complete alteration in my way of viewing the situation was first required. In order to achieve this it was necessary to compare sugar-water collectors and pollen collectors that were harvesting, at different distances, from artificial feeding stations of precisely known location.

To cause bees in the open country, where usually there is a rich supply of pollen at their disposal, to gather pollen from an artificial feeding station, one must first have a sufficient quantity of a suitable pollen and in the second place must make use of a trick. The choice of pollen is left to forager bees from other colonies during their natural harvesting flights. On their return to the hives we remove their loads mechanically with the "pollen traps" customarily used by beekeepers; these are wire screens that are placed in front of the hive entrance so as just to admit the bees but in doing so to strip from their legs the accumulation of pollen and collect it in a container below. In order that this pollen can be reharvested it must first be ground in a porcelain dish, dried in an incubator, and then reground. When the humidity is high it must be dried again as soon as the harvesting runs into difficulties. In getting the bees to collect pollen at an artificial feeding station, one places a shallow watch glass of sugar water in a petri dish the bottom of which is covered with pollen (Fig. 57). If now the sugar water is allowed to become exhausted, after some search one or another of the foragers will usually begin to harvest the pollen. To ensure suc-

FIGURE 57. In order to make bees gather pollen, their feeding dish is placed with pollen in a larger glass vessel and the sugar water is allowed to become exhausted.

cess it is recommended that a previous feeding of concentrated sugar solution be offered, so that newcomers appear. They are more easily induced to gather pollen than long-accustomed sugar-water collectors.

In July 1945 an artificial feeding station was set up next to the observation hive, and in the course of 10 days was moved step by step to greater distances; the behavior of sugar-water collectors and pollen collectors was observed. At distances of 10 m and 25 m the pollen collectors performed round dances, like the sugar-water collectors. At 50 m, both occasionally let some waggling be seen; at 100, 200, 300, and 500 m, the sugar-water collectors performed tail-wagging dances that differed in no way from those of the pollen gatherers (v. Frisch 1946: 24ff).

Once my attention had been directed to the situation, I was able to discern tail-wagging dances in unmarked bees that were returning from natural sources of food and that could be identified as nectar collectors because they offered little drops from their honey stomachs. Actually an occasion for such observations frequently is not given in our somewhat scanty food-producing areas because, with augmenting collection by the bees, the stores of nectar in the flowers soon become so meager that lively dances no longer take place.

Summary: The Tail-Wagging Dance

1. When the source of food is more than 100 m distant from the hive, the round dance is replaced by the tail-wagging dance. In addition to the data supplied by the round dance, the tail-wagging dance brings the hivemates information as to the distance and direction of the goal.

2. In the tail-wagging dance the bee runs through a short straight stretch and then returns in a semicircle—alternately to the right and to the left—to the starting point (Fig. 46). The straight portion of the course is given emphasis by lateral waggling motions of the body and by a buzzing noise. There are about 13–15 waggles (to right and left) per second. The noise results from repeated short vibratory episodes (about 30/sec); the vibrations have a frequency of about 250 cycles/sec (Figs. 47–50). Apparently the sound is produced by the thoracic musculature; the frequency of 250 cycles/sec corresponds to the wingbeat frequency. It is tempting to interpret this as movements indicating intention to fly off. The duration of sound production corresponds to that of the waggling movements. Tail-wagging dances with deviant patterns occur (Fig. 51). The dance followers are able by means of a squeaking sound to make the dancers halt and give out sugar water (Figs. 53–55).

3. The transition from the round dance to the tail-wagging dance is a gradual one, that takes place via either a figure eight or a sickle-shaped pattern (Fig. 56).

4. There is no difference between nectar and pollen collectors with respect to the form of their dances.

A. THE INDICATION OF DISTANCE

4. The Tempo of the Dance

When in the summer of 1945 I for the first time moved the bees' feeding station stepwise to greater distances, it struck me that there was a related change in the tempo of dancing. With the goal 100 m away, the dancer makes her circles to right and left hastily, with short waggles during the straight run. In the space of ¼ min she makes nine or ten circuits (1 circuit = a waggling run + a semicircle returning to the starting point). With the feeding station 500 m away, there are only six circuits in this time, and at 1500 m only four. The dance has slowed down, but has by no means become less lively. The interpolated waggling run comes more and more to dominate the picture, and this movement can be so vigorous that the invitation to fly out impresses the unprejudiced observer as being at least as emphatic and effective as the hectic circles of a bee dancing for a nearby source of food. With a stopwatch I measured the number of circuits per 15 sec for various distances of the feeding station. The first extensive experiment of this kind, during the period 7–23 June 1945, led in a northeasterly direction from Brunnwinkl through a somewhat elevated grassy valley to some 1500 m from the hive. Figure 58 shows the number of circuits per 15 sec (average values) for the individual stations of this journey.

In timing the dances, initially I counted the number of circuits per 15 sec. For short distances (rapid circuits) this can be done with some degree of accuracy. Half-circuits can be estimated satisfactorily. But we soon shifted to another procedure, the only one to use for greater

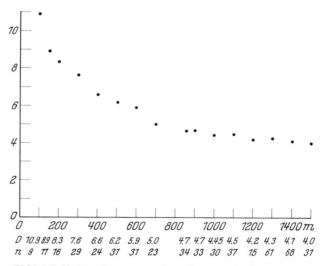

FIGURE 58. The dance tempo in a distance experiment over a range of 1500 m at Brunnwinkl, 7–23 June 1945: abscissa, distance to the feeding station; ordinate, number of circuits danced per 15 sec; D, average numbers of the circuits for n measurements.

distances: beginning always with the start of a waggling run, one records the time required for a definite number of circuits and later calculates the rate in circuits per 15 sec. In the interests of accuracy the preselected number of circuits should not be too small, but on the other hand one is limited by the duration of the dances; depending on the distance, we measured ten to two circuits.

The next question was whether the "distance curve" continued to follow a regular pattern in its further course, all the way to the limits reached in flight. As early as 2–16 July 1946 I ran an experiment that began at St. Gilgen and led northwestward through the gently rising meadows toward Fuschl to a distance of 3 km, at which the limits of the range covered in flight were reached, according to generally accepted opinion. And actually at this distance on two days of experiment favored by the finest weather only two of the numbered bees finally appeared (at 2800 m there had been 17), and only a single newcomer showed up. Nothing was to be gained by extending the journey farther. The dots in Fig. 59 show the results of this experiment.

On the same graph are entered as circles the values that M. Lindauer obtained 14 years later (28 August–3 September 1960) in an experiment in the Forstenrieder Park in Munich. This test extended for 4 km from the hive. Anyone with a clear conception of the performance of the dancers will regard such agreement with respect: in different years bees from differing, not closely related, colonies flew for kilometers over different countrysides, had to estimate and remember the distance flown, and on their return home were able, with their fine feeling for time, to dance at a tempo whose average values were in extensive agreement.

Have we reached the attainable limits with these distance experiments? What is actually the scope of the flying bee? Opinions in regard to this are at considerable variance.

According to v. Buttel-Reepen (1915:161ff) bees have a radius of action of 3–4 km around their hive, insofar as they are not impeded by

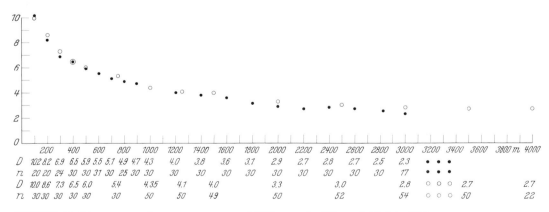

FIGURE 59. Distance experiments over ranges of 3000 m at St. Gilgen, 2–16 July 1946 (. . .) and 4000 m at the Forstenrieder Park in Munich, 28 August–3 September 1960 (o o o). For further explanation see Fig. 58.

high mountains. When the immediate surroundings are poor in flowers, the bees exceptionally may be enticed by fields of rape and other good food sources to collecting places 6 or even 7 km distant. The "bee baron" v. Berlepsch (1860:176) was the only one in his neighborhood to maintain bees of the yellow-banded Italian race, and hit on the good idea of looking for his naturally marked domestic animals in the surroundings. He saw them on flowers as much as 2–3 km distant, and one year during fine weather and a shortage of nearby food even up to 7 km. With the bees at her artificial feeding station, R. Beutler (1954) attained a maximum distance of 2650 m from the hive. According to Zander (1946:22), in springtime with a good source of food the bees' flight range is limited to a diameter of less than 1 km. Kickhöffel (*in lit.*), in experiments at the time when fruit trees were in bloom, even found an area of only 75 m radius. Herta Knaffl (1953) let the bees themselves inform her of the distance flown by means of the tempo of dancing on their return home from natural sources of food; she found that they harvest mainly from an area of diameter 1–2 km, but that part of them frequently visit food sources at distances up to 6 km, more rarely even as far as 9–10 km, and that this occurs more often when the nearby flowers are too strongly exploited by a large number of colonies. Peer (1955, cited by Lee 1959) found most of the foraging bees within 4 km, some as far as 6 km. W. R. Lee (1959) found marked bees harvesting from blueberries 4 km away, although the home hive was in a field of blueberries and yet other blueberries were nearby. John E. Eckert (1933) set up swarms of bees in the middle of a desert and reports that they flew as much as 13.5 km to the nearest watered areas with fields of flowers.

Thus the data in regard to the range of flight may vary from 75 m to 13.5 km. The explanation lies in the fact that weather and the condition of the food supply are of great influence. This afforded the opportunity, through capitalizing on favorable conditions and on technical tricks, of determining the shape of the "distance curve" over a yet greater extent.

This task was met by H. Knaffl and M. Lindauer, with the latter undertaking the journey with the feeding station. The experiment had an unexpected result. Between 12 and 22 September 1949 Lindauer went southward 12 km from the city limits of Graz through the Murfeld to the vicinity of Wundschuh. The last dances were to be seen at 11 km. At this distance the bees often performed only a single long tail-wagging run on the comb and then interrupted their dancing. From the originally very large group only two bees were still collecting actively at 12 km, but they were not dancing any longer. The results of the measurements are recorded in Fig. 60. With exception of the deviant value at 1000 m, the curve agrees with the earlier ones and shows a harmonious extension of them up to 11 km.

That during the journey the bees were led successfully with the feeding table to a distance of 12 km from the hive is to be attributed principally to the following circumstances; (1) level, free, open country without roads, woods, and groups of houses; (2) fine autumn weather;

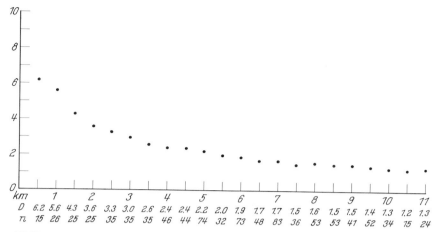

FIGURE 60. Distance experiment over a range of 11,000 m at Graz, 12–22 September 1949. For further explanation see Fig. 58.

(3) ample marking of the feeding station with color and odor and skilled advancement of the bees, conducted with sensitivity for their ability to follow; (4) after the 6-km mark was passed, feeding of 2M sugar solution with added honey; (5) the autumnal dearth of good natural food, and hence the great desire of the bees for something sweet. The shadow over this fortunate circumstance is that the feeding station was discovered again and again by strange bees from strong colonies in the vicinity; these quickly mob the sugar dish and drive off the experimental bees. But (6) before the beginning of the experiment, Lindauer had in 2 weeks' time given colored markings to all the 9000 bees in the observation hive, so that strangers from other swarms could be recognized at once and done away with. Without this precautionary labeling each and every newcomer to the feeding station would first have had to be marked and her appearing at the observation hive made certain of. At a great distance this takes so much time that an invasion of the feeding place is inevitable if a newcomer is from a strange colony situated nearby.

Additional distance experiments will be cited later in another connection. Here the four examples are sufficient to demonstrate the essentials. But at this time let us mention one other. From 1945 to 1947 I determined the tempo of dancing for distances up to 700 m with nine different colonies for five of them at intervals sufficiently different to permit the evaluation of curves. The latter were similar, but nevertheless differed somewhat according to the stock of bees, in that some colonies had a slower tempo of dancing than others (Fig. 61). This was confirmed later (e.g., M. Schweiger 1958:280, Fig. 6 and 281, Fig. 7). Bräuninger found stock-dependent differences in the dance tempo even in colonies that were queened with sisters. As an example I show in Fig. 62 the result of his measurements on two different colonies, one of which was observed in 1956, the other in 1957, at the same experimental location and at the identical flight range of 385 m. Since the

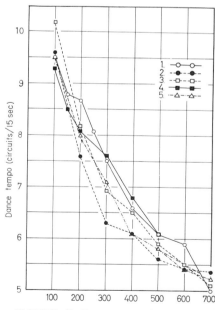

FIGURE 61. Dance tempo in five different observation colonies at flight ranges of 100–700 m: 1 and 2, 1945; 3, 1946; 4 and 5, 1947 (Brunnwinkl). Further details in v. Frisch 1948:14.

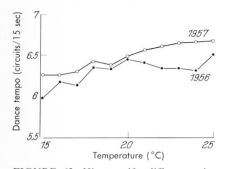

FIGURE 62. Hive-specific differences in dance tempo between two different colonies of bees; see Table 5. The values are arranged according to the temperature prevalent at the time of observation. Assembled by Bräuninger from his data.

tempo of dancing is accelerated somewhat by increasing environmental temperature (see pp. 75ff), the values were arranged according to the temperature prevalent at the time of each experiment (abscissa). One sees that at all temperatures in the range from 15° to 25°C the bees of one hive danced more slowly than those of the other. In a comprehensive evaluation the difference in dance tempo is highly significant ($P < 10^{-10}$). For the individual temperatures the values and the probable significance of the differences are given in Table 5.

If one regards the tempo of dancing as the signal that indicates to the hivemates the distance of the goal, one is interested first of all in the extent to which differences occur within the same hive. The individual values scatter appreciably (see pp. 70ff). But the averages too, which are given by the points in Figs. 58–60, may often deviate more or less, as one sees, in an irregular manner from the smooth course of the curve. Yet it is to be kept in mind that in these distance experiments the values for the points on a curve are obtained over several days of work, during which conditions do not remain constant. How this may influence the tempo of dancing we shall learn in the next section. There is more reason to marvel at the uniformity of the results than at the irregularities that occur.

In Fig. 63 are shown the combined results of the measurements available to me at present. In this curve are summarized as average values my own experiments concerning the reporting of distance together with observations by D. Bräuninger, H. Heran, H. Knaffl, and M. Lindauer, including the mean error and the number of observations. The calculation is based on measurements of 6267 dances (see Table 6).

TABLE 5. Hive-specific differences in the tempo of dancing; see Fig. 62. In order to eliminate the disturbing influence of temperature, only measurements obtained at the same environmental temperature are placed on the same line. Comprehensive evaluation of all the measurements shows a highly significant difference in tempo of dancing ($P \ll 10^{-10}$). After Bräuninger.

Temperature (°C)	Colony I		Colony II		Difference (circuits/15 sec)	P
	Tempo (circuits/15 sec)	n	Tempo (circuits/15 sec)	n		
15	5.97	108	6.26	327	0.29	10^{-9}
16	6.18	152	6.26	184	.08	0.10
17	6.13	204	6.30	366	.17	$3 \cdot 10^{-5}$
18	6.35	49	6.43	375	.08	0.24
19	6.33	87	6.37	542	.04	0.47
20	6.45	188	6.48	598	.03	0.34
21	6.40	178	6.56	453	.16	$3 \cdot 10^{-4}$
22	6.33	206	6.61	884	.28	$< 10^{-10}$
23	6.34	181	6.65	1043	.31	$\ll 10^{-10}$
24	6.31	107	6.66	811	.35	$\ll 10^{-10}$
25	6.50	165	6.67	660	.17	$8 \cdot 10^{-6}$

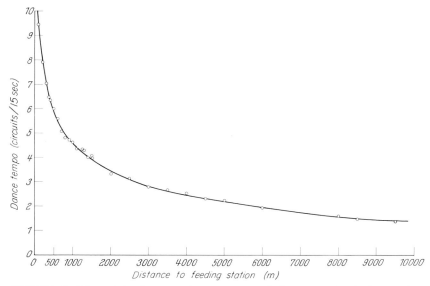

FIGURE 63. The relation between dance tempo and distance to the feeding station; average values from the experiments of Bräuninger, v. Frisch, Heran, Knaffl, and Lindauer. The curve is based on measurements of 6267 dances. Table 6 gives data for the points on the curve.

With the aid of this curve one is able—within the limits set by the variability—by timing the dance rates to determine the distance from which any desired bee has been harvesting. It is the hivemates who ordinarily profit from this possibility (see pp. 84ff), and they need no other measuring instrument than their own sense organs. A glance at the curve makes it clear that, as a result of its flattening, estimation of the

TABLE 6. Data for Fig. 63.

Distance to feeding station (m)	Number, n	Mean number of circuits/ 15 sec	Scatter, σ	Distance to feeding station (m)	Number, n	Mean number of circuits/ 15 sec	Scatter, σ
100	261	9.45	±0.90	1400	98	4.01	±0.29
200	781	7.90	.93	1500	128	4.06	.21
300	376	7.04	.60	2000	206	3.31	.30
385	400	6.49	.36	2500	120	3.13	.21
400	248	6.34	.51	3000	126	2.77	.27
500	514	6.01	.56	3500	146	2.65	.17
600	598	5.59	.42	4000	116	2.52	.19
700	119	5.07	.32	4500	163	2.30	.32
800	95	4.79	.39	5000	74	2.22	.20
900	463	4.73	.27	6000	73	1.93	.17
1000	344	4.62	.49	7000	83	1.71	.15
1100	66	4.34	.37	8000	53	1.62	.20
1250	450	4.33	.30	8500	53	1.46	.22
1300	61	4.30	.38	9500	52	1.36	.16

distance must become more difficult and less exact the farther away the goal. Thoroughly sensibly, communication works most precisely over the range that is most often frequented by the flying foragers.[6]

The results can be represented in another form by recording on the ordinate not the number of circuits in a given interval but the time required for a single circuit. In this way one obtains Fig. 64 instead of Fig. 63. We shall consider the shape of the curves later.

Lopatina, Kusnjezowa, and Pankowa (1958) dispute the statement that the number of circuits in the tail-wagging dance is dependent on the distance. Unfortunately they misunderstood me and confuse the number of circuits per 15 sec with the total number of circuits, that is, with the duration of the dance. Then indeed the relation with the distance does not exist.

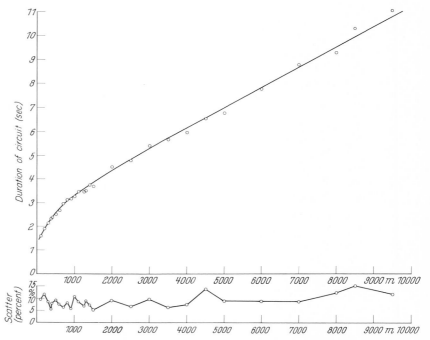

FIGURE 64. The relation between dance tempo and distance to the feeding station: a different graph of the same data shown in Fig. 63. Abscissa, distance to the feeding station; ordinate, duration (sec) of a dance circuit. *Below:* scatter (percent of the duration of a circuit).

5. The Influence of Internal Factors on the Dance Tempo

Even under uniform conditions the tempo of the individual dances is always somewhat variable. In order to give a picture of this, in Table 7 I have put together from Lindauer's experiment in the Forstenrieder Park (Fig. 59) the first 20 values measured at distances of 100, 500, 1000, 2500, and 4000 m.

[6]Concerning communication on horizontal combs relative to distance, see p. 137.

TABLE 7. Dance tempo (number of circuits per 15 sec) for the first 20 individual values measured for various distances in Lindauer's distance experiment (see Fig. 59).

100 m (10)[a]		500 m (8)		1000 m (6)		2500 m (4)		4000 m (4)	
Bee No.	Dance tempo	Bee No.	Dance tempo	Bee No.	Dance tempo	Bee No.	Dance tempo	Bee No.	Dance tempo
11	10.34	19	5.85	41	4.74	73	2.93	156	2.61
11	9.68	19	5.85	41	4.62	90	3.24	84	2.50
11	9.38	34	6.15	43	4.19	42	2.86	84	2.61
11	9.38	18	6.15	43	4.39	42	2.86	84	2.61
14	9.68	34	6.15	13	4.09	94	3.33	87	2.66
14	10.00	23	5.85	13	4.29	95	3.00	87	2.73
17	9.38	32	6.00	24	4.39	95	2.86	87	2.73
11	10.71	35	6.15	49	4.39	95	3.00	87	2.79
15	10.00	12	5.85	45	4.50	94	3.08	87	2.79
17	10.71	19	6.00	45	4.50	95	3.00	67	2.73
27	9.09	36	6.49	45	4.39	95	3.16	156	2.61
17	11.11	23	5.85	32	4.39	90	3.00	84	2.50
16	9.68	29	5.85	20	4.50	90	3.08	84	2.45
18	10.34	29	5.71	13	4.09	96	3.00	87	2.73
27	8.82	25	5.71	48	4.62	91	3.08	87	2.79
14	10.34	14	6.49	41	4.62	93	3.16	87	2.79
19	9.68	16	6.00	45	4.62	98	3.08	87	2.86
18	10.71	35	6.32	32	4.19	73	3.00	87	2.86
16	10.34	32	6.32	46	4.39	73	2.79	87	2.86
22	9.68	23	6.00	24	4.29	73	2.86	87	2.79

[a] Numbers in parentheses are the numbers of complete circuits timed, from which the dance tempo was calculated.

Since the numbers marked on the individual bees are given in the table, one may readily convince oneself that even the same specimen does not always maintain an identical tempo of dancing for a given collecting distance.

A telling picture of the variation in dance tempo with one and the same bee is given in Fig. 65, from measurements by D. Bräuninger in the Forstenrieder Park. All the values were obtained on 29 August 1958, a sunny day with high, rather even temperature and gentle wind

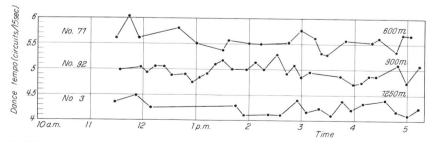

FIGURE 65. Variations in dance tempo in successive dances by one and the same bee, 29 August 1958, 11:20 A.M.–5:20 P.M., in the Forstenrieder Park; one bee each from the feeding stations at 600, 900, and 1250 m. From experiments by Bräuninger.

(calm from 4:00 P.M. on). Bees Nos. 71, 92, and 3 were from the same observation hive and were frequenting three different feeding stations, respectively 600, 900, and 1250 m from the hive. Each point on the curve refers to one dance. If with the longer dances several values could be determined successively, the point on the curve gives their average. No. 92 was an especially zealous dancer and provided the most data. From one dance to another the number of circuits per 15 sec shows differences of about as much as 0.4 circuit. Only a fraction of this variability can be ascribed to errors of determination. For according to tests the personal error of the observer in such measurements was at most 0.1–0.2 sec.

The variations in this case also are not to be traced to changes in external conditions during successive flights. They originate in the bee herself. For they appear to approximately the same extent when the tempo of dancing is measured several times in the course of a long dance. Thus with the zealous dancer No. 92 during the experiment of 29 August (Fig. 65), in the course of longer-lasting dances the time required for five circuits was measured several times in succession and converted by calculation into the number of circuits per 15 sec. The result is shown in Fig. 66. The differences of the measured periods within the same dance here amount to as much as 0.55 circuit/15 sec and average 0.18 ($n = 75$). At a flight range of 385 m Bräuninger in the experimental season of 1957 found on the average differences from 0.2 to 0.3 circuit/15 sec between successive periods of the very same dance.

In a closer examination of this variability in his extensive data he came to the conclusion that in a comparison of the first circuit with later circuits in the same dance no tendency toward an acceleration or retardation of the tempo is to be discerned. The variations succeed one another at random. In their extent they correspond approximately to the

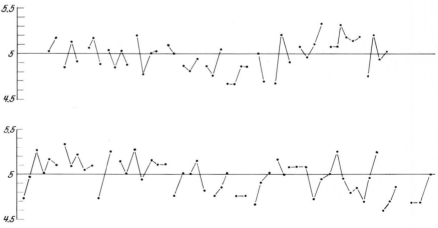

FIGURE 66. Bee No. 92, 29 August 1958 (see Fig. 65). Distance to the feeding station, 900 m. Variations in dance tempo within a single dance. Connected segments of the curves refer to a single dance. From experiments by Bräuninger.

alterations observed in different dances by the same bee (cf. Fig. 65). The differences seen between the two situations are so slight that their distinctness cannot be established.

Thus the dance tempo of a bee is not exactly equal for all circuits even in a single dance. The bee is no precision machine, but rather a living creature. Evidently she is unable to maintain the identical time for each circuit, but oscillates about the mean value striven for. H. Heran (1956:195, 196) reports the same phenomenon for phases of the same dance interrupted by pauses in the offering of food.

In addition there are undoubtedly individual differences in dance tempo among members of the same colony. On observing numbered bees attentively one will soon notice that there are both lively and phlegmatic dancers that regularly maintain a certain difference in the number of circuits per 15 sec. As an example Fig. 67 shows results obtained by Elisabeth Schweiger (1958). From a group of newly marked bees she chose five specimens that danced more rapidly than the swarm average and eight that danced more slowly than the average. With these bees she journeyed on the two following days to a distance that reached 600 m. The curves show the average dance tempos of the two groups for five different distances. The difference of the means is statistically significant ($P < 0.0027$). An important point is that these experiments dealt with bees of equal ages in relation to flight. For in addition to the individual predisposition, age can be a significant factor in the dance tempo.

Yet another example of clear-cut individual differences in dance tempo among bees from the same colony has been assembled for four different flight ranges in Table 8. The data are taken from Bräuninger's numerous records. Striking cases have been selected intentionally. For instance, at the distance of 385 m in 28 possible comparisons an individual difference was highly significant in 11 and insignificant in 17.

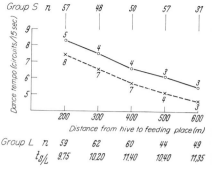

FIGURE 67. Individual differences in dance tempo among hivemates of equal flight experience. The curves show the average values of the dance tempo (circuits/15 sec) for a group of rapid dancers (S, upper curve) and a group of slow dancers (L, lower curve), at flight ranges of 200–600 m; n, number of individual observations, numbers of bees are indicated at the points on the curves. The differences are statistically significant. After Elisabeth Schweiger 1958.

TABLE 8. Individual differences in dance tempo (average number of circuits per 15 sec) with bees of the same colony; one example for each of four different flight ranges. The bees compared had the same flight experience and were of equal or nearly equal age (marked on the day of emergence), the temperature difference was at most 2 C deg, the difference in wind velocity at most 0.4 m/sec; P, probability that the difference in dance tempo was due to chance. After Bräuninger.

Distance of feeding station (m)	Bee No.	Dance tempo (date, September 1958)					Σn	P
385	41	6.34 (24),	6.24 (27),	5.89 (28),	6.09 (29),	6.15 (30)	128	10^{-12}
	78	6.90	6.54	6.42	6.61	6.46	77	
600	62	5.85 (2),	5.62 (3),	5.70 (4)			29	10^{-9}
	51	6.61	6.12	6.27			26	
900	24	4.33 (3),	4.38 (4)				15	10^{-11}
	85	4.87	5.40				16	
1250	7	4.56 (1),	4.46 (2),	4.15 (3),	4.30 (4)		21	6×10^{-4}
	2	4.71	4.62	4.54	4.57		30	

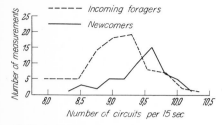

FIGURE 68. Comparison on 24 May 1953 between four newcomers (younger bees, $n = 54$) at the feeding station and four older bees ($n = 84$) that had been visiting it for some time. Statistical evaluation, $t = 3.40$. After Elisabeth Schweiger 1958.

Influence of age. The experiments (E. Schweiger 1958) are concerned with bees that were marked on the day of emergence (in the breeding box) and then put among the observation colony. Thus their age was known. They appeared at the feeding station 18–33 days after emergence. This time span of their absolute age was, however, without significance for the dance tempo. What was decisive was their age in relation to flight, by which is meant the number of days elapsed since their first appearance at the feeding station. For most of the experimental bees, this date is most probably the first day for harvesting in their lives; they had been under constant observation in the hive.

The newcomers at the feeding place, compared with bees that had been frequenting it for some days, had a remarkably rapid dance tempo. Figure 68 shows this for an experiment on 24 May 1953, with the feeding station at a distance of 200 m. On the abscissa is indicated the number of circuits per 15 sec, on the ordinate the number of measurements included under this value. The solid line is for four newcomers, that is, bees that appeared at the feeding place for the first time on the day of experiment, the dashed line for four bees that already had frequented the feeding place for some time.

The tempo of dancing diminishes rather steeply on the first day. It also becomes somewhat slower in the further course of individual life (Fig. 69). However, the first striking change occurs during a bee's last two or three days. This time may be designated the stage of senility. With some experience one notices also in the appearance and behavior of the bees during their collecting activity that their end is at hand. Their movements are hampered. Their dances often are irregular, so that no definite values for the rate can be obtained. If the dances nevertheless do seem continuous, yet they clearly are slower than on earlier days (Schweiger 1958). According to Lindauer (1948: 407, 408), the bees now surprisingly frequently perform dances ("Zittertänze") in which they tremble all over (see p. 282). Morphologically this stage is characterized by the onset and rapid progress of degenerative phenomena in the ganglion cells of the brain (Weyer 1932).

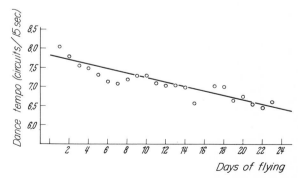

FIGURE 69. Relation between dance tempo and flight experience, from experiments of July and August 1955 in Munich; abscissa, number of days' flying; ordinate, number of circuits per 15 sec. The reduction in tempo is statistically significant. After Elisabeth Schweiger 1958.

At Heran's suggestion, Elfriede Schreindler (1964) repeated these experiments and improved the technique by joining, after a few days, to a collecting group another younger group. In this way bees of differing age could be tested simultaneously under exactly the same conditions. The diminution of dance tempo with increasing age was in general confirmed, yet exceptionally an opposite effect also occurred. From this it is concluded that the connection between dance tempo and age in relation to flight is affected by one or more other factors.

Influence of increasing experience. With bees that visited the same feeding place day after day E. Schweiger (1958) was able to discern a reduction in the scatter of the values indicated for distance. That is, when the same goal is visited persistently the information given as to its distance becomes gradually more exact. In Fig. 70 can be read from the abscissa how many days the bees already had been frequenting the feeding place. On the ordinate is plotted the degree of scatter (in fractions of a circuit) for the day in question, as the mean value obtained from all the dances measured on that day; the number of dances is given below the *x*-axis. One can see the tendency for the extent of scatter to diminish in the 3 weeks during which the same feeding station was frequented. Statistical calculation did not indicate certainty for a decrease in the degree of scatter, but the probability was about 99 percent.

6. THE INFLUENCE OF EXTERNAL FACTORS ON THE TEMPO OF DANCING

(a) *Temperature*

Chemical and physiological processes are temperature dependent. According to van't Hoff's rule (RGT rule), the reaction velocity often is doubled or tripled with a 10-deg rise in temperature. At 8°–10°C the bees begin to fly out and their zeal for foraging remains undiminished even at an air temperature about 30°C. Hence one is likely to wonder whether, over this wide range, the tempo of dancing is changed by the temperature.

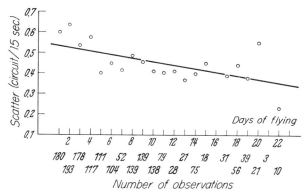

FIGURE 70. Reduction with increasing flight experience in the scatter of the distances indicated by dancing. From experiments by Elisabeth Schweiger 1958.

In this connection one must first recall that, although bees are by no means "warm-blooded" animals with a constant body temperature, in comparison with other poikilotherms they possess a respectable capacity for temperature regulation.[7] In the hive they maintain a constant temperature of 35°–36°C in the area of the brood nest. Outside the brood nest, and hence very often in the region of the dance floor, the temperature variations of the outside world make themselves clearly noticeable. The body temperature of the bee itself was measured by H. Esch (1960) with the help of thermistors.

Esch investigated the body temperature of individual bees at the feeding station, in the hive, and in flight. For measurement the thermistor was pushed into the body through a little hole in the thoracic integument. The procedure required only a few seconds. Without diminishing the precision of measurement the thermistor could also be glued to the outer surface of the integument with wax rosin. With such specimens, which ran around in the hive "on a leash," the body temperature could be followed over a long period.

On arrival at the feeding station the thoracic temperature lies about 10 deg above the outside temperature. Only after the air temperature exceeds some 25°C and the body temperature thus has reached about 35°C does the latter cease to rise and remain at this level. On cold days the body temperature sinks 2 or 3 deg at the feeding place, whereupon drinking is interrupted while the body is warmed up. In the hive Esch found thoracic temperatures of between 29°C and 38°C in dancing bees; in general these temperatures were higher than the values that had been measured at the feeding station. Among the followers of the dancer an increase of several degrees in body temperature was noted during the pursuit. They are "warming up for their coming task," as may be seen too in other bees preparing to fly out (Fig. 71). The abdominal temperature always is somewhat below that of the thorax.

Thus the collectors warm themselves up relative to the environmental temperature. But since they become warmer only relatively, their body temperature nevertheless changes with the temperature of the surroundings. The question is still justified whether the tempo of dancing is affected by the outside temperature.

There is no strong effect. For without an investigation carefully directed to the point it passes unnoticed.

Among the great number of dance measurements he made for other purposes, Heran (1956:198) had 34 pairs of mean values that were obtained—under otherwise identical flight conditions—at temperature differences of more than 4 C deg. In 25 cases the bees danced more slowly at the lower temperature, in 9 cases at equal rates, and in only 3 cases more rapidly than the bee at the higher temperature. Indications of an influence of temperature also are found in the work of Grete Werner (1954:471).

Clarification of the relation was provided via a comprehensive set of data amassed during 3 years' work by D. Bräuninger (1964). In Fig. 72,

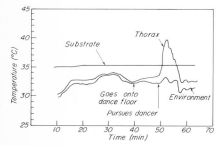

FIGURE 71. Example of warming up of a bee during pursuit of a dancer. The curves show the temperature of the substrate (measured in the wax of the dance floor), the thoracic temperature of the dance follower, and the environmental temperature (1–2 cm dorsal to the thorax). In 2.2 min the thoracic temperature rose 6 C deg and then, when the follower lost contact with the dancer, fell back to the starting level. After Esch 1960.

[7] In characterizing this difference, bees are contrasted with "homoiotherm" and "poikilotherm" animals as "heterotherm." They occupy an intermediate position.

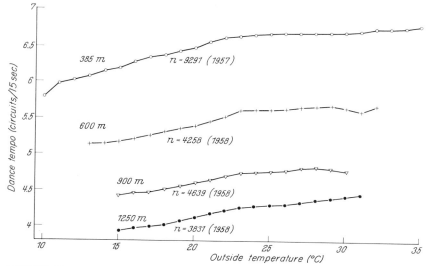

FIGURE 72. Dependence of dance tempo on the external temperature at four different flight distances. The number *n* of dances measured is given beside the curves. After observations by Bräuninger.

for feeding stations at 385 m (1957), 600, 900, and 1250 m (1958), all dance rates measured during the indicated year are listed in order according to the air temperature at the time of dancing, and are averaged. It is quite clear that on the grand average dancing is more rapid in warmer weather. This result is statistically significant.

Evident as is the change in dance tempo, its extent is just as slight. Whereas according to van't Hoff's rule a temperature rise of 10 C deg corresponds to a two- to threefold increase in reaction rate (thus as much as 200 percent), over this temperature range the tempo of dancing changes by only about 10 percent. Thus the temperature effect is in the main compensated, especially well in the range from about 20–30°C, as the flattened course of the curves shows and precise calculation confirms.

Since the temperature inside Bräuninger's observation hive could be regulated, its influence could be tested independently of the temperature outside. When the air temperature around the comb fell below 18°C there was no more dancing; in contrast the most zealous dancing occurred on one occasion when there was an unintentional overheating (to 44°C) and the wax in the upper part of the hive had already begun to melt.

Evaluation of all recorded dances has shown that the influence of the inside temperature is about equal to that of the outside temperature. This finding, too, is statistically significant. It may be demonstrated by Figs. 73 and 74. In Fig. 73 the results from one day of experimentation (29 September 1958) are shown. It was a sunny autumn day, with light wind and a nearly constant outdoor temperature. The bees were collecting from a feeding station 385 m distant. In the course of the day the temperature inside the observation hive was raised and lowered four times. It is plain how the tempo of dancing changed accordingly. Fig-

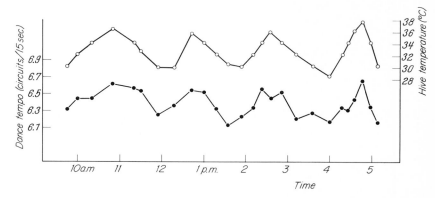

FIGURE 73. Influence of the temperature inside the hive on dance tempo; 29 September 1958, flight distance 385 m, 735 measurements. During the day of the experiment the hive temperature was raised and lowered four times (*upper curve, right-hand scale*); the dance tempo changed correspondingly (*lower curve, left-hand scale*). After Bräuninger.

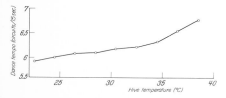

FIGURE 74. Influence of the temperature inside the hive on the dance tempo. Compilation of all (451) dances measured at an external temperature of 16°–17°C; abscissa, artificially regulated internal temperature of the hive; ordinate, number of circuits per 15 sec. Experiments of 1958; flight distance 385 m. After Bräuninger 1964.

ure 74 combines all the measurements made at an outside temperature of 16–17°C, assembled from all the available data and arranged according to the artificially regulated inside temperature.

The temperature of the food solution also enters the picture as a natural biological factor. According to Büdel (1956, 1959), in sunshine the inside of large goblet-shaped flowers, and hence their nectar, may be some 5–10 C deg warmer than the surroundings. With honeydew the same sort of thing is possible, since excess temperatures of more than 10 C deg have been recorded for leaves during strong sunshine and calm weather (Geiger 1961:278).

Bräuninger offered his bees sugar water at temperatures from 2° to 45°C. These limiting values were accepted only with hesitation. An influence of the temperature of the sugar solution on the tempo of dancing was demonstrable and was evident on statistical analysis. But it remained within modest limits. With the feeding place 385 m distant, an increase in the temperature of the food, from about 10° to about 35°C, caused an acceleration of the dances such as occurred when the air temperature rose from 10° to 15°C. However, it is worth noting that a single swallow of warm solution at the collecting station is enough to speed up the dance tempo recognizably.

When the air temperature rises from 15° to 25°C the approximately 10-percent increase in the dance tempo is minor in itself. But if one calculates in meters the error in indication of distance that could arise from it, one comes to the appreciable quantities of about 100 m for a feeding-station distance of 385 m and about 220 m for a distance of 1250 m. The accuracy of the response of the hivemates to the dances (pp. 84ff) is evidence against the existence of such an error-ridden system. Thus it is to be concluded that temperature-induced changes in the tempo of dancing are compensated for by the comrades, who are indeed subjected to similar conditions both inside the hive and when they fly out.

In contrast with Bräuninger, Esch (1964) was unable to find a connection between dance tempo and outside temperature over the range from 16° to 31°C. Esch's measurements cover only a few days at approximately the same average daily temperatures. In Bräuninger's work, too, a temperature dependence of the rate of dancing frequently could not be seen in the values from a single experiment nor in a few successive days, especially above 20°C. Esch conjectures that the tempo of dancing is dependent not simply on the factor of "outside temperature" but on the general meteorological circumstances that on the average are correlated with a certain outside temperature. But this interpretation is not wholly satisfactory, since in Bräuninger's experiments the dance tempo responded very clearly to short-term variations in the temperature inside the hive, and these certainly occurred with no change in the general meteorological conditions (see Fig. 73).

(b) The Wind

The influence of wind direction on the tempo of dancing became evident when I arranged my data from the years 1945–1947 according to the nature of the wind at the time of experiment. Although these matters were noted only approximately in my protocols, it was still evident that bees that had flown against the wind to the feeding station danced more slowly than when it was calm, and on the other hand that they danced more rapidly when they had flown downwind to the feeding place (Fig. 75). A headwind during the flight to the goal had the same effect on the dance as an increase in the distance, and a tailwind as a reduction. This gives us a glimpse into the mechanism whereby bees estimate distance (p. 109). But it was essential to determine the facts more precisely.

During his experiments that provided us with information about the effect of temperature, Bräuninger (1964) regularly noted the direction and strength of the wind. For measuring the wind velocity a cup anemometer was set up at a height of 2 m on the flight path. This height corresponded approximately to that at which the bees flew. But since the height they maintain is not constant, for example being nearer the ground in strong wind, and since further the strength and direction of the wind may change for brief intervals, this factor could only be approximated. Nevertheless, because of the large number of measurements (10,847 dances), the results emerging were clear.

First, all the dances recorded were assembled in groups of 45° according to the wind direction. Thus NE signifies a breeze coming from the sector NNE to ENE. For the foraging bees in the Forstenrieder Park NE wind was a tailwind on the flight to the feeding place (the flight path is shown in the sketch map, Fig. 89). In Table 9 for the three feeding stations in the Forstenrieder Park (600, 900, and 1250 m), the average values of the dance tempo in 1958 are given for each group of wind directions, together with the mean errors and the number of observations. For all feeding places, the dance tempo diminishes with a transition from tailwind (NE) to headwind (SW). With a sidewind (NW) the

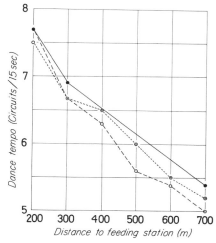

FIGURE 75. Dance tempo with a tailwind (———), in calm (- - -), and with a headwind (- - -) on the flight to the feeding place. Average values from experiments in the years 1945–1947.

TABLE 9. Dance tempo f (average number of circuits per 15 sec) and wind direction;[a] summary, without regard for other factors, from experiments in Forstenrieder Park, 1958. After Bräuninger 1964.

Distance of feeding station (m)	Tempo	Wind direction					
		NE	N	NW	W	SW	Calm
600	f	5.66	5.63	5.51	5.49	5.47	5.57
	sf	0.01	0.01	0.02	0.02	0.01	0.03
	n	1632	532	366	486	597	168
900	f	4.84	4.76	4.72	4.72	4.67	4.74
	sf	0.01	0.01	0.01	0.01	0.01	0.02
	n	1796	411	363	840	648	169
1250	f	4.38		4.25		4.21	4.30
	sf	0.01		0.01		0.01	0.01
	n	1521		495		622	201

[a] Wind direction: NE = tailwind, SW = headwind on the way to the feeding place, sf = mean error of average, n = number of observations

dances are slower than when it is calm (last column) and are closer to those in a headwind.

In evaluating the data further, measurements were compared that had been made at the same temperature. Thus the disturbing influence of air temperature was excluded. Figure 76a provides as an example the results obtained in 1958 for the feeding stations at 600 and 900 m at an environmental temperature of 24–26°C. Again a SW wind was a headwind and a NE wind a tailwind in the flight to the feeding place. The ordinate is the number of circuits danced per 15 sec. The slowing down of the dance tempo with a transition from tailwind to headwind is clearly evident.[8]

When the wind velocity is taken into account, the relation for tailwind and headwind in comparison with calm has the form shown in Fig. 76b. In these summaries taken from the great number of original data only the values for a narrow temperature range (23–26°C) have been singled out. Unfortunately, wind velocities appreciably in excess of 2 m/sec were represented too seldom for a statistical evaluation. But in the range under consideration the divergence of the curves, that is, the increasing effect of rising wind velocity on the tempo of dancing, is very clear, as indeed would be expected.

Within the framework of an as yet incomplete study, Heran and Lindauer were able in the summer of 1964 to carry out dance measurements in the vicinity of the Neusiedlersee (Austria) with almost pure head- and tailwinds and at much higher wind velocities; their results are shown in Fig. 77. All measurements were made between 29 August and 1 September and refer to the same observation hive. Each point on a curve is based on 60 measured dances. The wind velocity at the time of the measurements can be read from the arrows (b), the slight

[8] That the difference is larger at 600 m than at 900 m is a chance occurrence and is not confirmed by results for other temperature ranges and at other distances.

The Tail-Wagging Dance as Communication / 81

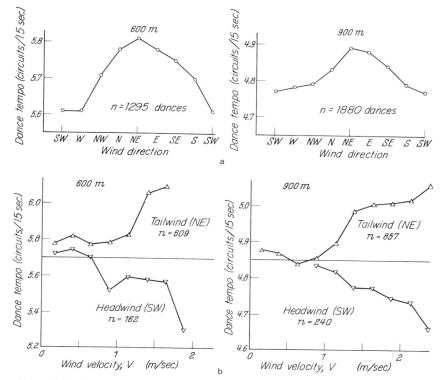

FIGURE 76. (a) Influence of wind direction on dance tempo: NE corresponds to tailwind, SW to headwind, on flight from hive to feeding place, at two different flight distances (1958); outside temperature, 24°–26°. (b) Influence of wind velocity V (abscissa) during tailwind and headwind in course of flight to feeding place; outside temperature, 23°–26° (1958). For comparison the dance tempo in still air is indicated by the thin horizontal line. After Bräuninger 1964.

sidewise components of the wind are given by the dashed lines, and (c) indicates the temperature. On the basis of the work of Bräuninger (1964) the tempo of dancing can be adjusted to a common temperature of 20°C. Then the curves are shifted slightly without an appreciable change in the over-all picture. Figure 77 shows the values actually measured. They form a good complement to Bräuninger's results, because they refer to shorter distances; also with such high wind velocities as were available at the Neusiedlersee the effect of tail- and headwinds is exceptionally clear. Note that the scale of the ordinate differs in Figs. 77 and 76. With weak tail- and headwinds the dance tempo, at a flight distance of 600 m, differs by about 0.2 (5.8 vs. 5.6 circuits/15 sec), and with high wind velocities, at a flight distance of 500 m, by almost an entire circuit (6.51 vs. 5.59 circuits/15 sec).

(c) Gradient of the Flight Path

If bees that on their way to the feeding place must overcome a headwind react with a decrease in the tempo of dancing, as though the goal had been moved farther away, one might suppose that overcoming a gradient would have a similar effect. This suspicion led to experiments that were carried out by Heran and Wanke (1952) and Heran (1956) at Mount Treuchtling at Tragöss in Styria and on the Schöckl at Graz.

Figure 78 shows the location (×) of the observation hive in the middle of a steep talus slope at Tragöss. From it one group of numbered bees was led up through the gully to the ridge and another down the rubble slope as far as the woods at the foot of the mountain. The average inclination of the slope amounted to something more than 30°, and the length of the entire stretch to about 1300 m. The training of the bees from the hive to the most distant feeding stations took from 28 August to 7 September 1950. After each 50-m stage some time was spent and the dance tempo was measured. Figure 79 shows that the bees that had to fly upward to the feeding place (dashed curve) clearly danced more slowly than those that flew an equal distance downhill to their goal (solid line). Although the scatter in the values was rather large, all differences are statistically significant except for the one at the 200-m stations.

Another place suited for such experiments was a bare outcrop on the Schöckl at Graz. The gradient there was a little less (25°), and the total length of the flight path was 800 m. The result of an experiment carried out there in September 1949 is shown in Fig. 80.

In these two instances the wind conditions were favorable to an extent rarely found on such slopes: mostly it was calm or the wind velocity was less than 0.5 m/sec. Only at the 200-m station in the first experiment (where the difference in rate of dancing is not statistically significant) was there a gusty updraft at velocities as high as 3.5 m/sec, and in the second experiment a strong steady upwind (2 m/sec) prevailed at 350 m, where the curves are seen to cross in Fig. 80. The updraft carries the bees aloft and hence has an effect opposite to that of the gradient.

From this it is clear that in such experiments on slopes—which make no small demands on the endurance and strength of the observer—the wind causes the most mischief. In actual fact it has thus spoiled many an experiment—if one understands this to mean that the curves obtained did not follow the expected course. Since the wind conditions always were recorded, deviations from the typical curves, such as those given in Fig. 80, can always be traced to other perturbing factors. Heran (1956) concerned himself with the latter, studied the effects of wind in the special form of thermal air currents on sunlit slopes, temperature differences, and shaded stretches. Since such influences do indeed make complications but change nothing in the fundamental result, they do not need to be discussed in detail at this time.

(d) Pharmaceutical Agents

With the effect of drugs we touch upon an area that may seem to go beyond the biological scope of this work. Yet these effects have relations to questions with which we shall be occupied further.

Marked bees that were frequenting a feeding place at a set distance were captured at the feeding station on the eve of an experiment and were fed in the laboratory with sugar water to which had been added

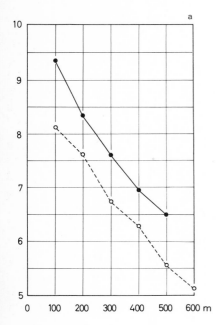
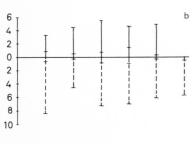
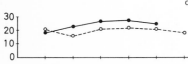

FIGURE 77. (a) Dance tempo in a strong tailwind (——) and strong headwind (– – –) on the flight to the feeding place; abscissa, distance to the feeding place; ordinate, number of circuits per 15 sec. Each point on the curves is based on 60 measurements of dances. Note the difference in scale of the ordinate from Fig. 76. (b) Wind velocity (m/sec) in the direction of flight; —— tailwind; – – – headwind; abscissa as in (a). The horizontal lines near the axis indicate the strengths of the lateral components of the wind. (c) Temperature (°C) at the time of the experiment; abscissa as in (a). After unpublished experiments by Heran and Lindauer at the Neusiedlersee, 1964.

quinine, thyroxin, or adrenalin. On the following morning they were released into the observation hive. In the experiments that followed they displayed in comparison with the normal bees a reduction in the dance tempo after feeding with quinine but an increase after feeding with thyroxin or adrenalin (Grete Werner 1954). As an example Fig. 81 shows the result of an experiment with the feeding station 300 m distant. Further experiments at flight ranges of 180, 700, and 950 m were conformable. The deviations in dance tempo from those of the normal bees are statistically significant.

This change in the tempo of dancing is not caused by a disturbance of the bees' time sense. For after they had been trained to a certain feeding hour they came in time as before, despite having been given quinine, thyroxin, or adrenalin.[9] Likewise the estimation of distance was not deranged. Grete Werner was able to show this by testing bees at distances of 40–70 m, thus at a range where transitions from the round dance to the tail-wagging dance are to be observed. The proportion of round and tail-wagging dances remained the same with treated and untreated bees. If there had been a change in the estimation of

---

[9] The influence of quinine and thyroxin on the sense of time was tested by M. Renner (1957) also. He reached the same negative result as Werner. For references to contrary reports, see Renner's paper.

FIGURE 78. View of the experiment site at Tragöss; × location of the observation hive.

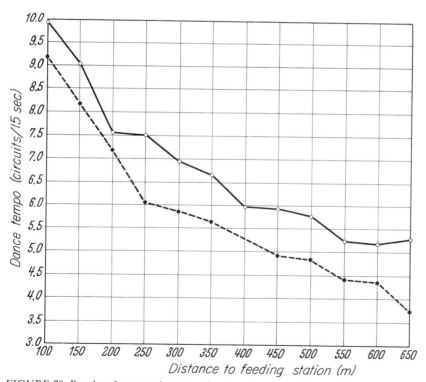

FIGURE 79. Results of an experiment on the dance tempo on the steep slope at Tragöss; abscissa, distance to the feeding station; ordinate, number of circuits per 15 sec; - - -, feeding station uphill; ———, feeding station downhill. After Heran and Wanke 1952.

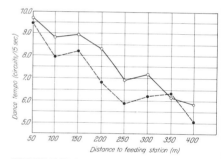

FIGURE 80. Results of an experiment on a steep slope at the Schöckl. Explanation as in Fig. 79.

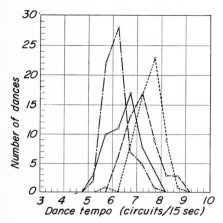

FIGURE 81. Influence on dance tempo (distance 300 m) of feeding quinine, thyroxin, or adrenalin: —·—·— quinine (50 mg percent); - - - thyroxin (5 mg percent); ---, adrenalin (100 mg percent); ——, normal; abscissa, number of circuits per 15 sec; ordinate, number of dances observed. After Grete Werner 1954.

distance, the proportion would have had to shift. The oxygen consumption of the bees was influenced demonstrably. After quinine had been given, oxygen consumption was reduced; after thyroxin and adrenalin, it was raised. Hence Werner reaches the conclusion that the change in dance tempo following administration of the drug is merely an expression of altered activity on the part of the bees. It was possible actually to demonstrate such a change in experiments on phototaxis: in the darkroom treated and untreated bees (with wings cut short) were made to run across a dark-covered table toward a source of light. The speed of running, compared with that of normal bees, was reduced by 12 percent after quinine had been given, but was increased by 12–18 percent after dosage with adrenalin or thyroxin.

7. How Accurately Can Newcomers Follow the Distance Indications? Stepwise Experiments (Stufenversuche)

A clear relation obtains between the dance tempo of the bees and the distance to the feeding place (Figs. 63 and 64). This relation is, however, influenced by internal and external factors (pp. 70–84), but within such narrow limits that nevertheless we are able with remarkable accuracy to read from the rhythm of the dancer the distance to the source of food at which she was active. The question now is whether the hivemates too understand the indication and how precisely they follow it.

The pure round dance shows no clear relation between the number of circuits per 15 sec and the distance to the goal. The dancer does change the direction of her circling, even during the same dance, often after even a half-circuit and again frequently only after from one to three complete rounds; thus the reversals succeed one another at differing intervals and their number in 15 sec is variable.

For example, in an experiment on 1 July 1945 (feeding place 10 m from the hive) during 12 round dances I counted from 6 to as many as 12 reversals per 15 sec. On 14 August 1964 from 9 to 10 A.M. I watched 60 dances that were carried out in relation to the same distance (10 m) and counted 5–11 reversals (Fig. 82, *left*). After the feeding place had been moved to 50 m, in 60 observations of dances between 10:45 and 11:45 A.M. there were 5–12 reversals (Fig. 82, *middle*). Thus the values still varied a lot, but they had shifted in the direction of a greater number of reversals per 15 sec. It now happened more rarely that a single dance would include complete circles or several rounds in the same direction, and this of course operated to reduce the number of reversals. With the feeding place 100 m distant, an indication of direction becomes clear, and with it the regular alternation of semicircles to right and left in almost all dances. Thus there arises not only the spatial but also the temporal orderliness of the tail-wagging dance. From 12:10 to 1:00 P.M., in the great majority of 60 dances 9–10 reversals per 15 sec were to be seen (Fig. 82, *right*).

To begin with, we accept as a fact that the bees informed by the tail-wagging dance are given the proper direction to the goal (pp. 129ff), and wish to ascertain whether they know its distance. In order

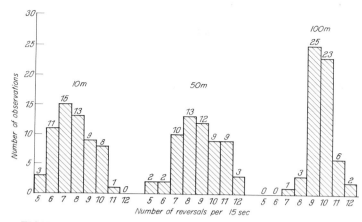

FIGURE 82. Number of reversals per 15 sec for distances to the feeding place of 10 m, 50 m, and 100 m. In the range for round dances the number of reversals (abscissa) displays considerable scatter, because frequently only a fraction of a circuit is made between two reversals, while often a complete circle or even more is made during the dance. At 100 m the orderly figure of the tail-wagging dance brings with it a more rigid temporal sequence.

to learn this we conduct stepwise experiments. The reason for doing so is simple.

At an arbitrarily chosen distance we set up a feeding station with meager food and without addition of a scent, which was visited by a group of numbered bees. During the experiment this group is given concentrated sugar solution with a scent added. Hence there now is dancing and the newcomers that are alerted go searching for the scent in question. Along the route from the hive to the goal and beyond, scent cards (without food) are set out at stepwise intervals. These cards are perfumed with the same scent as the feeding place and hence attract the newcomers as soon as they come near them, for the newcomers are seeking the scent they perceived on the dancer (pp. 46ff). At each scent card there sits an observer, who notes down the number of unmarked bees that come flying up and of those that settle. At the feeding place itself every newcomer is at once caught and killed. Thus the visitors to the scent cards have not yet found any food in connection with the specific scent offered; they come merely because of the information given by the dancer. Their distribution over the stepwise series of scent cards informs us where they are searching.

The behavior of the bees toward the scent cards is very expressive. They come flying slowly along near the ground and against the wind, searching in short zigzags, and hover over the card. They are recorded only when they have come within about 20 cm of it. As a rule they then settle on it and run about over the card, displaying unequivocally their interest in the source of scent. Since they are not captured and are unmarked, whether the same bee repeats its approach several times of course remains unknown. When the bee has disappeared from the observer's field of view, she is counted again on a new approach. That is no disadvantage, for the persistent searching of a given insect at a definite distance shows that it is just here that she is expecting something.

The proper preparation and performance of such an experiment is not simple. Often 12 observers are needed—for 12 scent cards. They must not only be able surely to distinguish a bee in flight from other insects, but they must also have the ability to concentrate on their task for hours at a time even when nothing happens. Careful selection of the helpers is a prerequisite to success. But this is threatened from other directions also, and it is necessary to point out some sources of error.

Of particular interest, naturally, would be the influx of newcomers at precisely the distance indicated by the dances, that is, at the feeding station itself. But in evaluating the results the number of flights to the feeding station must be left out of consideration. For here, because of the coming and going of the marked bees, which make use of their scent organ, quite different conditions prevail than at the observation cards. The additional attraction from the scent organ differs from experiment to experiment, and depends strongly on how intensively the organ is used and on the direction of the wind. Our practice has been to set out the closest scent cards 50 m before and 50 m beyond the feeding station; thus we are outside the effective range of the scent organ and yet near enough to the goal of collecting activity to evaluate the precision of the approaches.

In order that comparable conditions shall prevail at all stations, the scent cards must give off the scent at the same rate throughout the several hours of the experiment, and they must give it off at such a low intensity that only those bees that are searching in the immediate vicinity shall be attracted. The following arrangements have proved satisfactory. In a sheet of cardboard (15×15 cm) eight holes are punched and little shell vials (with 6-mm openings) are set into them. At the beginning of the experiment each vial is supplied with a few drops of the scent. We used oil of lavender, oil of peppermint, orange-blossom oil, and on one occasion linalool. To keep it from blowing over, the scent card is held by two blocks of lead lying on the ground (Figs. 83 and 84).

If during the filling of the vials a single drop of the essential oil falls beside them on the cardboard, the card cannot be used because the scent is intensified. In large-scale experiments there were about 100 vials to be filled at the observation stations before we could begin. Therefore from 1955 on we used another arrangement by which the filling and setting out of the scent stations can be done faster and more simply. The essential oil is put into a glass dish (30 mm in diameter and 15 mm high) so that the bottom is well covered. The dish is covered with a plate of glass or plexiglass in the center of which is bored a hole 8 or 10 mm in diameter to permit escape of the scent. A wire screen (2-mm mesh) fastened above it prevents the bees from forcing their way in and avoids displacement of the plate. The dish is placed on a short-legged wooden stand (15×15 cm), which is set on the ground (Fig. 85). In sunshine all such scent plates must of course be equally exposed.

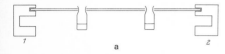

FIGURE 83. (*a*) Scent card held by grooved lead weights (1 and 2). (*b*) Detail: 1, cardboard sheet; 2, vial; 3, odorant.

FIGURE 84. Scent card with eight vials. Lead weights (K_1, K_2) prevent its being blown over. The wooden post on the left marks the location of the station.

FIGURE 85. Scent plate: a glass dish containing an odorant rests on a low wooden stand that is set on the ground during experiments in the open. A little of the scent escapes continually through the hole in the glass cover. A wire screen prevents the bees from forcing their way in. A newcomer has flown up and is taking an interest in the scent. Photograph by Renner.

The feeding table is about 60 cm high. For scenting the feeding place the arrangement shown in Fig. 21 is most satisfactory. When the feeding station is at a great distance an additional trace of the scent is added to the sugar water itself. This amount of scent must be properly measured out, so that it is sufficiently effective but will not inhibit the dances (cf. Kaschef 1957): 1 drop of essential oil to 1 liter of 2M sucrose solution.

In setting up the feeding station a group of numbered bees from the observation hive is led with the feeding table to the desired distance. In doing this, feeding should be done without a scent and with such a low concentration of sugar that dancing in the hive does not occur; this has to be rechecked constantly. Dances during the course of the journey alert newcomers to the flight path, but with repeated progression of the feeding place some of them do not find the goal and become a source of error when alerted again in the subsequent experiment. If the journey is extended to several kilometers, however, the support of a weak scent must be given from time to time. Then a different scent is chosen for the experiment proper.

The number of marked bees must not be too large lest the odor of the scent organ be too intense. On the other hand, the number must be proportioned properly to the distance flown. When a bee has to fly 4–5 km to the feeding place, she will appear there only about twice an hour. The following will serve as a rough guide: for a distance of 500 m, there should be 8 forager bees; for 2000 m, 30; for 4000–5000 m, about 50.

We laid out the flight path five times in the Forstenrieder Park at Munich, along a forest road that is closed to general traffic and that runs straight for 7 km, and twice in the marshlands around Erding along a perfectly straight road; in these instances the scent cards could be set out quite rapidly by using a car. After this was done the stock bottles of scents and the used pipettes had to be stored under tight cover.

Customarily the observation stations planned for were measured off and suitably marked on the day before the experiment.

The first three stepwise experiments (*Stufenversuche*) took place in 1947 and 1948 in Brunnwinkl and vicinity. They gave positive results, but mountainous terrain affords too little freedom of movement for this procedure, and conditions are insufficiently constant from place to place. For a quantitative evaluation experiments in flat country are preferable. Between 1949 and 1962 we carried out nine more stepwise experiments in the vicinity of St. Gilgen, Graz, and Munich. Two of them were unsuccessful.

I now report the seven successful stepwise experiments, arranged in order of increasing distance to the feeding station, followed by the two unsuccessful ones, which also are quite instructive.

1. At Gschwandt, near St. Gilgen, 3 September 1962; flight distance 300 m. From the former Gschwandt station of a short feeder line that was abandoned some years ago, the old railroad bed heads directly toward Zinkenbach. The rails have been removed; the roadbed, which rises only a little (about 30 cm) above the surrounding level fields (mowed at the time), has a cover of short grass. On 2 September an observation hive was set up on the roadbed (■, Fig. 86) and a group of numbered bees was conducted, by means of scentless feeding, to as far as 300 m from the hive. Not a single newcomer appeared. On 3 September, a warm, cloudless day, at first thin sugar solution was fed also, without scent. During the experiment proper (10:15 A.M.–12:45 P.M.) seven bees received 2M sucrose solution, with orange-blossom scent, at

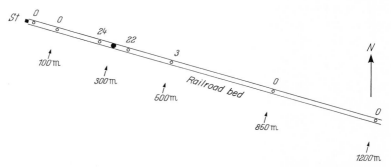

FIGURE 86. Stepwise experiment at Gschwandt (St. Gilgen) on 3 September 1962, 10:15 A.M.–12:45 P.M.: *St*, observation hive; ●, feeding station 300 m from the hive; ○, scent plates. The numbers of visits to each scent plate during the observation period are given.

the feeding place (arrangement as in Fig. 21). Scent plates (as in Fig. 85) with oil of orange blossom were set out in the middle of the road-bed 20, 100, 250, 350, 500, 850, and 1200 m from the hive.

At the feeding station, where the floral scent was supplemented by the attraction due to the scent organ, in 2.5 hr there appeared 80 newcomers; these were at once caught and killed. A gentle wind from NNW–NW prevailed. If the attractive effect of the scent organ had still made itself felt as a disturbing factor at the adjacent scent plates, 50 m distant from the feeding station, in view of the wind direction more flights would have been expected at the 350-m place than at 250 m.

The numbers of visits during the experimental period from 10:15 A.M. to 12:45 P.M. are noted in Fig. 86 and are shown graphically in Fig. 91. During the 2.5 hr, flights of newcomers were very regularly spaced and always were recorded only in the immediate vicinity of the distance indicated by the dancers.

2. Marshlands around Erding, near Munich, 19 July 1956; flight distance 450 m. The observation hive was set up on 18 July, 30 m south of the road (Fig. 87). Numbered bees were conducted along the road to a distance of 450 m from the hive; the final feeding station was placed in the meadow 30 m south of the road.

On 19 July food was offered from 7 A.M. on, and from 9 A.M. on the bees were trained to become accustomed to the experimental feeding dish. Neither during the journey nor at the final station did newcomers appear. From 9:20 to 9:45 seven scent plates were set out parallel to the road and 30 m south of it, all of them in the mowed field in fully flat country. At 9:50 the scent vessel at the feeding station was filled with oil of orange blossom and the weak, barely acceptable sugar solution was replaced by a 2M solution with added oil of orange blossom. It was visited by eight numbered bees. During the experimental period 88 newcomers flew in to the feeding station and were killed. The foraging bees were using their scent organs intensively. The weather was sunny and warm, the wind from E to NE and weak.

The numbers of visits to the scent plates during the experimental period from 9:50 to 11:50 are noted in Fig. 87 and are shown graphically in Fig. 91. Even the scent plates only 200 m before and beyond the feeding place received scarcely any attention. When one considers that a bee in free flight traverses a distance of 200 m in about 24 sec (v.

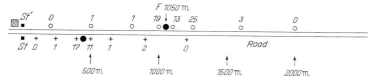

FIGURE 87. Stepwise experiments on the marshlands around Erding, near Munich. A country road running east and west is shown: *St′*, *St*, observation hives; o,+, scent plates, with the number of visits during the experiment; ●, feeding places. On 10 July 1956 a feeding place was set up north of the road at a distance of 1050 m; on 19 July another colony was set up south of the road (below it in the figure), with another feeding place, at a distance of 450 m.

Frisch and Lindauer 1955), it is quite remarkable that in the course of their searching flights the newcomers held so closely for all of 2 hr to the distance indicated in the dance.

3. Vicinity of Graz, 27 June 1949; flight distance 750 m. From the observation hive we journeyed along a straight stretch of the Graz-Köflach railway, over low, flat meadows and fields to the feeding station 750 m south of the hive.

No scent was used, but initially somewhat too concentrated food was given, so that 24 newcomers were taken over the first 350 m. Beyond this and up to the final station only 4 more came, and on the day of the experiment none whatever before it began. On 27 June from 10:25 to 11:20 A.M. the scent cards were set out with linalool. At the feeding place 2M sugar solution with linalool was offered at 11:15 A.M. It was frequented by 12 numbered bees. During the experiment 31 newcomers flew in to the feeding station and were killed. The weather was mostly sunny, the wind moderate from N.

The experiment ran from 11:15 A.M. to 12:45 P.M. The distribution of the scent cards and of the visits is shown in the sketch of Fig. 88 and the numbers are graphed in Fig. 91. Perhaps the use of linalool was the reason why in this experiment only about half the bees that definitely approached the scent cards settled on them.

4. Marshlands around Erding, 10 July 1956; flight distance 1050 m; the same territory as in the experiment of 19 July already described, but with a different colony. The hive, the route followed, and the feeding station lay 25 m north of the road. On 9 July we journeyed successfully with numbered bees, without a scent and without noteworthy dancing, to the final station at 1050 m.

On 10 July from 7:15 A.M. on, weak sugar solution was offered without scent. In 4.25 hr (up to the beginning of the experiment) only four newcomers appeared. At 10:40 training to the screened dish began for the bees at the feeding station; between 11:00 and 11:35 the scent plates with orange-blossom oil were set out. From 11:30 on, 2M sucrose with added orange-blossom oil was fed, the scent being added to the dish. The feeding station was visited by about 30 numbered bees; 132 newcomers were killed there during the 3-hr experiment. This time too the active use of the scent organ seems to have caused the large visitation. The sky was lightly overcast, with occasional sunshine; the wind was moderate, predominantly from N and NW, more rarely from the W.

Experimental period: 11:30 A.M.–2:30 P.M. In Fig. 87 are shown the locations of the scent plates and the numbers of visits. See also the graphical representation of Fig. 91. Almost all the newcomers flying to the scent plates settled on them and hunted about intensively.

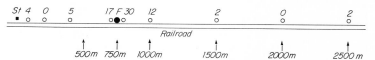

FIGURE 88. Stepwise experiment near Graz on 27 June 1949. *St*, observation hive; ●, feeding station at 750 m; ○, scent cards with the number of visits during the experiment.

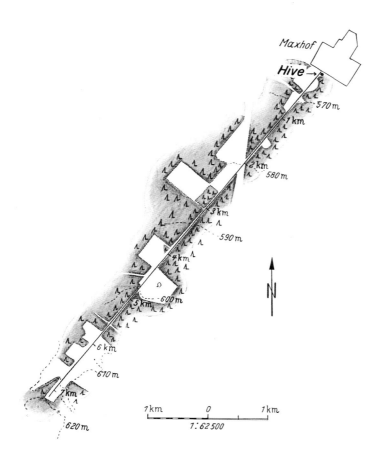

FIGURE 89. Sketch map of the experimental area in the Forstenrieder Park at Munich, on the southwest forest road. The figures in meters show the elevations above sea level.

5. Forstenrieder Park at Munich, 20 July 1952; flight distance 2000 m. The area for the experiment is shown in Fig. 89. From the "Maxhof" a gravel road, which is used only as a forest trail, runs southwesterly in a straight line for 7 km through pathways and clearings into the Forstenrieder Park (Fig. 90). Thanks to the cooperation of the Forest Service this road was useful to us in many experiments.

On 18 July the observation hive was set up near the Maxhof and on this day a journey was undertaken with numbered bees, with a 2M sugar solution but without scent, to a total distance of 1600 m; all newcomers appearing in the course of the trip were incorporated into the foraging group. On the next day only a small amount of feeding was possible because of rainy weather. On 20 July the feeding station was

FIGURE 90. Feeding station on the forest road in the Forstenrieder Park.

advanced to 2000 m by 10:00 A.M. Up until the beginning of the experiment (11:15 A.M.) no newcomers appeared. Between 10:30 and 11:45 the scent cards were set out with oil of lavender on the edge of the road. At 11:50 the bees were fed with 2M sugar solution and the scent vials were filled with oil of lavender. About 40 numbered bees were frequenting the feeding station, while during the experimental period 22 newcomers flew up and were captured. It was cloudless and warm, partly calm and partly with moderate wind of variable direction.

The experiment ran from 11:50 A.M. to 2:50 P.M. Figure 91 shows the results in graphical form.[10] It is noteworthy how sharply the large

---

[10] In this experiment newcomers appeared during the journey and were incorporated into the foraging group. In the course of the rapid trip some of them were lost from the group and consequently reappeared at various locations during the experiment. They were recognized by their markings and of course were not counted then as newcomers. Since in instances where the bee merely flew briefly about the scent card it was uncertain whether she had been marked, in this experiment only those bees that had settled down fully were taken into account, this being the situation with the overwhelming majority.

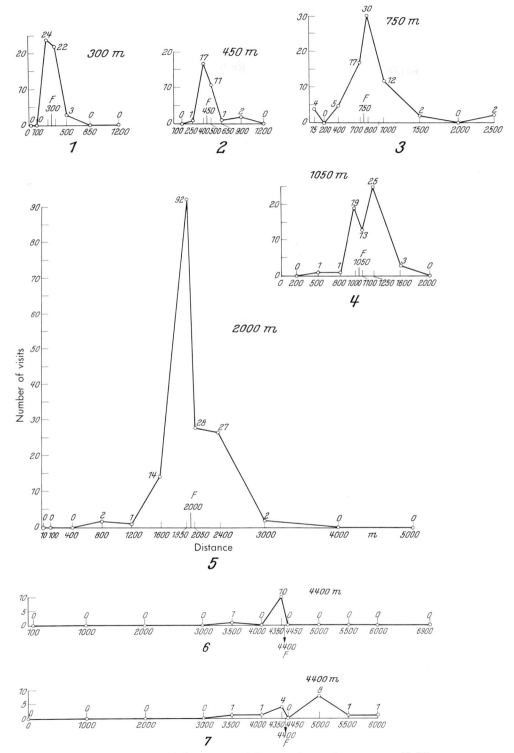

FIGURE 91. Graphed results of the seven stepwise experiments (see text, pp. 88–94); abscissa, distance (m); ordinate, number of visits.

number of approach flights is limited to the region between 1600 and 2400 m—a distance requiring less than 2 min for a bee in free flight. The dancers must have indicated this distance "convincingly." If the approach flights during the 3-hr experiment are taken for $\frac{1}{2}$ hr at a time, the constant searching at the announced distance is expressed over and over (Table 10).

6. Forstenrieder Park, 12 July 1953; flight distance 4400 m. The observation hive was set up on 8 June. In consequence of bad weather the journey to the great distance planned went very haltingly. On 7 July 3000 m was reached, and 4400 m at last on 12 July after more rain.

On 6 July a little lavender scent had been used between 2500 and 3000 m, but the journey was extended without scent thereafter and only a few newcomers appeared. Between 9:55 and 11:00 the scent cards were set out with oil of peppermint, and from 11:00 on the feeding station was perfumed with peppermint and 2M sucrose with peppermint in it was fed. The station was frequented by 40 numbered bees. During the 5-hr experiment nine newcomers arrived at the feeding station and were captured. The weather was sunny, the sky somewhat overcast at times; it was partly calm, partly with moderate winds of variable direction.

The experiment lasted from 11:00 A.M. to 4:00 P.M. It was continued for 5 hr because only a few newcomers appeared, on account of the great distance and because of the currently good supply of natural food. But these newcomers came with surprising precision (see the graphical representation of Fig. 91). Seven of the ten approach flights at 4350 m fell during the hour from 1:30 to 2:30 P.M. At this time the preceding activity in the hive was very lively.

7. Forstenrieder Park, 3 September 1954; flight distance 4400 m; repetition of the 1953 experiment. On 31 August the observation hive was set up in the Maxhof. As early as 1 September the 4400-m station was reached, without the use of scent.

It was necessary to use lavender scent to arouse more newcomers, and these were incorporated into the marked group. On 2 September there was scentless feeding with a rather weak sugar solution. On 3 September the scent cards were set out by 10:40 A.M. with oil of pepper-

TABLE 10. Stepwise experiment in Forstenrieder Park, 20 July 1952. Visits to the scent cards (and to the feeding station) during each half hour of the 3-hr experiment.

Time	Distance from hive (m)												
	10	100	400	800	1200	1600	1950	2000[a]	2050	2400	3000	4000	5000
11:50–12:20	0	0	0	0	0	0	0	(0)	1	1	0	0	0
12:20–12:50	0	0	0	2	0	3	7	(4)	1	4	0	0	0
12:50–1:20	0	0	0	0	0	4	17	(6)	2	6	0	0	0
1:20–1:50	0	0	0	0	0	3	30	(3)	2	2	0	0	0
1:50–2:20	0	0	0	0	0	1	24	(3)	17	7	2	0	0
2:20–2:50	0	0	0	0	1	3	14	(6)	5	7	0	0	0

[a] Feeding station.

mint. From 10:40 on the scent of peppermint and 2M sucrose solution to which peppermint had been added were offered at the feeding station, which was frequented by 70–80 numbered bees. During the 4-hr experiment nine newcomers appeared and were killed. There was a threat of thunderstorms, a condition unfavorable to the appearance of newcomers at great distances. Yet a storm did not break. After 1:20 P.M. it cleared to the southward, and was sunny. It was rather calm during the experimental period.

The experiment ran from 10:40 A.M. to 2:40 P.M. A graph of the results may be seen in Fig. 91.

A glance at the summarized outcome of all seven experiments brings up the question how closely the precision of searching matches the precision with which information is given. But this will not be discussed until the end of the next section (pp. 106ff).

The two unsuccessful experiments remain for consideration.

On 4 July 1954 we carried out a large-scale experiment in the Forstenrieder Park with the feeding station at 4400 m. I had set out the scent cards beyond the feeding station and was driving back to the hive in order to locate seven additional scent cards between it and the feeding station. In turning around the bottle of scent was tipped over, its stopper was loosened, and a few drops of oil of lavender leaked out onto the floor of the car. Since the car doors were opened at each observation station in order to set out a scent card, air heavy with perfume was enabled to flow out onto the road—a development all of whose consequences I did not realize immediately. The experiment ran from 9:55 A.M. to 1:25 P.M. The results were aberrant. As shown in Table 11, the observation stations at 3000 and 3500 m were most frequently visited. Both observers—without knowing of this event—repeatedly noted in their protocols that bees went hunting around on the road, several meters to the side of the scent cards, or that they approached the scent cards from the direction of the road. Apparently lavender scent that had flowed out of the car was still clinging to the ground there. Since the scent cards at 100, 1000, and 2000 m were scarcely noticed by the bees, or even not at all, the conclusion is tempting that in consequence of the indications received the newcomers were flying at considerable heights toward their goal and came close to the ground in their search

TABLE 11. Stepwise experiment in Forstenrieder Park, 4 July 1954, 9:55 A.M.–1:25 P.M. The lavender-scented feeding station at 4400 m was frequented by about 45 numbered bees; during the experiment 13 newcomers appeared and were captured. Following rain the previous night it was cloudless, and mostly calm; at other times there was light wind from the south or sometimes from the north.

Distance from hive (m)	100	1000	2000	3000	3500	4000	4350	4400[a]	4450	5000	5500	6000
Number of visits by newcomers	2	0	2	25	29	10	15	(13)	7	2	0	1

[a] Feeding station.

only when they were approaching the designated distance. This view is supported also by the fact that in general the scent cards between the hive and the vicinity of the goal were passed by unnoticed, although they lay on the flight course. In the instance here considered, when the bees came close to the ground they were caught in the cloud of strong scent and in part were impeded from flying onward toward the actual goal. This assumption is bolstered also by the second unsuccessful experiment.

This too took place in the Forstenrieder Park, on 16 July 1955, with the flight distance at 500 m. After a morning fog, fine warm weather prevailed. The experiment lasted from 10:30 A.M. to 1:30 P.M. with oil of lavender as the odorant. As a disturbing factor a heavy wind came up from the ESE (about 2–3 m/sec with frequent even stronger gusts). As is readily seen, in such a wind the bees fly nearer the ground, where the wind velocity is less. It is to be assumed that on this account they came within range of the odor of the scent cards closer to the hive and that the large visitation at all three observation posts between hive and feeding station is to be ascribed to this factor (Table 12). Beyond this, during the journey on the previous day as a result of too concentrated feeding the mistake was made of mobilizing newcomers along the way; no doubt some of these did not follow the whole distance and apparently were involved in the flights to the nearer stations during the experiment.

TABLE 12. Stepwise experiment in Forstenrieder Park, 16 July 1955, 10:30 A.M.–1:30 P.M. Twenty numbered bees were frequenting the feeding station at 500 m; during the experiment 46 newcomers flew in and were captured. Weather fine, but vigorous gusty wind from the ESE.

Distance from hive (m)	150	300	450	500[a]	550	700	1000	1500
Number of flights to scent cards	37	37	29	(46)	16	1	0	0

[a] Feeding station.

8. What Part of the Tail-Wagging Dance Is the Signal that Defines the Distance?

The stepwise experiments have shown that the newcomers alerted have been informed of the distance to the goal. They obtain this information from the tail-wagging dance of the foraging bees. Since the tempo of the tail-wagging dance is clearly related to the distance of the feeding place, we have considered the dance tempo[11] as the signal for

[11] Expressed as the number of circuits or reversals in 15 sec, or as the duration of a circuit; also referred to in some papers as the number of waggling runs or passages per 15 sec, or as the dance rhythm—all of them merely different designations for the same phenomenon.

the distance. But there remains to be investigated whether the definitive signal lies in the whole course of the dance or only in a certain phase of it, and whether the significant aspect is concerned with the duration or with some other variable element of this phase. In order to judge this, the elements included in the tail-wagging dance would have to be measured more precisely than is possible in observation with the naked eye. This was done by obtaining and evaluating motion pictures of the tail-wagging dances (v. Frisch and Jander 1957). A further basis was provided by electromagnetic registration of the dances (Steche 1957; Esch 1961, 1961a, 1964) and with microphonic recording (Wenner 1962).

For the technique of taking photographs see p. 24. They were first made in the Forstenrieder Park at Munich in 1953 in cooperation with the Institute for Scientific Photography (Göttingen) and were repeated utilizing our experience in 1954 and 1955. We chose a film speed of 60 exposures/sec, which suffices for a reliable count of the waggling movements made during the waggling run. In evaluating the photographs a manually operated scanning table with a "Cinetti" viewing apparatus was used. A motor drive is indeed more convenient but is not sufficiently regulable for the present purpose. The following results are based on pictures taken in 1955, in which the times were registered by simultaneously photographing a watch with a large second hand and a scale divided into tenths of a second. Insofar as comparisons can be made, the pictures from the two earlier years agree fully.

(a) *The Components of the Tail-Wagging Dance and Their Correlation with Distance*

Our idea was to seek the index of distance in whatever element of the tail-wagging dance might show the best correlation with the distance to the food source. Thus the various elements comprising the dance were to be compared in this respect.

With increasing distance the tempo of dancing slows down, that is, the number of circuits per 15 sec decreases and simultaneously the number of waggling movements made during the linear portion of the dance increases. The waggling is so rapid that with the naked eye the number of movements can only be approximated, not counted accurately. Our first endeavor was to compare, on the films, the precise number of waggles with the number of circuits per 15 sec. However, from the statistical standpoint this procedure is subject to the criticism that individual values (number of waggles per single run) are compared with averages (number of circuits per 15 sec). By such treatment the scatter of the "number of circuits" is reduced. Hence, for purposes of comparison we selected as a measure of the dance tempo not the number of circuits per 15 sec but the time required for *one* circuit.

The components of the tail-wagging dance that can be considered as an index of the distance may be grouped according to the nature of the signaling factor, as follows:

Factor used in signaling	Signal (see Fig. 92)
(a) Temporal rhythms during the movement	1. Frequency of the waggling movements[12] (= number of waggles/sec)
	2. Duration of a circuit (for the course $S + R$)
	3. Duration of waggling (for the course S)
	4. Returning time (for the course R)
(b) Numerical index	5. Number of waggles per linear run
(c) Length index	6. Length of the waggling run S
(d) Vibratory stimulus	7. Duration of sound production
	8. Vibratory frequency
	9. Temporal structure of the virbrations (duration of the vibratory episodes and intervening pauses)

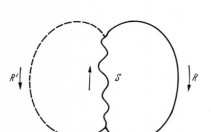

FIGURE 92. Pattern of the tail-wagging dance, schematic: S, waggling segment; R and R', return runs (as a rule alternately to right and left).

The last-named elements can be of value as signals only if the sound produced by the dancer during the waggling run (see pp. 57f) is perceived by the followers. Without doubt this possibility exists. Bees indeed have no true sense of hearing—if thereby one understands the perception of airborne sound—but they have a well-developed vibratory sense. They react clearly to vibrations transmitted through the substrate (Hansson 1945; Frings and Little 1957; Little 1959, 1962; Wenner 1962a).

Lindauer and Kerr (1958) were able to show that stingless bees (meliponids) produce buzzing sounds whose vibrations play a part in transmitting information to their comrades. Whether vibration is perceived by the honeybee with the antennae, the legs, or both together is an open question. First consideration is to be given to the antennae, which seek contact so conspicuously with the dancer when she is being followed; for according to Heran (1959) Johnston's organ in the antennae is most sensitive to vibrations from 200 to 350 cy/sec and the resonating frequency of the antennal flagellum lies at about 280 cy/sec, which corresponds well with the frequency (about 250 cy/sec according to Esch 1961a) of the sound produced. Autrum and Schneider (1948) succeeded with electrophysiological methods in showing a response of the subgenual organ of the honeybee to vibrations of the substrate, but its maximal sensitivity lies in a higher frequency range (around 2500 cy/sec).

Of the nine elements of the tail-wagging dance, three are excluded as possible indices of the distance because they show no relation to the distance of the feeding station. In our list these are signals Nos. 1, 8, and 9.

No. 1. In evaluating the photographic records of the waggling frequency in 1447 waggling runs we found an average of 13 cy/sec, that is, 13 waggling movements to right and left each second, with extremes of 10 and 16 cy/sec. Stepwise removal of the feeding station from 200 to as far as 4400 m had no influence on the frequency (v. Frisch and

---

[12] The term "waggling movement" always means a double waggle, that is, an excursion of the body to both left and right (= 1 cycle).

Jander 1957). With electromagnetic registration of the dances Steche (1957) found average values of 14–15.5 cy/sec for the waggling frequency, Esch (1956, 1956a), a mean of 15 cy/sec. The differences are perhaps to be attributed to the use of different colonies. In regard to what matters, all investigators are at one: the frequency remains unchanged with increasing distance. Hence the frequency cannot be the index of distance.

No. 8. For the frequency of the sound given off during the waggling run Esch (1961a) found a mean value of about 250 cy/sec, which varied insignificantly and independently of the distance during movement of the feeding place over intervals between 100 m and 4000 m.

No. 9. In the temporal structure of the vibratory episodes (their duration and the duration of the intervals between them) there seems to be a relation with the *quality* of the feeding place (p. 237). But no relation to the *distance* of the feeding place is indicated. The number (four) of excursions in a single vibratory episode, the duration of the individual episodes, the length of the intervals between them, and hence the frequency of succession of vibratory episodes (30 cy/sec) were approximately the same for all distances from 100 m to 4000 m (Esch 1961a:7, Table 2).

No. 7. According to Esch (1961a and personal communication), the duration of sound production is tied rigidly to the waggling phase (tested up to 3000 m). In comparison with the period of waggling there was never a longer period of production of vibration, and a shorter one only during aberrant dances that were found to be without an alerting effect. Hence the period of vibration corresponds to the waggling period, and does not require separate treatment here.

Thus our list has been condensed to the five points, Nos. 2–6: the duration of the circuit and of its components, waggling phase and return; the number of waggling movements per run; and the length of the waggling course. Each of these elements of the bees' dance shows some correlation with the distance to the feeding place. This relation is depicted in the curves of Figs. 93–95 (v. Frisch and Jander 1957).

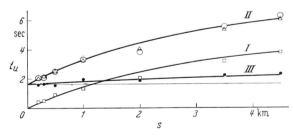

FIGURE 93. Curves showing duration of a circuit (II) and duration of waggling (I), based on photographic data (□, ○) and on direct observation (△). Curve III gives the duration of the return run as the difference of II and I. Abscissa, distance (m) to feeding station; ordinate, mean time (sec); see Table 13. From v. Frisch and Kratky 1962.

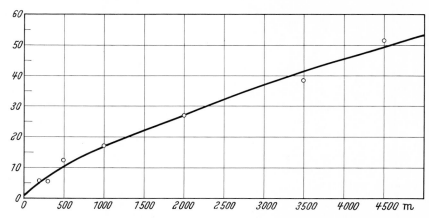

FIGURE 94. Curve relating the number of waggling movements to the distance, based on photographic data; abscissa, distance (m) to feeding station; ordinate, number of double waggles per run; see Table 14. From v. Frisch and Jander 1957.

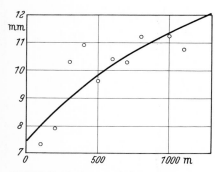

FIGURE 95. Curve relating the length of the waggling segment to distance, after measurements by Heran (1710 individual determinations); abscissa, distance (m) to feeding station; ordinate, length (mm) of waggling segment; see Table 15.

The mean values for constructing the curves of Figs. 93 and 94 were taken, for the different distances of the feeding stations, from the photographic data (Tables 13 and 14), and entered on a system of coordinates. The lengths of the waggling runs (Fig. 95) were not taken from our films, for a large number of measurements by Heran (1956:173) were available. They concern feeding places at distances of 100–1100 m (Table 15).

What interests us is a comparison of the error (of the value as signals) of these elements. For this two factors are determinative: the scatter of the single values for a given distance of the feeding place and the slope at this distance of the curve that relates the scatter to distance.

(1) With a given slope of the curve the error E in the indication of distance is proportional to the measure of scatter σ. If, for example, the scatter σ doubles, then the error in indication of the distance is twice as large (Fig. 96).

TABLE 13. Duration (sec) of a circuit, of the waggling period, and of the return run, for feeding stations at different distances. Measurements on a pure Carniolan colony.

Distance from feeding station (m)	Circuit		Waggling		Return	
	Mean[a]	n	Mean[a]	n	Mean[a]	n
200	2.09 ± 0.03	123	0.45 ± 0.01	123	1.64 ± 0.02	123
300	2.12 ± .03	89	0.50 ± .02	89	1.62 ± .02	89
500	2.53 ± .02	354	0.95 ± .01	357	1.58 ± .02	354
1000	3.32 ± .04	114	1.34 ± .03	114	1.98 ± .04	114
2000	3.78 ± .06	53	2.08 ± .06	55	1.70 ± .04	53
3500	5.59 ± .22	16	3.12 ± .14	16	2.47 ± .12	16
4500	6.30 ± .11	75	3.98 ± .09	70	2.32 ± .07	67

[a]The errors given are the ratio of the scatter σ of the individual measurements to the square root of their number.

(2) With a given degree of scatter σ of an element the error is inversely proportional, on the other hand, to the tangent of the angle of slope β of the curve. For instance, if the value of tan β is doubled, then the indication as to distance is only half as poor, or in other words twice as good (Fig. 97).

From (1) and (2) there follows the definition of the error E in the indication of distance:

$$E = \sigma / \tan \beta$$

Armed theoretically in this way, one can calculate the errors of the different elements of the dance and compare them. The slopes of the curves for the duration of waggling t_w, the duration of a circuit t_c, the time of returning t_r, the number of waggles n_w, and the length of the waggling run l_r are similar, that is, for any two curves the ratio tan β_1/tan β_2 is approximately constant for all distances.

For purposes of comparison the error of the duration of waggling E_{t_w} is taken as 100 percent. The relative error of the duration of a circuit $(E_{t_c})_{rel}$ then is:

$$(E_{t_c})_{rel} = \frac{E_{t_c}}{E_{t_w}} \times 100 = \frac{\sigma_{t_c} \tan \beta_{t_w}}{\sigma_{t_w} \tan \beta_{t_c}} \times 100 = 119 \text{ percent.}$$

Corresponding expressions give the relative errors of the other elements. Then the percentage values indicate how well the distance can be read from the element in question, in comparison with the duration of waggling.

An example will clarify how these relative errors are to be interpreted. With the feeding station 500 m distant the error of the duration of waggling amounts to about 200 m.[13] The error for the duration of a circuit then amounts to 119 percent of this = ± 238 m; in other words

[13] This error may seem large. But remember that the calculation is based on the scatter of *individual* waggling runs (see p. 97) and that in consecutive runs within a single dance the duration of waggling varies appreciably (see p. 72).

TABLE 14. Number of waggles per waggling run for feeding stations at different distances.

Distance of feeding station (m)	Mean number of waggles	n
200	5.95 ± 0.20	122
300	5.77 ± .24	89
500	12.4 ± .20	354
1000	17.0 ± .40	114
2000	27.0 ± .68	53
3500	38.3 ± 1.17	16
4500	51.4 ± 1.25	78

TABLE 15. Length of the waggling run for feeding stations at different distances (after data from H. Heran).

Distance of feeding station (m)	Mean length of waggling run (mm)	n
100	7.3 ± 0.21	111
200	7.9 ± .14	164
300	10.3 ± .17	179
400	11.0 ± .18	198
500	9.6 ± .14	177
600	10.4 ± .18	176
700	10.3 ± .15	177
800	11.2 ± .14	195
1000	11.2 ± .17	160
1100	10.7 ± .25	173

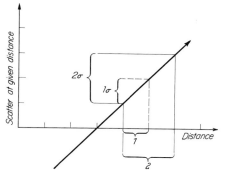

FIGURE 96. Dependence of the error in indication of distance on the scatter; see text. From v. Frisch and Jander 1957.

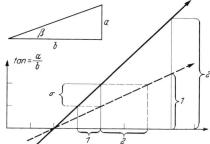

FIGURE 97. Dependence of the error in indication of distance on the slope of the curve relating distance to the index signal; see text. From v. Frisch and Jander 1957.

it is greater by 19 percent. The corresponding error for the number of waggles is greater by 29 percent, the error for the length of the waggling run is greater by 170 percent,[14] and the error for the time of returning by as much as 352 percent (Table 16).

Thus we reach the following conclusions:

Because of its poor correlation with the distance—as already expressed in the slight slope of the curve relating the two—the time of returning need not be given consideration. This is confirmed also by the behavior of the dance followers, who display but little interest in the return phase. Whereas they follow the waggling run in great excitement, during the return phase they often lose contact with the dancer. Not infrequently they remain awaiting her at the spot where the next waggling run is to begin. Besides, at great distances the return run may be replaced by a short S-curve (Fig. 51), or may be omitted altogether, which certainly would not be expected of a decisive indication as to the distance.

The values for the length of the waggling run are so widely scattered that they too are eliminated as a signal for the distance. This result of Heran's has been confirmed by our calculations (v. Frisch and Jander 1957:245, 248n). The length of the straight run would also be an unsuitable indication, since, because of the gradual transition at its beginning and end, it is not sharply distinguished from the rest of the course; nor could the dance followers judge it reliably from the beginning and end of the waggling movements, for one often sees that a dancer begins her waggling run even before the start of the straight run or continues waggling on for a little into the curve at its other end.

The information about the distance, expressed in meters, given by the number of waggles per run is about 29 percent worse than for the duration of waggling. This difference is not so great as to exclude the number of waggles as a usable signal. But the value of a signal is dependent not only on the accuracy of the information given but also on the precision with which it can be understood. Now the perception of phases of motion that are sharply distinguished temporally (duration of waggling or of a circuit), as in general the comprehension and remembering of durations and rhythms, is a widespread phenomenon in the living world. But the *number* of waggling movements would be a signal of a fundamentally different sort. That bees should be able to count as many as 50 or more waggling movements, that they should grasp differences in the number of waggles and be guided accordingly, is most improbable in view of all we know about the counting abilities of animals. This same consideration applies with reference to the number of vibratory episodes that occur in the course of a waggling run. (Regarding the relation between the number of vibratory episodes and the quality of the source of food, see p. 237.)

There now remains the necessity of discussing the duration of the

---

[14] In order to be able to compare Heran's measurements with our own we have treated them in the same way as the other elements. In doing so we limited ourselves to distances of 200, 300, 500, and 1000 m, since our films did not contain material for comparison with Heran's data at other distances.

TABLE 16. Comparison of error in the indication of distance by means of the several elements of the tail-wagging dance, as a percentage of the error of the duration of waggling.

Distance (m)	Duration of waggling,[a] t_w			Duration of circuit,[a] t_r				Time of return,[a] t_r				Number of waggles,[b] n_w				Length (mm) of waggling run,[c] l_r			
	σ	n	Error, $\dfrac{\sigma}{\tan \beta}$	σ	n	$\dfrac{\tan t_w}{\tan t_c}$	Error	σ	n	$\dfrac{\tan t_w}{\tan t_r}$	Error	σ	n	$\dfrac{\tan t_w}{\tan n_w}$	Error	σ	n	$\dfrac{\tan t_w}{\tan l_r}$	Error
200	±0.16	123	± 98	±0.30	123	0.8	150	±0.27	123	4	675	±2.22	122	0.1	139	±1.8	164	0.29	331
300	± .18	89	± 118	± .25	89	.8	111	± .23	89	4	511	±2.24	89	.1	124	±2.1	179	.29	343
500	± .27	357	± 204	± .42	354	.8	124	± .32	354	4	474	±3.69	354	.1	137	±1.7	177	.29	185
1000	± .29	114	± 276	± .47	114	.8	132	± .41	114	4	565	±4.22	114	.1	145	±2.2	160	.29	223
2000	± .41	55	± 565	± .45	53	.8	88	± .29	53	4	283	±4.92	53	.1	120				
3500	± .56	16	± 935	± .89	16	.8	127	± .48	16	4	343	±4.70	16	.1	84				
4500	± .73	70	±1350	± .92	75	.8	101	± .57	67	4	312	±11.2	78	.1	153				
	$(E_{t_w})_{\text{rel}}$, 100			$(E_{t_c})_{\text{rel}}$, 119 ± 8				$(E_{t_r})_{\text{rel}}$, 452 ± 55				$(E_{n_w})_{\text{rel}}$, 129 ± 9				$(E_{l_r})_{\text{rel}}$, 270 ± 39			

The quantities $\tan t_w$, $\tan t_c$, $\tan t_r$, $\tan n_w$, $\tan l_r$ are the slopes of the curves that relate the element in question to distance.
[a] See Table 13. [b] See Table 14. [c] See Table 15.

waggling segment and of an entire circuit. The duration of the whole circuit, that is, the tempo of dancing in the old sense, is only 19 percent less good as a signal than the duration of waggling. The difference is small and does not permit a decision between the two possibilities. But in favor of the duration of waggling there are—in addition to the best correlation with distance—yet other arguments. The dancer gives this phase special emphasis, both by means of the waggling movements and through the production of sound. The attention of the dance followers is directed above all to the waggling phase. With great distances it sometimes happens that the dance is limited to a single waggling run and that the return run is eliminated (see p. 66). It is to be noted also that the direction to the goal is signaled by the waggling run, and one is tempted to ascribe the indication of distance to the same phase of the dance.

Taking all these facts into account, we regard the acoustically emphasized duration of waggling as the index of distance.

The investigations concerning the influence of external and internal factors on the tempo of dancing, such as age of the bees (p. 74), temperature (pp. 75ff), and so forth, are based on measurements of this tempo, that is, of the duration of the circuit. One needs to examine whether these findings hold also in regard to the duration of waggling, or whether the latter is more constant. This work can be done more easily with modern recording techniques than on film.

Steche (1957), like ourselves, considers the waggling phase the significant index, but regards this as given by the number of waggling movements, without providing convincing support for this view (see v. Frisch and Jander 1957:261). Ribbands (1953a:155) also thinks that the bees might more readily perceive the number of waggling movements than the rhythm of dancing; but he offers no supporting arguments. Haldane and Spurway (1954:260) believe that at distances exceeding 700 m the number of waggling movements is a more precise index than the rhythm of dancing. According to our measurements this is not the case (Table 16).

(b) Experiments to Vary Components of the Dance Independently

We came to the conclusion that very probably the duration of waggling is the significant index of the distance—but we have not proved this. Proof is possible if one succeeds in testing only certain components of the tail-wagging dance and varying them. This was suggested by Haldane and Spurway (1954). Such experiments were done by Steche with models of bees that were caused artificially to waggle. He claims (1957a) to have achieved directed flights by this means, but his data are very inexact and later repetitions were unsuccessful.

Esch (1964) also, who extended the model experiments with improved technique, did not obtain decisive results but made some noteworthy incidental findings. As models bits of wood or paraffin-coated bees were used. Before they were used they were placed in the hive for some days and were thus impregnated with the odor of the colony. During the experiment he guided the models around on the comb in the dance pattern and made them waggle by mechanical means with a rotating magnet. Dance followers pursued them with interest (Fig. 98), but were not to be stimulated to seek out the feeding station. Nor was this accomplished when the model was caused, with the help of a second electromagnetic system, to emit the vibratory sounds. The model evidently lacked some significant characteristic without which it could not be taken seriously. But that it was recognized nevertheless as a dancer was shown, apart from the pursuit, in another unexpected manner: when a follower gave a peeping sound and the model did not thereupon halt, as a real bee does, in order to distribute sugar water (see p. 60), the model was attacked and stung. This was seen 30 times. On the same number of occasions Esch stopped the model at the sound of peeping. Then there was no attack. Apparently the anomalous behavior of the artificial bee revealed it as a stranger and it was treated accordingly.

Esch (1964) also experimented in another direction, which is promising and which, with sufficient time and patience, might lead to success. A feeding station was placed in the direction toward the sun and was shifted constantly as the latter moved. It was visited by a single bee. As soon as the bee had completed one waggling run upward on the vertical comb, without any view of the sun, she was shown the sun from below in a mirror. Thereupon she made an immediately consecutive waggling run toward the mirror image of the sun, that is, downward. At the conclusion of this run the sun was covered and the bee at once oriented herself according to gravity and made a waggling run

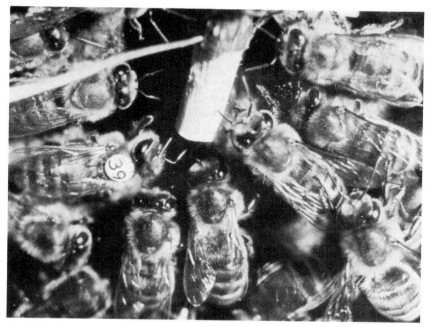

FIGURE 98. Dance followers pursue with interest a "dancing" wooden model. After Esch 1964. Photo H. Miltenburger.

upward. This was done several times in succession. Followers pursued the dancer during each waggling phase, and were stimulated by the abnormal dances too. The duration of waggling was identical whether the runs were oriented according to gravity or light. Since the waggling runs succeeded one another immediately, the time of return was omitted and the duration of a circuit was shortened correspondingly.[15] Now a stepwise experiment should provide a decisive answer: if the stimulated newcomers fly to the correct distance even after dances with reduced duration of the circuit, then it has been demonstrated that the waggling phase and not the total duration of the circuit is the significant index. Unfortunately the outcome is still unknown. In this situation the stepwise experiment can be carried out only with a single dancer at a time, since there must be certainty that all dances that have invited visits to the feeding station have been influenced, in accord with the nature of the experiment, by reflection of the sun. As yet it has been possible to carry out only one such experiment, and because of the single dancer only a few newcomers appeared. Repetitions were ruined by unusually unfavorable weather, and hitherto other circumstances have prevented Esch from continuing with this work.

(c) *Comparison of the Precision of Searching and the Accuracy of Distance Indication*

The stepwise experiments discussed in the preceding section have provided information regarding the precision with which the alerted newcomers seek out the indicated source of food. Now we ask what degree of accuracy is to be expected of these bees in consideration of the value of the duration of waggling as an index, and how well the two findings accord with one another.

Table 17 gives a résumé of the stepwise experiments. In the fourth column is shown the scatter (error in meters) of the indications of distance given by the dancers. These values for the scatter were calculated from the photographic data (v. Frisch and Jander 1957). In doing so the error E in respect to distance was first determined with the help of the relation $E = \sigma/\tan \beta$, as explained on p. 101. Graphical interpolation in Fig. 99 then yielded the error in respect to distance made by the dancing bees in the stepwise experiments here under consideration. In addition to this error in the duration of waggling, the scatter of the actual flights of our searching newcomers in these stepwise experiments is entered in Table 17 (column 5) and in Fig. 99. The comparison shows that the bees seek the goal within a narrower range than would be expected if they had been guided by the duration of waggling of a single waggling run. They come flying in more precisely than conforms with the degree of scatter of the individual determinations of the duration of waggling. From this we may conclude that they average together several waggling runs.

[15] With the feeding station 300 m distant the average duration of a circuit was measured as 2.26 sec ($n = 6009$) for normal dances, and as 1.47 sec ($n = 1205$) for dances shortened by means of a mirror. The duration of the dance phases was determined acoustically (technique: Esch 1961a).

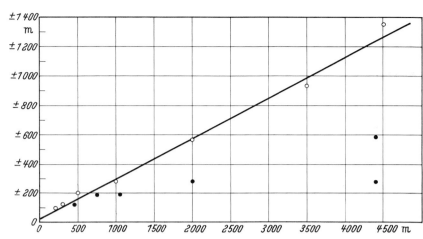

FIGURE 99. Relation between the distance to the feeding station (abscissa) and the bees' error (ordinate): open circles and curve, error of the duration of waggling if the latter is taken as an indication of the distance; solid circles, error (scatter) of the searching newcomers. From v. Frisch and Jander 1957.

Such ability may properly be ascribed to them. It is paralleled in other integrations with which we shall presently be concerned: apprehending the line of flight in experiments involving detours (pp. 173ff); signaling the true direction when there are sidewinds (p. 194); determining the distance when the way out and the way back differ (p. 117). It is a fact that before the bees fly out they follow closely not only several waggling runs in a single dance but also as a rule several distinct dances by the same or by different bees (Lindauer 1952). At the time of these observations Lindauer was able with 24 numbered bees to record all the pursuits of dancers made by them before they first flew out. On the average they had obtained directions from three or four dancers before they first set out themselves; in doing so, in each

TABLE 17. Comparison of scatter in the distances indicated by the dancers with the scatter in the flights of recruited newcomers in stepwise experiments (pp. 88–95).

Experiment number	Year	Distance of feeding station (m)	Scatter in indication of distance in waggling run (m)	Scatter in distances flown by newcomers (m)	Number of flights	Minimum number of waggling runs
1	1962	300	± 108	± 69	49	2.4
2	1956	450	± 150	±132	32	1.3
3	1949	750	± 215	±193	66	1.0
4	1956	1050	± 310	±170	62	3.3
5	1952	2000	± 580	±278	166	4.4
6	1953	4400	±1245	±280	11	20.0
7	1954	4400	±1245	±590	16	4.5
						Mean: 5.3

instance they followed closely after several (the precise number not counted) waggling runs. From their own dances after their first successful foraging flights it was to be deduced that 10 of the 24 newcomers had been harvesting precisely in the direction and at the distance signaled by the bees that had danced before them. On 11 occasions that was approximately the case, while on 3 the goal could not be determined because they danced but little and lethargically (personal communication from Lindauer on the basis of his old protocols).

The scatter of means diminishes in proportion to the increase in the square root of the number of individual determinations averaged. Hence one can calculate the minimal number of waggling runs the alerted bees must have averaged before they flew out (Table 17, last column). For the different experiments these values diverge greatly. Particularly conspicuous is experiment No. 6, in which only a few newcomers appeared, but with great accuracy. The calculation shows that on the average the bees must have followed the mean of at least five or six individual indications on the part of the dancers. But this is so under the assumptions that the newcomer receives the information free of error, takes its average without erring, and flies to the indicated goal precisely according to this mean. Only then should errors in the distance flown be attributed solely to errors in the dances. Now bees are not computers that work without making mistakes. Hence, since the assumptions stated certainly do not hold, more waggling runs must have been averaged than are indicated in the last column of the table. That also fits with Lindauer's observations. It is not possible to be more precise about these matters at this time. When the goal is far away, the fact that averaging has occurred becomes more conspicuous, as may be seen clearly from the graphical comparison of the scatter (Fig. 99).

Wenner (1962) likewise determined that the newcomers in our stepwise experiments flew in more accurately than is possible from the signal given in a single waggling run. But he doubts that the small scatter in these results can be accounted for by an averaging of the indications as to distance. He thinks that even if numerous waggling runs had been followed there could not be sufficient equalization of the differences that exist between the indices as to distance given by different bees. Here he overlooks the fact that the dance followers not only chase after several waggling runs made by a single bee but also pursue several different bees through their dances. He proposes another explanation: after the bees first settle they are said to fly about at random, and thereby improve their second choice appreciably. However, random flight cannot improve their precision.

It would be desirable to carry out a much larger number of stepwise experiments in order to retest the numerical findings, which for the present can be regarded only as approximations. An obstacle to doing so is the expenditure in time and personnel demanded for each stepwise experiment. Nevertheless the results to date are in accord with what might reasonably be expected, and do not suffer from intrinsic improbability.

9. How Does the Dancer Estimate the Distance?

If a bee is to announce the distance to the goal, this must be known to her. We enquire now as to the basis for this knowledge.

According to human custom distances are measured in meters, yards, miles, and so on. When the velocity of movement is known, time also is an ordinary index of distance: "The house is an hour's walk away." How a bee determines distance should be revealed by testing whether her announcement of it is related to the absolute distance or to the time required in flight. But neither of these proves to be the case.

(a) Bees Do Not Signal the Absolute Distance to the Goal

That a bee may correctly estimate by means of optical impressions the length of the course flown over is thinkable. For she is constantly taking note of the landscape rolling past beneath her.

This becomes evident when her flight is impeded by a headwind. Then she increases her flight speed and thereby compensates in part for her displacement by the wind; but with a tailwind she slackens her efforts and lets the motion of the air play a greater part in her progress (v. Frisch and Lindauer 1955). Heran (1955) succeeded in demonstrating that the terrain below contributes decisively to this regulation of flight velocity. He suspended a bee by the thorax and made her fly in one spot. In the darkroom she could see an illuminated, patterned background rolling away beneath her, as though she herself were going forward over a motionless region. She reacted to an acceleration of the background movement, that is, to an increase in her apparent flight velocity, with a decrease in her own exertion in flight (measured by the amplitude of the wingbeat).

Nevertheless, it follows from various kinds of observations that the indications of the dancer as to distance do not refer to the true length of the course.

1. When the foraging bee meets a headwind on her way from the hive to the feeding place, she signals, by means of a slower tempo of dancing, a greater distance than when it is calm, and with a tailwind a lesser distance than when it is calm (see pp. 79–81). Simultaneously this demonstrates that in her signaling of the distance it is only or predominantly her flight *toward* the goal that counts (see pp. 116ff).

2. When the foraging bee has to fly up a steep slope on her way to the feeding place, she indicates, by means of a slower tempo of dancing, a greater distance, and after a downhill flight a lesser distance, than in flat country (see pp. 81f).

3. When one loads the foraging bee with 55 mg of lead, which is so heavy that she can barely fly (Fig. 100), she continues harvesting, but she dances more slowly and thereby indicates a greater distance than an unladen comrade (G. Schifferer 1952). On the average 9.14 circuits/15 sec were counted for the unladen bees ($n = 398$) and 8.49 for those with weights ($n = 245$). The difference is statistically significant.

The weights were circular lead disks 2.5 mm in diameter and 0.85 mm

FIGURE 100. Bee weighted with 55 mg of lead (*B*), at the feeding dish. After Gertraud Schifferer. Photographs by Dr. Schick.

110 / The Dances of Bees

thick. They stick well to the bees if they are first glued to paper and the paper is then fastened to the thorax with shellac. The feeding station was 200 m from the hive. In contrast to Kalmus (1939), whose bees could no longer take off with about 50 mg, Miss Schifferer found that her foragers were still able to do so—even though with difficulty—and got along quite well after several flights. The bees, however, made use of a trick that is noteworthy with reference to the plasticity of their instincts: they decreased their intake of sugar water on the average from 47.75 to 38.35 mm^3, that is, by approximately 20 percent. In this way the overload was reduced from 55 to 43 mg.[16]

4. Increasing the air resistance by sticking on a strip of tinfoil (Fig. 101) resulted in a reduction of the tempo of dancing by the same extent as overloading (Schifferer 1952). Heran succeeded in confirming the effect of additional loading and increased air resistance, and extended these findings in experiments with bees whose flight had been impeded

FIGURE 101. At the feeding dish. The bee in the foreground carries a strip of tinfoil to increase air resistance during flight. After Gertraud Schifferer. Photograph by Dr. Schick.

[16] Gontarski (1935), in experiments that were similar but with a much smaller load (18 mg), found only an inappreciable decrease in the degree of filling of the honey stomach.

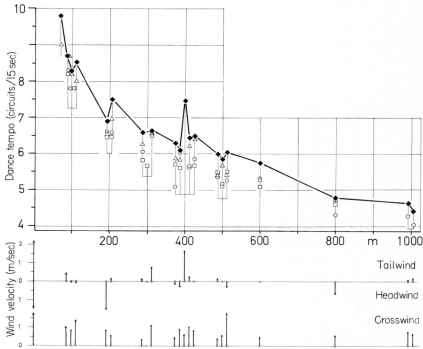

FIGURE 102. Dance tempo of normal bees (♦), compared with that of bees impeded in flight by lead weights of 35–50 mg (□), by attached strips of tinfoil (○), or by shortening the wings (△); ordinate, number of circuits per 15 sec; abscissa, distance (m) to the feeding station. Feeding with 2M sucrose. Symbols standing directly beneath one another refer to dances that occurred at the same time of day and under identical external conditions. The average wind during an experiment has been separated into its two components, tailwind (or headwind) and crosswind, which are shown below the abscissa for the outward flight. The bees whose flight was impeded signaled the position of the feeding station with slower dances than the normal foragers. Experiment by Heran and Scholze 1961.

by shortening the wings (Heran 1953 and *in lit.*). I am indebted to him for Fig. 102, in which he has summarized his results (in part unpublished).

In all four cases the indication of distance was changed, although the stretch flown had remained the same.

5. That the bee's estimation of distance is not determined through optical examination of the surface beneath her is confirmed by another observation: a correct indication of distance is given also when the foraging bees are flying over water where there is no possibility of their reading off from an inhomogeneous background the length of the course traversed.

As a suitable place for these experiments[17] we chose the "Narrows" of the Wolfgangsee; at this spot two opposite tributaries to north and south have each reduced the width of the lake with a delta of floodplain (Fig. 103).

On the morning of 4 September 1962 we set up on the south shore an observation hive (*St* in Fig. 103*b*) in a shelter as described on p. 11. The flight entrance, facing the water, was at the lake margin. As soon as the entrance was opened (6:25 A.M.), a feeding place was set up (Fig. 104) in one of the local long flatboats ("Traundl"), and within 3.5 hr we had succeeded in reaching the opposite shore with a group of foraging bees. The distance was 363 m.[18] The feeding station was kept in the boat, which was moored against the shore.

In order to lead the bees across the water as fast as possible, the feeding place must be identified conspicuously with large sheets of colored paper and with an effective scent (we used oil of lavender). Besides, the boat on the surface constitutes a good optical marker. Nevertheless one has to proceed very carefully initially in order to make the bees come along. After each advance of several meters we anchored until the foragers returned. Later, when they have become accustomed to such flights, one may go forward rapidly by stages of 20–30 m. Close attention is necessary to avoid breaking off the connection.

Beyond the close littoral zone the depth was between 10 and 30 m. The bottom was invisible. Under a cloudless sky the lake was at first smooth as a mirror, later slightly ruffled in part. The behavior of the dancers was unaffected by such a change. The dances gave the normal picture. From 9:50 to 11:37 A.M. an average value of 7.2 circuits/15 sec was obtained in 30 measurements. A tempo of dancing of 6.65 for the corresponding distance may be read from the curve depicting the grand average (Fig. 63). Thus in comparison with the average the tempo of these bees was slightly too fast, as if in flying over the water they had underestimated the distance slightly, yet the value still lies within the

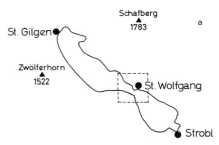

FIGURE 103. (*a*) The Wolfgangsee; the outlined square is enlarged in (*b*) the "Narrows"; *St*, observation hive; F_1 and F_2, feeding stations; altitudes in meters.

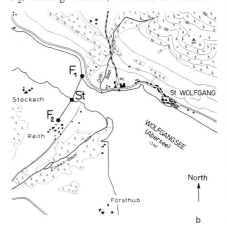

FIGURE 104. View from the south shore of the Wolfgangsee across the "Narrows." The boat (*Traundl*) contains a feeding station and numbered bees en route to the opposite shore, where × indicates the future location. In the foreground is the observation hive in the three-piece observation hut (p. 12), with the observer's entry open. The flight entrance is on the opposite side, toward the lake margin. During the experiment proper the lake was smooth; the picture was made at another time. Photograph by Rossbach.

---

[17] They were carried out in cooperation with M. Lindauer in 1962 and 1963.

[18] For the exact measurement of the distance across the lake we are indebted to Prof. Kneissl and his collaborator Dr. Messerschmidt. The latter determined with a geodimeter (an electro-optical distance-measuring device from the AGA Company, Stockholm) the distance of the hive from a point on the opposite shore; the distance of the feeding station from this place was easily calculated later.

FIGURE 105. View from the north shore of the Wolfgangsee (F_1, feeding station) across the "Narrows" to the observation hive St, 340 m distant, and F_2, the inland feeding station, likewise 340 m from the hive, in the opposite direction. See Fig. 103b.

range of scatter (at its upper limit) found with different colonies. A month previously (4–7 August) a dance tempo of 8.1 circuits/15 sec had been determined in 50 measurements with this same colony flying a distance of 200 m over land, as opposed to the value of 7.9 in the curve for the grand average. Thus the tempo of dancing of this colony lay somewhat, though only slightly, above the general average even in flights over land.

So it was demonstrated that the indication of distance given by the bees does not depend on estimation of the length of the course flown by means of optical perception of a structured background. But a more exact comparison of the dance tempo after flights over land and water was desirable.

Therefore we repeated the experiment the next year[19] and improved the procedure, by simultaneously making one group of foragers fly across the lake and another group from the same swarm fly inland an equal distance over level meadowland. The feeding station on the opposite shore lay not far from the previous year's landing place, this time at the end of a jetty built out into the lake (Fig. 105). The distance from the hive was 340 m, and the other feeding station was an equal distance inland (Fig. 103b). The water group of bees was trained to yellow plus oil of lavender, the land group to blue plus oil of orange blossom.

For four days experimentation was impossible because of cold, rainy weather; we struggled in vain to get the feeding stations going. In particular we failed again and again in attempts to cross the lake, because the few bees with which we managed to reach the other shore either stopped flying of their own accord or plunged into the water and drowned. In agreement with Heran and Lindauer (1963), we saw that often the bees flew unusually low over water, and in doing so frequently descended so far that they would hit the water and be unable to rise again. We assume that they are seeking optical contact with the underlying terrain and occasionally get down too far because they do not perceive the water surface. This happens in warm, sunny weather too, but is then compensated by the swift growth of the foraging group, while during the kind of weather prevailing the few bees that we had induced to fly back and forth across the lake were lost in this way.

On 9 September 1963 the weather improved, the temperature rose to 14°C, and the sun often shone weakly through the cloud cover. Between 10:00 and 11:25 A.M. we succeeded in measuring the dances of 20 bees from the water group and 21 from the land group. There was a brisk WNW wind (estimated at 2–3 m/sec) that was blowing similarly over the lake and the level meadow. The average dance tempo of the water group was 6.25 circuits/15 sec (scatter ± 0.38), that of the land group 6.6 circuits/15 sec (scatter ± 0.76); the difference of the means is 0.34 with an error of 0.19, and thus is far from being statistically significant.

[19] We used a racially pure Carniolan colony from the stock of Sklenar, Mistelbach, near Vienna.

On 10 September, with a cloudless sky and mostly mirror-smooth lake (the surface ruffled only temporarily here and there), we again succeeded in measuring 20 dances with each group and obtained an average of 6.9 circuits/15 sec (scatter ± 0.5) for the water group, 6.5 circuits/15 sec (scatter ± 0.7) for the land group. The difference of the means is 0.4 with an error of 0.2, and likewise is far from being statistically significant.

If one combines the two experiments, one obtains the same average dance tempo of 6.6 circuits/15 sec for the flights over water and over land. This time the value also corresponds quite closely with the general average of 6.65 from the curve of Fig. 63.

On 9 September the bees had danced somewhat more slowly after flying over water, but on the 10th somewhat more rapidly than after flying over land. On the 9th, waves were coursing obliquely to the line of flight; on the 10th the lake was smooth. Although the differences in the two instances are not significant, one is inclined to wonder whether they may not have a real basis and whether the smooth lake may have brought about an underestimation of the distance on the second day (cf. the 1962 results, p. 111). We regard this as improbable, because it is not to be assumed that the waves, which were moving obliquely to the direction of flight, would afford a surface pattern sufficient for estimating the length of the course flown. Through Heran and Lindauer (1963) we know that bees over water, even with waves, are not able to compensate, by turning the body obliquely, for a lateral drift induced by wind from the side, as they do at once when over land. They do not find the necessary optical clues in the waves (cf. pp. 186f).

Only a repetition of the experiment and the accumulation of more data will be able to clarify this wholly. For the present one may not with certainty exclude the possibility that perception of the structured surface beneath plays a modest part in estimation of the distance.

(b) The Indication of Distance Is Not Based on the Duration of the Flight

Considering the well-developed time sense of bees, one may readily suspect that they estimate the distance of the goal according to the time required for the flight. But this notion is refuted by the following observations:

1. In experiments on a steep slope (Heran 1956) bees that flew uphill to the feeding station clearly (and statistically significantly) danced more slowly than those whose feeding place was downhill—although the length of time occupied in flight was the same in both cases. This finding holds for the duration of the outward flight as well as for the return and for the entire trip (Heran 1956:193, Fig. 15, and 197, Fig. 16).

2. In G. Schifferer's (1952) experiments the duration of flight[20] with lead-weighted bees (see Fig. 100) was greater than that for unladen bees. By means of gluing on strips of tinfoil (Fig. 101) the flight time was increased significantly more. Despite this, the dance tempo was the same for the loaded bees and for those with strips of tinfoil (Table 18).

---

[20] Measured for the course from the hive outward to the feeding station.

TABLE 18. Dance tempo and duration of flight for normal and weighted bees, and for bees bearing strips of tinfoil (means). The differences $M_1 - M_2$, $M_1 - M_3$, $M_2 - M_3$, $M_1' - M_2'$, and $M_1' - M_3'$ are statistically significant. The precise agreement of M_2' and M_3' is to be regarded as a chance occurrence. After G. Schifferer (1952).

Bees	Dance tempo			Flight time		
	Symbol	(circ/15 sec)	n	Symbol	(sec)	n
Control	M_1'	9.14	398	M_1	25.62	330
Weighted	M_2'	8.49	245	M_2	28.19	189
With tinfoil	M_3'	8.49	166	M_3	32.52	186

3. If one does not insist on comparing flying times, but instead considers in a general sense the expenditure of time required for covering the course from hive to feeding station, a third argument is supplied from experiments by A. R. Bisetzky (1957). She made bees run on foot from the hive entrance to a feeding dish (for technique see pp. 115f). With these pedestrians she found the transition from round dance to tail-wagging dance when the goal was 3–4 m distant, whereas for free flight the point of changeover is at 50–100 m. The expenditure of time in traveling on foot for 3–4 m was about 50 sec, and for the flight of 50–100 m about 10 sec. Thus the same indication of distance is given for very different expenditures of time.

Wenner (1963) has a different view. According to him three kinds of experiments demonstrate that the bees' dance tempo is determined by the duration of the flight to the goal. He cites: (1) v. Frisch (1948), but there (p. 15) it is stated that the information given by the dancer "depends on the expenditure of time *or energy* needed for reaching the goal, rather than on the actual distance"; (2) Shaposhnikova (1958), but I find no pertinent experiment in this work; (3) Wenner (1962), where (p. 90) he states expressly that the expenditures of time and energy were not separable. In his own more recent work (1963) he finds that headwind and tailwind had but little influence on the bees' flight speed[21] and that hence the duration of flight could serve them as an index of the distance. But he provides no evidence of this. His opinion is refuted by Heran's and Schifferer's observations, cited above.

(c) *The Expenditure of Energy as a Measure of the Distance*

Inasmuch as there is no clear relation between the dance tempo of the bees and either the true distance or the duration of flight, we come to the conclusion that their estimation of distance must be based on the expenditure of energy for the flight.[22] That is a yardstick that truly is strange to man.

[21] In this connection see p. 190, note 63.
[22] Stumper (1955) also holds this opinion. It is only with his argument for it that one cannot agree. He rejects estimation based on the duration of flight because according to van't Hoff's rule the velocity of movements of insects should about double for a temperature increase of 10 C deg. According to our observations this certainly is not the situation with respect to the flight speed of bees.

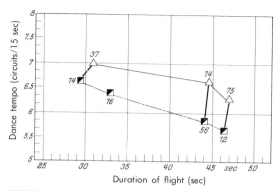

FIGURE 106. Dance tempo and duration of flight from hive to feeding station; △, bees with shortened wings; ◼, bees weighted with 40 mg. Differences of means on the heavy lines are statistically highly significant ($P < 0.0002$). After Heran.

This assumption does not arise merely for lack of another explanation. It also shows best agreement with the observations on loading the bees, on increasing their air resistance, with the effects of tailwinds and headwinds, and with the performance on steep slopes. In particular, among the manifold changing conditions on a steep slope, Heran (1956) found the dance tempo regularly well correlated with the expenditure of energy en route to the food source (in respect to calculation of the energy expenditure see Heran 1956:186). This result was confirmed by experiments with bees that were artificially impeded in flight during experiments in level country (Heran *in lit.*; Scholze, Pichler, and Heran 1964). Some of the bees had their wings shortened; others were loaded with 40 mg. The loaded bees danced more slowly (confirmed statistically) than those with shortened wings. The two groups flew at equal speeds and they met the same air resistance, but the energy expenditure was greater for the loaded bees, because they had to increase their lift accordingly (Fig. 106).

Further confirmation of this interpretation was given by the aforementioned experiments of A. Ruth Bisetzky (1957) taken together with the report by Scholze, Pichler, and Heran (1964). Bisetzky had placed in front of the entrance of the observation hive a runway (Figs. 107

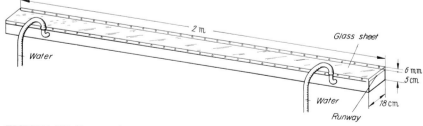

FIGURE 107. Runway in front of the entrance of the observation hive, covered with a sheet of glass 6 mm above the floor so that the bees must go on foot. Since the closed space becomes strongly heated when in sunshine, the metal sides are perforated and the running floor is the roof of a water-cooled metal box. The course can be lengthened by adding similar segments. After A. Ruth Bisetzky 1957.

FIGURE 108. The feeding dish underneath the glass cover of the runway for experiments on bees afoot, with a strip of filter paper moistened with scent. Colored stripes on the runway make it easier for the bees to hold to a straight line. After A. Ruth Bisetzky 1957.

and 108) that was covered at a height of only 6 mm with glass, so that the bees could not take flight. They ran on foot in this corridor to the feeding station several meters distant. If the distance was less than 3 m they performed round dances after their return home. If the interval was as much as 3–4 m there were some tail-wagging dances, just as after a flight of 50–100 m. Comparative measurements by Scholze, Pichler, and Heran now revealed the remarkable fact that the sugar consumption of a bee traversing 3 m on foot corresponds to that of a bee covering a course of 55 m in unimpeded free flight.

Unfortunately these experiments had to be limited to the distance at which there is a transition between the round dance and the tail-wagging dance. It would have been natural to make a comparison of greater distances of running with longer flights. But when the course for the bees afoot is lengthened further, the relation between distance and dance tempo becomes very irregular. That is quite understandable in view of the very unnatural conditions, but it puts an end to the investigator's calculations.

10. THE SIGNIFICANCE OF THE OUTWARD AND THE HOMEWORD FLIGHT IN THE INDICATION OF DISTANCE

When bees flying from the hive to the feeding station meet with a headwind, then its hindrance is balanced by a tailwind on the return. Despite this, the bees dance more slowly than when it is calm (pp. 79–81), as though the distance had grown longer. Correspondingly, dancing for a feeding place that is uphill is as though for a more distant goal on the flat (pp. 81f), and when the outward flight is assisted by a tailwind or when the feeding place is downhill there is more rapid dancing, as though for a shorter flight. From this it is apparent that relative to the indication of distance the outward flight is either solely determinative or else is given greater weight than the return. We shall now have to deal with this "either-or." The pertinent experiments have produced remarkably contradictory results.

In 1945, when the fact of the bees' announcements of distance was still a novelty, I did a few (unpublished) "transport experiments" in order to learn whether the information is based on the outward or on the homeward flight, or on both. Some numbered bees collected food at a feeding station 10 m from the hive, others at one 100 m away. Most of them were acquainted with both places. Those returning home from 100 m performed tail-wagging dances, the others round dances. On four occasions I carried a bee that had flown in to the 100-m station, swiftly and while she was still drinking, to the 10-m station. After her return home, she performed round dances. Six times I carried a bee that had come to the 10-m station to the 100-m station. Back in the hive they performed tail-wagging dances, with interpolated round dances. The information seemed to have taken account of both the outward and the homeward flights. Since the zone of transition from round to tail-wagging dances lies between 10 m and 100 m, I sought a more trenchant result by transporting bees over longer distances.

Numbered bees were collecting food at a station 600 m from the hive. On a bicycle I transported a bee that had flown in there toward the hive until she got ready to fly back home. Once the place at which she took flight was about 300 m from the hive, three times it was 200 m, twice 100 m, and once 10 m. Judged by the number of circuits per 15 sec, the dances in all cases were based on a distance situated between the place of feeding and that of taking wing homeward. The experiments were too few and too inexact for a clear decision. At that time other problems were more attractive and the question was deferred.

Heran and Wanke (1952) transported bees from a feeding station 300 m distant from the hive in the same direction but 200 m farther away. Most of the bees were so upset by this treatment that they did not dance. Among more than 1000 transported bees there were only 40 whose dance tempo could be measured. It corresponded to the length of the outward flight. However, this apparently clear result is by no means assured. One should recollect that the bees had become better accustomed to the shorter stretch through their previous outward flights, while the changed return course was new to them and probably could not immediately overcome the memory of the well-known shorter course homeward. This interpretation is favored by other analogous experiences, for example in experiments with a "moving feeding station" (p. 207).[23]

F. Otto (1959) succeeded in transporting bees still farther. With the help of a motor scooter he carried bees from a 600-m-distant feeding station over a 500-m course to a place 100 m distant from the hive. In a second experiment the bees were displaced from 100 m to a distance of 600 m. As with Heran and Wanke, the bees' readiness to dance was greatly disturbed at first. But Otto transported the same individuals over and over again, until they had learned the altered return route. Then their dance tempo indicated 350 m, the average between their places of arrival and takeoff. They had evaluated both outward and homeward flights equally, and had averaged them.[24]

The result is unambiguous. But it is in contradiction with the experience that under natural conditions, with flight on a steep slope or with headwinds or tailwinds, the outward flight is taken into account alone

---

[23] Shaposhnikova (1958) also transported bees that had flown in at a given distance to a different distance from the hive, while they were drinking, and found that their dances indicated the distance of the station to which they had flown. But she did not watch the dances, instead recording the number of newcomers that came in response to the dances to the original station and to the place from which the foraging bees had taken off. This indirect method is less precise than observing the dance tempo. Beyond this, Shaposhnikova seems to have overlooked a source of error: at the feeding station flown to, the foraging bees evert the scent organ, while the transported bees do not do so when they take off. In her experiments the preponderance of newcomers at the original station may well have been due in large part to the attraction caused by the scent organ.

[24] After transportation from 600 m to 100 m the dance tempo was 7.43 circuits/15 sec ($n = 40$); after transportation from 100 m to 600 m it was 8.16 circuits/15 sec ($n = 40$). These values approximately equaled those of a control group that was collecting at 350 m (7.55 circuits/15 sec). The differences among the three values are not statistically significant, but the values are significantly different from the dance tempo of untransported foraging bees that were collecting at 100 m (10.14 circuits/15 sec) and at 600 m (5.93 circuits/15 sec).

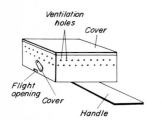

FIGURE 109. Carrying box, for moving bees while they are drinking. After Otto 1959.

or at least more strongly than the homeward flight. In order to clarify the contradiction we must examine more closely the behavior of the bees in the transport experiments.

Technically, care must be taken that the bees are not disturbed by the process of transporting them. In order that they may drink on calmly, very good food must be given: sugar water in the highest possible concentration. For moving them with a bicycle or motor scooter, protection against wind is essential. As a makeshift it is sufficient to place a glass cover over the little feeding table. It is better to feed the bees from the start in a ventilated box that can be closed and to transport them in this (Fig. 109).

A much more serious difficulty in the transport experiments arises from an aspect of bee behavior that was described some time ago by Wolf (1927). He placed a hive in a wasteland that had few markers for orientation, and set up a feeding station 150 m from the hive. When he shifted bees that had flown in to this station to a point to right or left of the hive or behind it and then let them take flight, they started out in the direction that would have taken them home from the accustomed feeding place; they flew about 150 m, then went hunting around, and finally, via detours and with great delay, found their way back to the hive (Fig. 110). But this tendency on the flight home simply to traverse in the opposite direction the entire outward flight makes itself felt by no means only in wasteland.

The foraging bees transported 500 m by Otto, once they took wing, first went the accustomed distance toward the hive, thus coming, like Wolf's bees (Fig. 110), to a wrong location, and from there hunted out the way home; this behavior found expression also in greatly lengthened times for the return flight. The return flight was not simply a reversal of the outward flight. In the course of frequently repeated displacements they learned the new way back. One can imagine that then this return flight was turned into an independent "outward flight, but one now toward the hive," and that consequently the homeward flight was given importance equal to that of the true outward flight in signaling the distance.

In my own transport experiments, also carried out over a distance of several hundred meters, the bees had found the way home quickly and without noticeable disturbance, possibly because they had collected previously in both locations or at least in the vicinity and were acquainted with the places. Although in consequence of this no retraining had taken place on the return course, evidently in giving their indication of the distance they took account of both the outward and homeward flights—in what proportion to one another cannot be evaluated from the relatively few observations.

But there is yet another way to learn something about the relative value placed on the outward and homeward flights. At my request Bräuninger[25] assembled from his experiments regarding the effect of meteorological factors on the dance tempo all cases in which, with

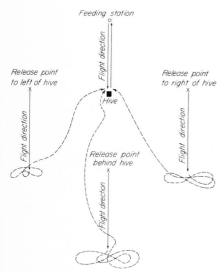

FIGURE 110. Location of the feeding station, and of the places where displaced bees took flight for home. The arrows show the direction of flight. After Wolf 1927.

[25] Sincere thanks to him here also for these time-consuming summaries and calculations.

wind velocity 1 m/sec (0.5–1.5 m/sec) and 2 m/sec (1.5–2.5 m/sec), there was headwind or tailwind for the bees on the flight to the feeding place.[26] Since the speed of bees in calm air is known, one can easily calculate the delay in a headwind or the acceleration in a tailwind.[27] This yields the apparent lengthening or shortening of the course flown and hence the dance tempo to be expected if the bees' report is based on the outward flight. From Bräuninger's observations we know the dance tempo that actually ensued.

The first line of Table 19 may serve as an example explaining this. With the feeding station at a distance of 600 m (column 1) the average dance tempo was 5.61 circuits/15 sec (column 2) in Bräuninger's experiments. The mean value has been calculated from the total number of Bräuninger's observations (column 3). In calm air the duration of the outward flight is 72.3 sec (column 4), the flight velocity being 30 km/hr = 8.3 m/sec. Against a headwind of 1 m/sec (column 5), the bee flies at a ground speed of $8.3 - 1 = 7.3$ m/sec, and hence needs 82.2 sec (column 6) to cover 600 m; consequently it is as if the course had been lengthened in still air to 682 m (column 7). For a distance of 682 m in calm air the dance curve yields 5.37 circuits/15 sec as an expected value, which is 0.24 less than for 600 m (column 8). The observed value was 5.51 circuits/15 sec, thus only 0.10 less (column 9). That is, the change resulted in only 42 percent of the amount to be anticipated if the signal was based solely on the outward flight (column 12).

The corresponding calculations for a headwind of 2 m/sec, for distances of 900 m and 1250 m to the feeding station, and for all of these with tailwinds, are to be found in the other lines of the table.

If outward and homeward flights were given equal weight by the bees, the dance tempo would have to be the same as in calm air. That is not the case. Thus the former experience is confirmed, that the dance tempo is slowed when a headwind prevails during the outward flight. But it is not slowed as much as it should be if only the outward flight were taken into account. Rather it is as if in reporting the distance the outward flight weighs more heavily than the flight homeward. The same is true when there is a tailwind on the way to the goal.

For the biologist, attention given exclusively to the outward flight would have been a more satisfactory finding. For in the first place the comrades are to find this outward course on the basis of the report; and in the second place it is only on the outward flight that the foraging bee, like the newcomer who responds to her signal, is unburdened; on the return flight she bears a variable load.[28] When the honey stomach is filled with 70-percent sugar solution, a crop capacity of 57 mm^3 means a load of 77 mg (Gontarski 1935). An added weight of as little as 55 mg clearly slows the dance tempo (Schifferer 1952). Thus, tak-

---

[26] Winds within 22.5° from (or toward) the goal were recorded as head winds or tail winds.

[27] When a head wind or tail wind is strong the bees compensate partially for their displacement. But such compensation first occurs at wind velocities of about 2 m/sec, so that it may be ignored in our present considerations.

[28] W. H. Thorpe (1949) has called attention to this point.

TABLE 19. Test of the theory that the distance signal is based solely on the outward flight. The theoretical dance tempo can be calculated (column 8) and compared with the dance tempo observed (column 9). The difference (column 12) is statistically significant ($P < 0.001$)[a] in all 12 cases. Hence it follows that the bees utilize not only the outward flight but also—to a slighter extent—the homeward flight. More detailed explanation in text. Observations and calculations by D. Bräuninger.

1	2	3	4	5		6		7		8		9		10	11	12
Distance of feeding station (m)	Normal dance tempo, calm (circ/ 15 sec)	Number of dances observed	Duration of outward flight, calm[b] (sec)	Wind		Outward flight		Length of course		Dance tempo[c] (circ/15 sec)				Number of dances	Observed difference from norm (percent of theoretical)	Difference between observed and theoretical dance tempo (circ/15 sec)
				Direction	Speed (m/sec)	Duration (sec)	Difference from calm (sec)	Apparent (m)	Difference from true (m)	Theoretical		Observed				
										Value	Difference from norm	Value	Difference from norm			
600	5.61	4258	72.3	Headwind SW	0.5–1.5	82.2	+ 9.9	682	+ 82	5.37	−0.24	5.51	−0.10	240	42 ± 12	0.14
					1.5–2.5	95.2	+22.9	790	+190	5.08	− .53	5.41	− .20	114	38 ± 8	.33
900	4.80	4639	108.4		0.5–1.5	123.3	+14.9	1023	+123	4.60	− .20	4.72	− .08	312	40 ± 9	.12
					1.5–2.5	142.9	+34.5	1186	+286	4.38	− .42	4.69	− .11	171	29 ± 6	.31
1250	4.31	3831	150.6		0.5–1.5	171.2	+20.6	1421	+171	4.16	− .15	4.24	− .07	258	47 ± 12	.08
					1.5–2.5	198.4	+47.8	1647	+397	4.02	− .29	4.23	− .08	129	28 ± 9	.21
600	5.61	4258	72.3	Tailwind NE	0.5–1.5	64.5	− 7.8	535	− 65	5.80	+ .19	5.66	+ .05	1000	26 ± 8	.14
					1.5–2.5	58.3	−14.0	484	−116	5.98	+ .37	5.69	+ .08	183	22 ± 9	.29
900	4.80	4639	108.4		0.5–1.5	96.8	−11.6	803	− 97	5.05	+ .25	4.84	+ .04	1262	16 ± 4	.21
					1.5–2.5	87.4	−21.0	725	−175	5.25	+ .45	4.97	+ .17	194	38 ± 5	.28
1250	4.31	3831	150.6		0.5–1.5	134.4	−16.2	1116	−134	4.46	+ .15	4.38	+ .07	1027	47 ± 7	.08
					1.5–2.5	121.4	−29.2	1008	−242	4.62	+ .31	4.40	+ .09	174	29 ± 7	.22

[a] Calculated according to F. Yates, *Sampling methods for censuses and surveys* (Hafner, New York, ed. 3, 1960).
[b] The flight velocity is taken to be 8.3 m/sec (v. Frisch and Lindauer 1955).
[c] According to the theory that the bees' distance signal is based solely on the *outward* flight.

ing the return flight into account could introduce more or less error into the signal for distance, depending on the load.

It seems as though, in the evolution of their communication of distance, the bees had struck out on a logically satisfying path without having followed it to its conclusion. Perhaps future investigators will find out why that is so. But perhaps there is behind it no more than that even a bee is not yet a perfected creature.

11. The Shape of the Curve for Distance

Our measurements have led to the construction of curves that visualize the relation between the distance to the goal and the tempo of dancing. If one plots on the abscissa the distance and on the ordinate the number of circuits per 15 sec, when distance increases the curve approaches asymptotically a parallel to the abscissa (Fig. 63). If as a measure of the dance tempo one takes the duration of a circuit, that is, the time required for a waggling run + return (Fig. 92), then one obtains a rising curve, concave downward (Fig. 64). We have seen that the dance tempo may vary from individual to individual, as well as from one colony to another, and that also it is influenced by external factors. In spite of the scatter introduced in these ways, the general course of the curves has been confirmed again and again in many thousands of observations.

It might have been expected that with increasing distance the dancer would have added a definite time to the duration of her circuit for each unit of increase. That would have yielded a linear rise in the curve for duration of circuit, but this is not realized. One might also have surmised that she would have responded to proportionally equal increases in the length of the course with equal extensions of the duration of circuit. That too is not the case. Thus we are confronted with the question of how to interpret the typical shape of the curve for distance.

Haldane and Spurway (1954) sought a mathematical formulation: They concluded that the number of circuits z in 15 sec depends on the distance s according to the following relation: $z = 4.76(3.95 - \log_{10}s)$. But for greater distances (over 3 km) another equation is said to hold; according to it the curve would reach zero at 9 km. Thus distances greater than 9 km could no longer be indicated. This conclusion is in contradiction with the facts. Also it seems unsatisfactory to base the curve on different relations in its initial and later course. The data on which Haldane and Spurway founded their derivation were at that time neither qualitatively nor quantitatively adequate for a statistical evaluation.

On the basis of the earlier measurements and his own more recent ones Heran (1956:201) made a calculation of the curve and, introducing two suitably chosen constants, obtained a linear relation. This formally successful attempt fails to satisfy only because it lacks an evident physiological basis.

Following a lecture that I gave in Graz in 1948, O. Kratky, a physical chemist, expressed the thought that in the course of their flight the bees might have "forgotten" a portion of the way traversed. The simplest assumption would be a gradual extinction of memory, by means of which during each unit of time a definite fraction of the path already traversed in flight would vanish into oblivion. This hypothesis, the foundations for testing which were insufficient at that time, we later re-examined in collaboration (v. Frisch and Kratky 1962).

Let it be said in advance that the expression "forgotten" is intended merely to designate in a readily comprehensible manner the actual central nervous process. On the one hand there is no doubt that in some form or other the length of the course traversed is remembered, and one can assume that the physiological changes in the central nervous system underlying the ability to remember decay to a certain degree during flight.[29] On the other hand it is not demonstrable that actually there occurs a gradual diminution in the impression retained of the part of the course traversed. It could be also that matters are exactly the opposite, in that the stretch flown over initially makes the strongest impression and that further segments of the course, under the influence of what has been traversed already, lose increasingly in emphasis—a concept reminiscent of the Weber-Fechner law. The mathematical relations are so similar in the two cases that at present no decision between these possibilities can be made.

Derivation of the mathematical relations.[30] A relation is here to be derived between the length of path traversed and the duration of the circuit,[31] based upon the idea that in each unit of time during the flight, the same fraction of the path s previously traversed is forgotten. Let us designate the remainder left in the memory as s^+, and let us assume further a constant flight speed (so that the distance flown and the time required are proportional); then the remnant of memory ds^+ after traversal during time t of an infinitesimal segment ds of the route is given by

$$ds^+ = ds \cdot e^{-ks}, \qquad (1)$$

where k is the fraction of memory that disappears per unit of distance

[29] The assumption of "forgetting" in the usual sense encounters difficulties, if the process is limited to the time of flying and nothing further is forgotten after the flight has ended. Whether matters are this way cannot be decided at present. At times a bee will repeat a dance much later without having flown out again in the meantime. Reports by Wittekindt and Wittekindt (1960) and by Wittekindt (1961) concerning the signaling of distance in such cases could be interpreted in favor of a further extinction of memory after flight had ended, but unfortunately are not adequate to clarify the problem. If no further extinction ensues, one might easily conceive that the central nervous processes that lie at the basis of the capability for remembering had been influenced in an inhibitory sense by the motor centers active during flight, which would amount to a memory decrease only while flight was in progress.

[30] I am indebted to my colleague Prof. Kratky for the following concise formulation for this book. For a more detailed exposition, see v. Frisch and Kratky (1962).

[31] On p. 104 we reached the conclusion that the determining signal for distance was to be seen in the duration of waggling. But data adequate for interpreting the curves are available only for the duration of a circuit (duration of waggling + return). Since the duration of a circuit is strongly correlated with the duration of waggling, and since a mathematical relation between the two could be derived from a series of experiments in which both the duration of waggling and that of a circuit were measured, it was possible to base the considerations that follow on the measurements of the duration of a circuit.

flown, a "constant of forgetfulness." The half-value distance σ, in the course of which an initially traversed segment of the route is half forgotten, is then of course

$$\sigma = (\ln 2)/k.$$

The very simple mathematical problem now consists of calculating the persistent trace of memory s^+ after a finite route s has been traversed:

$$s^+ = \int_0^s e^{-ks}\,ds = \frac{1}{k}(1 - e^{-ks}). \quad (2)$$

Now the bee has to indicate the magnitude of s^+. The simplest assumption is that of a waggling time proportional to this quantity:

$$t_s = K's^+. \quad (3)$$

Substituting s^+ from (2) we obtain:

$$t_s = \frac{K'}{k}(1 - e^{-ks}). \quad (4)$$

Now the duration of a circuit t_u is closely correlated with the duration of waggling. Therefore,

$$t_u = t_o + \frac{K}{k}(1 - e^{-ks}),$$

where t_o is a constant, the time for the return run if the goal is very near. In place of the constant K', we have another (K) that is larger by about 8/7.

We call K the slope constant. It defines the increase that the duration of a circuit would undergo if there were no "forgetting."[32]

If one chooses to assume that there is no process corresponding to "forgetting," but that instead the path traversed initially creates the strongest impression and that further segments of the path diminish increasingly in impressiveness owing to the force of what has already been experienced, one comes to very similar relations. On making the simplest assumption, that the impression created by a given element of the path is inversely proportional to the length of course already traversed, one reaches in the limit where the path is zero an illogical result, for there the degree of impressiveness should be "infinite." On taking the simplest imaginable way out of this difficulty, namely by adding a constant to the distance actually traversed, one obtains for the relation between length of path flown s and duration of circuit t the equation:

$$t = a \ln \frac{s + b}{b}. \quad (5)$$

Without enquiring more closely into the application of this equation, we

[32] Expressed graphically, it measures the slope of the initial tangent, the course of which would define the dependence of duration of a circuit on the stretch traversed, if there were no "forgetting."

may point out here merely that it differs from the one derived from the other concept ("forgetfulness") so little that an experimental choice is scarcely to be provided at present. We may add that equation (5) can be obtained from the concept of forgetfulness by making the assumption that the "constant of forgetfulness" varies in inverse proportion to the length of path flown, and hence decreases more and more as the path grows longer.

In what follows we shall hold to the first concept and indicate by quotation marks that we do not take "forgetting" literally and are at all times prepared to give it up for another interpretation whenever good grounds for doing so are provided.

When now one compares with the theoretical curve, calculated in accordance with equation (4), the measurements from the available distance experiments, which extend to more than 1000 m, one finds good agreement.[33] Figure 111 shows an example. The values measured in three series of experiments carried out by three different observers at three different places in three different years, can be interpreted by one and the same theoretical curve. Some scattering of the experimental data occurs in every effort of this sort.

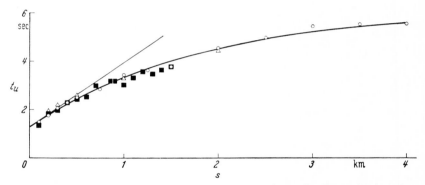

FIGURE 111. Theoretically calculated curve and experimentally observed values from three distance experiments: ■, experiment by K. v. Frisch in Brunnwinkl, 7–23 June 1945; △, experiment by W. Steche in Bonn (Steche 1957); ○, experiment by M. Lindauer near Munich, 28 August–3 September 1960; abscissa, distance (km) to the feeding station; ordinate, duration (sec) of dance circuits. From v. Frisch and Kratky 1962.

However, it was not possible to evaluate together all the experiments. In part they demand different constants. In this respect there are evident differences between different stocks even within the same race of bees. Here are confirmed and formulated more precisely the already long-known differences in the signaling of distance by different colonies. Regarding details reference is made to the more extensive publication mentioned above, and to Table 20, in which the results of all experiments that have been analyzed are summarized briefly.

[33] This agreement does not hold to the same extent for summarizing curves (for example, Fig. 64) in which the average values have been brought together from many experiments. This fact is explained by the circumstance that, as may be shown readily, with this kind of averaging another sort of curve is produced. Wenner (1962) obtained for the duration of waggling a curve that rose linearly up to distances as great as 1230 m and pointed out that further experiments would be necessary to test whether this will remain the case at still greater distances. Such experiments have been published and they show clearly that the curve is not a linear one (v. Frisch and Kratky 1962).

In the table are included the Indian species of bees studied by M. Lindauer (see p. 303). Their way of signaling distance is in principle the same as with our honeybees. But in the course of the curves, that is, in the values of the constants, they differ appreciably. Therein they give us a hint of the biological connections, as for that matter Lindauer (1956) already recognized.

As an example I take the little dwarf honeybee (*Apis florea* F.) (Fig. 112). In the initial portion of the curve the increase in the duration of a circuit is much steeper, the slope is about 14, and correspondingly with this bee the duration of a circuit lengthens very rapidly with an increase in distance. Evidently this is related to the shorter flight range. Whereas with our honeybee the flight range extends as much as 3 km from home and occasionally even much farther (see pp. 65, 66), with the dwarf bee it is limited to 300–400 m. Under such conditions the altered course of the curve is logical. For it implies a precise indication of distance within the principal flight range, and no imprecision in the data until distances are reached that are scarcely of practical significance. Over the shorter range of the dwarf honeybee an equal increase in the

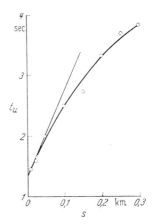

FIGURE 112. Indication of distance by the dwarf honeybee, according to experiments by M. Lindauer in Ceylon; abscissa, distance (km) to the feeding station; ordinate, duration (sec) of a dance circuit.

TABLE 20. Summary of the distance experiments that have been analyzed mathematically.

Observer	Fig. No.[a]	Species, *Apis*	σ[b] (km)	K[c] (sec/km)	t_0[d] (sec)	s_{max}[e] (km)	$K \cdot s_{max}$ (sec)	$K\sigma$ (sec)	$\dfrac{\sigma}{s_{max}}$
v. Frisch and Jander 1955 (publ. 1957)	4	mellifera carnica	2.0	1.95	1.65	3	5.9	3.9	0.67
v. Frisch and Lindauer 1954	7	mellifera carnica	2.8	2.00	1.58	3	6.0	5.6	.93
v. Frisch 1945	9 ■	mellifera carnica	1.25	2.70	1.28	3	8.1	3.4	.42
Steche 1957	9 △	mellifera carnica	1.25	2.70	1.28	3	8.1	3.4	.42
Lindauer 1960	9 ○	mellifera carnica	1.25	2.70	1.28	3	8.1	3.4	.42
v. Frisch and Lindauer 1954	10	mellifera carnica	2.57	2.1	1.57	3	6.3	5.4	.86
v. Frisch 1946	11	mellifera carnica	1.72	2.7	1.27	3	8.1	4.6	.57
Lindauer 1956	14	dorsata	—	3.12	1.32	≈3	9.3	—	—
Lindauer 1956	15	indica	—	6.94	1.35	0.8	5.2	—	—
Lindauer 1956	16	florea	0.24	14.1	1.32	0.4	4.9	3.4	.69

[a] Figure numbers of the graphs in v. Frisch and Kratky 1962, of which only Figs. 9 and 16 are reproduced here (Figs. 111 and 112, respectively).
[b] σ = half value of the distance in the course of which a segment of the route traversed at its beginning is half "forgotten."
[c] K = constant of the slope; it indicates for the initial tangent to the distance curve the increase in waggling time (sec) per kilometer of the route.
[d] t_0 = return time for short segments (minimal return time).
[e] s_{max} = maximal range of flight, approximately the greatest distance to which the food-gathering flights of the species in question usually extend.

flight path is of greater importance than with our honeybee, and the slope of the curve is correspondingly steep. Thus there becomes apparent a reciprocal relation between maximal flight range and the slope, a type of relation that is confirmed in other instances also. In Table 20 the product of the slope and the maximal flight range is found to be rather constant, whereas the slopes themselves vary over almost an order of magnitude.

On the other hand, the relation is very sharply curvilinear with *Apis florea,* and the half-value distance σ is extremely small, namely 0.24 km. It is almost an order of magnitude less than the average of this value for *Apis mellifera.* This strong degree of "forgetfulness" seems to be correlated with the shorter flight range, and the half-value distance for forgetting, σ, appears to go hand in hand with the maximal range of flight s_{max} (Table 20).

If on the one hand σ parallels the maximal flight range and on the other runs counter to K, then σ and K must also vary in opposite directions and their product $K\sigma$ should not depart far from constancy. Table 20 shows that this is the case.

The great differences that appear in a comparison of the slope constants and the half-value distance for forgetting vanish to a large degree when everything is related to the maximal flight range and this is taken as the yardstick of distance. At all events the phenomena then come close to agreement with the dictum: for equal fractions of the maximal flight range dancing is at the same tempo and like fractions are "forgotten."

The qualitative validity of the relations discussed is not open to question. In the present state of knowledge, the establishing of rigorous laws cannot be considered, because the scatter in the data leaves too much room for alternative interpretations. It might be rewarding to pursue further by experimental means the ideas that have been revealed. For it seems likely that the multiplicity of the ways of behaving fits into the framework of a uniform code.

Summary: The Indication of Distance

1. With increasing distance to the goal the dance tempo slows down: the number of circuits per unit time decreases (Fig. 63); the duration of the individual circuits increases (Fig. 64). The dance tempo indicates the distance of the goal. This regularity can be followed out to a range of 11 km. Different colonies may diverge somewhat in their tempo of dancing.

2. During a single dance the dance tempo of an individual bee is not identical for all circuits; it varies about a mean value striven for. With different bees individual differences in the dance tempo are seen. Newcomers to the feeding place dance relatively rapidly; with increasing age since the first flight the dance tempo customarily becomes somewhat slower. Observations over a period of several days' flight to the same feeding place suggest a decrease in the scatter of the indications of distance, as a result of greater experience.

3. External factors may influence the dance tempo.

(*a*) Dancing is somewhat more rapid when environmental temperature is high than when it is low (Fig. 72). But the increase is so slight that as a rule it is demonstrable only statistically; for a temperature increase of about 10 C deg it amounts only to 10 percent (in contrast with a 100–200 percent increase in reaction velocity according to the van't Hoff rule). The same increase as with increasing environmental temperature occurs when temperature is raised within the hive. Heated sugar solution at the feeding place has a similar but weaker effect.

(*b*) A headwind on the way to the feeding place acts like an increase in the distance: the dance tempo grows slower. A tailwind on the way to the goal speeds up the tempo of dancing (Fig. 77).

(*c*) Bees that fly up a steep slope to the feeding place clearly dance more slowly than those that fly downhill (Figs. 79 and 80).

(*d*) Feeding quinine slows the dance tempo; feeding thyroxin or adrenalin accelerates it. Since in the first case the oxygen consumption of the bee is diminished and in the other is increased, the phenomena are to be interpreted as an expression of altered activity. The latter is demonstrable in experiments on phototaxis.

4. Stepwise experiments ("Stufenversuche") show that the dancers' indications of distance are taken into account with remarkable precision by the newcomers aroused. In the stepwise experiments a group of numbered bees is fed concentrated sugar solution with added scent at a definite distance from the hive. Along the line to the feeding station scent cards with the training scent but without food are set out at given intervals from the hive to far beyond the goal. The number of flights of newcomers to these cards defines the region of search and the intensity of hunting at the corresponding distances (Fig. 91).

5. With increasing distance of the goal several elements of the tail-wagging dance vary (duration of a circuit, duration of the return run, duration of waggling, number of waggling movements, length of the waggling segment). We regard the duration of waggling as the index of distance, because this has the best correlation with distance. This interpretation is supported also by the facts that the duration of waggling is emphasized strongly by the dancer through her waggling movements and by the production of sound, that the interest of the dance followers is concentrated on the waggling phase, that in some instances the return run may be omitted, and that for very great distances the tail-wagging dance frequently consists of only a single waggling run. Experimental demonstration of the decisive importance of the duration of waggling is theoretically possible; preliminary experiments have been made.

The stepwise experiments reveal the degree of precision with which the newcomers search. A comparison with the accuracy of the indication of distance shows that their flights are made with more precision than would be the case if the newcomers were guided by a single waggling run. Hence it is to be deduced that they average some larger number of waggling runs. This agrees with observations of their behavior in the hive. Before the newcomers fly out, as a rule they follow after several dancers and many waggling runs.

6. The bees' estimate of distance is not based on optical perception of the structured terrain flown over, for the indication of distance is correct even after flights over mirror-smooth water. Also it is not based on the expenditure of time needed to reach the goal. On the contrary, it is the expenditure of energy that is decisive: with a headwind on the way to the feeding place, after flights up a steep hill, when loaded with a lead weight, if air resistance is augmented or flight is impeded by shortening the wings, too great distances are reported. Confirmation that distance is estimated in accordance with the required expenditure of energy is given also in comparisons of bees running on foot to the feeding place with those that fly to it: the former shift from the round dance to the tail-wagging dance after a trek of 3–4 m, the latter after a flight of 50–100 m; in both cases the consumption of sugar is approximately the same.

7. Bees that fly against a headwind to the feeding place have a tailwind on the return; nevertheless they announce too great a distance; thus in estimating the distance they place stress either solely or more heavily on the outward course. The choice between the two possibilities was provided by analysis of a large number of data on dances from experiments in which the velocity and direction of the wind were known. The outward flight is given greater weight than the flight homeward. The tempo of dancing lags about 30–40 percent behind what would be expected if only the outward flight were taken into account.

In apparent contradiction to this conclusion, bees that had been transported from their feeding place to a nearer or more distant location announced the mean distance of the two courses, and thus set equal value on both outward and homeward flights. The explanation is that initially the displaced bees found the way home only via detours, and had to learn the proper return course anew. In this way the return flight, as being an "outward flight, but one now in the direction of the hive," acquired the significance of a true outward flight toward the feeding place.

8. The characteristic form of the distance curve (Figs. 63 and 64) may be comprehended by making the assumption that in the course of flight the bees "forget" in each unit of time a definite fraction of each segment of the path already traversed (instead of an extinction of the memory other central nervous processes may be assumed instead). By assigning a "constant of forgetfulness" for the fraction of recollection that disappears for a unit of the journey and a "slope constant" for the increase at the beginning of flight in the duration of a dance circuit per unit of path flown, one can calculate a relation between the length of the path and the duration of a circuit that agrees well with the curves obtained experimentally (Fig. 111). For many colonies, somewhat different constants must be chosen, and in this necessity the well-known hive-specific differences are more incisively expressed. Appreciable differences in the constants are found with other species of bees. In this phenomenon biological connections become clear. Thus, in correlation with its short flight range the slope constant is large and "forgetfulness" strong in the dwarf honeybee (Fig. 112). This implies a precise indication

of distance over the principal range of flight and a poorer performance (flattened curve) at distances that scarcely are of practical importance. In general the dictum seems to hold: for equal fractions of the maximal flight range dancing is at the same tempo and a like fraction is forgotten.

B. THE INDICATION OF DIRECTION

12. First Hints of the Mode of Indicating the Direction of the Goal

In the summer of 1944 a few very simple experiments led to a result that was just as unexpected as it was thrilling. I was feeding a group of numbered bees at a greater distance than previously, some 200 m from the hive. At different places in the vicinity scent cards were distributed with the same scent that characterized the feeding place. As soon as a high concentration of sugar was offered, the newcomers aroused came to the scent cards too, but they appeared with marked preference at the feeding place itself and at those scent cards that were set out on a line with it. On the basis of earlier experiments (1923) I was convinced at that time that in response to the dances the entire surroundings were explored equally. To myself I accounted for the influx toward the neighborhood of the feeding place by the attraction produced by the scent organs of the foraging bees for their comrades searching in the vicinity. A control experiment was designed to confirm this. In the sketch of Brunnwinkl (Fig. 113) the location of the beehive (*St*) is shown. Now two feeding places were set up, in different directions. One numbered group was collecting food in the hilly but yet open country 300 m NNE of the hive (station *A*), the other to the southward, 200 m distant in a direct line, on a narrow, stony footpath that was bounded on one side by cliffs and large forest, on the other by a lake (station *B*). Here no great number of newcomers wandering about at random would be expected.

Following some "meager feeding," 2M sugar solution with the scent of lavender was next offered on the afternoon of 17 September 1944 in the open country at *A*. The consequence was that within 1 hr 178 newcomers had to be captured, the first just 11 min after good feeding was begun. The number of flights to the lavender-scented cards (without food) is shown by the figures in parentheses in Fig. 113. The preference for the vicinity of the feeding place is unambiguous.

Anxiously we started the companion experiment on 18 September and at *B* offered 2M sugar solution scented with thyme. To my astonishment here too there appeared unmarked newcomers after no more than 10 min; they came in crowds and without interruption along the narrow path, to a total scarcely experienced before. During the hour's observation at the feeding place it was necessary to kill 362 newcomers. The number of flights to the scent cards may be seen from the figures in brackets in Fig. 113.

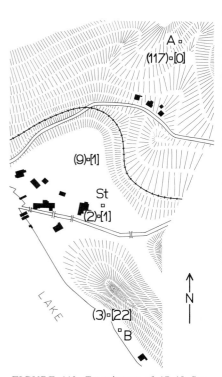

FIGURE 113. Experiment of 17–18 September 1944, Brunnwinkl; *St*, observation hive; *A*, *B*, feeding stations for two groups of numbered bees from this hive. 17 September: feeding at *A* with 2M sugar solution scented with lavender; the number of flights of newcomers during 1 hr to the lavender-scented cards is given by the figures in parentheses. 18 September: feeding at *B* with 2M sugar solution scented with thyme; the number of visits during 1 hr to the thyme-scented cards is given by the figures in brackets. To the feeding station itself there came on 17 September 178 newcomers (to *A*), on 18 September 362 newcomers (to *B*).

Had the foraging bees everted their scent organ on the flight from the hive to the feeding place and thus laid down through the air a perfumed trail that was followed by their comrades? This suspicion was testable—and proved to be false (see pp. 222f). Autumn arrived and the riddle remained unsolved until the cold months were past and the bees were active again.

With the beginning of new experiments it was conspicuous in the hive that the numbered bees coming from a known feeding place all kept to the same direction in their waggling runs, whereas foragers returning at the same time from natural sources were mostly waggling at other angles. If the sugar-water dish was situated north of the hive, on the vertical comb the waggling run was directed downward at noontime; if food was being offered to the south, dancing was exactly opposite, in the upward direction. As in succeeding hours the sun moved on clockwise the direction of the waggling run was shifted counterclockwise. Thus it became clear that the direction toward the goal was indicated by the straight waggling run, with reference in fact to the position of the sun and in a peculiar indirect manner. On the vertical comb in the dark beehive a straightforward signal of direction according to the sun's position is not possible. This difficulty is removed by the dancer's signaling the direction toward the sun by a waggling run upward, and indicating any desired angle right or left from the sun by waggling runs at a corresponding angle to right or left of the vertical direction upward. She transposes her signaling of direction from the realm of vision to that of the gravitational sense.

In order to learn in what portion of an ordinary beehive are carried out the dances that till then I had seen only in the shallow observation hive with its abnormal arrangement of combs, under the open sky I took one comb after another out of an ordinary beehive. The numbered dancers, which were frequenting an artificial feeding place, were scarcely disturbed. Then the thought occurred to me: What would they do if one held the comb horizontal, so that a transposition to the direction of a plumb line is impossible? They did not hesitate a second, but continued their dances on the horizontal surface, and now pointed with the direction of their waggling run straight at the feeding place, as we ourselves might indicate the goal with upraised arm. When the comb was rotated like a railroad turntable, the bees maintained their direction, like the needle of a rotated compass. If the sun had moved forward, despite this their waggling runs pointed unchanged to the feeding place. At that time I was no little astonished at the cleverness of the bees that mastered so smoothly an unusual situation—but basically I was wrong. Rather it was the dancers on the vertical comb that deserved admiration. For what I now saw for the first time was much the easier solution of the problem of indicating a direction to the comrades. No doubt this is also the more ancient, phylogenetically older form of indicating direction in the honeybee stock (see p. 304). Consequently it is appropriate to begin a more detailed exposition with the more easily comprehensible relations on a horizontal dance floor.

What can be read here in a few minutes became clear to me only during the course of a couple of months in the summer of 1945. For the principle just described, according to which direction is signaled, is obscured for the still inexperienced observer in a downright treacherous manner by conditions whose nature was at first a riddle. How were bees able to take a stance on the horizontal comb in relation to the sun when this was invisible to them? Why at times on the vertical comb did they point more than 40° to the right or left of the expected direction? Why were the dances on the opposite side of the comb oriented differently? How was it that the signal for direction was altered, with the goal the same as before, when the observation hive was rotated so that the flight entrance faced another point of the compass? All of this was confusing[34] and would have been hopeless if the fundamental principle, indication of direction according to the position of the sun, had not been so prominent.

13. THE INDICATION OF DIRECTION ON A HORIZONTAL SURFACE

Occasionally dances on a horizontal surface are seen on the flight board in front of the hive when the weather is very warm and the inhabitants "camp" in great numbers before the entrance. Many of the foragers then distribute their loads even before entering and dance under the open sky. The same thing may happen with a swarm cluster when scout bees are making known by their dances the location of a favorable nesting site (pp. 269ff). Thus in no wise are we dealing with an unnatural occurrence.

In order to be able to study at will dances on a horizontal comb, I laid the whole observation hive with its combs over horizontally (see the tiltable hive, pp. 8ff and Fig. 3). It is remarkable how little bees are disturbed by this maneuver. Difficulty arises only because bees that want to get out have a hard time finding the hive entrance when the protective cover is open. Their tendency to go toward the light drives them against the glass window of the observation hive, and in this position too gravity does not guide them downward to the exit. So restlessness develops, but may be overcome quickly by covering the hive temporarily with a dark cloth. As far as other conditions are concerned, the hive can be left for hours in a horizontal position without any recognizable disruption of normal goings on. In October 1947, as a check I left a hive on its side for 11 days. The bees went about their foraging activity as though nothing had happened. Chauvin (1960) even kept a hive for several months in such a position that the combs were horizontal and reports that the deposition of eggs and storage of honey continued normally on their upper and under sides.

In the horizontal tail-wagging dance, dancing bees from an artificial feeding place point with the straight segment of their run quite precisely in the direction of the goal. As an example I cite an experiment of 11 August 1959 (observer, M. Lindauer).

[34] The explanation will be found on p. 134 and on pp. 196f.

132 / The Dances of Bees

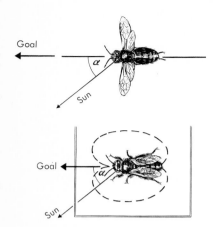

FIGURE 114. The principle whereby direction is indicated by means of dancing on a horizontal surface. During the waggling run (*bottom*) the bee takes such a position that she sees the sun at the same angle α as during her previous flight to the feeding place (*top*).

The observation hive was on a level stretch of heath in the Schleissheim Forest near Munich. Numbered bees were foraging 200 m directly east of the hive, which lay horizontal for the duration of the hours of observation. The upper part of the hive was removed (cf. Fig. 7), so that the bees had a free view of the sky on all sides. It was nearly cloudless. The sun was shining into the hive, but in order to prevent excessive heating the hive was shaded between observations. The direction of the waggling runs was determined with the protractor (Fig. 26). Table 21 shows the divergence from the direction to the goal for all dances during the observation time (8:43 A.M.–1:47 P.M.). On the average the dances pointed 0.21° to the left of the goal (scatter $\sigma = 0.96$, mean $\bar{x} = 1.07$, $n = 95$).

How the bees accomplish this is explained in Fig. 114. During their flight from the hive to the feeding place they take note of the position of the sun. We designate as the angle of the sun the angle between the compass direction to the goal and the compass direction of the sun's position (in astronomer's terms: the angle between the azimuth of the goal and the azimuth of the sun, angle α in Fig. 114, *top*). The bee keeps this angle in her memory. After her return home she sets herself at the same angle to the sun during her linear waggling run (Fig. 114, *bottom*). Thus she shows the dance followers what angle to the sun they must themselves maintain in order to fly in the direction of the goal. The compound eye, with its thousands of slightly divergent and very exactly oriented ommatidia (Fig. 115; cf. also Figs. 410 and 411) is an excellent protractor for such purposes. Some other arthropods, too, know how to profit from it.

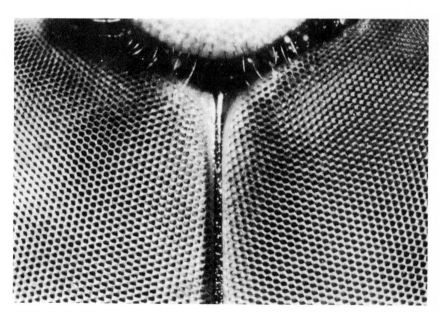

FIGURE 115. Compound eye of a dragonfly, partial view from in front. At the top is the vertex of the head; below it is the line of separation between right and left eyes. See also Figs. 410 and 411. Photograph by Hans Bamberger.

TABLE 21. Indication of direction on a horizontal surface with a free view of the sun and blue sky; experiment of 11 August 1959; observer, M. Lindauer.

Time	Divergence (deg) Left	Divergence (deg) Right	Time	Divergence (deg) Left	Divergence (deg) Right
8:43	0.5		11:18	2.0	
8:48		0	11:21		0.5
8:50		0.5	11:23	1.0	
8:53		1.5	11:26	1.5	
8:57		1.5	11:28	2.5	
9:00		0.5	11:30	1.5	
9:03		0.5	11:34	2.5	
9:05	0.5		12:00		0
9:07		2.0	12:05	1.5	
9:10	1.5		12:10	5.0	
9:13		1.0	12:11	1.5	
9:17		0	12:16	1.5	
9:20		4.5	12:21		0
9:22	1.5		12:23	0.5	
9:24	2.0		12:25		0.5
9:25		0	12:26		1.0
9:27		0.5	12:34	2.5	
9:29		0	12:35	4.5	
9:31		1.5	12:37	1.5	
9:33	1.0		12:38	1.5	
9:35		0.5	12:41	0.5	
9:38	1.5		12:44	1.0	
9:40	0.5		1:03	0.5	
9:42	1.0		1:03	0.5	
9:44		0	1:06		1.5
10:03		2.0	1:08	0.5	
10:05		0.5	1:09		1.5
10:07		1.5	1:11	1.5	
10:09	0.5		1:13		1.0
10:10		0	1:16		2.0
10:14	1.5		1:19		0.5
10:16		0	1:20	0.5	
10:19		0.5	1:23	1.5	
10:21	1.0		1:25		1.0
10:24	0.5		1:27	0.5	
10:26	0.5		1:28		2.5
10:29		0	1:30		0.5
10:31		0	1:33		0.5
10:33		0	1:34	2.5	
10:37		1.0	1:35		1.5
10:39		0	1:36	1.5	
10:41		2.0	1:38		1.0
10:44		0.5	1:40		0.5
11:03	0.5		1:42	1.5	
11:08		0	1:43	1.5	
11:12		1.5	1:46		0
11:14	1.5		1:47		1.0
11:16		0			

This manner of using the eyes was discovered by Santschi (1911) in desert ants. He marveled that on the sand, which affords scarcely any optical clues for orientation and on which scent trails quickly disappear, these ants could make linear scouting journeys and then return home without going astray. He found the explanation to be that they use the sun as a compass by maintaining during their wandering the angle initially taken in relation to it. For the return trip they alter their angle of orientation by 180° and thus are guided home by the sun. Santschi provided the elegant demonstration of this behavior with his classical "mirror experiment": when he shaded from the sun an ant that was on her way home laden with booty and by means of a mirror displayed the sun on her other side, she turned about through 180° and ran off in the wrong direction. Von Buddenbrock (1917) recognized that such "light-compass movement" occurs quite generally in insects and today we know that it is widespread throughout the entire animal kingdom. In bees this type of orientation was demonstrated by Wolf (1927). In his experiments (already mentioned on p. 118), in a wasteland poor in optical clues, the bees were displaced, while drinking, from one feeding place to another location; when they took flight they maintained the direction relative to the sun that would have taken them home from the accustomed feeding station (Fig. 110). By means of this type of orientation bees achieve no more than many other insects. But only the bees have carried developments to the point of using the solar compass not only for their own direction but also for showing the way to their hivemates with its help.

A bee collecting nectar or sugar water that has returned home first looks for recipients for her burden. While she is distributing the contents of her honey stomach (Fig. 28), she sits with her head in any chance direction. Only thereafter does she turn, with a tentative circuit, into the proper direction toward the goal and begin her first waggling run.

At first I was completely unable to comprehend how the bees were able to orient according to the sun when it was hidden from them by the wooden roof of the observation stand. The greatest variety of explanatory experiments were tried and discarded. The general distribution of light intensity was not the criterion, nor could magnetic influences be demonstrated; penetrating long-wavelength radiations were without effect. After manifold efforts in such directions it was sobering to realize that under certain conditions no more than shading the sun with a sheet of paper destroyed orientation; "under certain conditions"—namely when the dancer was unable to see any blue sky. It turned out that the bee can orient just as well with respect to the polarized light of the blue sky, whose direction of vibration depends on the position of the sun, as with respect to the sun itself.

When no blue sky is visible, only the sun's rays that fall on the eye determine the position taken by the dancer. If the sun is then concealed, she may not stop dancing; but from this moment on the dances are disoriented, the waggling runs pointing irregularly in all possible directions (Fig. 116). Disoriented dances always take place if a dancing

FIGURE 116. Three disoriented dances. The direction of the waggling runs was traced during the dances.

The Tail-Wagging Dance as Communication / 135

bee is unable to orient herself while performing the waggling run. She reacts differently if she cannot orient herself according to the sun on the flight to the goal (for example, with the feeding place on the radio tower, p. 167). Then she performs round dances in the hive. In the first instance she strives in the dance for a direction that she cannot find; in the second she does not know what direction to signal.

Esch (1964) with the help of a probing microphone recorded disoriented dances on a horizontal comb also. The time of a circuit was shorter than for normal dances on a vertical comb surface, but the duration of waggling changed only slightly with increasing distance to the feeding place (Fig. 117). Disoriented dances inform the hivemates properly as to neither the direction nor the distance to the goal.

In the observation hive that lay in a horizontal position all day long the returning foraging bees used to run to the lower surface of the comb, where it was darker. There there were only disoriented dances. Neither sun nor blue sky was to be seen there. When one sets the hive up in a fiberboard shelter, with diffuse illumination on the upper side of the combs also, *all* dances are disoriented. In spite of this they have an arousing effect. But the newcomers search in all directions (see pp. 153f).

In the shelter one can easily convince oneself that actually the sunlight, and the azimuth of incident light, determine the orientation of the dances. Even with a pocket flashlight, shone onto the comb from the direction of the position of the sun, one can immediately transform the disoriented dance into a correct one, directed toward the goal. If the light is allowed to fall on the comb from the opposite direction, the waggling runs point away from the feeding place. By positioning the lamp in any desired direction, one can make the bees dance correspondingly. They always maintain in relation to the lamp that angle that they took with respect to the sun during their flight to the feeding place. (v. Frisch 1948:40).

For this effect, light in our range of visibility suffices for the orientation of the dancer. Orientation is not impaired by exclusion of the ultraviolet (Schott filter GG_{13}). Heat radiation has no orienting effect (v. Frisch 1948:41; Heran 1952:198ff).

In the fiberboard shelter a light held in the hand can be brought only approximately into the desired direction. For testing this more precisely we used instead a canvas tent that had a hole for a "sun pipe" (Fig. 118). The latter could be moved circumferentially in two directions and could be set at the desired elevation and azimuth (Fig. 119). For the dancing bees then what was visible through the open tube, depending on the plan of the experiment, was either the sun, the blue sky, or a closed cover, or—in the present case—an artificial light source. The canvas tent screened off the blue sky beside the tube.

The sunlit tent has to be sufficiently heavy; if it is significantly brighter on the side toward the sun, uniform brightness inside must be provided by suitable shading, since unilateral illumination can distort the dances.

In front of the outer opening of the tube there was fastened a sheet of cardboard in the center of which a circular hole 18 mm in diameter

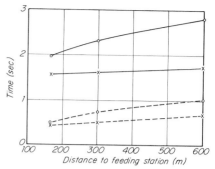

FIGURE 117. Durations of a circuit and of waggling in normal and disoriented dances at different distances to the feeding place: ○——○, circuit time of oriented dances; ×——×, circuit time of disoriented dances; ○- - -○, waggling times of oriented dances; ×- - -×, waggling times of disoriented dances. After Esch 1964.

FIGURE 118. The observation hive is laid flat; the side walls of the fiberboard hut, the protecting roof, and the roof support (see Figs. 5 and 6) have been removed. A canvas tent restricts the bees' view of the sky to the small circular portion they can see through the "sun pipe." Photograph by Dr. Renner.

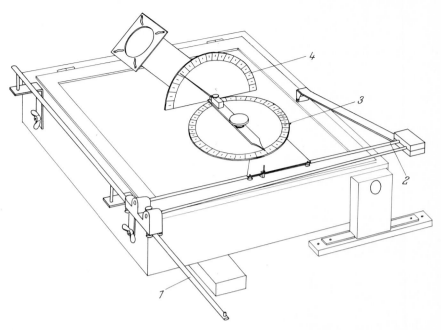

FIGURE 119. The observation hive in a horizontal position, with the "sun pipe." The latter can be moved on the rails (1 and 2) and fastened above the dance floor of the comb; 3, arc for setting azimuth; 4, arc for setting elevation.

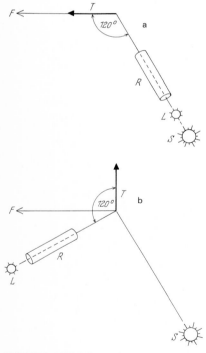

FIGURE 120. Effect of artificial sun on direction indication: F, direction of the feeding place (west); S, position of the sun; L, artificial light, visible to the dancer through the tube R; T, direction of dancing. (*a*) Lamp in the direction of the sun, dances properly oriented toward the west; (*b*) lamp displaced 90°, dances point 90° wrong, toward the north.

was cut. Translucent paper was stretched over the hole, which was lighted from without.[35] Thus the dancers looked at an artificial sun. At the lower end the tube could be closed with a shutter.

When the tube was closed the bees performed disoriented dances. If the artificial sun was visible to them in the direction of the true position of the sun, they danced toward the west, where the feeding place was, 200 m away. If, keeping the elevation constant, I displaced the artificial sun clockwise by 90°, the dancers pointed precisely northward (Fig. 120). Thus they maintained relative to the artificial light the angle that had pointed to the feeding place in respect to the sun. Exactly the same happened if I changed the inclination of the tube to 25° less or to 24° or 40° more than the actual elevation of the sun (experiments of 18 and 20 September 1960).

Hence for the correct orientation of the bees according to the position of the sun it is not essential that the sun's image fall on the very same ommatidia as during the flight to the feeding place. But it must fall on ommatidia of the same dorsoventral circumference. Ommatidia that look out upon the same compass direction apparently are coupled in respect of their afferent nervous connections. The *elevation* of the sun is without effect, or of subordinate significance. It may be shifted to a great extent without disturbing the orientation of the dancing bee, whereas the slightest azimuthal displacement of an artificial light source evokes a corresponding change in the direction of dancing.

[35] Either with a 200-W bulb or with a daylight bulb (Osram HWA 300, 160 W).

As regards the indication of distance I have noticed no difference between dances on horizontal and on vertical comb surfaces. Measurements made from time to time yielded the same tempo of dancing on both. Thus for a distance of 200 m to the feeding place 50 dances each were observed on a horizontal and a vertical comb and averages of 7.95 and 8.076 circuits/15 sec were determined (4, 6, and 8 August 1962). A prerequisite is that bees dancing on a horizontal surface shall be able to see the sun or polarized light from the sky. If that is not the case, the dances are disoriented. They lose their regular pattern and along with it the regulated duration of circuit and the rigid relation between the duration of waggling and the distance to the goal (cf. p. 135).

14. The Indication of Direction on the Surface of a Vertical Comb

As a rule the bees dance inside the beehive on the comb, which hangs vertically. Then they cannot indicate the angle of the sun directly by the direction assumed during the waggling run. That is technically impossible on the vertical surface. Also in the dark hive they lack the view of the sun as a point of reference. Under these conditions they transpose the angle of the sun from the sensory realm of visual perception to that of the perception of gravity, according to the following formula: the direction toward the sun is indicated by waggling runs in an upward direction, and any angle to right or left of the sun's azimuth is shown by a corresponding angle to right or left of the upward direction (Fig. 121).

How they might have come upon this solution seemed at first incomprehensible. But then it turned out that such transposition is a widespread facet of arthropod behavior that was by no means first evolved in order that they might communicate with one another. Even where biological significance is lacking, when the light is eliminated an angle of orientation to a light source can be transferred automatically to gravity. Thus an important element for the construction of a "language of the bees" has been available from the beginning. We shall deal with this fact later (pp. 326f).

The indication of direction is best observed with the aid of the protractor described on p. 26, in a fiberboard shelter (Figs. 5,7) with diffuse light. Blue sky might distort the dances strongly. Care must be taken that the inner walls of the shelter that are visible to the dancers are illuminated uniformly by the diffuse light. In order to compare the dance angles of the bees with the solar angles, we used azimuthal curves from which the azimuth of the sun at the experimental location could be read exactly, within about 0.5°, for every day and every minute.[36] This degree of precision was sufficient for our purposes. The momentary position of the sun was not known to the observer. Comparison with the azimuthal values was never made till later.

[36] We are indebted to Dr. Jahn (Munich Observatory) for constructing the curves.

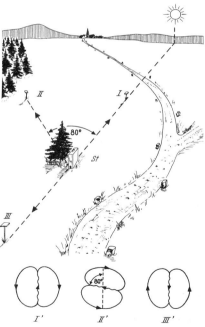

FIGURE 121. Three examples of the indication of direction on a vertical comb surface: *St*, beehive; I, II, III, feeding stations in three different directions; I′, II′, III′, the corresponding tail-wagging dances on the vertical comb.

When the foraging bee has reached the hive on her return home, she runs upward on the vertical comb surface to the region where the bees that take the food from her are waiting and where as a rule the dances also are performed (cf. p. 36). With sources of food that were more than 2 km distant, Boch (1956) noted that the dancing areas regularly were farther from the hive entrance the greater the distance to the place where food was obtained. This is so not only of the observation hive with only one or two superimposed combs, but also of the normal arrangement with several successive combs, and is true not only of artificial feeding places but also of natural food sources. This observation was confirmed by R. Götz (with Lindauer, not yet published), and he was able to demonstrate it even for shorter flights also. Figure 122 shows how, according to his observations on a normal beehive (with the combs in the "cold-weather arrangement"), with increasing distance to the feeding places the dancing areas were moved stepwise upward and backward from the hive entrance. This spatial separation of the dancing areas for bees gathering food near at hand and from far away may facilitate their arousing the appropriate foraging groups.

The dancing area apparently is not only defined topographically by its location relative to the hive entrance but seems also to bear a local sign, possibly of an odorous nature. For when R. Götz closed the entrance of an observation hive, rotated the hive 180°, and opened another entrance on the opposite side, which now faced toward the front,

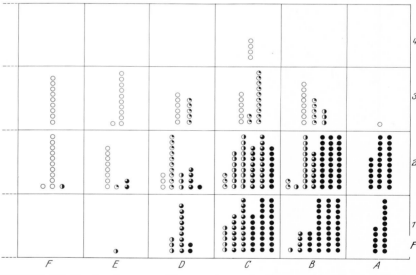

FIGURE 122. The dancing areas of bees on the comb of a normal beehive, for differing distances of the feeding station; F, hive entrance. The comb is divided into square fields. For each dance the field was recorded, but not the precise point of dancing within the field. Each circle designates one dance for a feeding place at a particular distance: ●, 50 m; ◐, 300 m; ◓, 1000 m; ◔, 3000 m; ○, 5000 m. The hive contained ten combs in the "cold-weather arrangement" (comb surfaces parallel to the sidewalls of the beehive). The actual dances took place on the middle combs, which hung behind the hive entrance, and in the figure have all been projected onto a single comb surface. With increasing range of flight the dance areas are moved in steps backward and upward away from the hive entrance. After unpublished experiments by R. Götz (Frankfurt am Main).

the homing foraging bees, which previously had been dancing in the vicinity of the entrance, now ran all the way through the hive to the back; and thus their dances were still performed on the region of the comb that formerly had served them as a dancing area. In the course of hours or days the dances once again gradually approached the hive entrance. These experiments are being continued.

When in one experiment the bees were fed 6 km distant from the hive, Boch observed a definite shifting of the dancing areas in correspondence with the temporal change in the direction of dancing (Fig. 123). It looked as though the foraging bees were influenced in their course as soon as they ran onto the comb by the direction of dancing that presently they were to maintain (Boch 1956:163).

In order to see whether anything similar could be detected also with bees collecting food from nearby sources, I made a few observations in Brunnwinkl in August 1964. At the same time I wished to learn how the bees behave when their dances are directed downward. Then indeed any influence on the direction of their entering, which in the natural course of events has to be obliquely upward, was not to be anticipated.

The feeding place was 200 m west of the hive. Since dancing by no means always occurs for the first place where food is distributed, I did not make note of the areas of dancing as Boch had done, but instead I measured with the protractor the direction of entry. In most instances these determinations went well because when the bee runs in she customarily holds clearly to a definite direction. When exceptionally she followed a zigzag course, the run was not taken into account. The measurements are combined as sketches in Fig. 124.

The individual figures have been arranged in succession according to the time of day, independently of the day of observation. The solid arrow in each case gives the average value of the observed directions of dancing for the time interval in question. The dashed arrows show the scatter of the dances (inclusive, of course, of scatter due merely to the change in the sun's position during the observation time). The directions of entry (individual values) are indicated by the peripheral points, that is, the directions of entry correspond to the lines connecting the center of the circle to the points. One sees that there is no relation to the direction of dancing, when it is obliquely downward. Naturally the home-coming bees then head upward into the region where food recipients are waiting. In essence, that still remains the case when the angle of dancing rises above the horizontal, yet it seems soon to gain a certain influence over the bees running in (Fig. 124, *7*). For angles of dancing nearer the vertical, a connection between the direction of entry and the later direction of dancing was evident (Fig. 124, *9–18*). Simultaneously the scatter in the direction of entry diminished. When the direction of dancing had passed the vertical, the connection was loosened, and the earlier condition of a rather large degree of scatter to the upper right was restored (Fig. 124, *19, 20*).[37]

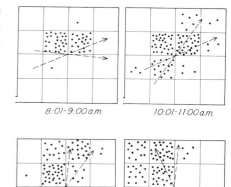
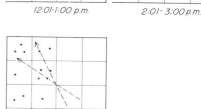

FIGURE 123. The dancing area of bees from far away (feeding station 6 km from the hive) is shifted on the comb in the course of the day. Each point designates the location of an observed dance. The dashed arrows indicate the limits within which the signaling of direction by the bees varied during the period of observation. Hive entrance at the lower left of the diagrams. After observations made by Boch, 9 August 1952.

---

[37] The direction of entering may also be aimed a few degrees to the left (for instance, *3, 4*, etc. in Fig. 124), because after the bees have passed through the hive entrance they often run straight ahead for a short distance while they are still below, before they climb up onto the comb.

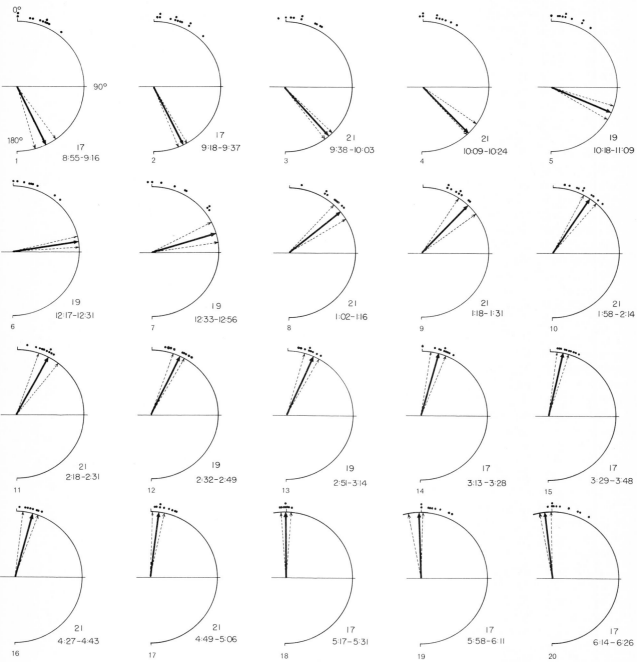

FIGURE 124. The direction of entry to the comb in relation to various directions of dancing; feeding station 200 m west of the hive. The diagrams are arranged according to time of day; the date (August 1964) and time are shown in each diagram. The direction of dancing (heavy arrow, average value; dashed arrows, range of scatter) is shifted counterclockwise in the course of the day. The direction of entry in each instance corresponded to the line connecting the center and the points on the arc; each point refers to an observed dance. Insofar as was permitted by the general location of the dancing area, the direction of entry was influenced by the later direction of dancing (diagrams 9–18); that is, when the bee enters she already has in mind the subsequent direction of dancing. For purpose of illustration the sites of entry to the comb have all been concentrated into a single point. In actuality they vary over a space of several centimeters from the hive entrance.

Thus it has also been confirmed for nearby sources of food (in our example 200 m) that when the foragers enter the hive they already have in mind the direction of their forthcoming dances. This participates in determining the direction of their entry, insofar as that is possible and so long as they are not thereby deflected too far from the area of the comb where the food recipients are sitting, into other regions where they would not be freed of their burden.

If the foraging bee meets a recipient on the dance floor or even further below, she may initially remain at rest in the direction of her running, while she is distributing the food. But in turning toward additional recipients or in circling about in search of them, she assumes during the distribution of food the greatest variety of positions. Thus if one records in general the positions occupied during the distribution of food one finds—just as with horizontal dances—no connection with the direction of dancing.

Not until the start of a waggling run does the bee wheel in a curving course into the correct angle. Although in doing this she must transfer from memory the visual angle into the direction with respect to gravity, she strikes the right value with astonishing precision. Table 22 will give some notion of this. It contains measurements of 35 dances that were made on 6 September 1959 between 11:11 A.M. and 12:03 P.M. with an observation hive at Aich (not far from Brunnwinkl). The feeding place was 270 m due south of the hive. The changes in the solar angle that resulted from the progressive movement of the sun are taken account of so rapidly that within a few minutes they are reflected in the altered angle of dancing (column 4). In order to make that even clearer, in Table 23 I have summarized the solar angles for intervals of 2° and have set beside them the average values for the angles of dancing.

In this case the dances were aimed close to the zenith. But that is not the reason for their good orientation. In Table 24 are brought together the last dances of the experimental day (4:45–5:29 P.M.), when the sun was almost due west and the solar angles lay between 80° and 88°. In the first observational series the scatter σ amounts to 1.14 (mean $\bar{x} = 1.74$, $n = 35$), in the second 1.24 ($\bar{x} = 1.55$, $n = 29$). Since during the afternoon the distance of the sun's position from the (southerly) feeding place grows greater and greater, in Table 24 the solar angles and angles of dancing increase progressively, whereas in the morning (Tables 22 and 23) they decreased.

Not always do the angles of dancing agree so well with the solar angles. Often discrepancies of 10°–15° in one direction or the other are found. These apparent errors depend on the direction of dancing and follow a definite pattern. Only to us do they appear as a "misindication." The bees are directed appropriately to the goal. For the present we shall leave this phenomenon aside and will discuss it later (pp. 196ff).

Another type of discrepancy is represented by the dances that show divergence. Often one sees that the waggling run deviates from the correct line, for instance to the left, whereupon the next waggling run deviates toward the right to the same degree. The direction of dancing fluctuates in regular alternation about the proper value. It is not hard

TABLE 22. Example of the signaling of direction on a vertical comb surface; experiment in Aich, 6 September 1959, 11:11 A.M.–12:03 P.M.; feeding station 270 m from the hive.

Time	Bee No.	Solar angle, right (deg)	Angle of dancing (deg)	Error (deg)
11:11	29	20	22.5	+2.5
11:14	10	19	16.5	−2.5
11:15	20	18.5	17.5	−1
11:16	19	18	14.5	−3.5
11:17	15	17.5	14	−3.5
11:19	10	17	13.5	−3.5
11:20	11	16.5	15.5	−1
11:21	29	16	18	+2
11:22	23	16	14	−2
11:24	20	15	15	0
11:26	13	14.5	10	−4.5
11:29	10	13	11	−2
11:30	29	13	12	−1
11:30	16	13	14	+1
11:32	11	12	13	+1
11:33	25	12	8	−4
11:35	29	11	13	+2
11:36	15	10.5	9	−1.5
11:37	11	10.5	9.5	−1
11:38	23	10	8	−2
11:43	11	8	9.5	+1.5
11:44	29	8	7.5	−0.5
11:46	13	7	5	−2
11:47	16	6.5	4.5	−2
11:48	23	6.5	5	−1.5
11:49	12	6	4	−2
11:50	15	6	6	0
11:51	16	5.5	6	+0.5
11:52	11	5	6	+1
11:54	29	4	7.5	+3.5
11:56	23	4	5.5	+1.5
11:58	29	2.5	3.5	+1
12:00	17	2	1	−1
12:02	11	1	1	0
12:03	29	0.5	2	+1.5

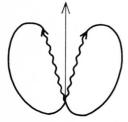

FIGURE 125. Tail-wagging dance showing divergence. The bisector points the direction to the goal.

to set the protractor on the bisector, which corresponds to the angle to the goal (arrow in Fig. 125).

Such diverging dances are seen above all when the feeding place is only a short distance away. Here they are the rule (see p. 61, Fig. 56). The farther removed the feeding place, the smaller the angle of divergence, until finally at a flight range of several hundred meters the waggling runs often are superimposed; but this does not have to be. In the protocols I find dances with divergence noted even at a range of 2000 m. The variability is considerable. In a single experiment often

TABLE 23. Solar angle (2-deg classes) and angle of dancing, from Table 22.

Time	Solar angle (deg)	Angle of dancing, average (deg)	n
11:11–11:15	20–18.5	18.8	3
11:16–11:20	18–16.5	14.4	4
11:21–11:26	16–14.5	14.25	4
11:27–11:31	14–12.5	12.3	3
11:32–11:37	12–10.5	10.5	5
11:38–11:42	10– 8.5	8	1
11:43–11:48	8– 6.5	6.3	5
11:49–11:53	6– 4.5	5.5	4
11:54–11:58	4– 2.5	5.5	3
11:59–12:03	2– 0.5	1.1	3

TABLE 24. Indication of distance on a vertical comb; experiment in Aich, 6 September 1959, 4:45–5:29 P.M.; feeding station 270 m from the hive.

Time	Bee No.	Solar angle, left (deg)	Angle of dancing (deg)	Error (deg)
4:45	22	80	84	+4
4:46	15	80	78	−2
4:47	17	80	82.5	+2.5
4:49	16	80.5	80.5	0
4:51	12	81	84	+3
4:52	29	81.5	80.5	−1
4:53	13	81.5	81.5	0
4:54	16	81.5	82	+0.5
4:57	29	82	81.5	−0.5
4:58	23	82.5	84.5	+2
4:59	15	82.5	82	−0.5
5:01	16	83	82	−1
5:07	15	84	84	0
5:08	16	84.5	86.5	+2
5:08	19	84.5	88.5	+4
5:11	22	85	87.5	+2.5
5:12	20	85	87	+2
5:13	23	85	89	+4
5:14	29	85.5	86.5	+1
5:15	11	85.5	85.5	0
5:16	19	86	88.5	+2.5
5:17	16	86	87	+1
5:19	29	86.5	86.5	0
5:20	21	86.5	88	+1.5
5:22	25	87	88.5	+1.5
5:23	16	87	88	+1
5:25	19	87.5	89.5	+2
5:27	29	88	88.5	+0.5
5:29	23	88	90.5	+2.5

144 / The Dances of Bees

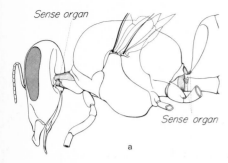

FIGURE 126. (*a*) The honeybee's gravitational sense organs at the articulations of the thorax with the head and the abdomen. (*b*) Areas of the joints under higher magnification; head and abdomen have been pulled somewhat apart from the thorax, in order to render visible the concealed locations of the sense organs. After Lindauer and Nedel 1959.

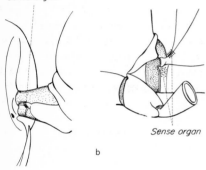

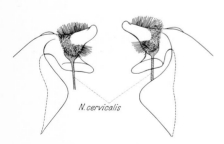

FIGURE 127. The sclerotized processes of the prothorax with the sensory hairs and their nerve strands. After Lindauer and Nedel 1959.

many bees will dance with divergence, others without. I have not devoted much attention to the conditions underlying the occurrence of diverging dances. Perhaps a lot more could be discovered about them.

Transposition of the solar angle into the direction relative to gravity presupposes the existence of effective organs of gravity perception. Such were long sought in vain in the bee.

Since there was some question whether the antennae might not be the seat of a gravitational organ, I checked the dances of bees whose antennae had been amputated. Unilaterally antennectomized specimens danced somewhat uncertainly at first and with increased scatter, but without permanent disability. After amputation of both antennae orientation in the hive was at first (for a few hours) greatly disturbed. It was necessary, by temporarily closing the opaque protective cover, to help the bees that had been operated on to find their way in and out, since otherwise they were irresistibly distracted by the light from the side. When they resumed traveling between hive and feeding place, after several flights they again began to dance, but with complete disorientation. The suspicion that with removal of the antennae they had lost their sense of gravity was quickly abolished: they also danced disorientedly on a horizontal surface. It seemed that control over the movements of their own body had been impaired. Kalmus (1937) and Wolf (1926) had already called attention to the latter. However, with the few bees that continued to frequent the feeding place after bilateral loss of the antennae, this disturbance gradually diminished and normal dances with well-directed waggling runs were again to be seen after even a few hours. My best example is a bee (No. 40) whose right antenna I had cut off at the scape on 21 August 1955 and the left one on 26 August; 3–4 hr later during an hour's observation she performed nine dances, most of them of long duration and well oriented. Only in part were the dances marked by strong divergence and by temporary uncertainty.

Thus the bees are able even without antennae to orient in relation to gravity in their dances. But I have never seen an antennaless bee show the slightest interest in a dancing comrade. For reception of the information contained in the dances the antennae are essential.

Lindauer and Nedel (1959) deserve the credit for having discovered the sensory organs for perception of gravity in the bee. They are located in pairs on the joint between head and thorax and between thorax and abdomen. In Fig. 126 head and abdomen are slightly torn away from the thoracic tagma in order to show the concealed sense organs. In Fig. 126*b* these parts of the body are shown in greater magnification. In its normal position the head rests on two sclerotized processes from the prothorax; they protrude into the foramen magnum and are covered with sensory hairs (Fig. 126*b*). In the figure the hairs are merely indicated; actually they form considerable fields of bristles, as Fig. 127 makes evident. They are in contact with the opposing sclerotized parts of the occiput. The center of gravity of the normally positioned head (Fig. 126) is lower than the joint with the thorax. Hence the head acts like a pendulum. When the bee is facing upward on the vertical comb, gravity tends to pull the ventral part of the head against the thorax, and the reverse when she is facing downward (Fig. 128). In both cases the weight of the head is supported by opposing parts of the hair fields. During transition from one position to the other there is exerted on the sensory cushion a shearing force that will affect maximally different definite regions of the bristle fields, according to the nature and intensity of the obliqueness of the bee's positon. Perception of the body's position depends on the nervous registration of the stimulation pattern. Thurm (1963), by means of morphological and electrophysiological investigations, succeeded in explaining the mode of action of the bristles. The bristle fields at the joint between thorax and abdomen work according to the same principle. They are stimulated especially when the abdomen is bent downward or to the side.

FIGURE 128. The force of gravity, acting at the center of gravity of the head, tends to rotate the head toward the ventral or the dorsal side of the thorax, depending on the position of the bee.

It has been possible experimentally to demonstrate the correctness of these views. If by fastening a small lead weight to a bee's head one displaces its center of gravity upward, the head, obeying gravity, hangs in a direction opposite to the normal one and the bee displays her deception convincingly, by confusing upward with downward. Lindauer and Nedel succeeded also in severing the nerves that supply the head organ. Some bees thus treated flew back and forth between hive and feeding place for several days. They also started dancing, but they could not carry out any oriented waggling runs. If the observation hive was laid horizontal and a view opened to the sky, the dances were at once oriented properly. Thus the responsiveness of the bees was not impaired in general, but they were merely unable to take a proper position relative to gravity.[38]

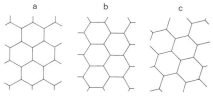

FIGURE 129. (a) Normal comb pattern. (b) Less frequent but naturally occurring comb pattern. (c) Unnatural comb pattern, rotated 15° relative to (a) or (b).

The precise maintenance of the size of the angles raises the question whether the dancer may not be assisted here in ways other than by the information from the sense organs for gravity perception. One could imagine that perhaps she gains a frame of reference for the direction of running from the unusually regular pattern of the comb. But an orientation in relation to the arrangement of the cells is unlikely, because the side walls of the cells are not always vertical (Fig. 129a); there occurs also an arrangement with horizontal cell walls above and below (Fig. 129b). Zander (1947:44, 45) designated such variations as rare. According to Rademacher (1960) they are frequent. Both he and O. W. Park (in Grout 1949:90) supply examples of both arrangements in photographs of combs that—without center walls bearing a preimprinted pattern—had been constructed by the bees in the course of their natural building activity. One would have to assume also that the dancers could alter their ways of reacting in accordance with the local pattern. That is quite improbable. Observations made in 1945 also gave evidence against such behavior.

When I lifted out of a normal hive a comb with dancing bees on it and, retaining the vertical position of the comb, rotated it by 90° in such a way that its lateral margins became the upper and lower margins,

---

[38] Regarding the mutual influence of the organs of the head and abdomen and the (subordinate) cooperation of the bristle fields on the legs, see Lindauer and Nedel (1959) and Markl (1962, 1963).

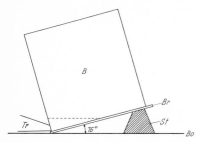

FIGURE 130. Observation hive *B* tilted obliquely; *Tr*, flight funnel; *Bo*, baseboard of the stand; *Br*, movable board; *St*, wooden support; dashed line, horizontal ledge for placement of protractor.

the dancers kept to their original orientation relative to gravity, although the comb pattern was turning beneath their feet. When signaling direction they oriented themselves not relative to the comb but with reference to the force of gravity (v. Frisch 1946:41, 42).

Complete certainty was provided by a few experiments that I carried out in Brunnwinkl in the summer of 1961 in cooperation with Lindauer. First we set up two observation hives beside one another and separated by 3 m. From each colony a group of numbered bees was harvesting, each at its own feeding place 200 m to the west. One group was dancing on a normal comb (cell pattern in Fig. 129*a*), the other on a previously inserted comb with a cell pattern like that of Fig. 129*b*. The dances, observed simultaneously, gave no basis for any influence of cell pattern.

Considering the possibility that the bees were familiar with both cell patterns, because both occur under normal conditions, we exchanged the dancing comb of one colony for a comb that was rotated only about 15° instead of about 30° (Fig. 129*c*). That is definitely an abnormal cell pattern. But even so, despite immediate observation, no influence on the direction of dancing could be seen.

Since in these experiments dances from two different colonies were to be compared with one another, we altered the arrangement as follows in order to eliminate colony-specific differences. The screw connection of the observation hive with the floor board of the stand was loosened. By pushing a suitable wooden support beneath it the hive could now be tilted 15° and set upright again as desired (Fig. 130). Thus it was possible to test the dances of the same bees at intervals of a few minutes with a normally and abnormally oriented cell pattern. No difference appeared. We were able also to make the shift during a single dance and were convinced that the dance was continued in an unchanged direction. Thus the comb pattern has no recognizable influence on the direction of dancing. The exactness of the orientation of the dances is to be ascribed to the sense of gravity.

15. Dances on an Oblique Comb Surface

On a horizontal surface the dancer takes her position with immediate relation to the sun; her waggling run points directly at the goal. On a vertical surface she transposes the visual angle to one related to gravity. What does she do when she dances on an oblique surface? That is no purely theoretical question. By no means rarely one sees that a bee is dancing on the obliquely rounded lower edge of a free-hanging comb or on a strongly undulant comb surface without being in any way bothered and without the least sign of any distraction among the followers. But above all the rounded swarm cluster provides a broad field for oblique dances, where the scout bees point out to their shelterless folk the direction to a new home (see p. 269).

By tightening the wing nuts (5, Figs. 3*b* and 4) one can fix the tiltable observation hive in any desired oblique position. The angle of inclination can be read from the scale. If the dancers on the oblique comb have a view of the blue sky, so that they are able to read the sun's position from the polarized light of the heavens, then the tendency to take position with reference immediately to the light comes into competition with taking position according to the gravitational angle. We are already acquainted with this latter phenomenon from dances on the vertical comb (p. 134). But now there is added the complication that for the proper positioning of the dancer different weight is given not merely to light, according to the nature of the visible heavens (clouds, polarization in the blue sky), but also to the force of gravity, according to the angle of inclination of the obliquely positioned comb. Thus different directions of dancing occur in accordance with circumstances.[39] It is to be assumed that the dance followers, which in all instances are exposed to the same conditions, properly grasp the intricate indications of the angle. An analysis of the success of the dances when they are influenced by the visibility of blue sky has been made hitherto only for the vertical position of the comb (p. 212), yet we know that the dances on the freely suspended swarm cluster are only rarely oriented precisely vertically or horizontally and in spite of this are highly successful in sending the swarm to its goal (pp. 269ff).

In what follows we ignore such complex interrelations and test the behavior of dancers on an oblique comb in the fiberboard shelter, where orientation in reference to light was excluded. We have three series of experiments from 1948 and one from 1961. Observations were made in uniformly diffuse light. The feeding place was 200 m distant from the hive.

If the surface of the comb is inclined at 30° to the horizontal, the dancer has no difficulty; at 15° she is still well oriented; and at an angle of inclination of 10° the indication of direction is still quite good, yet the dances are obviously less sure, the waggling movements less energetic; the scatter of the indications of direction increases, as does the angle of divergence—often the alternate waggling runs form an angle of 40–60° with one another. This increased oscillation around the correct orientation with respect to gravity is no doubt an expression of greater difficulty in finding this orientation. With this degree of obliqueness there occur also scattered very aberrant waggling runs, and not infrequently probing circular runs are interspersed among the dance figures. With an inclination of 5° above the horizontal the uncertainty becomes great. Disoriented dances are frequent. With the majority of dancers the waggling runs still are approximately correct. But apparently the perception of gravity by the bristle fields is now near the limits of its capacity. The data in Table 25 give a typical picture of the decreasing exactness in this region of the indications of direction.

Orientation with respect to gravity on oblique surfaces has been investigated also with some other insects. The staphylinid beetle *Bledius bicornis* Grm. builds vertically downward into the sea sand the tubes it inhabits. If enclosed between tilted glass plates it digs downward as

---

[39] These differences were incomprehensible to me at the time of the first communication about dances on an oblique comb surface (v. Frisch 1948:42), because then I was not yet acquainted with the influence of polarized light from the sky.

TABLE 25. Dances on an oblique comb surface, 15°, 10°, and 5° above the horizontal; Brunnwinkl, 12 September 1961.

Inclination of comb (deg)	Bee No.	Time	Solar angle, right (deg)	Angle of dancing (deg)	Error (deg)
15	38	9:56	132	127	− 5
	33	9:57	132	128	− 4
	31	9:58	131.5	140	+ 8.5
	30	9:58	131.5	130	− 1.5
	38	10:01	130.5	129	− 1.5
	40	10:02	130.5	135	+ 4.5
	33	10:04	129.5	127	− 2.5
	30	10:05	129.5	134	+ 4.5
	41	10:06	129	134	+ 5
	27	10:07	129	125	− 4
10	33	9:35	138	153	+15
	36	9:36	138	149.5	+11.5
	38	9:38	137.5	135	− 2.5
	31	9:39	137	146	+ 9
	40	9:40	137	142	+ 5
	27	9:41	136.5	131	− 5.5
	41	9:42	136.5	140	+ 3.5
	36	9:42	136.5	148	+11.5
	40	9:42	136.5	148	+11.5
	33	9:45	135.5	146	+10.5
5	38	10:13	127	139	+12
	35	10:16	126	Disoriented	
	33	10:19	125	135[a]	+10
	38	10:22	124	100	−24
	41	10:22	124	Disoriented	
	31	10:24	123.5	114	− 9.5
	33	10:26	123	131[a]	+ 8
	35	10:27	122.5	Disoriented	
	35	10:29	122	130[a]	+ 8
	31	10:30	121.5	138[a]	+16.5

[a] With disoriented runs.

steeply as possible. It can still find the steepest direction at an inclination of 20° above the horizontal, but no longer at 10° (Bückmann 1954, 1962). The carabid *Dyschirius nitidus* Dej. under such conditions can find the direction upward on a surface of only 3° inclination (Bückmann 1955). Ants (*Formica rufa* L.) also are able to find the upward direction on a surface that is inclined only 3° above the horizontal (Markl 1962).

The beetles and ants mentioned thus seem to equal the bees as regards the efficiency of their gravity sense. But they had the simple task of finding their way upward on an oblique surface. The bees had to maintain a definite angle to this direction on the oblique surface, an angle that was prescribed for them by the position of the sun and that then had to be transposed according to the gravitational sense. The ability to do this when the angle is only 5° is a respectable accomplishment.

16. INDIVIDUAL DIFFERENCES IN THE INDICATION OF DIRECTION, AND THE INFLUENCE OF AGE

In Sec. 5 on the indication of distance we mentioned individual differences in the dance tempo. That the vivacious rhythm of the dances leaves some room for the play of personal organization is not surprising and fits in with the picture of individual variation that frequently is exhibited to the observer by bees eager to dance and others sluggish about it, by dancers that seem precise or inaccurate, lively or phlegmatic.

In indications of direction individual differences seem less conspicuous. And yet they exist. When comparing different bees of the same colony, that were observed during several days while they were frequenting a feeding place 200 m distant, Elisabeth Schweiger (1958) found a statistically significant individual difference in 20 out of 109 instances that could be compared, that is, in 18 percent of the cases. In the tempo of dancing the differences were appreciably more frequent (52 percent). But the general notion that specific individuals should deviate regularly in a given sense in their indication of direction seems strange at first. The significance of such "errors" of transposition of the visual to the gravitational angle will be considered again when I return to the discussion of "misdirection" (p. 212).

If the indications of direction by one and the same bee are compared at different ages, there becomes evident a decrease in the scatter as she grows older. In Fig. 131 are combined the results obtained with 42 bees from two different hives. On the abscissa is shown the mean age of the specimens during the time of observation, on the ordinate the scatter of the indications of direction.

We have already become familiar with the same phenomenon in relation to the indication of distance (p. 75). There it is the more readily comprehended, because one can imagine that a constant distance will be judged more accurately after often-repeated flights. But the indication of direction does not remain constant; it shifts continually with the position of the sun, and after each flight the bee has to signal a different angle. This raises the suspicion that with advancing age performance improves without respect to the specific goal. Perhaps the same is true for the indication of distance. The question whether bees with greater flight experience also signal a *new* goal more precisely will have to be clarified by further experimentation.

17. COMPARISON OF THE EFFECTS OF ROUND DANCES AND TAIL-WAGGING DANCES

In the tail-wagging dance the direction-giving waggling run is repeated after each turn to the right or left. In this way there is intro-

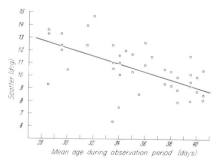

FIGURE 131. The scatter in indication of direction diminishes with advancing age; observations on 42 bees from two colonies in 1955. After Elisabeth Schweiger 1958.

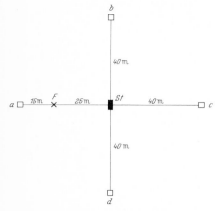

FIGURE 132. Experimental setup for testing at what distance to the feeding place a *directed* search begins: *St*, hive; *F*, feeding station; *a*, *b*, *c*, *d*, scent cards.

duced into the dance the rhythm on which the indication of distance is based. In the round dance the turns succeed one another at irregular intervals, sometimes after ½ or ¾ of a complete circle, at other times only after 1, 1½, or 2 complete circles to left or right. Consequently the round dance contains a signal neither for distance nor for direction.[40] The prerequisite for such signals is developed only with the institution of regular turns, when oriented semicircles or a course shaped like the figure 8 appear, with the goal at distances of 25–50 m. At this point the indication of distance is still distorted at first by the frequent interpolation of complete, often multiple, whole circles, until the tail-wagging dance becomes fully regular at about 100 m (cf. p. 61).

In order to learn where in the transitional region between round dance and tail-wagging dance an indication of direction becomes recognizable to the dance followers, I tested the direction in which searching was done by the newcomers flying out, with the feeding station at distances of 10, 25, 50, and 100 m. An example chosen at random from these series of tests (v. Frisch 1948:17) will make clear the experimental arrangement (Fig. 132). The observation hive (*St*) is in the middle of the level meadows at Aich. At *F*, 25 m west of the hive, a small group of marked bees is fed with sugar water to which has been added a trace of star-anise oil. The foragers arouse their hivemates. Where will these search? We use the already frequently mentioned device of attracting them to scent cards. These (without food but each with a drop of star-anise oil) are set out in all four compass directions 40 m from the hive (*a*–*d*).[41] During a whole hour, while 2M sugar solution is being offered at *F* and the dancers are sending out their comrades in search of star-anise fragrance, all flights to the scent cards are recorded by four observers. Thereupon the feeding station, with the sugar concentration reduced so that arousal does not continue, is moved to another location, a new experiment is conducted with 2M feeding, and so on.

In two experimental series of this kind (1–3 August 1946) we proceeded so that the feeding station was set up 10 m eastward, then 25 m eastward, 50 m northward, and 100 m northward during the first series, and 10 m westward, then 25 m westward (as in Fig. 132), 50 m southward, and 100 m southward during the second series. The scent cards were always 15 m farther from the hive than the feeding station.

In this way the feeding station was set up twice at each distance, on the second occasion always in the opposite compass direction. The combined result of the eight experiments is shown graphically in Fig. 133. Each group of four adjacent columns refers, as indicated below it, to the scent cards set out in the four compass directions. The height of the columns and the numbers below them show the number of flights to each during an experiment. The black column refers to the scent

[40] Regarding direction-oriented round dances see p. 304.
[41] Along the line of flight of the foraging bees, it would have been better to put the scent cards in front of the feeding station instead of beyond. But in some of the experiments the distance to the feeding station was too small for this. It was more important to keep conditions constant. In particular, the distance between the feeding station and the scent card situated on the same line should remain identical, so that any distortion due to the scent organs of the foragers might be kept as constant as possible.

card that was on the same line as the feeding station, the white column to the scent card in the opposite direction, and the hatched columns to the two scent cards situated at right angles to the line of flight of the foraging bees. It is clear that when the feeding station is 10 m distant hunting is fairly even in all directions. But at 25 m there begins to appear an oriented search, which is emphasized more strongly at 50 m, until finally at 100 m the scent cards that are off the track are scarcely heeded any longer. Hence at 25 m the dance followers already take note of an indication of direction, which for us is recognizable only when we pay the closest attention and then by no means in all dances; in agreement with our own perceptions, the indication of direction becomes clearer and clearer for the bees, too, as distance increases to 100 m.

Thus the bees themselves have made known to us that in their "language" round dance and tail-wagging dance mean different things: an instruction to search round about the hive in all directions, or an instruction to fly out in the direction of the waggling run.

But the experiment does not reveal whether the search made by the flying newcomers aroused by round dances is confined to the distance that is indicated by the round dance. In order to test this, on 4 September 1963 we set up a pure Carniolan colony in the level meadowland at Aich and let nine marked bees collect 2M sugar solution at the feeding station F (Fig. 134) 10 m southwest of the hive, between 9:45 and 11:45 A.M. The food dish was placed above a scent-bearing watch glass that contained oil of peppermint (arrangement as in Fig. 21). Observation showed that the bees were making only round dances.

During the 2 hr of observation there flew to the feeding place, favored by the short distance and by the great activity of the scent organ, 278 newcomers; these were killed. Throughout the experimental period the wind remained weak and varied repeatedly in direction (NNW, SSW, S, NE).

Scent plates with oil of peppermint (as in Fig. 85) were set out in two approximately opposite directions at 50, 100, 200, 300, and 450 m from the hive, and two more cards 20 m from the hive approximately to the north and to the east. The number of flights of newcomers during the 2-hr experiment is given in Fig. 134 and is also pictured there by dots. In spite of the persistent, intensive recruiting dances, the area of searching is limited predominantly to a range of 50 m diameters; at 100 m there appear but few newcomers and at 200 m and beyond none. With the goal distant between 50 and 100 m the pattern of the tail-wagging dance develops more and more clearly from the round dance (cf. p. 61). Thus the newcomers aroused covered obstinately only that area that was indicated by round dances. The dance followers interpret this form of dance as the sign of a nearby source of food (for exceptions, see pp. 252f).

That these gestures are interpreted correctly can be shown in yet another way. Of two marked groups from the same observation hive, one was collecting from a feeding station 12 m from the hive, the other from one 400 m away, both of them without an added scent. The bees

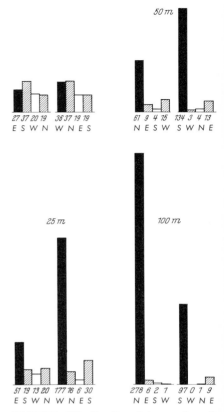

FIGURE 133. Results of two series of experiments on 1–3 August 1946. The feeding station (F in Fig. 132) was placed alternately at 10 m, 25 m, 50 m, or 100 m from the hive. The numbers of visits to the scent cards (a–d in Fig. 132) are indicated by the height of the columns and the numbers below them: black column, scent card on the line to the feeding station; white column, scent card in the opposite direction; hatched columns, scent cards at 90° to the line of flight. Beneath each number of visits is recorded the compass direction (from the hive) in which the scent card was placed.

of the former group were performing round dances, the others tail-wagging dances. After a rather long interruption in feeding at both stations both groups were sitting inactive on the comb; only scouts were still making inspections from time to time. Now the dish at the distant station was filled with sugar water, while there continued to be no food offered at the nearby station. Each contact between a forager returning to the hive from the distant feeding place with bees of either the distant or the nearby group was recorded and its result noted. Contact between the dancer and a numbered bee was considered successful when the latter appeared at her[42] feeding place within 5 min. Then, after food had been offered at both stations and a new pause in feeding had been made, a reciprocal test was run with food at the 12-m station and with none at 400 m.

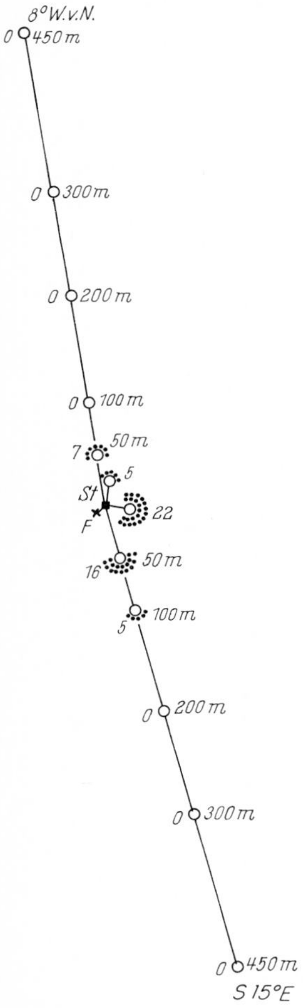

The outcome of the two experiments may be summarized as follows. The bees of the identical group (that is, those collecting from a distance who came in contact with a waggle dancer and those collecting from nearby who came in contact with a round dancer) were aroused successfully 11 times in 12 instances[43] observed, but bees of a strange group only 12 times in 32 observed meetings. Thus by coming together with a dancing bee 92 percent of the bees from the identical group were induced to seek out their feeding place, but only 37.5 percent of the bees from a strange group, for whose feeding place the other form of dance would have been the proper one. Anyone who is surprised that even so more than a third of the resting foragers were induced to fly out by a dance that signaled another distance may recall that, even without a dance, contact with a forager returning home successful occasioned a revisiting of the feeding place in 39 percent of the cases, in contrast with 89 percent when contact was with a forager that danced (p. 30). In terms of percentages the relation between unsuccessful and successful contacts was almost identical in the experiments just discussed. Despite the numerically modest number of data there is good reason to suppose that for resting foragers a dance that does not fit the accustomed distance has only as much force of arousal as does contact with a food-bearing returning bee that does not dance at all. Repetition of the experiment at another time had the same result (v. Frisch 1946:14–17, 1948:15).

Round dance and tail-wagging dance are intrinsically distinct dance forms. I now wished to see how, after a pause in feeding, bees react to the dances of another group when both feeding stations are located in the range for waggling but are in different directions. The experiment was made on 10 and 11 September 1960 in Brunnwinkl. The feeding places for the two groups were 200 m west of the observation hive and 200 m 10° east of south (Fig. 135). Feeding at both was without added scent. In order to be sure that no source of error was introduced by training to individual human odor, the personnel caring for the two

FIGURE 134. Experiment in Aich on 4 September 1963: *St*, observation hive; *F*, feeding station 10 m southwest of the hive (with peppermint scent); ○, scent plates in various directions 20 m, 50 m, 100 m, 200 m, 300 m, and 450 m from the hive. Numbers and dots show the number of visits. By means of *round dances* the newcomers are induced to explore the *nearby* surroundings, up to as far as the distance at which tail-wagging dances replace the round dances.

---

[42] She always flies first to the feeding place from which she is accustomed to collect.

[43] The relatively few contacts with bees belonging to the same group are to be explained by the fact that the latter start collecting food and that hence the number of them on the comb declines rapidly.

feeding stations exchanged positions hourly. After an interruption in feeding at both places, on the 10th sugar water was offered at F_W and on the 11th—in the reciprocal experiment—at F_S, while on each occasion the other vessel remained empty. The informants who returned home successful promptly mobilized their group comrades in the customary way. We are interested primarily in how the bees of the strange group, whose feeding place was at the same distance but in a different direction by 100°, behaved on contact with the tail-wagging dancers. The direction of dancing did not point to their feeding place. Would they remain cool to the invitation, or would they nevertheless let themselves be induced to take a look at their vessel?[44] During the two experiments on 19 occasions I saw a bee of the resting group come in contact with a dancing bee from the other group that was already harvesting again. Only 3 of these 19 bees came within 5 min to their feeding place; in 16 cases the contact was without success. Sometimes they ran along for a few turns, sometimes they showed immediate lack of interest. Particularly impressive were two bees that turned decisively away from the dancer—one of them not until she had run after her for a few circuits—as if after the sudden realization: "That is none of my affair." Thereafter they remained at rest on the comb. They had understood and heeded the indication of direction.

The experiments previously described, with feeding stations in the zone of transition from round dance to tail-wagging dance, have shown that the newcomers do not undertake an oriented search until oriented dances appear among the foragers. The dependence of oriented search on oriented tail-wagging dances can also be demonstrated convincingly in another way. On a horizontal comb with diffuse illumination the tail-wagging dances are disoriented (Fig. 116) because although the bees know the direction to the feeding place in relation to the sun they have no point of reference by which to guide themselves in dancing (see pp. 134f). Under such conditions dancing lacks orientation in the dark also. This may be seen, for example, in the fiberboard shelter in red light (not visible to the bees). By no means is there any reduction in the eagerness to dance, nor equally any diminution in the interest of the dance followers. They trip along in the wake of the disordered dances just as vivaciously as after definitely oriented waggling runs.

In the level meadowland near Aich we set up a tiltable observation hive in the little hard-fiberboard hut (Fig. 7) and on 30 August 1962 established a feeding station 250 m south of the hive. During the journey unscented sugar water of low concentration was offered as food, and no arousal occurred.

On 31 August, a warm, sunny day, the hive inside the fiberboard shelter was laid horizontal. From 9:57 A.M. to 12:27 P.M. the feeding place was perfumed with oil of anise in a screened scent dispenser

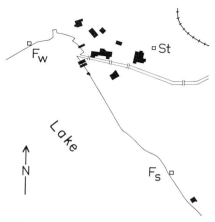

FIGURE 135. Sketch of the experimental layout at Brunnwinkl on 10–11 September 1960: *St*, observation hive; F_w, feeding station 200 m west of the hive; F_s, feeding station at 200 m 10° east of south.

---

[44]A third possibility, that they fly to the strange feeding place indicated to them by the dancer, is never realized at once. The foraging bee remains loyal to her accustomed feeding place—for how long is an individual matter and varies greatly according to external conditions also. When the interruption at the accustomed feeding location lasts for hours and days and lively dances reveal a profitable source of food elsewhere, part of the foragers gradually are converted.

(Fig. 21), and the ten numbered bees received 2M sucrose there. Before the experiment began scent plates (wooden stands, vessels with oil of anise, glass covers pierced with 8-mm holes, as in Fig. 85) were set out to the N, E, S, and W (most of them displaced slightly to one side as dictated by conditions in the field; see Fig. 136); those to the N, E, and W were put at 100, 250, and 400 m from the hive, those to the south at 100, 200, 300, and 400 m (because the feeding station was at 250 m).

The experiment falls into two parts. During the first 1.5 hr (9:57–11:27 A.M.) the hive remained horizontal; its wooden cover was usually closed, but on several occasions it was opened momentarily in order to observe the dances. In the diffuse light inside the fiberboard shelter they were disoriented and pointed irregularly first in one direction and then in another. They must also have been disoriented with the cover closed, in the dark. In Fig. 136 the position of the scent plates is given and every visit by an unmarked newcomer during the 1.5 hr of experiment is depicted by a dot. One sees that the newcomers hunt about in all directions after the disoriented dances, while the number of bees

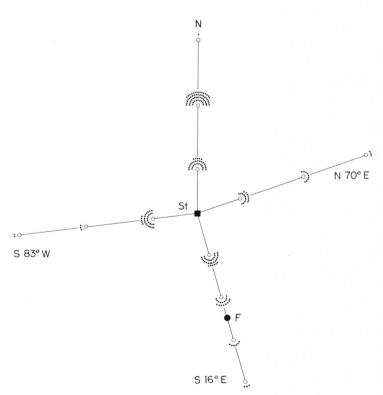

FIGURE 136. Experiment at Aich on 31 August 1962: *St*, observation hive; *F*, feeding station. To the south, scent plates, o, were located 100 m, 200 m (feeding station at 250 m), 300 m, and 400 m from the hive; in the other directions 100 m, 250 m, and 400 m from the hive. While the darkened hive lay horizontal for 1.5 hr (disoriented dances) the newcomers went hunting in all directions. The number of dots beside each scent plate shows the number of visits.

declines rapidly with increasing distance[45] (see also Table 26). Thus they do not know what direction to hunt in. But even in the disoriented dances there is an indication of the *distance* of the goal. This is evident in a comparison with their behavior following round dances (see Fig. 134).

At 11:27 A.M. all that happens is that the horizontal hive is set upright, so that the comb surfaces have their normal vertical position and the dancers, oriented according to gravity, are able to designate the direction. Now the second portion of the experiment begins (11:27 A.M.–12:27 P.M.). Of course there are at first searching newcomers still under way that had been aroused while the disoriented dances were going on. These do not have to be taken into account for more than half an hour. In order to exclude this time of transition, in Fig. 137 (for comparison with Fig. 136) only the visits during the second half hour (11:57 A.M.–12:27 P.M.) are entered. The success of the indication of distance is striking. In Table 26 there have been noted also the visits during the initial half hour after the hive was set upright. Closer study of the protocols shows that after as little as 10 min the stream of newcomers turns quite predominantly in the direction of the feeding place and dries up in other directions.

TABLE 26. Number of flights to the scent plates; experiment of 31 August 1962 (cf. Figs. 136 and 137).

Distance of scent plates (m)	North			20° N of E			16° E of S			7° S of W		
	a	b	c	a	b	c	a	b	c	a	b	c
100	30	5	1	17	3	1	31	34	35	30	8	1
200	—	—	—	—	—	—	17	36	40	—	—	—
250	57	14	0	9	0	0	Feeding station			3	2	0
300	—	—	—	—	—	—	6	62	68	—	—	—
400	1	0	0	3	5	1	3	22	31	2	1	1

[a] Combs horizontal, dances disoriented (1.5 hr).
[b] During the first 0.5 hr after the hive was placed upright, oriented dances.
[c] During the second 0.5 hr after the hive was placed upright, oriented dances.

As early as 1947 I carried out several experiments of this kind, though with many fewer observers and scent cards. The basic result was already clear at that time: undirected arousal with disoriented dances (the hive dark, horizontal), directed arousal with oriented dances (the hive dark, vertical).

The same success as comes from setting the hive upright is obtained with it horizontal, provided the wooden cover is removed so that the dancers can see blue sky and read off the position of the sun from the

---

[45] The preference for the northerly direction is to be explained, in agreement with much experience, as due to the wind's being from the north at the time of the experiment. The reduction in visits to the northern 100-m scent plate in comparison with the 250-m plate is accounted for with no difficulty by the presence of an apple orchard that forced the bees to fly aloft between the hive and the 100-m station.

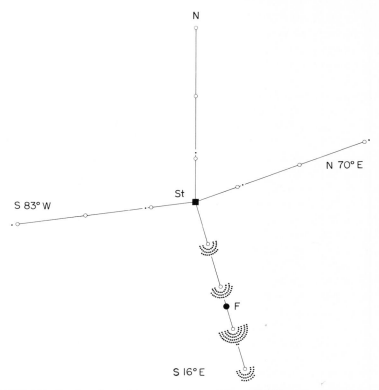

FIGURE 137. Continuation of the experiment of FIG. 136. When the observation hive is set upright, the bees are able to orient according to gravity in the dark hive and perform directed dances. Within about 15 min the stream of newcomers turns southward into the direction toward the feeding station. The figure shows, after 0.5 hr transition, the visits made during the second 0.5 hr after the hive was placed upright.

polarized light. Then the horizontal dances are oriented correctly at once and the newcomers fly out to the goal as surely as when the combs are vertical (experiment of 25 August 1947).

18. How Precisely is the Indication of Direction Followed by the Newcomers? Experiments in a Fan-shaped Pattern

The experiments just discussed show that the indications of direction by the dancers are followed by the hivemates informed. The precision with which this occurs can be clarified by experiments in a fan-shaped pattern. In their arrangement these resemble the stepwise experiments pp. 84f), except that the scent cards are all set out at one distance in a fanlike distribution. Thus one defines the cone of scatter within which the newcomers swarming out search.

In the course of the years we carried out 12 experiments with the fan-shaped pattern. From the first test (Brunnwinkl, 27 September 1947; v. Frisch 1948:18, Fig. 12) it could already be gathered that the majority of the specimens aroused moved in a direction that deviated by no more than about 15° to the left or right of the goal. But the mountainous terrain was unfavorable.

Consequently two further experiments (18 and 19 July 1948) were made on the flat meadowlands at Aich. It was demonstrated clearly that after the feeding place was shifted 15° to the left—with conditions otherwise unchanged—the stream of newcomers changed its direction correspondingly. At the same time it became evident in these preliminary trials that the wind is an unwelcome source of distortion. Even a gentle breeze coming obliquely against the hunting newcomers may bring the scent from an odor card at the side even to those bees that are maintaining the indicated direction quite precisely, and may lead them off in that direction (Fig. 138). This has been confirmed over and over again, and can obscure the hoped-for clear experimental result, as the following examples will show. Even under natural conditions something of the sort may happen, but then it does no harm to the bees. For when the specific floral odor for which they are looking is brought to them from the side, they find there no phantoms deceptively set out, but instead food-giving flowers of the same kind as those from which the dancer had harvested.

The three experiments so far mentioned served first of all for working out a useful technique. What else they led to is shown by the results presented in Figs. 139–144. First I shall discuss three experiments in the completely level meadows and marsh at Erding near Munich.

An observation hive (*St* in Fig. 139) was set up on 29 August 1955 beside a shed, and a feeding station *F* was established 250 m to the southwest; in doing this—just as in the stepwise experiments—during the journey and subsequently at the goal feeding was meager and unscented up until the beginning of the experiment, in order to avoid a premature arousal. At distances of 200 m from the hive and at angular separations of 15° seven scent plates were placed on the ground (wooden stand, a glass dish with oil of lavender as in Fig. 85, the glass cover of the dish pierced with an 8-mm hole to permit escape of the scent).

Starting at 12:40 P.M., 1M sugar solution was offered at the feeding station (50 m farther from the hive than the scent plate in the same direction) and at the same time the watch glass beneath the food dish was supplied with oil of lavender. Twelve numbered bees were collecting. Thus with their dances they sent the newcomers out searching for lavender perfume. The number of visits of unmarked newcomers to each of the scent plates in course of the succeeding hour can be seen in Fig. 139*b*. There was a strong, gusty west wind. A deflection toward the direction of the wind has been noticed principally with weak, even airflow and less when the wind was turbulent.

In this experiment the scent plates were separated from one another by only about 50 m. Hence the searching bees could stray rather easily into the field of scent given off by a neighboring plate. Therefore we immediately added a second experiment with a new group of bees, and established the feeding station—this time to the east—at a distance of 600 m. The scent plates, again at angular separations of 15°, were 550 m from the hive and each 140 m from the next. Preparations went so well on the afternoon of 29 August that at 10:55 A.M. on the 30th it was possible to offer 1M sugar solution in the presence of lavender

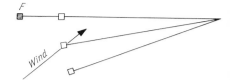

FIGURE 138. Portion of a fan-shaped experiment: *F*, feeding station; □, scent cards. Searching newcomers flying out in the indicated direction may be deflected by an air current toward the scent cards at the side.

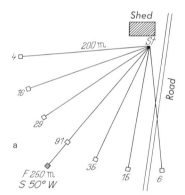

FIGURE 139. Fan-shaped experiment near Erding on 29 August 1955. (*a*) *St*, observation hive; *F*, feeding station 250 m from the hive 40° west of south; □, scent plates (with oil of lavender) 200 m from the hive at angular separations of 15°. The numbers show the number of visits during the hour of observation from 12:40 to 1:39 P.M. (*b*) Graphical representation of the results. The black column refers to the scent plate on the line to the feeding station.

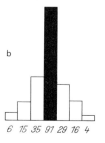

158 / The Dances of Bees

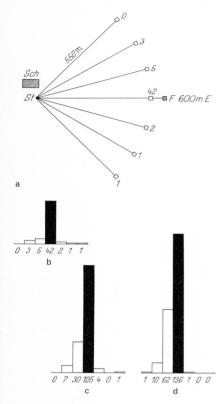

FIGURE 140. Fan-shaped experiment near Erding on 30 August 1955, 10:55 A.M.–1:25 P.M. (a) *St*, observation hive; *Sch*, shed; *F*, feeding station 600 m east of the hive; , scent plates (oil of lavender) 550 m from the hive at angular separations of 15°. The numbers show the number of visits during the first 50 min after the start of the experiment. (*b–d*) Graphical representation of the visits during the first, second, and third 50 min of the experiment: black columns, scent plates on the line to the feeding station.

scent. On account of the great distance we let 20 numbered bees do the collecting and also added some oil of lavender to the sugar water. Figure 140 shows the distribution of visits during the first 50 min of the experiment. As was to be expected, the increased distances between the scent plates improved the appearance of the numerical results. The slight preference for the first scent plate to the north of the direction to the goal is ascribed to the weak, constant ENE wind, which was interrupted by rather short periods of calm. This time we continued feeding and observations for 2.5 hr. In the graphic representation of the results in Fig. 140 the numbers of visits are shown separately for the first, second, and third 50 min. During the entire time the newcomers hunted persistently in the indicated direction. Even their "disappointment" at the empty scent plates caused scarcely any increase in the scatter.

We were curious as to how the result might look at even greater distances. On this account in September 1956 we again brought an observation hive to the same place. Poor weather delayed the experiment. I was able to be present only for the preparations; the actual conduct of the trial was taken over by M. Lindauer. On 16 September 1956 the final distance of 1250 m from the hive was attained. The scent plates were 1200 m from the hive at angular separations of 10°, each 220 m from the next. Unfortunately it stayed cold, and owing to a strong, rather constant west wind (estimated at 4–6 m/sec) only three bees frequented the feeding station regularly and only a few newcomers came flying out. They kept close to the ground, as is observed again and again when the wind is so strong. From 11:55 A.M. on concentrated sugar solution was offered in the presence of orange-blossom fragrance. The meager result may be seen from Fig. 141. The relatively strong visitation of the scent plates to the left of the goal probably is to be ascribed to the westerly wind, and in connection with this the smaller degree of angular separation (only 10°) should also be considered. The number of flights is small, but the performance of these few bees is noteworthy. With a sidewind that tended to sweep them from their course at a velocity of about 5 m/sec, most of them kept with a precision of about 10° to the indicated direction, over a distance of 1200 m. This presupposes that they swiftly recognized and compensated for the lateral drift—a capacity with which we shall be occupied again.

For the six remaining experiments the presentation of the results and a few short indications will suffice. Two of them[46] precede the fan-shaped experiments at Erding just discussed.

On 25 September 1948 a group of numbered bees at Aich was fed between 11:20 A.M. and 12:20 P.M., in the presence of lavender perfume, 250 m from the hive. Figure 142 shows the setup and the visits to the scent plates during the first half hour; Fig. 142*b* gives a graphical representation of the results for the period in question and Fig. 142*c* shows the visits during the next half hour. Gentle wind from the W and WSW during the entire experiment caused an evident deflection to the

[46] At that time the equipment described on p. 86 (Figs. 83 and 84) served as scent cards, but with only three scent vials. At the feeding station a few drops of the scent were placed on the filter paper spread beneath the feeding dish.

W. In the second half hour the number of flights was greater, but their distribution was not significantly different.

On 27 September 1949 we repeated the experiment in the broad, level valley meadow of the Zeppezau (district of St. Gilgen). There the bees found surprisingly good natural sources of food for the time of year, which had an unfavorable influence on the recruitment of newcomers for our feeding station. In the 1.5 hr of experiment (9:10–10:40 A.M.), with fine, almost calm weather, the flights amounted to those recorded in Fig. 143.

The last four fan-shaped experiments took place in Aich on 4 and 5 September 1959. Their technical execution corresponded to that of the tests at Erding. The feeding station was 270 m distant from the hive, the scent plates were 20 m nearer the hive, their angular separation was only 8°, and the distance of the plates from one another was only 34 m.

In the first experiment on the morning of the 4th (Fig. 144a, a') we made the mistake of adding the scent in too great concentration at the feeding station. Since, moreover, the nearest scent plate was only 20 m away (it was 50 m in all previous experiments), apparently searching bees from the circle of influence of this plate were deflected in rather large numbers to the feeding station itself, to which an unusual number of newcomers came flying and had to be killed. At all events that is the most likely explanation for the fact that this time the scent plate in line with the goal was visited less often than the two neighboring plates. Repetition of the experiment in the afternoon (Fig. 144b, b'), with a reduction in the intensity of the odor at the feeding station,[47] gave a better looking result, as expected. The precise maintenance of the direction to the goal on this occasion is all the more noteworthy because during the experiment there prevailed a strong NNE wind with gusts estimated at more than 6 m/sec.

Immediately thereafter we shifted the feeding station southward, with scentless feeding, and during the early and late afternoon on 5 September 1959 there followed the last two fan-shaped experiments, with the identical arrangements used previously. The first test again produced outstandingly well-directed visits (Fig. 144c, c'). Moderate wind from the NE may have been the cause of the six flights to the scent plate 8° east of the midline. Shortly before the second experiment the wind shifted to the south. It remained weak and thus bore the scent toward the bees flying out, but according to the protocol its direction shifted repeatedly between SE and SW in the region of the three scent plates situated to the west (Fig. 144d, d').

In the fan-shaped experiments at Aich, in spite of the smaller angular separation, the newcomers flew to the laterally situated scent plates less often than in most of the earlier instances. That may have been due to our using selected dancers and excluding, before the experiment began, any that were sluggish about dancing or that deviated from the norm in their indications of direction (see p. 149).

---

[47] Of course all apparatus concerned with the scent was cleaned carefully or replaced in the meantime. Feeding between the two experiments was with weak sugar solution and without a scent.

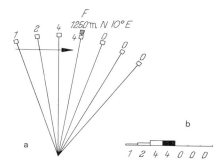

FIGURE 141. Fan-shaped experiment near Erding on 16 September 1956. (a) F, feeding station 1250 m from the hive 10° east of north; □, scent plates (oil of orange blossom) 1200 m from the hive at angular separations of 10°, with the number of visits in 2 hr, from 12:05 to 2:05 P.M.; the arrow shows the wind direction. (b) Graphical representation of the results.

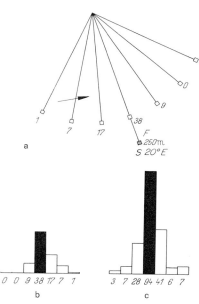

FIGURE 142. Fan-shaped experiment at Aich on 25 September 1948. (a) F, feeding station 250 m from the hive 20° east of south; □, scent plates (oil of lavender) 200 m from the hive at angular separations of 15°, with the number of visits between 11:20 and 11:50 A.M.; the arrow shows the wind direction. (b, c) Graphical representation of the number of visits, from 11:20 to 11:50 A.M. and from 11:50 A.M. to 12:20 P.M., respectively.

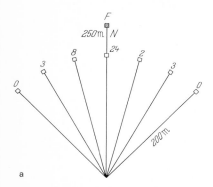

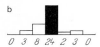

FIGURE 143. Fan-shaped experiment in the Zeppezau on 27 September 1949. (*a*) *F*, feeding station 250 m north of the hive; □, scent plates (oil of lavender) 200 m from the hive at angular separations of 15°, with the number of visits from 9:10 to 10:40 A.M.; almost no wind. (*b*) Graphical representation of the results.

160 / The Dances of Bees

It is worth while to devote a little thought to the chain of accomplishments that take place between the flight of the harvesting bee and the directed flight of the newcomer. In her dance the harvester transposes the solar angle that is proper to the goal into the gravitational angle—even when with a sidewind she has to set herself obliquely against the lateral drift and hence sees the sun at a different angle (see p. 194). The dance followers understand the angle correctly. In doing so their eyes are no help to them. In the dark hive the eyes are functionless, and the senses of smell and touch take over the transmission of information. The dance followers grasp the proper angle with astounding precision, although most of them stand oblique to the dancer—especially when she has a large following (Fig. 52). Beyond this they take note of the floral scent that clings to the dancer and of her announcement of the distance by means of the dance tempo; in doing the latter they observe several dances and average the result. If then they fly out they go in search of this scent at the corresponding distance, transposing the gravitational angle back into the visual angle and scarcely letting themselves be borne aside from the direction striven for even by a violent crosswind. As for anyone to whom the feeling of reverence for nature's creations is foreign—it might well dawn upon him here.

19. Dances when the Sun is in the Zenith

The heavenly compass loses its efficacy when the sun is in the zenith. Then a direction is neither indicated by nor to be found from its position. Bees that live between the Tropics of Cancer and Capricorn must fall into this difficulty at noontime twice a year. Their behavior under such circumstances should provide information as to whether the position of the sun is solely determinative of their indications as to direction.

Lindauer (1957) used an expedition to Ceylon for clarification of this problem. At his location in Paradeniya the sun passed through the zenith at 12:06 P.M. on 9 April. As long as 3 weeks beforehand he himself was no longer able to make out the southerly direction at noon from the position of the sun—with the sun still 8° from the zenith. The bees' eyes proved to be superior.[48] Even at noon they still pointed with well-oriented dances toward an artificial feeding station after the sun had already approached within 3° of the zenith. But on 2 April, with the sun 2.5° from the zenith, they first gave evidence of their "embarrassment"—at all events in an unexpected manner, which almost prevented a clear decision: from 11:45 A.M. to 12:20 P.M. they took a midday siesta and stayed at home. Lindauer found a way out. During the morning he gave the bees but little food, with numerous interruptions. Then toward noon he served them a full meal, whereupon they became so eager that they kept on harvesting. Their eagerness to dance was less than usual, but nevertheless during those days altogether 87 tail-wagging dances were observed; these were disoriented.

---

[48] For these experiments Lindauer used the Indian bee (*Apis indica* F.), which is so like our honeybee that originally it was considered merely a subspecies of it.

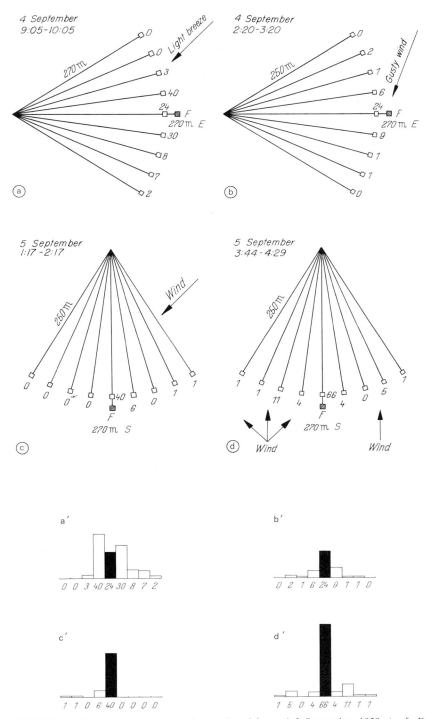

FIGURE 144. Four fan-shaped experiments in Aich on 4–5 September 1959. (*a–d*) *F*, feeding station 270 m from the hive; □, scent plates (oil of orange blossom) 250 m from the hive at angular separations of 8°, with the number of visits. (*a′–d′*) Graphical representation of the results.

In addition there were a few dances with a hint of orientation, but the waggling runs pointed so inexactly that they could not be measured (Lindauer *in lit.*, on the basis of a re-examination of his data). Before and after the time indicated the dances were well oriented without exception.

Particularly significant is an experiment on 10 April, the first day after the sun passed the zenith, with the hive laid horizontal. The bees had a free view of the sky. Around noon, from 12:02 to 12:16 P.M. (from 5 min before to 9 min after the maximum elevation of the sun), Lindauer saw 14 dances, all of which were disoriented. The sun was 1°–2° from the zenith.

Thus it is shown convincingly that in indicating direction the bees are oriented solely by the position of the sun; and further that a solar zenith angle of 2°–3° suffices for their recognition of the compass direction.[49] This is the quantitative evidence for the statement (p. 132) that the compound eye is a superb protractor. The value fits well with the findings of del Portillo (1936) and Baumgärtner (1928) that in the upper portion of the honeybee's eye adjacent ommatidia diverge by about 2° or 3°. Thus a lateral displacement of the sun's image by the width of one ommatidium from the single facet that looks directly upward is enough to enable the bee to recognize the direction of the sun's deviation from the zenith. From this it may be deduced further that the flying bee takes extremely accurate note of the position of her body relative to the horizontal.

Lindauer's experiments in Ceylon with *Apis indica* were repeated by D. A. T. New (1961) in Jamaica with *A. mellifera*. At the time of the tests there were only meager natural sources of food, hence the bees were eager for sugar water, and thus there occurred spontaneously what Lindauer had to achieve by means of an artifice: the marked groups of bees frequented the feeding station without interruption even when the sun was near the zenith. New did not watch the dances; he paid attention to the newcomers that were aroused by the dances. Toward noon, when the sun was only 1°–3° from the zenith, they came in definitely smaller numbers. But since any came at all he decided that there was some mechanism for transmitting information independent of the sun's position. But this conclusion is not convincing because there might have been bees aroused earlier that were late in finding the feeding station. New himself discusses this possibility and discards it because within even a few minutes almost no more newcomers came after he had captured and removed the harvesting bees at the feeding station; hence those arriving during the critical minutes must have been aroused just previously, when the sun was at its height. But he overlooked the fact that in catching the foragers he also diminished the attractive odor from their active scent organs, through which the newcomers searching about were led very effectively from the surrounding area to the feeding station.

[49] A little amphipod crustacean, *Talorchestia mortensii* Weber, is not able to orient according to the sun's position unless it is at least 5°–9° below the zenith (Ercolini 1964).

In a continuation of the experiments (New and New 1962) the dances were watched on the vertical comb. Many bees stopped dancing or danced disorientedly, as Lindauer had described, as soon as the sun approached the zenith. But even under these conditions a few performed well-oriented dances. In explanation it is assumed that these bees, by whom the position of the sun was no longer to be defined, continued to calculate it according to their "internal clock." That this assumption is correct seems highly probable to me from other observations of New and New. One must realize that when the arc of the sun is to the south the solar angle shifts clockwise in the course of the day, but counterclockwise when the arc is to the north. This shift, which occurs when the sun traverses the zenith, was not made in precise synchrony by the bees. Many were late in changing over to the different basis of indicating direction; others made the shift several days too soon. Thus in doing this they were not guided by their immediate visual perceptions, but instead calculated the course of the sun, and this apparently not according to their own experience but on an inherited basis—otherwise it would not have been possible for many of them to anticipate the forthcoming change.

In Lindauer's observations on the horizontal dance floor no skill in calculation was able to bridge over the critical minutes, for the dancers could not refer to gravity for determining their position. Under these circumstances the dances were disoriented without exception.

20. No Indication of Direction Upward or Downward

Food-giving flowers may be found on the ground or high in the crown of a tree; in the mountains the route to them may lead uphill or downhill; it would not have been surprising if the "language of the bees" disclosed an expression for the height of the goal also. Yet no such expression was demonstrable.

Negative findings are less forcefully convincing than a positive result. So we had to vary our experiments over and over again for a period of 7 years before we were sure of the situation. In order to attain suitable differences in elevation we used a balcony on our country house, the roof of the institute in Graz, gorges and cliffs, the bridge at Echelsbach, and a radio tower. Anyone interested in the difficulties of providing evidence will find the details in v. Frisch, Heran, and Lindauer (1953). Here it suffices to report the principal experiments.

In August 1951 we took an observation hive up the Schafberg and set it at the foot of a northward-facing cliff (at St_1, Fig. 145). On the 24th a feeding station was established (without a scent) at F_1, 85 m distant from the hive. Our plan was to establish a second feeding station at F_2, directly above F_1, and by alternate feeding at F_1 and F_2 to test whether the newcomers would search preferentially below or above according to the source of the dancers. As in the stepwise and fan-shaped experiments, appropriately placed scent cards were to give information about the distribution of the newcomers as soon as vigorous

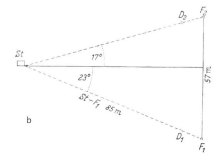

FIGURE 145. (a) Rock wall running from east to west on the Schafberg: St_1, observation hive on 25 August 1951; F_1 and F_2, locations of the feeding stations; St_2, observation hive in the experiments of 15–19 August 1951; f, location of the feeding station. (b) Experimental arrangement on 25 August; D_1 and D_2, scent cards.

recruiting on behalf of one of the two stations had been induced after addition of a fragrance and with succulent feeding.

In order to set up the second, elevated feeding station we planned to lead part of the harvesting bees aloft from F_1 by hauling up a hanging feeding table. But as we climbed the ridge, searching newcomers were already present in numbers. They had been sent up there by the bees collecting below. That of itself argues against there being a signal for elevation. Viewed from the hive the difference in altitude corresponded to a 40° angle. From the fan-shaped experiments we know that customarily no newcomers appear at scent cards that deviate from the goal by 40° horizontally. The 40° difference in elevation seemed to be meaningless. This first impression was strengthened by the following observations. When food was offered below and there was an interruption in feeding above, the dances of the bees collecting below kept on arousing also the other foraging group, who continued obstinately to examine over and over the empty dish above—and conversely the group below was also aroused by the bees collecting above. But never was a bee that sought food in vain at the accustomed place converted by the dances of the other group to their collecting spot—as happens often when one of two feeding places at different azimuths is profitable and the other is not. Apparently any difference in the manner of dancing was as little evident to the bees as it was to us, whether the harvesters came from above or from below. In agreement with this is the fact that newcomers came swarming above as well as below whenever food was offered at either of the two places. For a quantitative comparison the site was not well chosen, because the collecting place on the ridge was freely exposed to the sun while the lower one lay constantly in the shadow of the wall.

A repetition of the experiment at the Echelsbach bridge in May 1952 afforded better conditions. About 20 km from Oberammergau the highway from Augsburg crosses the deep-cut Ammer valley. The roadway is 76 m above the valley floor. For the observation hive we chose a site halfway up the slope (St, Fig. 146). Thence we journeyed with a marked group of bees up the slope and across the bridge to its center, where the feeding station F_2 was established on the railing. To the other feeding place F_1, exactly beneath F_2 on the valley floor on the banks of the Ammer, we had journeyed downhill with a second group of bees.

On 27 and 28 May 1952 we carried out four experiments. On two occasions the numbered group above on the bridge was fed with concentrated sugar solution in the presence of peppermint fragrance, as was done twice with the other group on the valley floor; and each time, during an hour, we recorded the visits to two peppermint-scented cards that had been set out on the line of flight 10 m nearer the hive (as in Figs. 83 and 84, without food). The scent card D_2 was opposite the feeding station on the other railing of the bridge (Fig. 146b), and D_1 was on the valley floor an equal distance from the feeding station. The number of visits of newcomers to the two scent cards was as follows:

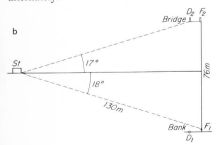

FIGURE 146. (a) The Echelsbach Bridge: St, location of the observation hive; F_1 and F_2, locations of the feeding stations. (b) Experimental arrangement on 27–28 May 1952: D_1 and D_2, the two scent cards, below on the river bank and above on the bridge railing, respectively; F_1 and F_2, the feeding stations, supplied with sugar water alternately.

May		D_1	D_2
27	2:00–3:00 P.M. (feeding above)	156	0
27	3:30–4:30 P.M. (feeding below)	108	0
28	11:30 A.M.–12:30 P.M. (feeding above)	197	0
28	1:00–2:00 P.M. (feeding below)	120	0

Hence it made no difference whether the dancers were striving to arouse interest in the upper or the lower place. In both instances the newcomers went hunting on the valley floor; during those 4 hr not a single one came to the scent card on the bridge.[50] The foragers from the bridge railing evidently gave no directions to fly upward. This was expressed also in the fact that—as previously in the experiment on the Schafberg—the dancing group always provoked the other crowd too, whose feeding dish was empty, to steadfast searching. That here the higher scent card remained unheeded, whereas it was visited assiduously on the Schafberg, is easily understood. There the newcomers following the indicated direction kept in touch with the terrain even when going upward along the face of the cliff, whereas for those here the bridge was freely suspended in the air, where bees normally have nothing to look for. Both on the Schafberg and at the Echelsbach bridge it was consistently demonstrated that the direction of the goal is indicated only by the azimuth and not by the elevation.

These negative results raise the question whether an indication of elevation may not be perceptible in extreme situations nevertheless. If the route from the hive to the feeding place runs vertically upward or downward, the goal does not lie in any compass direction. In this situation a signal for difference in elevation should emerge clearly, if there is such a thing.

Initially we carried out experiments of this sort on the Schafberg, on the same cliff with which we became acquainted in Fig. 145, in the oblique experiment. At that time, the observation hive was set up at St_2 and a group of numbered bees was gradually led from here up to f (Figs. 145 and 147) on a hanging feeding table. The feeding station was precisely above the hive entrance (difference in elevation 53 m). To our surprise the bees signaled a direction in their dances. It pointed into the cliff. As the sun's position progressed, the angle of dancing shifted so that the direction continued to point into the wall. The observation hive was in the shadow of the cliff. When we tilted the hive, the dances on the now horizontal comb surface were visibly and convincingly pointed directly into the rock wall (the bees were orienting themselves by the polarized light of the blue sky; see pp. 380ff). The dances differed in no way from the customary dance form. Nothing hinted at any difference in elevation. And yet the behavior was meaningful. The newcomers sent forth ran at once into the wall and were

---

[50]That they were able to find their way up there is demonstrated by isolated flights to the feeding place itself, attraction toward which was exerted not only by the oil of peppermint but also by the scent organ of the foraging bees. During the 2 hr of observation while there was feeding on the bridge two newcomers came to the feeding station there, and a few others during feeding outside of the time of experimentation.

FIGURE 147. Experiments on 15–19 August 1951. The hive is below at the foot of the cliff, St_2 in Fig. 145. The feeding table has just been raised by means of the hoisting boom (to f in Fig. 145). The second boom with the bucket carried notes between upper and lower observers.

FIGURE 148. Experiment on 7 September 1951: rock wall running from NE to SW; St, observation hive; F, feeding station.

able to pursue the indicated direction farther only by flying to the top of the cliff, where they arrived in large numbers at the feeding station.

At a later time in the reciprocal experiment we placed the observation hive above on the ridge, and by roping down a hanging feeding table we set up the feeding station directly beneath the hive entrance, 53 m below, at the base of the somewhat overhanging wall. This time the dancers pointed equally decisively away from the cliff. Again they were successful with this behavior. They sent the newcomers out over the abyss, and since there is nothing for bees to fetch out in the free air they flew downward and in large numbers swiftly reached the feeding station.

In order to learn whether the dances actually were directed in relation to the wall or in relation to other factors unknown to us, we repeated the experiments on a cliff with a different course (one that instead of running from east to west ran from NE to SW). Again the dances indicated the direction toward the wall when the feeding station was above, and away from the wall when it was down below the hive (Fig. 148).

Hence there was no indication of elevation, but a directing signal was given in reference to the horizontal plane, and under the conditions that prevailed this signal led to the goal. The behavior probably can be explained by the fact that in flying upward the bees' heads were directed toward the wall while they faced away from the wall on the downward flight. It was hard to see the position of the body during flights next to the wall, but in a series of cases support for this notion was found. In dancing the bees then indicated the compass direction assumed by the body during flight to the feeding place.

In order to deprive the bees of the rock face as a point of reference, we put the observation hive on the concrete foundation of a radio tower and on 1 May 1952, with the help of a hanging feeding table, brought a marked group up to the platform 50 m above the hive entrance (Fig. 149). Now when the bees collecting above returned home they performed round dances exclusively. Thus it was confirmed that the indication of direction in the earlier experiments had depended on the position taken relative to the rock wall. The open latticework of the radio tower (Fig. 149*b*) afforded no corresponding frame of reference. The collecting bees were seen now inside the tower, now outside, making their way aloft without assuming any fixed stance relative to the framework.

In the radio tower we had set up a colony of the Italian race of bees. On a horizontal flight course Italian bees point the direction to the goal at a distance of only about 10 m (see p. 294). But after flying 50 m upward they performed round dances. We had finally hit upon an experimental arrangement in which they could not indicate any direction to their comrades. They did not show them any compass

FIGURE 149. (*a*) *St*, location of the observation hive inside the metal framework of a radio tower; the feeding station is 50 m above it on the platform. (*b*) View upward inside the radio tower.

direction, but they also did not tell them that they should fly upward. That fact the newcomers revealed to us unambiguously through their behavior. They flew out and went hunting about—on the ground in the meadows around the radio tower. The feeding station was scented with oil of bergamot. Cards scented with the same oil were set out in the field around the tower, and numerous newcomers flew to them. The round dances were a signal to examine the ground around the hive. But on that fine, warm day in May not a single newcomer found the way up to the feeding station on the platform of the radio tower, although the ten numbered bees that we had led up there on the preceding afternoon danced most vigorously for 4 hr. They sent their hivemates astray—their ability to communicate broke down when faced with the unaccustomed task.

In order to carry out the reciprocal experiment, we went again to the Echelsbach bridge, on 16 June 1952, and set up the Italian colony on the bridge railing. A feeding table was roped down and a feeding station established 76 m below directly beneath the hive entrance (Fig. 150). In this instance too only round dances were to be seen. They induced the hivemates to go hunting round about. But the situation was now different from the experiment at the radio tower. The newcomers swarming out descended and many found the feeding dish. That they were making an undirected search was shown in a test with three scent cards that—like the feeding place—were supplied with oil of anise. One was below on the river bank, 10 m to one side of the feeding station and thus 76 m beneath the bridge. During the 1-hr period of observation it was visited by 180 newcomers. The other two scent cards were on the same bridge railing on which the observation hive stood, 76 m east and west of it. During the hour of observation they received 42 and 39 visits, respectively. The preference for the lower scent card is comprehensible. Ground covered with vegetation is more attractive to bees in search of food than is the iron bridge railing.

In a control experiment we convinced ourselves that the same bees that, when fed 76 m vertically beneath the hive, had by means of round dances sent their comrades forth undirected, when fed on the level 76 m from the hive performed oriented dances and dispatched the newcomers directly to the goal.

Thus we sought in vain in the "language of the bees" an expression for the direction upward or downward. This will hardly cause surprise. Flowering trees develop the same flowers on their branches near the ground as in their crowns, and the gradual spread of the foraging bees over the entire tree results automatically once the tree is found. Where the structure of the terrain is such that flight leads upward or downward, indication of the compass direction to the goal is ample, because the newcomers hunt near the ground and are guided correspondingly by its course. Thus it comes about that in our experiments the ability of the bees to communicate failed only where it was given an altogether unbiological task.

FIGURE 150. Experiment on 16–17 June 1952. The observation hive is on the far side of the upper railing on the Echelsbach Bridge (at F_2 in Fig. 146), directly above the feeding station F on the left bank of the river.

21. The Significance of the Outbound and Homebound Flights for the Indication of Direction

The bees' indication of direction refers to the outbound flight to the feeding place. The dances on a horizontal surface illustrate that convincingly. The waggling run points in the direction to the goal. But it does not follow from this that the solar angle perceived on the return flight is without significance for the indication of direction.

The return flight might be given less weight by the bees, so that the outward flight would be the deciding factor for the direction of dancing. Another possibility is more likely, namely that by means of a central nervous switch the solar angle perceived during the return flight is shifted by 180°, into the sense of the outward flight, for the purpose of indicating direction, just as the solar angle perceived on the outward flight can take over guidance of the return flight only by such a switching maneuver (see p. 118 and Fig. 110).

As regards the indication of *distance,* we came to realize that the dancer takes account of the return flight also, but gives the outward flight greater weight (p. 119). The question could be solved by considering the part played by wind direction. But with respect to the indication of *direction* the wind is of no help, because tailwind and headwind on the flights out and back do not have opposite effects on the direction of flight. Only the method of experimental displacement is left. If one carries the dish with the drinking bees to one side, so that they have to maintain a different solar angle on the flight back to the hive, then it should become clear whether the return stretch too is taken into account in indicating direction.

In these experiments F. Otto (1959) ran into the same difficulty as with displacement of the bees to a different distance (see p. 118). When they took wing the displaced bees paid no attention to the change in place; they made a simple reversal of the outward flight, which under normal conditions would have taken them back to the hive but which, after the displacement, led them astray (Fig. 151), so that they got home via detours and with loss of time. Only after repeated hourlong displacements from the site to which they flew (*H*, Fig. 151) to the site of return takeoff (*R*) did they learn to strike out directly from there on the route to the home hive. They had to be trained to the different flight path. Under such circumstances, Otto found that in their dance the bees indicated the bisector of the angle between the place to which they flew out and the place from which they flew back.

The performance of the experiments may be clarified by an example. The observation hive (*St*, Fig. 152) stood in a level meadow. Marked bees flew out to *H*, were displaced to *R* while they were drinking, and had learned to fly directly back to the hive from there. The matter of "misdirection," which varies with the direction of flight and with the time of day (pp. 205f), had not yet been clarified at that time. In order to exclude it as a source of error, adjacent to the place flown out to a second feeding station *K* was established for control bees that were not displaced. Their dances were watched at the same time as those of the

FIGURE 151. Expected and actual outcome of a displacement experiment. Marked bees fly from the observation hive *St* to the feeding station *H*, are displaced while drinking to the place of return *R*, and are expected to fly directly home from there (dashed line). In actuality, being oriented according to the sun, the bees performed the reverse of the outward flight (solid line). They had to be trained to the other route home.

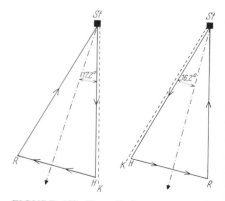

FIGURE 152. Two displacement experiments: *St*, hive; *H*, place flown to; *R*, place of return; *K*, feeding station for the (undisplaced) control bees; angle of displacement, 30°. The indication of direction by the displaced bees (—·—·—·—) corresponds approximately to the bisector of the angle between the place flown to and the place returned from. After Otto 1959.

displaced bees. In this way one learns the degree of "misdirection" prevalent at the time and can take it into account in the calculations. The dashed and dotted arrow indicates the direction in which and the extent to which the indication of direction by the displaced bees diverged from that of the control group; thus it shows the influence of the altered return flight. The difference amounted to 17.2° (6 experiments, 32 dances), or approximately half the angular displacement of 30°. In the reciprocal experiment with other, newly trained bees (Fig. 152b), the places flown to and returned from were interchanged and a control group was collecting food adjacent to the new place flown to. The indication of direction by the displaced bees differed by 16.2° from that of the control bees (Fig. 152b; 3 experiments, 16 dances). In both cases the bisector of the angle was 15°.

Since the displaced bees indicated quite precisely the bisector of the angle between the direction from hive to place flown to and from the hive to the place of return, the suspicion expressed initially, that the solar angle perceived on the return flight is transposed by 180° and is taken into account in indicating the direction in the sense of the outward flight, is confirmed. Only by making this assumption does the observed indication of direction become comprehensible.

Increasing the angle of displacement from 30° to 60°, 90°, or 120° required tedious training for the return flight, but led to no different result. The bees continued to indicate the approximate bisector of the angle.

If the site of takeoff for the return flight was moved in a direction opposite to that of the place to which the bees flew out, there was no room for compromise in the indication of direction. The dancers were required to decide in favor of the direction to the place flown out to or the place of return. This is what they did in fact, though the decision between the two possibilities depended in a peculiar way on the direction of flight and on the position of the sun.

Training the bees to return from the opposite direction was particularly tedious. In order to make it easier, the feeding stations were set up only 20 m from the hive and Italian bees were used, so that the indications of direction were clear even at this short distance. Figure 153 shows the observation hive in the countryside. Sheets of colored cardboard on the ground marked the flight course, in order to simplify for the bees the return flight in the inverted direction. When they flew out to the southern feeding station (F_S, Fig. 154), they were displaced in a large semicircle (in order not to disturb the other bees frequenting F_S) to the northern feeding station F_N, and after a corresponding amount of training flew home to the hive from there. At all times of day they signaled the direction of the place flown to.

If the bees were fed at the northern site F_N and displaced to the southern station F_S, they too signaled the direction of the place flown to if the sun was more than 30° east or west of the line of flight. But if the sun approached the segment of the return flight, the dances signaled the direction of the place of return (F_S). On the vertical comb

FIGURE 153. Displacement experiment; angle of displacement 180°. Sheets of yellow cardboard laid on the ground mark the flight line for the bees. Aerial photograph, after Otto 1959.

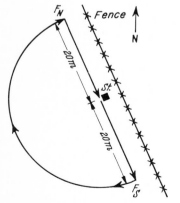

FIGURE 154. Location of the feeding station in a 180°-displacement experiment, here from the southern feeding station F_S to the northern one F_N. After Otto 1959.

they pointed upward. Both on the flight out and on the return the bees had flown away from the sun and despite this signaled the direction toward the sun, that is, the position of the feeding table from which they had returned to the hive. In this way they once again instructed us that the solar angle perceived on the return flight, transposed by 180°, is taken account of in the dance, in the direction of the outward flight.

The results may be summarized in the statement that when there is a displacement of 180° the position of the sun during the outward flight generally is determinant; only when the bee flies away from the sun on the return flight (and on the outward flight) does she signal the place returned from; thus the position of the sun during the return flight, transposed by 180°, then decides the direction signaled (see Otto 1959). Unfortunately I do not know an explanation for this odd behavior.

The clear and simple finding, that the direction of the bisector is indicated after a lateral displacement, obtains only when the clues for orientation on outward and return flights are equivalent. When one of the two courses is marked more conspicuously by landmarks than the other, its solar angle is given greater weight in formulating the indication of direction. Otto found this out by chance. In one series of experiments the observation hive was only about 2 m removed from a high iron fence, along which one of the two flight paths ran (Fig. 155). If the bees were fed at the western feeding station and displaced to the fence for their return flight, the dances signaled the feeding place in the direction of the fence, that is, the place of return (in doing which the solar angle perceived during the return flight once again was transposed 180° in the dances). But if the outward course lay along the fence and the bees flew home from the western feeding station, where their route was marked out only by yellow sheets of cardboard on the ground, the dances gave indications quite close to the place flown out to (with deviations of 4°–10° toward the place of return).

The same effect as that of the iron fence was produced by installing a set of wooden palings 1.5 m high and 5 cm wide at intervals of 20 m along the path of the return flight. If the course of the outward flight was unmarked, or indicated only by pieces of yellow cardboard on the ground, the dances pointed toward the spot from which the return flight had taken place. By means of further variants any desired intermediate in the indication of direction could be achieved. If, for instance, the outward flight led across the meadow without any striking marks of reference, whereas the return course was designated by pieces of yellow cardboard (12 × 23 cm) laid on the ground at 50-cm intervals, the dances pointed in a direction that was fairly precisely midway between the bisector and the line to the place of return.

Another possible means of giving different weights to the routes of outward flight and return is by changing the length of the course. If, for instance, the bees, after flying out a distance of 100 m, are displaced laterally (by an angle of 30°) and at the same time carried nearer the

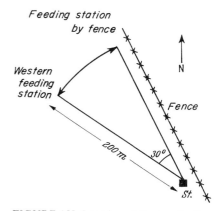

FIGURE 155. Location of the two feeding stations in the experiment by the fence. After Otto 1959.

hive so their return path is only 50 m long, then—when outward and return courses are equally marked—the bees signal a direction that lies in between the bisector and the direction on the outward flight.

Thus, by favoring one of the two flight paths by means of clues for orientation or by differential regulation of their length, one can shift at will the dancers' indication of direction from the objective of the outbound flight to the site of return. With stretches of equal value both outward and homeward flights have equal weight. But let it be remembered that the aberrant return course had to be imparted to the bees by training them. It is possible that only by this means did it acquire significance in the indication of direction. The experiments say nothing about whether the perceptions made during the return flight are used in indicating direction under normal conditions also, where the return is the simple reversal of the outward flight.[51]

Yet by means of two experimental arrangements, thought out by M. Lindauer (1963), we reached a consistent and conclusive solution of the problem.

1. The Indian bee (*Apis indica* F.) and our honeybee are so closely related that they were formerly regarded as races of the same species. They also agree in the principles of their "language," yet the Indian bee performs tail-wagging dances with a clear signal of direction when the feeding place is no more than 2 m distant. In these experiments the feeding station was at the end of a passage the roof and walls of which were of glass. The bees had first to fly 3 m southward from the hive and then an additional 3 m eastward in the passageway (Fig. 156). After they returned home they signaled southeast. If the passageway was darkened, so that while on this stretch the bees could not see the sky, then on their return home they pointed to the entrance to the passageway, to the south. If they are allowed to fly home directly from the end of the darkened passageway they have no difficulty in finding the way to the nearby hive (Fig. 156b), but in spite of the divergent return route the dances point southward, that is, in the direction of the outbound flight.

2. Other experiments were carried out with our own honeybee. By means of two tricks, it was possible with them to observe an indication of direction after lateral displacement without training them to the new return flight path. As soon as an aroused newcomer made her first appearance, she was moved in a closed box with the feeding dish to B, 150 m west of the hive (Fig. 157); from there she quickly found her way back to the home hive. In Otto's experiments the bees had to be trained to the altered return route (p. 169). These were by then experienced foragers that were already familiar with the outward and return

---

[51] That the return flight is learned independently of the outward flight is demonstrated also by an observation that at first glance seems astonishing and that reveals quite clearly how little the displaced bees take in the situation. If the return route is, for example, marked conspicuously with a fence, with their dances they send their comrades to the site of return. But scarcely has their dance ended in the hive when they themselves fly straight away to the objective of the outward flight. The independently learned return route influences the indication of direction, but the bee seems to have no knowledge of that and sticks to her accustomed outbound path.

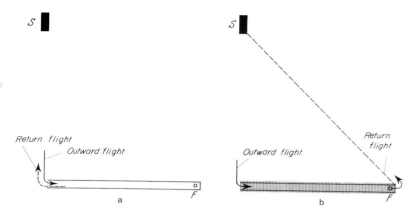

FIGURE 156. Experiments with *Apis indica*; S, hive. (a) To reach the feeding station F the bees fly southward 3 m, then 3 m eastward in a glass passageway; the return flight is over the same route. The direction they indicate is southeast. (b) Outward flight identical, but with the passageway darkened; the return flight is directly from F to S. The direction indicated is south, uninfluenced by the return flight. After Lindauer 1963.

paths. But in the present instance a newcomer was making the outward flight for the first time, never having flown the return, although she was no doubt acquainted through preliminary orientation flights with the surroundings of the hive. Such a bee will find the way home relatively easily in spite of being displaced. As a rule a newcomer dances only after several foraging flights. In order to avoid any possible training to the altered direction of the return flight, Lindauer wished to see dances indicative of direction after the very first return home. This was achieved by a further trick: care was taken that the swarm was suffering acutely from a shortage of food. Under such conditions the forager will sound the alarm on her very first return. With her dances she pointed unambiguously in the direction of the goal of the outward flight, toward A. The direction of the return flight was ignored.

With these observations our suspicion is confirmed that the bees in Otto's experiments averaged the directions of outbound and return flights only because they had to learn the return route anew. If such training for the direction of the return flight is eliminated, the direction of the outward flight alone determines the angle of dancing.

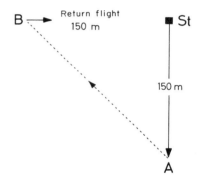

FIGURE 157. Displacement experiment: *St*, hive; *A*, feeding station 150 m south of the hive. A newcomer flying to *A* is displaced to *B*. After she returns home she signals in her dancing the direction to *A*. From experiments by Lindauer 1963:161–162.

22. Detour Experiments

(a) First Observations and Preliminary Experiments

In one experiment regarding the indication of direction, during the summer of 1945, the flight line led from the hive to the feeding station in a gently curving arc around a spur of wooded slope. The dances signaled a direction that corresponded approximately to the chord of this arc. This occasioned a repetition of the experiment with a more strongly curved flight path. We found a suitable location about 1 km from St. Gilgen, where the highway from Salzburg then ran in a sharp curve around a steep wooded slope.

174 / The Dances of Bees

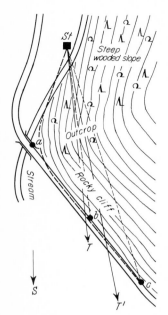

FIGURE 158. Detour experiment near St. Gilgen, July 1946: *St*, observation hive; dashed curve, flight path of the foragers frequenting feeding station *b* or *c*; *T*, *T'*, average indication of direction by bees collecting at *b* or *c*, respectively. Distance $St-a = 55$ m; $a-b = 50$ m; $b-c = 45$ m.

On 20 July 1946 an observation hive was set up (at *St*, Fig. 158) and a feeding station with marked bees was first moved to *a* and then on to *b*. On the following days as well, the foraging bees continued to fly around the mountain spur the indirect course indicated by the dashed line. Now it was unequivocally clear that their dances referred neither to the first nor to the second leg of their flight path, but to the air line between hive and feeding station, and hence to a direction they had never traversed. In 11 experiments between 20 and 25 July, 12 to 20 dances were observed on each occasion and their mean was calculated. The direction signaled corresponded exactly with the air line once; five times it deviated to the right (westward) and five times to the left (eastward). The grand average differed from the direct line by 2° to the west (arrow *T* in Fig. 158). In three experiments the feeding station was moved ahead along the second leg to *c*. The average direction indicated lay 6° west of the air line (arrow *T'* in Fig. 158). In the two cases the deviation from the direction of the first leg of the course flown was 24° or 20° and from the direction of the second leg 30° or 34°, respectively. Further details are to be found in v. Frisch (1948).[52]

It is not worth while to go further into the details of this experiment because the data are not precise. At that time it was not yet known how greatly the direction of dancing may be deflected by the view of the blue sky (see pp. 197ff). Hence this source of error was not eliminated. Also the technique of measurement was not yet fully developed.

On 28 July 1947 the bees of an observation hive in Brunnwinkl visiting their feeding place 100 m distant had to fly in an extensive arc around the curved margin of a wood. Once again the direction of the air line was signaled with all clarity.

(b) Experiments on the Schafberg

We looked around for a more favorable terrain and found an ideal place. Brunnwinkl lies at the foot of the Schafberg. From its 1780-m peak there runs toward the ESE a long rocky ridge that ends in a sudden precipice, the "Devil's Bite" (Figs. 159 and 160). On 21 August 1950 we took up an observation hive and established it on the strip of grass beneath the northeast wall (at *St* in Fig. 161). With a group of marked bees we journeyed around the rock spur and on the morning of the 22nd reached the feeding station F_1 (Fig. 161; × in Fig. 159). Most of the forager bees kept thenceforth to the route by which we had led them; a few shortened it somewhat in that they chose to pass over via the notches S_1 or S_2, but they did not fly on over the high ridge. In six experiments between 22 and 24 August (10–12 dance measurements each) the direction signaled was to the left (eastward) four times and to the right (westward) twice, on the average 1.25° east of the air line from hive to feeding station (Fig. 162). Since the bees were prevented from seeing the blue sky during the observations, the deviations

---

[52] In the reference listed there is given as the result of all experiments (feeding stations *b* and *c*) "an average deviation of 9° from the air line." The difference from the values cited above is explained by the fact that at that time in calculating the mean I summed all the errors without regard for whether the deviation lay to right or left of the air line.

The Tail-Wagging Dance as Communication / 175

FIGURE 159. Terrain of the detour experiment on the Schafberg; ×, feeding station in 1947. The observation hive was on the other side of the ridge at approximately the same elevation. At that time the bees flew over the ridge instead of out around it as was intended.

FIGURE 160. The same site, seen from ESE. The white spots (pieces of linen) at 1 and 2 indicate the position of beehive and feeding station in the 1950 experiments. The human figure at 3, beside the precipice at the end of the ridge, shows the relative dimensions.

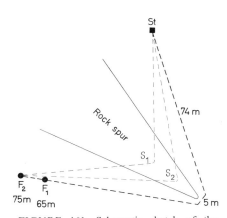

FIGURE 161. Schematic sketch of the detour experiment on the Schafberg: St, observation hive; F_1, F_2, feeding stations 65 m and 75 m, respectively, from the precipice at the end of the ridge; heavy dashed line, flight path around the outcrop; light dashed line, shortened flight path taken by some bees over the notches S_1 and S_2 (see also Fig. 159).

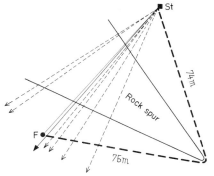

FIGURE 162. Detour experiment; indication of direction on the *vertical* comb surface, 22–24 August 1950; dashed arrows, the mean directions indicated in each of six experiments (10–12 dance measurements each); solid arrow, mean value for all six experiments; difference from the air line, 1.25°.

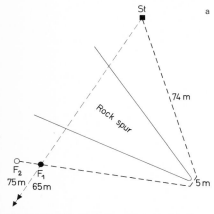

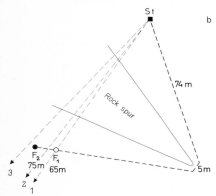

FIGURE 163. Detour experiment; indication of direction on the *horizontal* comb surface. (*a*) 22-23 August 1950: dashed arrows, during two series of experiments (10 measured dances each), on both occasions the bees on the average pointed directly to the feeding station. (*b*) When the feeding station was removed to F_2, after 0.5 hr the bees were still pointing to the earlier site F_1 (arrow 1). Arrows 2 and 3 refer to the indications of direction (each the average of 10 measured dances) about 3 hr after the feeding station was shifted, and again on the following day.

recorded as errors in the individual experiments did not arise as deflections due to the polarized light. But the "residual misdirection" (*Restmissweisung*) (see p. 204), which had not yet been explained at that time, may have caused the high degree of scatter in the six experimental series. That the mean corresponds so exactly with the air line must be regarded as due to chance.

This misdirection does not occur with dances on a horizontal surface. We had put the colony into a tiltable hive (Fig. 3) and in five experiments on 22, 23, and 24 August observed dances on the horizontal comb; in these the bees in shadow were orienting by the direction of vibration of the polarized light of the sky and were pointing directly at the goal. At first the feeding station was at F_1. In each of two experiments 10 dances were measured. In both cases the average direction signaled corresponded precisely with the direction of the air line (Fig. 163*a* and Table 27). The feeding station was then shifted 10 m farther, to F_2 (Fig. 163*b*). One-half hour later the bees signaled a direction 6° left of F_2, still corresponding exactly with the direction to the earlier accustomed feeding station F_1; about 3 hr after the feeding place had been moved the deviation amounted to only 3.5° to the left. On the next day it lay 2.5° to the right of the line to the goal (Table 27). With horizontal dances one does not need to make a conversion. The bees point out directly and convincingly the air line to the feeding place, a line that they do not know from their immediate experience.

Their behavior is sensible. If they were to indicate the direction of the first or second leg of their flight path they would send their comrades astray. To indicate alternately first the one and then the other stretch and in addition to tell which leg is to be flown first and which afterward lies beyond their capacity, and even if they were capable of doing this the task would be insoluble where the flight path is not bent at an angle but runs in a smooth arc. That they are able from the different solar angles perceived in flight to derive the solar angle of the air line and signal it is wonderful enough.

The experiments on the Schafberg, with their long, sharply angled flight path, were intended also to answer the question of what distance was signaled in such situations. I should add here that from 7 to 13 July 1947 we already had an observation hive on the Schafberg and with the same arrangements as in 1950 (see Figs. 159–161) had con-

TABLE 27. Indication of direction on a horizontal comb surface; detour experiment on the Schafberg, August 1950.

| Feeding station | Experiment No. | Day | Time | Average deviation (deg) of direction of dancing from— | | | Number of dances measured, n |
				Air line	First leg	Second leg	
F_1	101	22	2:05–2:20	0	52 r	66 l	10
F_1	106	23	11:25–11:40	0	52 r	66 l	10
F_2	110	23	1:00–1:10	6 l	52 r	66 l	14
F_2	116	23	3:16–3:24	3.5 l	54.5 r	63.5 l	10
F_2	120	24	9:40–9:46	2.5 r	60.5 r	57.5 l	10

ducted a group of bees around the "Devil's Bite," only at that time we had chosen a somewhat longer, hairpin-shaped stretch around the rocky spur and it developed—contrary to our estimate—that the route over the ridge was a shorter way from the hive to the feeding station (Figs. 164, 165). The bees found that out, and after a few hours it was already noticeable that they were not flying over the circuitous route on which we had led them but instead were following the shorter detour over the ridge, in the direction of the air line (Fig. 164). By doing this they crossed out our calculation. As far as the indication of direction was concerned the experiment turned out to be wasted. But it could still be turned to account with reference to the communication of distance. On the 8th and (after an interruption by rain) on the 11th the dance tempo after a linear flight of 100 m (feeding station F_1, Fig. 164) was 9.3 circuits/15 sec ($n = 31$) and on the 13th at a distance of 150 m (beyond the ridge) 8.5 circuits/15 sec ($n = 27$). The difference is statistically significant ($p < 0.0027$). The corresponding values from the over-all average curve (Fig. 63) are in good agreement: for 100 m, 9.45, and for 150 m, 8.65 circuits/15 sec. Hence the indication given as to distance referred to the route actually flown and not to the air line, which was 90 m long.[53]

I have already reported briefly on this (v. Frisch 1948). It has been possible to retest the result on two further occasions, during which, however, the bees kept to the desired course around the outside of the ridge.

In the experiment of 1950 on the Schafberg the distances on the flight legs were changed so that we achieved our intention relative to the indication of direction. This has been reported on pp. 174–176. Now in our measuring we found 154 m for the entire way around, about 133 m for the abbreviated distance over the notches (Fig. 161), and some 80 m for the air line. The dance tempo for flying the detour was found in four experiments to be 10.0 circuits/15 sec ($n = 43$). For a linear path of 80 m the same bees danced 10.64 circuits/15 sec ($n = 30$). The difference is significant ($p < 0.0027$). Thus they indicated a distance greater than the length of the air line.

In a third experiment at the same place in 1956, to which I shall return presently in another connection, the length of the detour flown was about 160 m, while the length of the air line from hive to feeding station was reckoned at 82 m. The dance tempo for the detour was 8.3 circuits/15 sec ($n = 15$). Two days thereafter, when the hive had been brought back to the valley, a group of bees (in part the same bees used in the detour experiment) was conducted with their feeding station from a distance of 200 m toward the hive, in steps, until the dance tempo for the detour experiment had been reached. That occurred at a distance of 170 m (8.37 circuits/15 sec, $n = 30$). This distance agreed closely with the length of the detour.

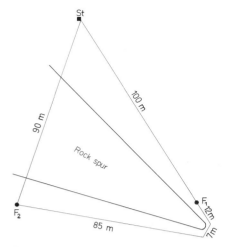

FIGURE 164. Detour experiment on the Schafberg, 1947: St, observation hive; F_1, F_2, feeding stations; air-line distance St–F_2, 90 m; over the ridge, about 150 m (see Fig. 165).

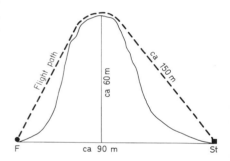

FIGURE 165. Detour experiment on the Schafberg, 1947: the flight path over the ridge.

---

[53] The lengths of the air line and of the route over the ridge were determined with the help of a measuring tape, compass, and protractor (aligning sights on the summit of the ridge), that is, trigonometrically. On a later occasion, when one of the observers went up the ridge, it was possible to check and confirm by direct measurement the elevation of the ridge above the foot of the wall (60 m).

Thus it was confirmed that only the indication of direction refers to the air line, while the approximate length of the detour flown is signaled by the communication as to distance.

Now we ask: How on the basis of this information do the comrades reach the goal? That they find it is demonstrated by the newcomers, who appeared in numbers at the feeding dish even in the detour experiments. I first assumed that, following the indicated direction, they would fly against the wall and then proceed laterally around the obstacle until, having traversed the announced distance, they reached the goal (v. Frisch 1951). But how could they tell whether to detour to left or right? Could it be that they flew directly over the ridge? Such doubts impelled our third experiment on the Schafberg (1956). This time the flight path of the newcomers was to be observed. The plan was simple: Three scent plates with the same scent that had been added at the feeding station should permit us to decide whether the aroused newcomers when confronted with the obstacle flew to left or right or possibly on over the ridge. The work was not altogether easy, because climbing the ridge is more difficult than it appears to be from Fig. 159. My student Böllinger, an experienced mountaineer, accompanied by a guide, took over this observation post.

On 17 August 1956 a colony was set up at the same place as in 1950 and on the 18th, in sunny weather, a group of numbered bees was led over the customary route around the spur of the ridge. At 12:00 M. the last feeding station was reached (see Figs. 159, 160, 166) and the numbered bees now received concentrated sugar solution with the addition of orange-blossom scent. The foraging bees flew over the detour as in 1950. Meanwhile three scent dishes with oil of orange blossom had been set out (glass covered, with holes 8 mm in diameter), two of them at the foot of the wall 50 m left and right of the intersection with the air line, and the third on the summit of the ridge where Böllinger meanwhile had occupied his observation post (Fig. 166). The experiment ran from 12:00 M. to 1:30 P.M. Mostly it was calm, but frequently there were gentle winds of variable direction. During the 1.5 hr of observation a single bee flew to and settled on the scent dish to the right at the foot of the wall, and two on the dish to the left, but the one on the ridge was visited by 19 bees, 8 of which settled on it. About 12 marked bees were frequenting the feeding station; during the experiment 23 newcomers appeared here and were captured.

Contrary to our expectations, the bees aroused thus held to the direction of the air line and did not bypass the obstacle to one side but instead flew over it. The observer on the ridge did not see a single one of the marked foragers pass by. Hence he was able to confirm that the latter flew out around the ridge. These two findings can be recorded as new evidence for the fact that the bees informed go hunting for the route independently, on the basis of the information they have received, and do not simply fly out in the wake of the foragers.

(c) Experiments with Italian and Indian Bees

Every undertaking on the Schafberg meant an appreciable expenditure of time and material preparations. Besides, it happened that a sud-

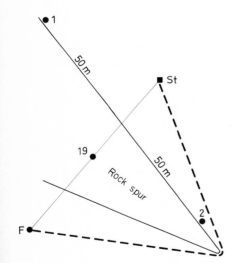

FIGURE 166. Detour experiment on the Schafberg, 18 August 1956: *St*, observation hive; *F*, feeding station; dashed line, flight path of the foragers. Two scent plates (black dots) were set out at the foot of the rock wall, 50 m to each side of the air line from hive to feeding station, and one on the summit of the ridge in the direction of the air line. The adjacent numbers show the number of visits to each.

FIGURE 167. Siemens building in Munich-Solln, seen from the north side.

den change in the weather put an end to the experiment already begun and to the life of the colony of bees. But we wanted to check once again the behavior just described and sought a simpler arrangement. Instead of Carniolan bees we took into the observation hive a colony of Italian honeybees, which perform direction-indicating dances for distances of as little as 10 m; and in place of the mountain range was substituted an 11-story (35-m) building of the Siemens Company in Solln near Munich (3–9 Leo Grätz Street). In the summer of 1961 everything for the experiment was ready, but just as the weather became favorable I had to go away and M. Lindauer undertook to supervise the work. Figures 167 and 168 show the north side of the building. Here, 8.5 m away from the wall and 25 m distant from the narrow west side, the observation hive was set up on the morning of 8 August 1961, and a group of numbered bees was led, with scentless feeding, around the west side to

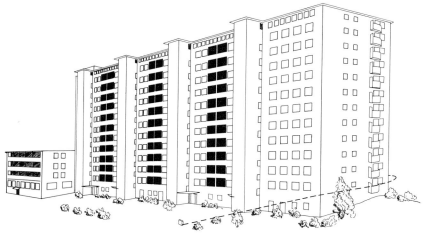

FIGURE 168. Siemens building from the north side: dashed line, flight path of the foragers from the observation hive to the feeding station on the south side of the building; height of the building, 35 m; length, 74 m.

the feeding station *F* (Fig. 169). This was reached at 10:30 A.M. During the experiment it was frequented by 15 bees. From 10:54 A.M. to 12:59 P.M. they were fed with concentrated sugar solution with orange scent. In order to obtain data on the flight path of the newcomers aroused, two orange-scented scent plates (as in Fig. 85) were set out on the grass below, 25 m west and east respectively of the hive, and three additional ones were put on the flat roof (Fig. 169). It was warm and sunny with gentle variable wind, gusty on the ground. All the foraging bees took the detour around the west side of the building. Frequently they flew from the feeding station obliquely upward, but in doing so they rounded the building not higher than the second or third story. This is in agreement with the fact that not a single one of the marked foraging bees was seen by any of the three observers on the roof. Since these bees had a large white spot on the abdomen they were of striking appearance in flight. The number of unmarked newcomers that settled on the scent plates during the experimental period is given in Fig. 169. By far the largest number of approaches (20) was recorded at the station on the roof in the direction of the air line. Of the five flights to the eastward scent plate on the roof, four fell in the initial period of observation, when a gentle east wind was carrying the scent toward the middle of the roof (cf. our experience during the fanwise experiments, p. 157). During the experimental period 90 newcomers appeared at the feeding station and were killed.

Thus the experiment brought full confirmation of the experiment on the Schafberg, that the newcomers aroused fly over the obstacle in the direction given by the dancers, and thus reach the goal.

The occasion was used also to note once again the indication of direction by the dancers that flew along the detour. A quick inspection at 11:00 A.M. showed that it corresponded to the direction of the air line. Between 12:43 and 1:01 P.M., with the sky obscured and a vertical comb surface, 11 dances were measured with the protractor (Table 28). On the average they signaled 5.6° west of the air line (Fig. 169). The difference from the first leg of the flight path amounted to 77°, from

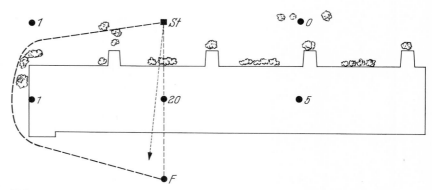

FIGURE 169. Detour experiment at Siemens building 8 August 1961: *St*, observation hive; *F*, feeding station; dashed curve, flight path of the foragers; black dots, scent plates, the adjacent numbers showing the number of visits; dashed arrow, indication of direction on the vertical honeycomb (average of 11 dances, 12:43–1:01 P.M.).

The Tail-Wagging Dance as Communication

TABLE 28. Indication of direction on a vertical comb surface (individual values); detour experiment at a skyscraper in Solln, near Munich, 8 August 1961.

Time	Angle of dancing (deg)	Deviation (deg) of direction indicated from—		
		Air line	First leg	Second leg
12:43	9 l	1.5 r	78.5 l	81.5 r
12:49	5.5 l	8 r	72 l	88 r
12:51	7 l	7.5 r	72.5 l	87.5 r
12:53	10 l	5.5 r	74.5 l	85.5 r
12:54	13.5 l	2.5 r	77.5 l	82.5 r
12:55	11.5 l	5 r	75 l	85 r
12:56	9 l	7.5 r	72.5 l	87.5 r
12:57	10.5 l	6.5 r	73.5 l	86.5 r
12:58	14.5 l	3 r	77 l	83 r
12:59	10 l	7.5 r	72.5 l	87.5 r
1:01	11.5 l	7 r	73 l	87 r

the second leg 80°. As in earlier detour experiments, a striking feature was a strong divergence in the dances (p. 142).

By using the Italian race of our honeybee it was possible to limit the detour experiment to a smaller area and to simplify it in this way. Lindauer (1963) went a step further in this direction. He used Indian bees, which perform direction-indicating dances when the distance is no more than 2 m. This was already spoken of on p. 172 in another connection. The result of the experiment shown in Fig. 156a is to be evaluated as a detour experiment also. The bees that first had to fly 3 m southward, then through the glass passageway 3 m eastward to the feeding station *F*, and after drinking to return by the same route, indicated in their dances the direction of the air line.

The experiment had a different outcome when the first leg of the rectangular course led through the glass passageway (Fig. 170). In this case the part of the course that was in the passageway was not taken

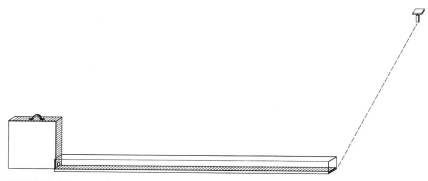

FIGURE 170. Experiment with *Apis indica*. In order to reach the feeding station the bees first fly to the south through a glass passageway and then, after a 90° bend, to the east. In indicating direction, only the free flight to the east is taken into account, while the passageway is regarded as part of the hive. After Lindauer 1963.

into account in the indication of direction. When the passageway pointed to the south and the feeding station was east of the exit from the passageway, the dances signaled the eastward and not the southeasterly direction, which would have corresponded to the air line from the hive entrance to the goal. Thus the bees were regarding the passageway as still a part of their home. For this reason it was also occupied by guard bees and kept free of rubbish. That outside of the hive sensory impressions during flight through the passageway are heeded just as strongly as in free flight, whereas they are not taken into account within the home area of the hive, is a noteworthy accomplishment of the nervous centers responsible for orientation (Lindauer 1963:172-173).

(d) The Biological Aspect

With respect to the natural biological circumstances the following experiments by Lindauer on our honeybee are of special interest; they are concerned exclusively with the behavior of bees that, aroused by dancing, came for the first time to the feeding place, and with their dancing after the first return home. In consequence of an acute shortage of food they began to dance right after the first successful flight for nourishment (cf. p. 173). If a newcomer who had flown to A was displaced inside a closed box to B (cf. Fig. 157), from where she flew home with a full honey stomach, her dance—as already mentioned on p. 173—signaled the direction to A, that is, the direction of her outward flight. Her displacement to B was not taken into account, as was expressed also in the fact that on her next flight out the bee again hunted for the objective A. Quite different was the result when the newcomer was compelled to take an active part in the displacement from A to B. Lindauer accomplished this by scaring the bee off after she had landed at A and had imbibed a short test drink, enticing her then to follow along on a slow journey with the feeding table. Along the way she was repeatedly allowed a taste at the feeding dish, but only at the final point of displacement B was she permitted to drink her fill. Under these conditions on her return home she signaled the direction to the site of the return takeoff B, that is, the direct route to the goal. The flight traversed actively in a new direction was included in her calculation. That emerged also from the fact that these bees found their way home from B more rapidly than previously displaced bees, apparently via a direct route, and that on their next outward flight they were hunting for the place of return B.

That the capacity for integrating the direction of the air line from an angular flight between the hive and the feeding place is used by the bee not only in her indication of direction but also for her own flight home was demonstrated by Heran and Lindauer (1963:52ff) in the following experiment. A numbered bee had flown back and forth all day long between her hive, set up on the lake shore, and a feeding station 248 m distant on the opposite shore; in doing this she flew along a floating plank bridge (Fig. 180). On the next day the feeding station was displaced 20 m along the shore after each visit by the bee. Figure 171 shows the result. For the first four displacements, up to the

80-m site, the bee in her outward flight always struck out on the route across the plank bridge; and not until the fifth displacement did she begin to leave this guide line, deviating to the right. But the return flight followed the air line every time. She must have found out this direction by considering the two segments of the outward flight in relation to the celestial compass. The environment around the lake afforded no landmarks that would have allowed her to steer directly toward the hive from the other shore (v. Frisch and Lindauer 1954:248).

Similarly too a scout bee that during an exploratory flight has sampled a flower here and there without much success and ultimately, perhaps after many detours, has discovered a profitable source of food, will find the shortest way home and point with her dance directly to the worthwhile goal (Lindauer 1963:161).

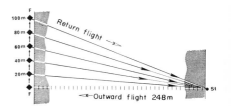

FIGURE 171. Direct return flight following an angular outward flight: St, hive; F, feeding station. Fuller explanation in text.

(e) *Detour Experiments with Bees on Foot*

Still another method was applied by A. R. Bisetzky (1957) by letting the bees travel on foot between the hive and the feeding station. The results that emerged from the experiments relative to the indication of distance have already been discussed on p. 114 (see also Figs. 107 and 108, showing the course). But the actual objective of the work was detour studies.

Let it be recalled that after a trip on foot bees perform tail-wagging dances when the feeding place is only 3–4 m distant. In comparison with bees in flight there appeared a very broad scatter in their indication of direction. Probably that is to be ascribed to the unnatural circumstances. Despite this the average values were fairly correct.

In commencing a detour experiment the bees were led gradually through a corridor, for instance one that was composed of two legs meeting at right angles, from the hive entrance to the end of the second leg. This training was subject to difficulties, especially at the angle points, and numerous individuals failed to master it. But with the others dances were observed after the return home. Even under these abnormal conditions the dances signaled the approximate air line to the feeding place.

As an example of such an experiment and in clarification of the technique Fig. 172 will be discussed. In this instance the air line from the hive entrance to the feeding station ran directly southward. The average direction of dancing is given by the arrow T_1. The individual values are entered as dots on the inner circle (each dot represents three measurements). In order to exclude the "residual misdirection" as a source of error, a second colony was set up adjacent to this one, with the corridor directed straight south. Its dances were noted simultaneously with those of the experimental colony (arrow T_2, and points on the outer arc). They inform us as to the prevailing degree of misdirection. The same degree of misdirection may be assumed for the experimental colony. Consequently a deviation of 25.5° west of the air line is obtained for this particular indication of direction.

In the further experiments the deviation from the air line was in general not always so large, but it lay regularly toward the open side of the angle (or of the arc, see Fig. 174) of the corridor. Bisetzky in-

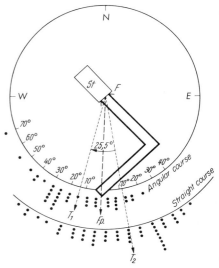

FIGURE 172. Detour experiment with bees on foot, angular corridor: St, observation hive; F, hive entrance; Fp, direction to feeding station in the corridor (exactly south); T_1, mean of directions indicated; T_2, mean for the control colony (not shown) that had a *straight* course to the south. The dots indicate the individual values, each dot representing three measurements of a class. After A. Ruth Bisetzky 1957.

FIGURE 173. Observation hive with an attached semicircular corridor. After A. Ruth Bisetzky 1957.

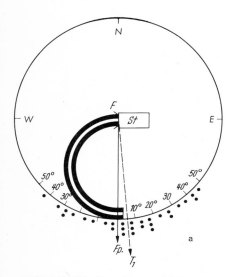

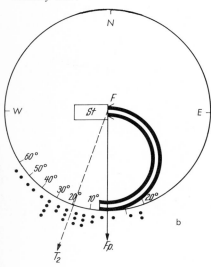

FIGURE 174. Experiment with two opposed semicircular corridors: St, observation hive; F, hive entrance; Fp, to feeding station, exactly south. (a) Westerly corridor; T_1, average of the directions indicated, $6.7° \pm 2.9°$ E of S ($n = 86$); difference from the air line not statistically significant ($P = 0.02$). (b) Easterly corridor; T_2, average of the directions indicated, $19.4° \pm 2.4°$ W of S ($n = 80$); difference from the air line statistically significant ($P < 0.0002$). Deviation from the air line was in the direction of overstressing the final segment of the course in both cases. After A. Ruth Bisetzky 1957.

terprets this behavior to mean that bees give more weight to the second half of the course than to the first. If their estimate of distance is based on the amount of energy expended for the way traversed (pp. 114ff), the misdirection could be ascribed to increasing fatigue during the unaccustomed long march on foot. Bisetzky speaks of their setting an excess valuation on the final stretch, a plausible even if unproved explanation.

In yet other instances the foragers had to walk to the feeding place through a semicircular corridor (Fig. 173). The results agree in principle with those from the angular course. Figure 174 shows two experimental examples.

According to our conception, the task is even more difficult for the bees on foot when the corridor makes a closed circle or a hairpin turn, or is S-shaped. Unfortunately this part of the work could not be completed and would require new experiments. So let us here merely refer to Fig. 175 and its legend, which indicate an accomplishment by the little wayfarers that should be retested and, in case it is confirmed, the path already begun should be pursued. This path is indeed a thorny one and demands great endurance and attentiveness.

We return to the simpler experiments with an angular course, which were varied in different ways and have yielded a certain degree of insight into the mechanism of indicating direction in the detour experiment.

If one covers the first leg of the corridor so that the bees can see neither sun nor blue sky (Fig. 176), in dancing they signal the direction of the second leg (Fig. 177). One might imagine that under these conditions they regard the darkened leg as being still a part of the hive.[54] But if one leaves the first leg open and covers the second one, the

---

[54] That is also the case in the absence of shading when the passageway is a continuation of the flight channel used by all bees (see p. 182). But in Bisetzky's present experiments the hive entrance was the normal site for takeoff, and was also known as such to the bees used in the experiment. In training them to the corridor they were conducted from the hive entrance into the shallow passageway, which for them did not belong to the familiar home area.

dances indicate the direction of the first leg. Thus in their indication of direction the bees refer only to that segment of the route in the course of which they could use the celestial compass.

Important too is a further experiment in which the final segment of the second leg was open (Fig. 178). The air line to the feeding station ran to the south. A control colony with a straight southerly corridor danced out a direction 5° west of south (T_2, Fig. 178). The bees collecting in the angular corridor pointed on the average in the direction T_1 (Fig. 178), 21° south of the direction of the first leg. The open final segment of the second leg clearly exerted an effect, but evidently the darkened part of the course was ignored. If it too had been open, bees, according to all our other experience, would have pointed even further southwestward of T_2, in consequence of the "excess emphasis on the final segment traversed."

With these results the conception is confirmed that in the detour experiment the bees derive the angle of dancing appropriate to the air line by a kind of integration of the solar angles perceived on the different segments and of their respective lengths (v. Frisch 1955). I regard this in the same sense as has been formulated clearly by Jander (1957:194) in his experiments on optical orientation in ants: "By integration is understood the central nervous reduction of complex spatial and temporal patterns of sensory perceptions to one resulting datum." The learning processes studied by him in ants, which, as the ants orient to a source of light by random search or a compulsory simpler path, lead to their maintaining the air-line direction between starting point and goal, bear the closest relation to the accomplishments of the bees in the detour experiment. Spiders too (the funnel-web spider, *Agelena*) that have come upon their prey via a detour are able on the basis of the optical impressions received to strike out in the direction of the air line (Görner 1958). What there was to be said about bees differs insofar as they—under the stress of circumstances—keep to the angular course themselves, but inform their comrades of the "calculated" air line and thus guide them successfully to the goal.

Hannes (1959 and 1961) disputes the idea that bees are able to

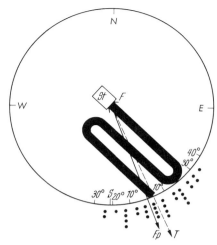

FIGURE 175. Detour experiment: *St*, hive; *F*, hive entrance; *Fp*, air line to feeding station at the end of the S-shaped corridor; dots, individual values; *T*, average of the indicated directions ($n = 31$), $3.9° \pm 2.6°$ eastward. The difference from the air line is not significant ($P = 1.5$). After A. Ruth Bisetzky 1957.

FIGURE 176. Detour experiment with corridor bent at right angles, the initial segment covered so that in this part of the course the bees see neither the sun nor the sky. After A. Ruth Bisetzky 1957.

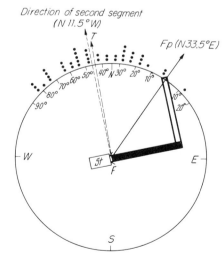

FIGURE 177. Result of the detour experiment with the initial segment of the bent corridor covered: *Fp*, air line to feeding station; direction of second (open) segment, 11.5° W of N; *T*, direction indicated (mean), $11° \pm 3.5°$ W of N; dots, individual values ($n = 42$). After A. Ruth Bisetzky 1957.

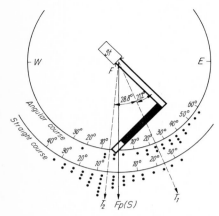

FIGURE 178. Result of a detour experiment with the initial segment (direction SE) of the bent corridor uncovered, and the second segment (direction SW) partly covered: Fp, air line to feeding station (due south); T_1, indication of direction given after proceeding through the bent corridor on foot, mean value 21° S of the uncovered segment ($n = 69$); T_2, direction indicated by a control colony (not shown) with a *linear* course to the S on foot, 5° W of S ($n = 98$); dots, distribution of individual values, each dot representing three measurements. After A. Ruth Bisetzky 1957.

"calculate" the air line from the direction and length of the segments of the path. In his view the direction of dancing in the detour experiment comes about through a kind of "mixture" of the neuronal learning effects during the segments—a statement that simply clothes the phenomena in other words.[55] The expressions "central calculation" and "integration" are much used in neurophysiology today (see B. Hassenstein 1960). Of course that must not deceive us into overlooking the fact that for the present the actual intimate phenomena in the nervous centers are unknown to us.[56]

23. THE INDICATION OF DIRECTION IN A CROSSWIND

When bees are subjected to a crosswind on the way to their goal, they do not let themselves be driven from their flight path. They compensate for the lateral drift by orienting themselves with the long axis of the body oblique to the wind (Fig. 186). Thus they act like an oarsman guiding his boat across a stream. The stronger the crosswind, the more oblique is the stance of the bees. This is not merely a theoretical requirement. Under favorable conditions we were able to observe it directly.

It is a prerequisite to such behavior that direction and strength of the lateral drift be noted. This is possible partly on the basis of visual perception. In calm air the bee sees the ground below passing by in a direction from front to rear. She can recognize a lateral drift by the oblique course of the ground beneath and compensate for it in such a way that once again she moves toward the goal. Of course the ground below now again runs obliquely past the long axis of her body. Hence —according to feedback principles—she must include her oblique stance in the calculation. Its amount she can read from the solar angle.[57]

In laboratory experiments Heran (1955, 1956) demonstrated a visual regulation of flight in accordance with the speed of movement of the pattern beneath.

That the visual perception of the substrate plays a significant part in compensating for lateral drift was also shown by Heran and Lindauer (1963). In a level terrain without conspicuous landmarks they trained bees to proceed across a water surface from one lake shore to the other (flight path, 248 m). The water affords no landmarks for the bees. In a crosswind they were able to compensate for lateral drift only in the immediate vicinity of the shore, then they were carried to one side, and not until they reached the other shore were they oriented by the landmarks as to the location of the feeding station (Fig. 179a). A floating wooden "bridge" of raw planks 4 m long and spaced 3 m apart, stretching from one shore to the other (Fig. 180), did not enable the

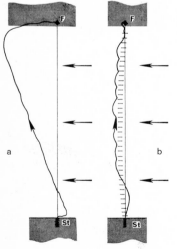

FIGURE 179. (a) Flight path of a bee flying in a crosswind (arrows) over rippling water: St, hive; F, feeding station. (b) Scalloped flight path along a plank "bridge" (see Fig. 180); maximal distance from the flight path to the edge of the "bridge" about 5 m. After Heran and Lindauer 1963.

[55] His other polemical remarks are answered by the data communicated in this section.
[56] Mittelstaedt (1961, 1962) has attempted to provide a theoretical feedback model for such steering of a course and related problems. But this treatment is in contradiction with various facts (Schöne 1962, Jander 1963) and my feeling is that the effort is premature.
[57] When neither the sun nor polarized light is visible to the bees, both can be replaced by distant landmarks, which are essential anyway for orientation under such conditions.

FIGURE 180. View of the course over water with a plank "bridge": at the right, from the hive; at the left, from the feeding station. Smooth water on the right, rippling water on the left. After Heran and Lindauer 1963.

bees to compensate by taking an oblique position. They continued to be driven to the side, but followed a scalloped course to the lee of the plank bridge (Fig. 179b). From this it is to be deduced that they had perceived the row of planks as a visual clue for the course, but that they did not at once find the proper compensatory stance; they would have to "fly their way into" this position while they were being displaced laterally, and the width of the plank bridge was insufficient for this.

But antennal perception of the air stream also is involved in steering. The flow of air against the flagella is registered by Johnston's organ (Heran 1959). Neese (1965) discovered that in addition sensory hairs, very neatly arranged at the angles of the ommatidia (Fig. 181) without protruding into the field of vision, take part in the perception of air flow. He made a tiny scalpel, the cutting edge of which was formed of fragments of a razor blade, with which he could at will remove the bristles from the whole eye or from certain parts of it without damaging the surface or interfering with foraging. In very strong wind (over 6 m/sec) bees without eye bristles no longer found the way to their accustomed feeding place, 660 m distant, which continued to be frequented by their normal comrades. Even at lower wind speeds the bristleless bees were driven aside in a remarkable manner by gusts. In analysis of and compensation for the wind relations evidently perception of movement relative to the stationary ground and perception of air flow work most closely together, even though at present the details of the regulatory processes are not exactly understood.

If now in consequence of her compensatory oblique stance a bee in a crosswind sees the sun from a different angle than when it is calm, where her body axis is headed toward the goal, the question arises:

FIGURE 181. Bee's head from in front: (right) the hairs at the boundaries of the ommatidia are aligned precisely with the optical axis of each ommatidium; (left) the hairs have been shaved from the eye. After Neese 1965.

188 / The Dances of Bees

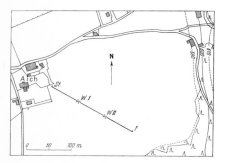

FIGURE 182. Experimental area and arrangement for the tests with crosswinds, 1953: *St*, observation hive; *F*, feeding station 210 m from the hive; W_I and W_{II}, the two anemometers, 70 m and 140 m from the hive, respectively. In the experiments of 1954 the locations of hive and feeding station were interchanged.

after she returns home will she point out in her direction-indicating dance the solar angle that she actually has seen, or the angle to the flight path that she would have seen in calm air? Experiments with Lindauer that we carried out in 1953 and 1954 in the level meadows at Aich brought the answer (v. Frisch and Lindauer 1955).

The experimental terrain is sketched in Fig. 182. In the summer of 1953 we set up the observation hive at *St*. A group of marked bees were frequenting the feeding station *F*, 200 m from the hive. The line of flight to the feeding place was so chosen that the typical wind in fine weather (out of the N to NE) came from the left, while under other conditions and less often wind from the right (a south wind) was to be counted on. In the summer of 1954 we exchanged the location of hive and feeding station, so that on the outbound flight the wind came mostly from the right. During the morning hours it usually was calm. In order to know for each bee whose direction of dancing was measured the wind relations to which she was subjected during the outward flight and the return, we had set up on the flight path two anemometers[58] (W_I and W_{II} in Figs. 182 and 183). The direction and velocity of the wind were always read as exactly as possible when the bee whose dancing was to be observed after her return home was passing the anemom-

---

[58] Obtained from R. Fuess, Berlin-Steglitz. The direction and strength of the wind were determined 2.5 m above the ground, which corresponded approximately to the height at which the bees were flying. In strong wind they customarily flew somewhat lower, and in a calm somewhat higher.

FIGURE 183. View from anemometer W_{II} (in right foreground) toward anemometer W_I and the observation hive *St*; see Fig. 182. In the left foreground are the meters of anemometer W_{II} for wind direction and velocity.

eters on her outward flight to the feeding station and on her return flight. In order to know just when this time would be we had to know the bees' speed of flight under different wind conditions.

Where the length of the route is known, the speed of flight is given by the times of takeoff and arrival. Simple as this procedure is, the statements in the older literature are just as contradictory. The reason probably is that many observers have recorded as the time of arrival the moment of settling at the goal. But the bee often hovers about the food for a long time beforehand. Then overlong times of flight are measured. Consequently a special observer stood where he could note with certainty the true time of arrival at the feeding station of the bee whose takeoff from the hive was signaled to him. In calm air the speed was found to be 8.2 m/sec or about 30 km/hr for the flight from the hive to the feeding station.[59] This value is in good agreement with the findings of other observers for other colonies, at other places and (in part) at other distances, insofar as they have paid attention to the true times of arrival (Table 29). For the return flight the bees need somewhat longer on the average. That is connected with the fact that with their burden they not infrequently stop to rest. It is noteworthy that their air speed is reduced in a tailwind and increased in a headwind.[60] In the former case the bee exerts herself less and takes advantage of the wind driving her toward the goal; in the second she compensates partially by an increased expenditure of energy for the hindrance of the wind (Fig. 184)[61] Yet this compensation is not begun until the wind speed reaches about 2 m/sec. With weaker winds the bees simply let themselves be carried, perhaps because they do not notice them.

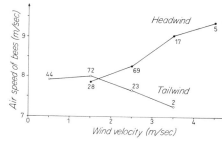

FIGURE 184. The air speed of bees in a tailwind and in a headwind, during feeding with 1M sucrose. Numbers beside the points on the curves are the numbers of observations on which the average values are based. The differences in air speeds at wind velocities of 2–3 and 3–4 m/sec are statistically significant ($P < 0.0002$ and $P = 0.001$, respectively).

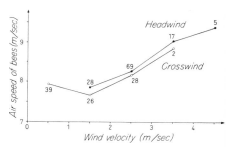

FIGURE 185. The air speed of bees in a headwind and in a crosswind, during feeding with 1M sucrose. For further explanation see Fig. 184.

[59] After feeding with 2M sucrose. As the sugar concentration is reduced, speed of flight is diminished.
[60] Under "tailwind" and "headwind" we have grouped all experiments in which the wind direction did not differ by more than 45° from the line of flight.
[61] This was noted long ago by O. W. Park (1923). Heran (1956) came to the same conclusion.

TABLE 29. Average flight velocity in winds of less than 0.5 m/sec on the flight from hive to feeding station; n, number of observations; concentration of the sugar solution, 2M or 1M.

Observer	Place	Distance flown (m)	Flight velocity (m/sec)	(km/hr)	n
G. Schifferer (1952)	Oberpfaffenhofen/Obb.	usually 200	7.7	27.7	213
H. Heran (1956)	Graz	200	8.2	29.5	81
v. Frisch and Lindauer (1955)	Aich near St. Gilgen	210	8.2	29.5	76
R. Boch (1956)	Near Munich	6000	7.7	27.7	177

190 / The Dances of Bees

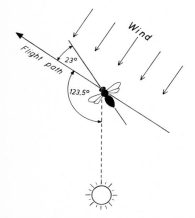

FIGURE 186. Compensatory oblique position of a bee in a crosswind from the right: true solar angle, 123.5°; angle of drift, 23°; apparent solar angle, 146.5°. See Fig. 187.

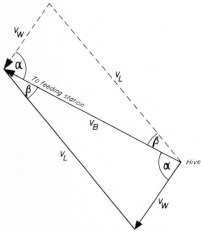

FIGURE 187. Example for calculation of the compensatory oblique attitude that corresponds to the angle of drift β; 19 August 1954, bee No. 1, 12:00 M. (see also Fig. 186): feeding station 30° N of W; $V_B = 8.4$ m/sec; anemometer I, NE 3.4 m/sec; anemometer II, NNE 3.5 m/sec; whence $V_W = 3.45$ m/sec, $\alpha = 86.25°$. Consequently $V = 8.88$ m/sec and $\beta = 22.65°$. The drawing is to scale, 6.3 mm = 1 m. For the bee the solar angle was increased by 23°, but her angle of dancing was 3° too small.

In a crosswind[62] the air speed is increased almost as if in a headwind (Fig. 185), as Bräuninger (1964) was able to confirm.[63]

These facts defined the prerequisites for our experiments. For each individual observation we proceeded as follows. At the hive the takeoff of a numbered bee was registered with a stopwatch. On the basis of our knowledge of the flight speed the observers posted at the anemometers were given a signal by whistle to take a reading at the time the bee flew past. The observer at the feeding station learned the bee's number via the field telephone and in the same way announced the moment of her departure, so that both anemometers could be read at the proper time during her return flight. Finally the angle of dancing of this forager was measured in the hive. The observer did not know the value expected. Only afterward was the sun's azimuth read from our curves and thus for each dance the error, that is, the deviation of the dance from the correct solar angle at the time of the dance, was determined.

What interests us in each individual observation is the amount of displacement of the bee by the wind. For from it is derived the compensatory oblique stance and hence the angle relative to the sun's position that she had to maintain during flight. Figure 186 shows a bee flying toward the feeding station in a crosswind from the right. We know (see Fig. 187):

V_B, the speed of translation of the bee (over the ground) = distance flown/time required = 210 m/x sec; for x was substituted, on the basis of our experiments on flight speed under various wind conditions, the proper average value for the given circumstances;

V_W, the wind speed (mean value of the readings on both anemometers at the time of the bee's passage);

α, the angle between the wind direction (average of the directions read on both anemometers) and the direction to the feeding station.

In Fig. 187 these values are shown to scale (6.3 mm = 1 m) for a

---

[62] The wind direction more than 45° to one side of the line of flight.

[63] In his study of speed of flight Wenner (1963) says that hitherto there have been no reports on the influence of wind velocity in individual observations of bees. That is not correct. What he wants of a theoretical nature is exactly what we have done (v. Frisch and Lindauer 1955). He supposes that as yet nobody has paid any attention to the time that is lost during takeoff and landing. In order to control these factors especially a trained observer was assigned in our experiments. Our report says: "Failure to take into account the true time of arrival at the end of the flight course certainly has led earlier observers to assume too low a speed of flight."

For the unburdened flight outward to the feeding station Wenner gives a speed of 7.5 m/sec (we found 8.2 m/sec). With a headwind (from 0.5 to 4 m/sec) according to his data the bees compensated for some 90 percent of the drift. In our experiments the bees began to increase their air speed only when the wind speed reached about 2 m/sec, and at 2–3 m/sec were compensating for about 25 percent of the drift. We regard our values as more exact, if only because Wenner did not measure directly the important initial rates in calm air but instead interpolated them from a regression line—though it has not been proved that the regression concerned actually is linear. His measurements of time spent in flight arouse our doubts. Thus, for instance, analysis of his Fig. 1, p. 28, shows that more than one-third of the bees that flew to the feeding station in a headwind arrived there definitely earlier than in a calm. Hence they had unequivocally overcompensated for the drift. During the 196 flights with a headwind we measured in 1953 that never happened once.

randomly selected example, and the distance V_B flown in 1 sec is shown. The graph gives the angle of drift $\beta = 23°$.[64] The bee must set herself obliquely toward the right against the wind at this angle in order to fly straight to the goal. Thus the (apparent) solar angle was 23° larger for her than it would have been during flight in still air. Her error in dancing was $-3°$, that is, her angle of dancing was 3° less than the true solar angle (cf. Fig. 186). Obviously her indication of direction referred to the solar angle that she would have seen under calm conditions.

In this as in other examples from the protocols it is taken for granted that the impressions that determine the bee's indication of direction are those she receives on the outward flight to the feeding station. The justification for this assumption was given on pp. 172f. In spite of this we carried through the calculations for the wind relations not only with the outward flights but also with the return flights, and with both considered together. No difference emerged from the results.

Out of the 16 series of experiments from the two years I next cite two examples. On 18 August 1954 between 2:22 and 3:50 P.M. dances were measured for 20 bees. The individual values for the first ten dances are given in Table 30 as an example from the protocols. The graphical representation in Fig. 188 affords a broader view. The solid line connects the values for the compensatory oblique stance determined for each of the 20 bees, arranged according to increasing wind speed (hence according to increasing obliquity of stance). In consequence of their oblique position the bees saw a larger solar angle (in correspondence with Fig. 186). In this series of experiments the angles of dancing were larger than the true solar angle (the measured angles of dancing are connected by the dashed line; for each measurement the times are given beneath the abscissa). The beginning portion of the curve gives the impression that the dancing bees wished to indicate the solar angle that they had to maintain in order to compensate for the drift. But the fact that at higher wind speeds (to the right in the figure) the indication

[64] The angle of lateral drift can also be calculated; see v. Frisch and Lindauer 1955:383.

TABLE 30. Crosswind experiment, 18 August 1954; experiment No. 100; first half, sunny, 20°C.

Bee No.	Departure time	Time of dancing	Angle of dancing, right (deg)	Solar azimuth (deg)	Solar angle (deg)	"Error" (deg)	Wind direction and velocity (m/sec)			
							Outward flight		Return flight	
							Anemometer I	Anemometer II	Anemometer II	Anemometer I
11	2:18	2:22	78	230	70	+8	NNE 1.4	NNE 1	N 2	NNE 0.3
10	2:21	2:24	72	230.5	69.5	+2.5	NNE 0.7	NNW 1.5	0	NW 0.7
2	2:30	2:33	72	233	67	+5	NE 2.5	N 2.7	NNE 3.5	NE 2.7
19	2:32	2:36	68	234	66	+2	NE 2.7	NNE 3.5	N 2.5	NNE 2.5
17	2:35	2:39	69	235	65	+4	NNE 3.3	NNE 3.5	NNE 3.5	NE 2.5
10	2:37	2:41	74	235.5	64.5	+9.5	NE 2	NNE 2.5	NE 2	ENE 1
11	2:40	2:43	76.5	236	64	+12.5	NE 4	NNE 3.5	NNE 3.5	NNE 3.8
16	2:42	2:45	73.5	236.5	63.5	+10	NE 4.5	NNE 3.5	N 3	N 3.5
14	2:47	2:50	77	238	62	+15	NNE 2.5	E 2.5	N 4	N 2.2
6	2:49	2:53	73.5	238.5	61.5	+12	NNE 1.7	NNE 3.5	N 3	NNE 3.3

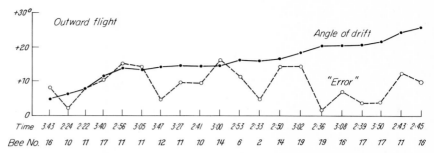

FIGURE 188. Comparison between angle of drift and angle of dancing, experiment of 18 August 1954. See the partial protocol in Table 30. The 20 individual observations are arranged in order of the strength of drift on the outward flight to the feeding station (solid line). The "errors" in the angle of dancing are connected by the dashed line. Below the abscissa are shown the time of the dance and the number of the bee. The degree of misdirection does not increase with increasing drift (with increasing obliquity in the bee's attitude).

of direction does not follow the increasing lateral drift is in contradiction with this interpretation.

This latter is refuted completely by the results of other experiments, an example of which is shown in Fig. 189. This time the wind came from the opposite side. In order to compensate for the drift the bees had to head obliquely toward the left and hence saw the solar angle as smaller. Nevertheless here too the angles of dancing were larger than the solar angles.

Thus in both cases, in spite of opposite wind directions, we found the angle of dancing too great. Such "errors" in the indication of direction ("misdirection") have occupied us on many previous occasions. At the time of our cross-wind experiments their genesis was still unclarified. Yet we knew that they were a consequence not of imprecision, but of systematic deviations that have a complex diurnal course influenced by various factors. Today we know that the deviations appear as "misdirections" only to us, whereas to the hivemates they indicate the location of the goal correctly. This will be discussed in the next section. Here we shall merely point out that the angles of dancing shown in the curves are far from being as "erroneous" as they seem.

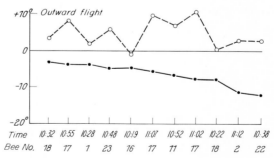

FIGURE 189. Comparison between angle of drift (solid line) and angle of dancing (dashed line); Experiment on 23 August 1954; wind south. The compensatory oblique attitude *diminished* the solar angle, but the angles of dancing were *too large*.

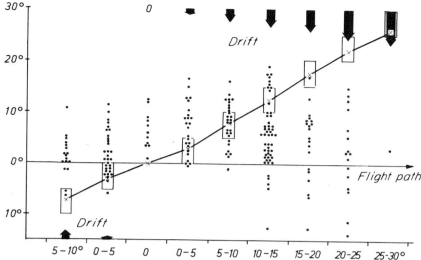

FIGURE 190. Comparison of angle of drift with angle of dancing: all observations from summer 1953 regarding the indication of direction in a crosswind. The crosses on the curves indicate the strength of the drift, and thus the obliquity adopted by the bees in compensation for it. The black arrows show from what direction the wind was blowing; their length is approximately proportional to the wind velocity. Each dot represents a measured dance and shows the direction and magnitude of the error (see ordinate); it is correlated with the angle of drift measured during the outward flight of the individual bee. The rectangles bound for each point of the curve the area in which the values for the respective group would be expected to fall if misdirection were caused by the oblique position of the bee.

Instead of further examples, I give in Figs. 190 and 191 a summary of all the 293 measurements from both years of experimenting: 199 measurements of the angle of dancing from 1953 (Fig. 190) and 94 from 1954 (Fig. 191). Again they have been put in order according to the amount of drift. The 0° line refers to the flight path, which ran in

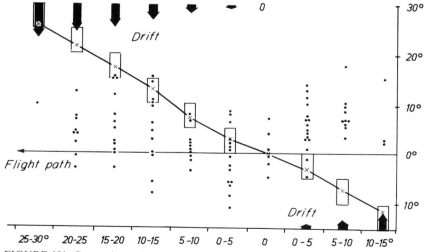

FIGURE 191. Comparison of angle of drift with angle of dancing: all observations from summer 1954 regarding the indication of direction in a crosswind. For explanation see Fig. 190.

opposite directions in the two summers. The black arrows above and below show whether the wind was coming from right or left; their lengths are approximately proportional to the force of drift. The small crosses connected by the curves show the direction and extent of the obliqueness (combined in classes of 5° each) adopted by the bees in compensation for the wind conditions given. The points indicate the size of the "error" for each dance angle measured and are correlated each time with the amount of drift (crosses on the continuous curve) that was determined for the outward flight of the bee in question. That these "errors" in the angle of dancing are not to be understood as actual mistakes in the indication of direction has already been mentioned. Thus we do not attach any great importance to individual differences between the angle of dancing and the angle of drift. As a significant result it is obvious that there is no relation between the indication of direction and the amount of drift, for the measured values do not follow the curves shown by the compensating oblique position of the bees. The dancer points out the same direction to her comrades whether the wind blows from left or right, strongly or weakly, or not at all.

This can be confirmed numerically. The correlation coefficient r was calculated (Gebelein and Heite 1951), with the following results: summer 1953, 199 pairs of values, $r = 0.08$; summer 1954, 94 pairs of values, $r = 0.03$. Here $r = 1$ signifies complete positive correlation, $r = -1$ the opposite, and $r = 0$ no correlation. Both of our values lie far below the maximal probable limit of 0.21 or 0.3, respectively (Koller 1940). Hence there is no correlation between the angles of drift and dancing.

This result seems to me of great interest. For it means that the bee in a crosswind later reduces the solar angle that she has actually seen in consequence of the oblique position of her body during the flight to the feeding station to the angle that she would have seen on a flight in calm weather. That is biologically important, for the wind can shift extremely rapidly. When the comrades aroused set out, the wind will not infrequently blow otherwise than it did during the preceding foraging flight. Then they would be led astray by an indication based on the solar angle actually seen by the dancer. But if the true azimuth of the goal is given them, they will fly the proper course by compensating on their own for such drift as corresponds to the wind then prevailing. The accomplishment gains in significance from the fact that strength and direction of the wind may vary in different segments of the flight course, and the extent and duration of these variations would have to be taken into account by the dancer in reaching her final result.

Thus the bee is capable of maintaining a solar angle (or of indicating it by her dance in the hive) that she has not actually seen but has deduced indirectly from the information provided by her sense organs. We have become acquainted with the same ability in the detour experiments (pp. 173–186), and also in the dance followers, which for the most part receive the indication of direction in the tail-wagging dance from an oblique position at one side (p. 160 and Fig. 52).

The indication of direction in a crosswind could be interpreted also to suggest that the bee averages the solar angles perceived during outward and return flights, in doing which she would of course have to transpose the impressions received during the return flight to the direction of the outward flight. For this interpretation it would be necessary to make the (improbable) assumption that, under the special conditions pertaining to a crosswind, the return flight too would have to be taken into account and with the same weight as the outward flight.

In order to clarify how this is meant we may begin with the simpler relations that obtain in calm air. There is no doubt of the fact, already mentioned on several occasions, that the bee is able to strike out in the direction of the return flight according to the position of the sun, by transposing the solar angle of the outward flight through 180°. Figure 192 shows this. During the outward flight the feeding station F was 50° to the right of the sun (Fig. 192a). The bee finds the direction of the return flight by hunting for the hive 130° left of the sun (Fig. 192b). Now we assume during the outward flight a cross wind from the right, compensated by adopting a stance 20° on the bias. Thus the bee flies to the feeding station in a bodily position 70° to the right of the sun (Fig. 192c). On the return flight she has to head 20° left against the wind, her body points 150° left of the sun (Fig. 192d). Transposed to the direction of the outward flight, the location of the feeding place is deduced as 30° to the right of the sun. As the average of 70° and 30° the bee finds the correct line to the feeding station at 50° to the right of the sun.

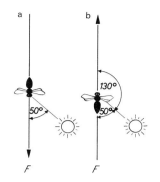

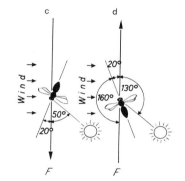

FIGURE 192. Actual and apparent solar angle during outward and return flights: (a) and (b) in calm; (c) and (d) in a crosswind. Further explanation in text.

Such an interpretation is formally possible, but it probably is not right. Apart from the difficulty already noted, that the return flight would have to be taken into account just as the outward flight is, it is more plausible to base the bee's behavior in a crosswind on the same capabilities that she has displayed so unambiguously in the detour experiments and in the pursuit of the dancers. Also the former considerations would add nothing to our understanding. Whereas in calm air the transposition imagined can be interpreted relatively simply (Fig. 192), to the effect that on the return flight the bee shifts to the hinder region of the opposite eye the solar angle that she had before her on the outward flight, with the body in an oblique position (Fig. 192c, d) this simple transposition does not lead to the goal; rather, the oblique position has to be included here in the calculation, by some sort of integration process. Besides, as in the other case, her performance is complicated by the fact that the variable strength and duration of the wind over the flight course would have to be included correspondingly in the calculations.

Hannes (1959:7) disposes of the problem in a radical manner with the thesis that in our experiments the indication of direction does not follow the air line to the goal in any way and has no causal connection with it. In his opinion the dancing bees point out the solar angle that they actually saw from their compensating position during the outward flight. That the angles of dancing measured by us conform in no way to the oblique positions assumed by the bees at various times (see Figs. 190 and 191) does not bother him. The indication of direction by the dance is in fact alleged (p. 46) to be "the result of a kind of 'mixture' of the neurophysiological learning effects of those solar angles that the bee in question had 'experienced' (and 'learned') not only during the last preceding flight out to the source of food but also during several such outward flights." This contention is refuted by the following facts:

1. In some of our experiments the angles of dancing deviate from the solar angles in the opposite direction from the one that—if Hannes were right—should have prevailed after the preceding flights. Thus the angles of dancing for the experiments reproduced in Fig. 189 were too large during the entire hour of observation. In accordance with the compensatory stance they should have been too small. During the morning when the bees began their flights to the feeding station the wind relations were identical with those during the observations of the dances.

2. In another series of experiments, shown by v. Frisch and Lindauer (1955) in their Fig. 14, for $\frac{3}{4}$ hr the angle of dancing likewise was larger, and the angle of drift smaller, than the solar angles; before the experiment started this day was calm. The arguments whereby Hannes (1959:58, 59) seeks to reconcile these protocols with his conception are therefore to no purpose.

3. For our 16 series of experiments I have now compared the "errors" in the angles of dancing, that is, the deviations of the points in Figs. 190 and 191 from the 0° line, with the available observational data regarding "misdirection" that are presented in summary form in the circular diagram of Fig. 212. The result of this comparison leaves no doubt that the deviations observed correspond to the "misdirections" that were to be expected under the various conditions obtaining. For the recipient of the information they do not represent any misdirection, but guide her correctly to the goal. Only apparently do the direction-indicating dances in Figs. 190 and 191 show so great a scatter and so strong a deviation from the 0° line. More information about this will be provided in the next section, especially on pp. 212f.

24. "Misdirection"

That in pointing out direction bees refer to the position of the sun and indicate a goal that is in the direction of the sun by an upward tail-wagging dance, and one that lies to right or left of the sun by dancing at a corresponding angle to the right or left of the vertical, was clear in principle as early as 1945. For clearly the diurnal course of the sun was met by a shift of the direction of dancing in the opposite sense. But often there were discrepancies in the "corresponding angle of dancing." Under some conditions the size of the angle of dancing might deviate by some 40° from that of the solar angle. These were no chance inaccuracies, for the "misdirection" showed a systematic diurnal course. Figure 193 gives an example. The feeding station was north of the observation hive. In 16 experiments that I carried out at various times of day in fine weather between 31 August and 14 September 1945, the "misdirection" was very large in the morning and toward evening, and at noon became practically nil (Fig. 193). But it was confusing that it differed in colonies with different locations, and even in a single colony when the direction of flight was altered, and in fact even with an identical direction of flight and at a single time on the other side of the honeycomb. By no means always were the angles of dancing too large, as in Fig. 193; they might be just as much too small, or too big in the morning and too little in the afternoon, or vice versa—with which not all the possibilities were exhausted. This occasioned many experiments and much vain headscratching before it became evident, in the summer of 1947, that location of the hive, direction of flight, and the side of the honeycomb are wholly indifferent in themselves. All that matters is in what direction and at what angle the sun stands relative to the bee dancing on the vertical comb surface. That depends not only on the position of the sun, but also on the position of the comb surface in space and on the position of the bee in her tail-wagging dance (which differs, for instance, under otherwise identical conditions, on the two sides of the comb).

It was not easy to come to the idea that this "misdirection" depends on the angle of the sun relative to the dancing bee; for in the experiments the sun indeed was not visible to the dancer. Either it was on the other side of the hive, or it was obscured.

But now it became evident that marked "misdirections" occurred only when the dancers—after removal of the protective wooden cover of the hive—had a free view of blue sky. At that time this was the first indication that bees perceive the direction of vibration of the polarized light of the blue firmament and are able to read the position of the sun from it.

But even with an unbroken cloud cover and in the closed fiberboard shelter there remained some "misdirection" of smaller magnitude. This "residual misdirection," too, had a systematic diurnal course. As in fact remained hidden for a long time, it is connected with the force of gravity.

When the relations had been clarified, the light-dependent "misdirection" lost its miraculous halo and interest in it declined to the level suitable for a technical and readily avoidable experimental error. In order to exclude it, it was necessary to measure the dances by diffuse light, preferably in the closed fiberboard shelter. The principles underlying the residual "misdirection," dependent on the force of gravity, have now become understandable, but in their details still leave many nuts to be cracked.

(a) Light-dependent "Misdirection"

I still have a vivid recollection of a certain observation, although it was made 15 years ago. The sky was cloudy; the dancing bees were pointing the direction to the goal quite accurately; then very swiftly a patch of blue sky came into their field of view and just as quickly they changed their direction of dancing by about 40°. Evidently their reading the position of the sun from the blue sky had caused the marked shift.

We know that a bee performing the tail-wagging dance on a horizontal surface sets herself directly into the proper solar angle, while on the vertical comb surface she transposes it to the direction relative to

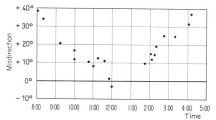

FIGURE 193. An example of "misdirection" and its diurnal course in dances on a vertical comb in view of the blue sky; experiments in Aich, 31 August to 14 September 1945; flight direction north. The "misdirection" is the difference between the solar angle and the angle of dancing. The points on the curve are average values of the angles of dancing measured during the time of observation (usually 20 min), referred to the azimuth of the sun at the midpoint of the period of observation.

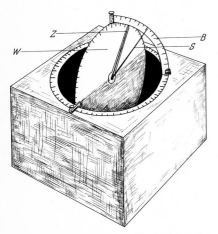

FIGURE 194. Device for determining the projection of the sun's position on the vertical honeycomb: W, "honeycomb plate"; B, 90° arc for setting the pointer S at the azimuth and elevation of the sun. When the pointer Z is set on the shadow of the pointer S it indicates the angle desired.

gravity. The suspicion was aroused that, when the position of the sun was perceived while she was on the vertical comb surface, her endeavor here too to adopt directly the proper visual angle came into competition with her tendency to maintain the correct angle relative to gravity, and led to a compromise solution that appears to us as "misdirection."

In order to test this idea one had to know what angle the bee on the vertical comb would strive for in taking a position relevant directly to the sun. For this it was necessary to know the position of the sun projected onto the vertical comb surface at the time of the dance. This purpose was achieved with a device (Fig. 194) designed by Professor Hans Benndorf and constructed in the Physical Institute of the University of Graz. In using it, the apparatus had to be set up in a darkened room. In the top of a wooden box is a circular hole with a scale. In this opening a circular plate, the "honeycomb plate," likewise bearing a scale, is mounted so that it can turn about a vertical axis. By means of the clamp screw the plane of the plate is fixed at the compass direction corresponding to the position of the comb in the observation hive. The 90° arc B is adjusted to the sun's azimuth at the time of the dance. The pointer S, with a ball joint at the center of the honeycomb plate, is set at the sun's elevation at the time of the dance. Hence it points to the sun. At the same height as the top of the box and at right angles to the honeycomb plate a lamp is placed at a distance of a few meters. The pointer Z should be set so that its direction corresponds to that of the shadow of pointer S. It indicates the direction of the sun, projected on the comb surface. The values found with this device agreed within 1°–2°, as was satisfactory for our purpose, with a subsequent calculation for which I am indebted to Dr. W. Jahn (Munich Observatory). Table 31 and the graphs are based on these calculated values.

With the aid of this device, in the winter of 1947–48 I made a supplementary check of the 163 available experiments concerning "misdirection," and in many instances found it to be true that the direction of dancing falls about midway between orientation with reference purely to gravity and that with reference to light. Yet certain values might also deviate appreciably from this expectation (v. Frisch 1948).

The data obtained at that time were inexact in several respects: first, the protractor was still imperfect; second, I did not evaluate the dances singly, but instead took average values for the angle of dancing and for the solar angle, over periods of about 20 min; in the third place, I still did not know of the existence of "residual misdirection," the main source of the imprecision noted.

Recognition of these circumstances led to new experiments (v. Frisch 1962) in which I used the improved protractor (Fig. 26), took each dance for evaluation individually, and eliminated the "residual misdirection" by the following procedure. In regular alternation ten dances were measured with a free view of the blue sky and ten in the fiberboard shelter in diffuse light. Subsequently the position of the sun at the moment of dancing was determined for each dance. The angle between the azimuth of the feeding station and the azimuth of the sun is the solar angle at the time of dancing. Through the observations in

TABLE 31. Experiment on 30 August 1961, 2:26–2:36 P.M.; blue sky visible to the bees, in the W and NW.

Time	Solar altitude (deg)	Solar azimuth (deg)	Solar angle[a] (deg)	Zenith angle[b] (deg)	Bee No.	Individual residual misdirection[c] (deg)	Expected angle of dancing			Observed angle of dancing			
							By sun (deg)	By gravity (deg)	Mean (deg)	Value (deg)	Deviation from expected direction		Deviation from mean (deg)
											By sun (deg)	By gravity (deg)	
2:26	41.5	228.5	41.5 r	37 r	36	−13	78.5 r	28.5 r	53.5 r	52.5 r	−26	+24	−1
2:27	41	229	41	37	35	−11.5	78	29.5	53.5	53	−25	+23.5	−0.5
2:27	41	229	41	37	30	−11.5	78	29.5	53.5	51	−27	+21.5	−2.5
2:28	41	229	41	37	31	−10	78	31	54.5	47	−31	+16	−7.5
2:28	41	229	41	37	34	−13	78	28	53	47	−31	+19	−6
2:29	41	229.5	40.5	37	36	−13	77.5	27.5	52.5	51	−26.5	+23.5	−1.5
2:31	40.5	230	40	37	25	−5	77	35	56	55	−22	+20	−1
2:32	40.5	230.5	39.5	36.5	27	−10	76	29.5	52.5	48.5	−27.5	+19	−4
2:33	40.5	230.5	39.5	36.5	32	−11	76	28.5	52	46.5	−29.5	+18	−5.5
2:36	40	231.5	38.5	36.5	31	−10	75	28.5	51.5	52	−23	+23.5	+0.5

[a] The angle between the solar azimuth and the azimuth of the feeding station.
[b] The angle between the direction of the zenith and the direction to the sun projected onto the vertical honeycomb surface.
[c] Determined for the bee in question, with the light of the sky shut off from her, in immediately preceding or succeeding dance measurements.

the fiberboard shelter we learn what deviation in angle of dancing from the solar angle is shown by the bee in question,[65] that is, her "residual misdirection" at the time of the experiment. This can now be included in the calculations for the observations with a free view of the blue sky.

Table 31 gives an example from the protocols of one series of experiments. In order to explain it let us take the first line. Bee No. 36, according to determinations made shortly before or afterward, was showing at the time a "residual misdirection" of $-13°$; that is, with a solar angle of $41.5°$ to the right (r) her angle of dancing, oriented according to gravity, was $28.5°$ to the right ("expected angle of dancing according to gravity" in Table 31). Now we need to know her direction of dancing on the vertical comb when oriented purely by the sun. With our measuring device we find that the direction toward the sun projected onto the comb surface is $37°$ to the right of the zenith ("zenith angle" in Table 31). If the dancer perceives the position of the sun she must head $41.5°$ to the right of this direction, whence is derived the "expected angle of dancing according to the sun" (Table 31) of $37° + 41.5° = 78.5°$ to the right. In the experiment with a view of the blue sky she danced $52.5°$ to the right of the vertical. The mean of the expected angles of dancing according to gravity and the sun is $53.5°$ to the right. The angle of dancing was only $1°$ different from this mean.

More plainly than the table, Fig. 195 shows for this experiment with what surprising precision the dancers maintained the direction (dotted line) midway between the expected directions of dancing according to light and to gravity. At the right below, in the frame, the mean of ten individual observations is shown. The experiments on the following afternoon gave a concordant result. During the observations the direction of dancing to be expected was always unknown to me. All the experiments were made in "ignorance."

That a bee guided by the polarized light of the blue sky should, when on a horizontal surface, head toward the goal (see pp. 382ff) is easily understood. She need merely seek the position of the body in which the polarization pattern perceived coincides with what she saw with the same parts of her eyes shortly before, on her flight to the feeding place. But in contrast therewith, when she dances on a vertical comb surface the polarization pattern is displaced onto other parts of the eyes. That in spite of this in her compromise she takes the solar angle into account with such exactness presupposes a precise mechanism of regulation. Still, in the experiments discussed hitherto she had behind her back not only the blue sky but the sun, the latter concealed from the bee. I was curious to know how the bee would react if she saw blue sky above her while the sun was below—a situation that never occurs during her flights in the open country.

With this idea in the back of my mind, I set up the observation hive in such a way that the comb surface was in the north-south plane. The dance floor was the comb surface that faced westward. Hence in the morning the sun was ventrad of the dancers, on the other side of the

---

[65] There are individual differences.

The Tail-Wagging Dance as Communication / 201

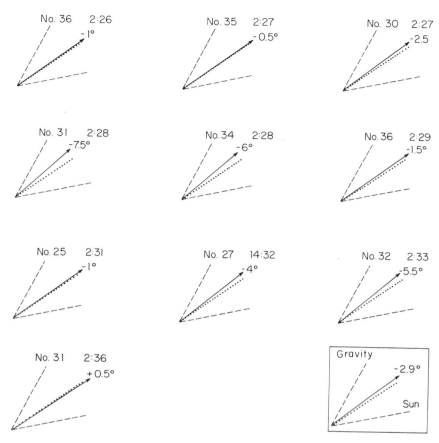

FIGURE 195. Ten dances on a vertical comb surface, 30 August 1961, in sight of the blue sky, but not of the sun; graphic representation of the series of experiments of Table 31: dashed lines, expected direction of dancing for orientation to gravity and to light; dotted line, mean of these two directions; arrow, direction actually danced. The deviation of the angle of dancing from the mean value, the time of observation, and the number of the bee are given for each picture. In the frame are averages of the ten single observations.

comb. Under these conditions too the bees pointed no less exactly midway between the directions that would have corresponded to orientation by gravity and by light, respectively (Fig. 196). Out of eight additional series of experiments in the mornings (ten dances each), seven were concordant; only in one instance, for unknown reasons, the direction of dancing was shifted somewhat in the direction of orientation by light.

The performance of the bee in this experimental arrangement is surprising—and continues to be so even when one recalls that, although she never has seen the sun beneath her in her flights in the open, still the task is not unbiological. A swarm of bees hanging free in a tree crown is informed by the dances of scouts that have found a suitable dwelling about its location (see pp. 269ff). In doing this the dancers that are active on the shaded side of the swarm must be guided by the polarized light of the sky, while the sun is below their ventral side. This

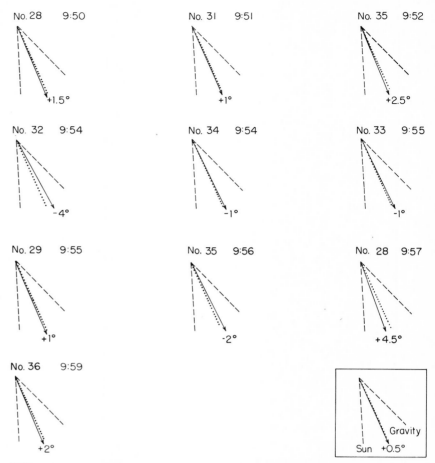

FIGURE 196. Ten dances on a vertical comb surface, 5 September 1961, in sight of the blue sky, the sun on the opposite side of the comb, ventrad of the bees. Further explanation as in Fig. 195.

capability may be an ancient heritage, from a time when the honeybee too—like her more primitive relatives even today—used to build her nests under the open sky.

In a series of experiments in which not only blue sky but the sun too was made visible to the dancers, they oriented themselves according to the light. The force of gravity was beaten by the sun (Fig. 197).

When a polarization pattern was no longer recognizable through a thin, complete cloud cover, but the sun was still shining through perceptibly, the directions of dancing were in between orientation toward light and toward gravity. During the succeeding series of experiments the cloud cover had thickened. For us the position of the sun was just barely detectable. Now the dances were oriented entirely according to the gravitational force. Thus the sun seems gradually to lose in influence as the cloud cover becomes heavier.

In orienting by the polarized light of the sky the bees behaved otherwise. Whether a small or a large area of blue sky remained visible to

The Tail-Wagging Dance as Communication / 203

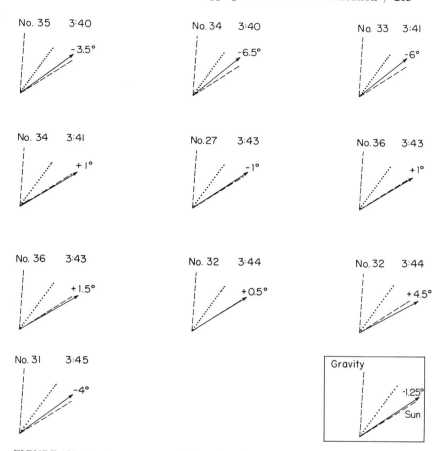

FIGURE 197. Ten dances on a vertical comb surface, 31 August 1961, in sight of the sun and the blue sky. The deviation of the angle of dancing from precise orientation to light is shown. Further explanation as in Fig. 195.

them, whether the sky was clear blue or moderately overcast, whether the polarization pattern seemed more or less contrasting in accordance with the changing position of the sun, had no influence on the orientation of the dancers—as long as the plane of vibration still remained recognizable.

The obvious question whether the newcomers aroused are led astray by the "misdirection" was examined as long ago as 27 September 1949 in a fanwise experiment. An observation hive was set up in the Zeppezau, a broad, meadowed valley not far from Brunnwinkl, at the base of the southern slope, and on the 26th a feeding station (without scent) was established on the opposite side of the level valley floor 250 m due north of the hive (Fig. 198). The sugar concentration was kept so low that no newcomers were aroused, so that there was no preliminary training to the direction of the station. On 27 September seven scent plates[66] with oil of lavender were put out 200 m from the hive at 15° intervals, after which 2M sugar solution on a lavender-scented base was

[66] As in Figs. 83 and 84, but with only three scent vials each.

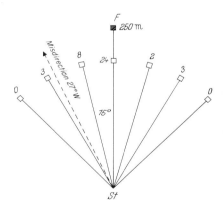

FIGURE 198. Fan-shaped experiment on 27 September 1949, for testing the question of whether the newcomers are led astray by the "misdirection": St, observation hive; F, feeding station 250 m from the hive. The numbers adjacent to the seven scent plates (200 m from the hive) indicate the number of visits to each. Period of observation, 1.5 hr.

offered at the feeding station between 9:10 and 10:40 A.M. During the period of experimentation it was cloudless with a weak east wind, calm at times. The flights to the observation sites are indicated in Fig. 198 by the adjacent numbers.

During the observation time the dancers had a free view of the blue sky, from the dance floor on the northward-facing comb surface. The average misdirection for 29 dances measured in the experimental period amounted to 27° west. If the newcomers had followed this direction, they would have had to deviate to the west from the line of flight of the foragers by two observation sites (dashed line in Fig. 198). But most of the visits led correctly toward the feeding station.

From this it is evident that the newcomers informed, who indeed are subject to the same influences as the dancers, compensate for the deflection when they transpose the indication back into the angle of flight relative to the sun. This conclusion later was confirmed by the great accuracy with which the dances on a hanging swarm lead to an often deeply hidden goal (p. 271). Here one meets the already mentioned case in which under natural circumstances blue sky comes into competition with the force of gravity.

(b) *"Misdirection" Due to the Force of Gravity*
    ("Residual Misdirection")

Now we shall observe the dances on a vertical comb with the sky invisible, in diffuse light. Any deflection due to the blue of the sky is impossible. The marked "misdirections" discussed in the previous section have therefore vanished. We shall now have to deal with the remaining smaller "misdirection."

In order to understand what follows, one must keep clearly in mind the nature of our dance measurements. A feeding place that lies toward the sun is indicated on the vertical comb by dances pointing vertically upward. If after some time the sun has progressed 10° clockwise, the direction of dancing will have changed 10° counterclockwise and will point out the goal as being 10° to the left of the sun. Now it is possible, beginning with 0° for the vertical direction, to count off counterclockwise (thus to the left) through 360°. Some authors have done so. But we count the angles of dancing only as far as 180° and distinguish between dances that are inclined 1°–180° to the left or to the right. That corresponds logically to a goal that is situated to the left or right of the sun. That it also has logic in evaluating the observations is revealed by a glance at Fig. 199 (with which indeed we anticipate our conclusion, that the "residual misdirection" depends on the sense organs for gravity perception).

In the left-hand diagram a northern feeding station F is assumed. In the morning with the sun in the southeast (I), it is 150° left of the sun and correspondingly is to be indicated on the vertical comb by an angle of dancing 150° left of the direction upward (right-hand diagram, expected angle of dancing I'). Too large a dancing angle is observed (in the example $+10°$). For a corresponding afternoon position of the sun (left-hand diagram, II; the feeding station 150° to the right of the sun)

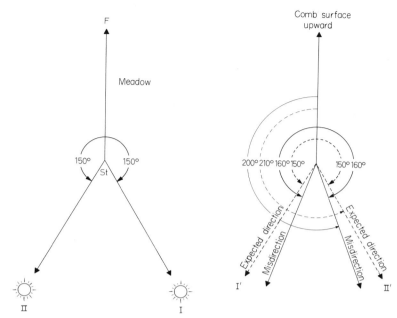

FIGURE 199. Diagrams explaining how angles are measured: (*left*) *St*, observation hive; *F*, feeding station (north); I and II, position of the sun in the morning and afternoon, respectively; (*right*) the indications of direction expected from the left-hand diagram (dashed lines) and the "misdirections" (each 10°, solid lines), on the vertical comb. We reckon the angle of dancing through 180° right and left from 0° for the direction vertically upward, which yields a positive value for the misdirection in both instances; with continuous reckoning over 360° (counterclockwise) there would be a positive misdirection on the left and a negative one on the right.

the dance is directed toward the lower right (right-hand diagram, II′) and the "misdirection" likewise amounts to +10°. Physiologically this is a satisfying result: with symmetrical orientation to right and left the "errors" due to the force of gravity correspond. If we were to traverse the angles of dancing from 1° to 360° the dancer, orienting symmetrically to gravity, would announce the angle as being too large toward the left and too small toward the right, as may be seen from the two solid arcs.

Data. Experiments in earlier years had shown that with the "residual misdirection" the systematic diurnal course also changes with the direction of flight. In order to obtain some insight into the relations we (v. Frisch and Lindauer 1961) measured, from morning to night, on different days and in different years, with different colonies, with different directions of flight and in different environments, the dances of bees that came from an artificial feeding station. As examples, in Figs. 200–203 there have been selected experiments in which the feeding station was due south, north, east, or west of the hive.

First let us consider Fig. 200. The feeding station was in level terrain 200 m south of the observation hive. On the abscissa are entered the times of day, on the ordinate the angles of dancing. For each bee observed after her return home the dance measurement was recorded

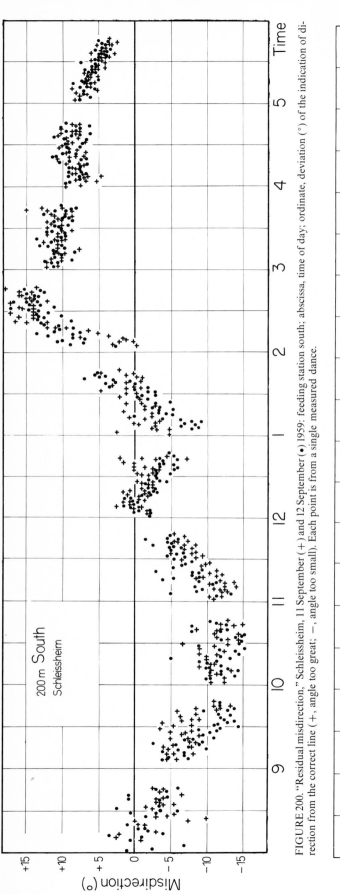

FIGURE 200. "Residual misdirection," Schleissheim, 11 September (+) and 12 September (●) 1959: feeding station south; abscissa, time of day; ordinate, deviation (°) of the indication of direction from the correct line (+, angle too great; −, angle too small). Each point is from a single measured dance.

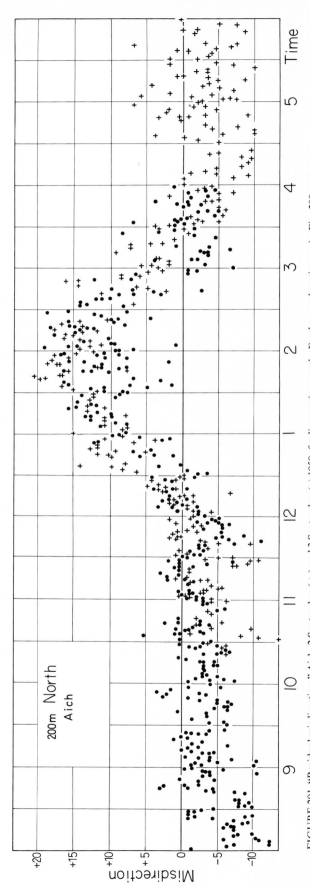

FIGURE 201. "Residual misdirection," Aich, 2 September (+) and 3 September (●) 1958: feeding station north. Further explanation as in Fig. 200.

separately. Angles of dancing too great are recorded above the 0° line, those too small below it.[67] The experiment was performed on 11 September 1959 from about 8 A.M. to 6 P.M. (crosses in the figure) and was repeated on the 12th (dots in the figure). It is evident that the values coincide. Hence the striking diurnal course cannot be ascribed to chance.

The diurnal course is not, however, dependent solely on the time of day. With a northern feeding station (Fig. 201, experiments in the level meadows at Aich on 2 and 3 September 1958) the "misdirections" at these same times of day are different. Thus in Fig. 200 there appears a sudden rise in "misdirection" between 2:00 and 3:00 P.M., but in Fig. 201 a swift decline during the same hour. In the late afternoon in Fig. 200 the angles are too large, in Fig. 201 too small. Here, too, repetition of the experiment on two successive days (crosses and dots in Fig. 201) produced concordant data.

Anyone who perchance might think that the difference between the north and the south experiments could be attributed to the difference in years or in the experimental location may be referred to Figs. 202 and 203. In Fig. 202 the feeding station was 200 m east of the hive (Schleissheim, 24 August 1959), in Fig. 203 200 m west (Schleissheim, 27 July 1959). The courses of the "misdirections" in these experiments are almost mirror images of each other and in both instances are quite different from those with a north or south flight direction.

Apparently the "misdirection" is related to the daily course of the sun, which results in a constant change of the direction of flight in accordance with the sun's position. If that is the decisive factor, then the "misdirection" should remain constant when the sun stands still. Of course we cannot halt the advance of the sun, but we can let the feeding station travel with it. The first result of such experiments—begun in cooperation with Lindauer and then extended by him alone—was an astonishing one.

Experiments with a traveling feeding station ("motionless sun"). If the feeding station is to lie in the direction of the sun throughout the day, then it is sighted in initially in this direction over the hive entrance and displaced thereafter, with constant checking every minute or two, in conformity with the change in the sun's position. The feeding table has to be marked conspicuously with scent and color so that the foraging bees will readily find it again despite its shift in location. In the evening it is brought back from the west to its eastern starting point, if a repetition of the experiment is intended. If the feeding station is to stand at a constant angle to the left or right of the sun, the sighting bar on the hive is joined at the desired angle with a second one. The experimental terrain was free and clear on all sides.

---

[67] In recording the angles, formal difficulties arise when the expected direction of dancing and the observed direction of dancing lie on opposite sides of the vertical. If the deviation was equal on both sides, for instance, the expected angle 2° (or 178°) to the right with an observed angle of 2° (or 178°) to the left, we have omitted the observation. When the angles were unequal, such marginal cases have been treated as though the bee had comprehended correctly the larger angle (no matter whether relative to light or gravity) and had localized the smaller angle in the wrong direction. Whether in doing so we hit upon the proper course remains uncertain. The over-all result is not affected by these individual cases.

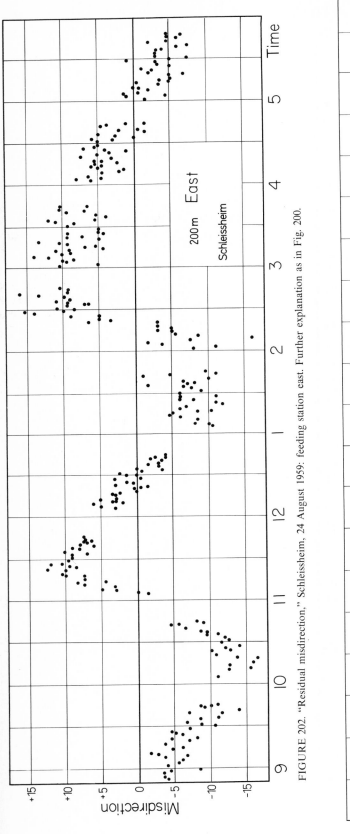

FIGURE 202. "Residual misdirection," Schleissheim, 24 August 1959: feeding station east. Further explanation as in Fig. 200.

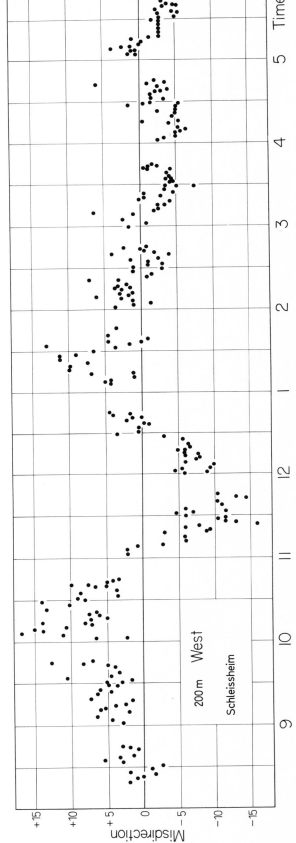

FIGURE 203. "Residual misdirection," Schleissheim, 27 July 1959: feeding station west. Further explanation as in Fig. 200.

Experiments had shown that no "misdirection" ensues when the feeding place lies in the direction of the sun (for example, in Fig. 200 the direction south at noon). We anticipated that from early morning till late afternoon there would be no "misdirection" if the goal remained in the direction of the sun during the entire time. But there occurred a leftward "misdirection" of appreciable magnitude (Fig. 204). The deviation was in the same sense whether the feeding station was kept at a constant angle to the left or to the right of the sun's position. Thus these experiments did not help us toward a comprehension of "residual misdirection." Probably the result comes from the fact that we had created thoroughly unbiological conditions and that the bees, fully accustomed to a constant location of their feeding places—whether flowers or feeding stands—and likewise familiar with the diurnal motion of the sun, after some little lapse of time went hunting to the left of the sun's position for a goal that they had discovered in the direction of the sun. With the traveling feeding station this tendency, combined with a continuous process of learning the way to a new location, leads to a compromise such that the flight path lags behind the movement.

This interpretation seems plausible and is supported by experiments that Lindauer (1963) made, in which the feeding stand followed the course of the sun by jumps, only every $\frac{1}{2}$ hr. At first the dancers kept pointing to the old feeding place, but gradually they shifted the solar angle to the new situation (Fig. 205). We have already come upon the same phenomenon when the feeding place was moved in a detour experiment (see p. 176).

Full confirmation was supplied by a repetition of the experiment with bees that Lindauer had reared without any view of the sky. As he had succeeded in showing previously, bees have to learn during the first days of their life the diurnal course of the sun (see pp. 364f). The young bees grown without any sight of the sun had not yet learned this and hence were not tempted to hunt to the left for the feeding station as the sun progressed. And now the anticipated result finally appeared. There was no "misdirection"; from early till late the dancers, with little scatter, pointed vertically upward, that is, toward the sun (Fig. 206).

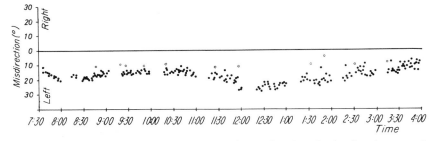

FIGURE 204. Experiment with a traveling feeding station, kept in the direction toward the sun (4 August 1961). During the entire day there is strong misdirection to the left. The indication of direction lags 1 hr and more behind the westward-moving goal; the open circles show the initial dances of bees that had been marked $\frac{1}{4}$–$\frac{1}{2}$ hr previously as newcomers to the feeding stand; their misdirection is less, since they indicate the location of their recent initial visit, which is still near the current location. After Lindauer 1963.

210 / The Dances of Bees

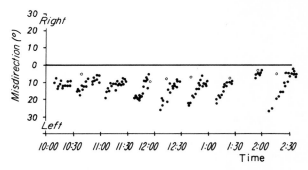

FIGURE 205. Like Fig. 204, but with the feeding stand moved only once each 0.5 hr (11 August 1961). The times of movement are marked with dashes on the abscissa. Note the sudden jump and gradual leveling off of the misdirection after each shift of the feeding station. After Lindauer 1963.

Thus it was eventually shown indirectly that a determining element in "misdirection" is to be seen in the direction of flight relative to the sun.

No "misdirection" in dances on a horizontal comb. Matters were taken a step further when we laid the hive over horizontally and gave the bees a free view of the sky, so that they could orient their dances directly in accordance with the sun and the blue sky.[68] Although we were watching older flight bees, which were familiar with the motion of the sun, there was no "misdirection" (Fig. 207, flight direction east, experiments on 11 and 21 August 1959). The scatter about the correct value was random and remained within narrow limits. There are concordant experiments with a southward flight direction, from 5 and 8 July 1959. Further observations in July and August 1962 (direction of flight, west) were put at my disposal by Neese: in a total of 69 dances the average

[68] In order to avoid overheating the hive and disturbing the bees the sun has to be excluded from time to time by shading.

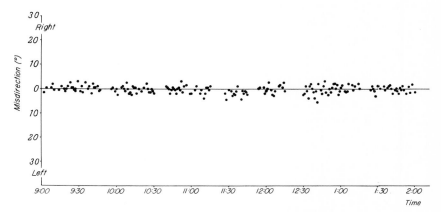

FIGURE 206. Experiment with a traveling feeding station, kept in the direction toward the sun (1 September 1961). The bees were reared *without sight of the sun,* and there was *no misdirection.* From early till late the dances are scattered evenly and to only a slight extent around the zero direction (upward). After Lindauer 1963.

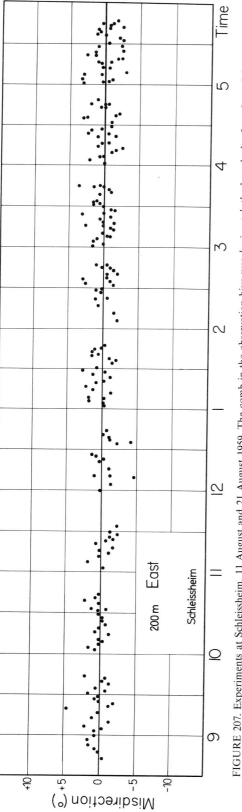

FIGURE 207. Experiments at Schleissheim, 11 August and 21 August 1959. The comb in the observation hive was *horizontal*; the bees had a free view of the sun and the blue sky, and there was *no misdirection*.

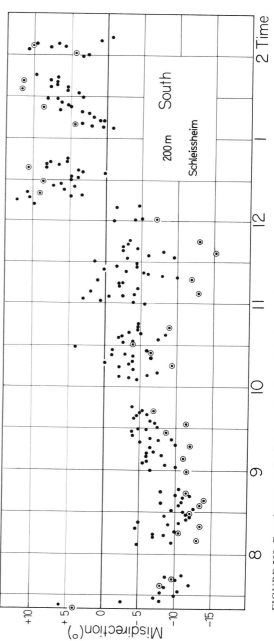

FIGURE 208. Experiments on "residual misdirection," Schleissheim, 6 July 1959; feeding station 200 m south. Bee No. 146 (circled points deviates in her indication of direction beyond the range of scatter shown by other bees.

"misdirection" was 0.4°, toward the right of the feeding station; the greatest "misdirection," observed on only one occasion, was 5.5° to the right.

One comes to the conclusion that the "misdirections" in the dances on the vertical comb surface arise from transposition of the visual angle to the gravitational angle.

Individual differences. The amount of "misdirection" can vary from one individual to another. Some specimens even reveal themselves as pronounced deviators. Figure 208 gives a typical example. The feeding station was 200 m south of the hive (6 July 1959). In the morning the bees were signaling too small angles. The indications of direction by bee number 146 (circled dots) lay at the lower limits of the area of scatter or even lower; but in the afternoon, when the angles of dancing were too large, they are found at the upper limits of the range of scatter, or above. That means that in both instances the bee signaled beyond the range of scatter a direction to the left of the goal, as might be expected if for some reason or other the sense organs for gravity perception of the left side were dominant. With other bees we saw deviations to the right. Such a tendency to right or left may—but does not always have to—persist in an individual for days at a time. In one case we had the opportunity to see that even an extremely aberrant dancer stayed quite within the normal limits of scatter on a horizontal comb. That too agrees with the assumption that her deviant performance on the vertical comb was connected with a peculiarity of her gravitational sense organs.

Apart from these rare rather strong deviations there are fairly frequently in the indication of direction systematic differences of small extent (see p. 149 and Schweiger 1958), a fact that leads to an increase in the scatter in experiments with larger groups of bees. Hence the curves obtained look nicer if one limits oneself to a small number of individuals for the measurements. Since the magnitude of scattering also differs with individuals, where possible we tested several bees before the beginning of an experiment and used only those that gave clear-cut indications of direction. The difference is made evident by a comparison of Figs. 200 and 201. In the first of these experiments the measured values were obtained on both days of observation from 7 or 8 good dancers; the values for the two experiments shown in Fig. 201 came from 20 or 21 bees selected at random. Each of the two additional figures, Figs. 202 and 203, is based principally on the values obtained from 16 bees after the exclusion of poor dancers.

Are the bees themselves also "misdirected?" Before discussing how these "errors" arise in the dances on the vertical comb surface, we enquire whether bees are "misdirected" at all when we measure a "misdirection." We have already answered the question in the negative as regards "misinformation" due to the light (p. 203), but it remains to examine it in relation to the "residual misdirection" (*Restmissweisung*).

To clarify this, fanwise experiments were carried out in September 1959 with an observation colony in Aich at a time of strong "misdirection." On 4 September 1959 a group of marked bees were collecting

270 m east of the hive. The feeding station was scented with oil of orange blossom. Somewhat nearer the hive and at angular separations of 8° scent plates were laid out during the experimental period, with so little oil of orange blossom that newcomers searching about were attracted from no farther than a few meters distant.[69] A half hour before the beginning of observation strong positive misdirection already prevailed; during the experiment between 2:20 and 3:20 P.M. it amounted on the average to +12°. If the comrades informed had adhered to so great an angle relative to the sun's position they would have had to fly out in the direction of the dashed arrow in Fig. 209. But the distribution of the visits, which are recorded adjacent to the scent plates shown in Fig. 209, does not reveal any influence of "misdirection."

At that time we did four fanwise experiments, two with the direction of flight east, two with the direction of flight south, one of each with positive and the other with negative "misdirection." In no case was there a deflection of the newcomers to the side indicated by the "misdirection" (v. Frisch and Lindauer 1961:589). Thus it is clear that the indication of direction is correct to the bees and that the "errors" that ensue after transposition of the visual to the gravitational angle are felt as such only by ourselves.

Under what conditions does "misdirection" occur? Now we would like to know how these apparent mistakes come about.[70] Although there were at our disposal 33 usable experiments with feeding stations toward 9 different points of the compass, of which we have given here only 4 examples in Figs. 200–203, we long sought in vain a reasonable basis for the changing diurnal course of the curves—until the following consideration proved fruitful.

The misdirection changes when the line of flight to the goal is changed (see Figs. 200–203). But with a shift in the direction of flight, under otherwise identical conditions, there comes a change in the direction of dancing on the vertical comb. Hence it might be profitable to arrange the observations not in the order of the time of day of the dances (as in Figs. 200–203) but rather according to the direction of dancing.

Figure 210 refers to such a representation of the experiment of 27

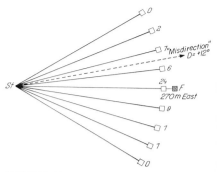

FIGURE 209. Fan-shaped experiment with strong "residual misdirection": *St*, hive; *F*, feeding station. Nine scent plates were set out, at angular separations of 8°. The adjacent numbers show how many newcomers settled on them during the 1 hr of observation. Dashed arrow: line of "misdirection"; this had no effect.

---

[69] For the scent dispenser at the feeding station see Fig. 21; scent plates as in Fig. 84 were placed on low wooden stands (Fig. 85).

[70] There are two dissertations that bear on this question. E. Palitschek v. Palmforst (1952) observed in Graz—in an independently conducted study—that in their indication of direction both on a vertical and on a horizontal comb bees almost always point past the goal to the left ("left-hand misdirection"). We were unable to confirm this. In his experiments the feeding station was invariably southwest or southeast of the hive. With these directions of flight there actually is mostly a left-hand misdirection in dances on a vertical comb, but with other directions of flight right-hand misdirections very often occur. Palitschek's complicated attempt at explanation for the occurrence of misdirection starts from a basis of left-hand misdirections and thus is founded on premises that by no means always are exemplified. A. Djalal (1959) worked in the neighborhood of Munich. We have repeated his experiments in part at the same places and time of year and have become convinced that the enormous scatter in his measured values is a consequence of inadequate technique. On two important points we agree with his results: that the misdirection is small when the direction of flight is toward the sun or away from it, and that the elevation of the sun has no influence on the misdirection. He found no explanation for the cause of misdirection.

214 / The Dances of Bees

No.	Time	n
1	8:19– 8:45	14
2	9:01– 9:25	13
3	9:28–10:03	15
4	10:05–10:35	20
5	10:37–11:03	9
6	11:06–11:26	13
7	11:28–11:45	12
8	12:01–12:09	7
9	12:11–12:28	11
10	12:30–12:45	11
11	1:07– 1:12	4
12	1:14– 1:35	12
13	1:37– 2:03	7
14	2:05– 2:32	20
15	2:33– 3:09	15
16	3:10– 3:45	28
17	4:02– 4:39	25
18	4:41– 5:05	7
19	5:09– 5:35	18
20	5:37– 5:50	11

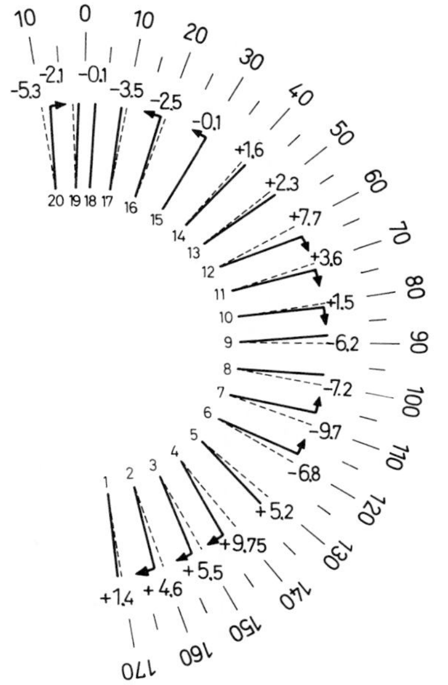

FIGURE 210. Results of the experiment on 27 July 1959 (see Fig. 203), arranged according to the direction of dancing. The time course of the experiment is to be followed counterclockwise; dashed lines, expected directions of dancing; solid lines, observed directions of dancing, combined in means for each 10°. For each stage the amount of misdirection is given; its tendency to shift in the course of the experiment is shown by the arrows. Time intervals and the numbers of the individual observations are given for each stage in the list at the left. Further explanation in text.

July 1959 (flight direction west), the results of which, arranged in order of the times of day, we have already become acquainted with in Fig. 203. For better understanding think of Fig. 210 as being projected onto the vertical comb surface of the observation hive, so that the 0° point of the circular scale points upward, the 180° point downward (the scale in degrees is given by the numbers on the outer margin). The feeding station was to the west (azimuth 270°). At the beginning of the experiment the sun was somewhat south of east (azimuth 100°). Thus the location of the feeding station was 170° to the right of the sun. Hence the bees ought to dance 170° to the right of the direction upward. This situation at the beginning of the experiment is indicated in the circular diagram by the dashed line at number 1 (the numbers on the inner margin refer to the order in time); here as elsewhere the dashed line indicates the expected direction of dancing. The solid line shows the angle of dancing actually observed: it was 1.4° too large (mean for all dances whose expected values lay between 165.1° and 175°). In the further course of the experiment the direction of dancing shifts counterclockwise, as already emphasized. Next one finds indicated in this way at 160°, by means of the solid line, the average "misdirection" for all dances whose expected value lay between 155.1° and 165°, and so on in 10° steps.[71]

For the number of each stage, the time of observation and the number of the individual data (n) are given in the list beside the figure. The figure shows for the different directions of dancing how the dances deviated from the directions expected according to our way of interpretation (the angular deviations for each are shown in the middle row of

[71] In the diagrams we have departed from the use of 10° steps only at 0° and 180°, where the angles are divided into intervals from 0° to 5° or 175° to 180°, respectively, to right and left, in keeping with our separate evaluation of the dances directed to left and right.

numbers of the circular diagram). And now a rule is discernible. In the lower quadrant, the angles of dancing are too large at first and deviate from the expected direction toward the vertical; then they grow too small and deviate toward the horizontal; after this has been passed (at a direction of dancing of 90°) they approach it from the other side and again become too large, then in the upper region once again deviate with too small angles toward the plumb line, as also happens from the other side once this has been exceeded. The same rule held true for experiments with corresponding opposite directions of flight and therefore dances toward the left, that is, in the other half of the circular diagram.

This shift between too large and too small angles of dancing in the different sectors did not always appear so regularly. But almost without exception the tendency for a corresponding change in the angle of dancing was clear. Figure 211 will show how that statement is intended. The diagram refers to the experiment of 2–3 September 1958 with the direction of flight northward. The measurements of the dances, arranged according to the time of day, have already been presented in Fig. 201. At the beginning the expected angle of dancing was 150° to the left (No. 1 in Fig. 211) and by noon it reached 180°. In this sector, in analogy with Fig. 210, dance angles too large were to be anticipated. In actuality they were too small. But a tendency toward the vertical is expressed in the decrease of the "misdirection" from $-5.6°$ to $0°$. In its further course the performance corresponds to that of Fig. 210.

Similar deviations and irregularities were frequent in the experiments. We believe that they are to be attributed to disturbing factors that as yet we do not recognize[72] and whose clarification remains the

---

[72] Among other things one might imagine an influence of the varying elevation of the sun, since the elevation of the source of light can have significance in the light-compass orientation of fish (Braemer 1960; see also 1961), of dung beetles (Birukow and De Valois 1955), and of ants (Jander 1957:191). In the indication of direction by bees only the sun's azimuth and not its elevation is determinant (p. 136). The "misdirection" too is independent of the elevation of the sun. From our experimental protocols we have compared measurements of dances for which the solar angles were equal but the elevations different. Thus a southern feeding station is in the direction of the sun at noon, a western one in the evening, while the sun is at its highest position in the one instance and is nearing the horizon in the other. In spite of this there is no difference in the "misdirection." We made corresponding comparisons for solar angles of 30°, 60°, 90°, 120°, and 150°, both to the right and to the left. Here too no relation could be seen with the same azimuth and differing elevation of the sun between the latter and the magnitude of the "misdirection." Details, with quantitative data, will be found in v. Frisch and Lindauer (1961:591, 592).

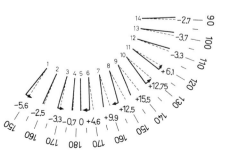

FIGURE 211. Results of the experiment on 2 September 1958 (see Fig. 201), arranged according to the direction of dancing. Further explanation as in Fig. 210.

No.	Time	n
1	10:29–10:57	12
2	10:59–11:23	20
3	11:24–11:49	24
4	11:50–12:03	8
5	12:04–12:15	11
6	12:17–12:38	17
7	12:43– 1:06	16
8	1:09– 1:35	12
9	1:39– 2:09	22
10	2:11– 2:43	12
11	2:45– 3:25	24
12	3:28– 4:12	26
13	4:13– 5:02	34
14	5:03– 5:56	37

216 / The Dances of Bees

task of further research, but that in the pattern followed by the "errors" in Fig. 210 there emerges the rule that lies at the basis of the "misdirections." We reached this conviction because this rule is exceptionally clear in the general averages from all experiments (Fig. 212).

In explanation let us remark that in Fig. 212 we have gathered all experiments except those that were spoiled by rain or heavy clouds. We do not enquire where they were done, at what time, with what flight distances and directions, but we arrange all the data, 6759 measured dances from 33 experiments, exclusively in accordance with the expected direction of dancing on the vertical comb surface. Thus, for instance, for the expected direction of dancing 10° to the left there are included from all the experiments all the dances in which the value expected lay between 5.1° and 15°. The "misdirection" was calculated for each dance and the mean of the values was entered in the diagram.

One sees that the indications are quite precise when the dances are directed upward or downward or approximately horizontally to left or right. In the upper and lower sectors the "misdirections" tend toward the vertical, in the lateral sectors toward the horizontal; the boundary between the two tendencies lies at about 60° to either side of the vertical. Thus pictorially expressed, the vertical exerts greater attraction than the horizontal. That is expressed also in the fact that, if we follow the "misdirections" counterclockwise in correspondence with the diurnal movement of the dances, after the horizontal direction of dancing is passed a tendency toward a reversal of the deviation does become evi-

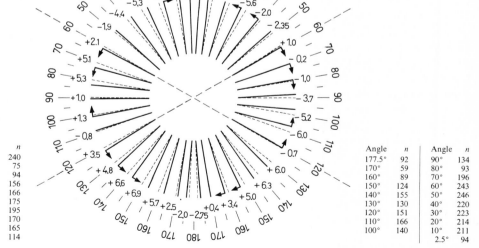

FIGURE 212. Over-all average values from all usable experiments regarding "misdirection" (33 experiments with 6759 measured dances), arranged according to the angle of dancing. The direction of dancing is to be followed counterclockwise, as are the numbers (n) of the individual values in the lists at left and right. For further details see the legend of Fig. 210.

dent (see the arrows in Fig. 212), although there is no very strong excursion, whereas the excursions toward the vertical from both sides are very clear.

Thus there is displayed an obvious relation between the direction of dancing on the vertical comb and the "misdirection." There arises the question of its physiological basis.

The physiological basis of "misdirection." We first suspected that the "misdirection" came about as an error of translation in transferring the visual angles to gravitational angles, since the bees use differing physiological scales for visual angles and gravitational angles (v. Frisch and Lindauer 1961).[73] Hence it was of interest to learn whether these angular deviations constituted a peculiarity of bees or whether one was dealing with a widespread phenomenon.

With some other insects, too, there occurs a transposition of visual angles (perceived with the body in a horizontal position) into gravitational angles (after the surface beneath is tipped into a vertical plane), though without the biological significance that is present in the bee dances (see pp. 326f). Unfortunately, little attention has been paid in these experiments to how precisely the angles are maintained. However, Jander (1960) studied the behavior of the caddis fly *Limnophilus* in this respect. When placed on a horizontal board in the dark room, most of the specimens ran obliquely toward a light source visible at horizon height, in doing which they held to a definite—individually differing— angle to right or left of the light (menotaxis). Some of them also ran away from the light in a corresponding manner (Fig. 213a, *left*). If the light was turned off and at the same time the board was tipped through 90°, so that the surface was vertical, the caddis flies transposed the visual angle into an orientation to gravity, not directly but in accurate proportion (Fig. 213a, *right*). The angle with respect to gravity was related to the visual angle as 2:3 (Fig. 213b). With directions of running approximately vertically upward or downward, the gravitational angle therefore deviated, as with our bees, in the direction toward the vertical. Since Jander's caddis flies only exceptionally ran at right angles to the line of illumination, we do not know whether they also behave on the vertical surface like the bees when the horizontal direction of running is approached.

According to Inge Tenckhoff-Eikmanns (1959), the beetle *Tenebrio molitor* L. transposes the visual angle into the gravitational angle approximately in the ratio 1:1. For the ladybird *Coccinella septempunctata* L. according to her statement (1959:329) the same is true. But her protocols (Fig. 13, p. 330) show clearly that, after transposition from the visual to the gravitational sense, the angles deviate on the average toward the vertical. This holds true even more strongly with the poplar-leaf beetle *Melosoma populi* L., where even Tenckhoff-Eikmanns speaks of "a sort of misdirection" relative to the path in light (p. 331 and Fig. 14). In all cases the "misdirection" is in the same direction as with bees.

[73] V. Holst (1950) succeeded in demonstrating comparable relations between orientation to gravity and to light in fish, and Schöne (1954) in crustaceans.

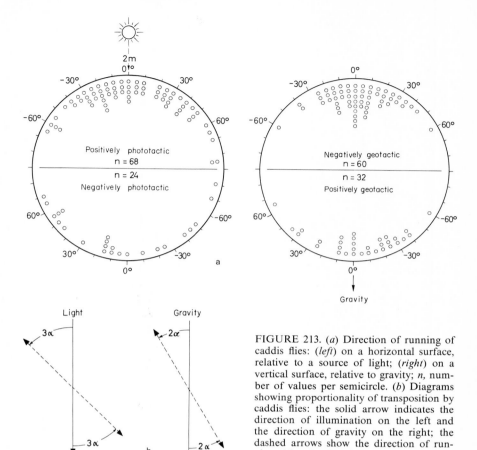

FIGURE 213. (a) Direction of running of caddis flies: (*left*) on a horizontal surface, relative to a source of light; (*right*) on a vertical surface, relative to gravity; *n*, number of values per semicircle. (*b*) Diagrams showing proportionality of transposition by caddis flies: the solid arrow indicates the direction of illumination on the left and the direction of gravity on the right; the dashed arrows show the direction of running. After Jander 1960.

The same seems to hold for the water strider *Velia*. According to Birukow (1956), on dry land these insects always run toward the south and in doing so orient themselves by the sun, maintaining in the morning an angle, corresponding to the time of day, to the right of the sun's position and during the afternoon one to the left. In a dark room they adopt the same course toward a lamp, varying with the time of day, as they do toward the sun out of doors. On a vertical surface in the dark they transpose the visual angle to the gravitational sense, and in the morning run upward to the right of the vertical, straight up at noon, and to the left of the vertical in the afternoon. "The amplitude of the change in course, however, is invariably less during orientation to gravity than during orientation to light . . ." (Birukow and Oberdorfer 1959:695). Because of the variability in the angular orientation of *Velia* and the great scatter in the experiments, a reevaluation of these findings, which are mentioned only briefly in the paper, would be important. A systematic investigation of such behavior in other insects would be desirable.

These fragmentary observations raise the suspicion that, in transposing visual to gravitational angles, deviations, such as we have described as "misdirections" in bees, are a widespread rule.

In interpreting the phenomena, we have recently progressed a step forward through investigations by Markl (1964) on ants. In them he found errors of geotactic orientation, similar to those of bees—but occurring during orientation purely to gravity, without any transposition. In view of this, our earlier assumption that errors in "translation" were concerned becomes very improbable for bees too.

As a surface for his ants (principally *Formica polyctena* Förster) to run on, Markl used a vertical circular board (arena) with a central opening behind which was the nest (Fig. 214). The ants could be trained to find offered food after leaving the nest opening by maintaining during their course a definite angle relative to gravity. During this performance regular deviations were observed in the directions run from the line to the goal; and these were not altered by long training. Their average values from numerous experiments are shown in Fig. 215. Altogether the ants were trained to 14 different directions. These are given as dashed lines inside the circle. The solid lines inside the circle indicate the average direction actually chosen in leaving the nest. In the upper semicircle these deviate upward from the line to the goal, in the lower semicircle downward from it; thus in both cases they tend toward the vertical. These deviations cannot be "errors of translation" because there was no transposition in these experiments. The matter is one of a genuine mistake in orientation: the ants were unable to reproduce properly the angle that accorded with their experience. They would have run past the goal were it not that, like the bees, they had a good estimate of distance, so at the proper place they would begin to search in loops.

These mistakes of the ants, with their tendency toward the 0° direction when trained obliquely upward and toward the 180° direction when trained obliquely downward, are so reminiscent of the performance of bees that a common basis must be sought. There is an exception, however, in that deviations toward the horizontal, when this direction is approached, are not to be recognized with ants but are distinct with bees—even though weaker than the upward and downward trends (see Fig. 212).

For the results of his investigations with ants Markl supplies an analysis that I subscribe to, and not only for the ants; it both makes the corresponding behavior of other insects comprehensible and in particular affords an explanation of the main elements of "residual misdirection" among bees.

Bees and ants (like other arthropods) possess an inborn fundamental orientation toward stimuli from light and gravity. This is expressed, for example, as positive phototaxis and negative geotaxis in ants when they leave the nest and is found with polarity reversed among specimens returning homeward. It is limited to a few preferred directions (positive, negative, transverse) and is the original manner of orientation from which menotaxis, the setting of the body, on the basis of experience, at the chosen angle relative to the source of stimulation, is to be

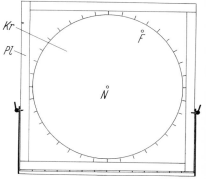

FIGURE 214. Vertical surface for training ants to a direction relative to gravity. The nest is behind the surface. From the nest entrance N the ants emerge onto the plate Kr (diameter 60 cm). At the periphery of this circular area a feeding vessel F is placed. The ants must learn to maintain the proper angle relative to gravity while seeking the food. The circular arena is sunk below the surrounding frame, and the entire square is bordered besides with lucite Pl coated with paraffin oil so as to be impassable for the ants. After Markl 1964.

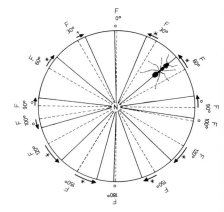

FIGURE 215. Results of training to gravity. In the course of the experiments the ants were trained to 14 different directions relative to gravity; the dashed lines inside the circle (arena Kr in Fig. 214) show the different directions at which food F was offered to the ants emerging from the nest N. The continuous lines are the directions (average values) actually taken (without food) by the trained ants. The arrows on the margin indicate the direction of deviation: *, deviation statistically highly significant; +, weakly significant; o, not significant. After Markl 1964.

220 / The Dances of Bees

derived (see Jander 1963). When ants learn to hold to a definite direction relative to a source of stimulation while they are seeking a goal, this inborn fundamental orientation becomes noticeable in the first instance as a deflecting factor (Jander 1957). If we assume that in Markl's attempts to train the ants to a given gravitational angle the latent fundamental orientation exerts a persistent influence on the direction of running, then it becomes evident that a deflection toward the vertical occurs when the direction is obliquely upward or downward. That bees behave somewhat differently, in that with them a deflection toward the horizontal also is noticeable, can be ascribed to a transverse geotaxis. But why such a thing should exert an effect in the "misdirection" of bees and not in Markl's experiments with ants is an open question. However, with regard to the positions adopted spontaneously relative to the direction of vibration of polarized light ("oscillotaxis" in Jander's terminology), we know that often only one, in other cases several, of these basic directions occur (see p. 407f). To imagine a similar situation with relation to geotactic orientation is by no means misleading.

Markl tested his ants not only as they departed from the nest but also on their return. Here the deviations from the expected directions are somewhat different (Fig. 216). If the return route from the food to the nest entrance[74] leads obliquely upward, a deviation from the direction of the goal is not significant, whereas in the lower sectors deviations toward the vertical become striking. This finding provides support for the conception that error in the position taken is to be attributed to the influence of the fundamental orientation. For the fundamental negative geotactic tendency of the departing ants has been reversed into a positive geotactic trend in the ants returning, and this is

[74] In the figure the nest entrance, actually central, is drawn on the periphery, in order to facilitate comparison of the deviations with those in Fig. 215.

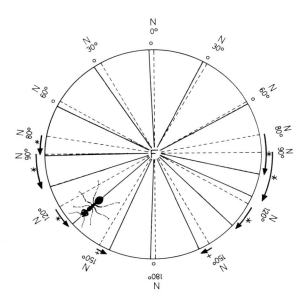

FIGURE 216. Deviations of ants from the expected direction of running on returning home from the feeding place to the nest. Explanation as in Fig. 215; for the sake of better comparison with this figure the nest has been drawn on the periphery and the feeding place at the center. The actual direction of running from the feeding place to the nest relative to the direction expected is of course indicated correctly in each case. After Markl 1964.

expressed in the upper sector as a decrease and in the lower one as an increase in deviation due to the fundamental orientation.

If after these findings we once again consider the "residual misdirection" of the bees, as it is expressed in the comprehensive representation of all observations (Fig. 212), we need no longer think in purely pictorial terms of the "attractive influence" of the vertical and the horizontal (see p. 216). These deviations can be ascribed to the concurrently operative fundamental orientation. With the bees they do not lead to any error in orientation on the part of the comrades informed because they are retransposed according to the same rules to the position with respect to light by the dance followers and thus are compensated.

According to Markl (*in lit.*), he has meanwhile succeeded in training *bees* to run at a given angle to the vertical. The bees flew from their hive to the entrance of a corridor that led into a dark room and ended in the center of a vertical arena (as in Fig. 214). The bees had to learn to maintain, from this point onward, the angle to the direction of gravity that would lead them to the peripheral feeding vessel. With bees such training is not easy. Nevertheless, skillful techniques, for which reference is made to the original publication,[75] produced success.

Like the ants, the bees in these experiments deviated in a definite manner from the angle to which they had been trained. The average values from all observations are given in Fig. 217 at the right for the outward runs and at the left for the return runs. The arrows on the margin make clear the direction of the deviation and the symbols indicate the statistical significance, as in Figs. 215 and 216. Here there is seen, in the outward runs to the source of food, not only a deviation

[75] *Zeitschrift für vergleichende Physiologie* 53, 328–352 (1966).

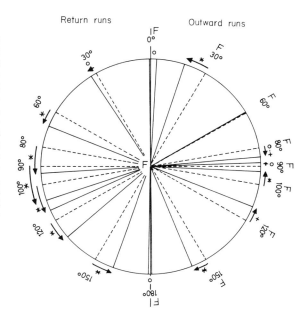

FIGURE 217. Results of training bees to gravity: dashed lines, directions of training; solid lines, directions of running (average values) by the trained bees; right semicircle, on the outward route to the food; left semicircle, on the return to the entrance to the arena. Here, as in Fig. 216, for the sake of better comparison the location of the food is drawn as central, that of the actually central entrance is shown as peripheral. The arrows on the margin indicate the direction of the deviation: *, deviation statistically highly significant; +, weakly significant; ○, not significant. Total number of observations: 967 outward runs, 1072 return runs. After Markl (unpublished).

toward the vertical in the runs obliquely upward or downward but also a deviation toward the horizontal in runs that approached this direction, as was not the case with ants. But as with the ants, in the return runs a tendency toward stronger deviation downward is characteristic; this decreases the trend toward the vertical in the upper sector and increases it below, while the horizontal direction no longer manifests attraction.

With respect to "residual misdirection," of greatest interest is the performance during the outward run to the source of food, for dancing refers of course to the direction of the route outward to the goal. There it is seen that the incorrect courses adopted by bees trained to a given gravitational angle agree, in all the angular sectors, with those deviations that we have described as "residual misdirection" for the dancers. Hence the explanation of the incorrect courses taken by the bees in the geotactic runs is also the key to understanding their "residual misdirection."

In analogy with the ant experiments these incorrect orientations have already been interpreted to the effect that the latent fundamental orientation takes part in the menotaxis and deflects the runs toward the directions known to be preferred in fundamental orientation. As with the ants, the differing behavior of the bees in the return runs was ascribed by Markl to their dominantly positive geotactic reactions when they are in the mood to return home. Jacobs-Jessen (1959:629, 630) demonstrated this positively geotactic basic mood (in combination with negative phototaxis) in returning bees. But during the outward runs the positive and negative tendencies of fundamental orientation were activated by the animals' being in a mood to search for food; and with bees the transverse geotactic tendency as well. Thence arises the incorrect orientation seen in the geotactically guided outward runs; and that also would be the gist of "residual misdirection."

25. The Role of the Scent Organ and Floral Odors with Distant Sources of Food

When the arousal of newcomers by dances was already known, but not the indication of direction by them, Lineburg (1924) proposed the theory that the foragers flying to the hive kept their scent organ everted and laid through the air a fragrant trail that was followed by their hivemates flying out. He does indeed note that such a trail in the air can have little stability, but he recollects both the white streaks left by skywriters, which under some circumstances remain visible for a long time, and the possibility that the newcomers follow the foragers returning to the goal at a very brief interval.

Originally I was of the same opinion and knew of no other explanation for the fascinating phenomenon that a few minutes after feeding was begun newcomers came streaming in great crowds to good feeding places even at a distance (see p. 129). I must now report the experiments that at that time, in 1944, raised my doubts of this conviction.

In Fig. 113 is shown the location of two feeding stations, *A* and *B,* that had been established in opposite directions from the observation hive. It has been reported earlier that numerous newcomers always flew in to the station at which food was offered, whereas the other feeding place, at which there was an interruption in feeding, lacked any attractiveness (cf. the experiments of 17 and 18 September 1944, p. 129). On 19 and 20 September such experiments were repeated, but in these the scent organ of the foraging bees was put out of action by covering it with shellac (for technique, see p. 53). On the 20th at feeding station *B,* on the narrow trail between the wooded slope and the lake, 200 m from the hive, 2M sugar solution with added fragrance of peppermint was offered for 1 hr. The marked bees arriving hovered about the feeding dish before settling, as they did customarily on other occasions. One had the impression that they were trying—in vain!—to evert the scent organ. That the shellac covers remained tight was constantly rechecked. This time there could have been no perfumed trail through the air. Nevertheless during the hour's observation 80 newcomers appeared at the feeding station, and 30 at the scent card set out nearby. In the opposite direction two approach flights were recorded at *A,* and none at the site halfway to it. On 19 September, when sugar water with added lavender scent was offered in the open countryside, 300 m from the hive, the influx of newcomers moved just as decisively in this direction, despite the covering of the foragers' scent organ. Today we know that the direction is indicated by the dancers and that the newcomers who—following this message—approach the goal are attracted from close by by the floral scent (or here by the fragrance of the ethereal oil).

This last fact was also known to me at that time and I wished to ascertain what happens when not only the scent organ but also the scent at the feeding station is eliminated. Therefore the two experiments were repeated on 22 and 23 September 1944, the scent organ of the foraging group was again sealed, and no scent was added at the feeding stations. At the adjacent observation sites little watch glasses with some honey were set out instead of scent cards. As was expected, the number of newcomers remained small, but nevertheless the ratio of flights toward the feeding station frequented by the foraging bees was as $10:1$ or as $19:0$, respectively, in comparison with the opposite direction.

Thus the newcomers find the direction even when the formation of a "scent trail" is surely prevented. Such a scent trail would of course be destroyed very rapidly by every breath of air, and Lineburg's suspicion that the newcomers fly out close behind the foragers has been shown to be incorrect.

The detour experiments, too, have taught us that the bees informed pursue the direction indicated by the dancers and not a trail of scent left by them. For they fly out in the direction given them, rather than around the obstacle over the detour on which the foragers are going back and forth (see pp. 178 and 180).[76]

[76] Under quite different circumstances, namely when bees are proceeding on foot to a feeding station located within the hive, a scent trail to the goal may be laid. Such an instance is described by Jacobs (1925:47). In a small observation hive with four combs the weak colony occupied only the two anterior ones. A feeding vessel in the lower part of

When the sites to be harvested are far away, neither the scent of the odorous gland that is used by the foragers before they settle nor the fragrance of the flowers flown to can build for the newcomers a bridge from the hive to the goal. Only after they have been led by the indication of direction and distance into the neighborhood of the feeding place do these perfumed aids become effective. The scent of the glands attracts them directly to the spot where foragers have had success. The specific odor of the feeding place was to be perceived in the hive on the body of the dancers (see p. 32) and now guides the newcomers unequivocally to the proper source. By means of experiments with ethereal oils and natural flowers we had already learned that the fragrance that the forager carries home from the harvesting site in her body hair is of importance for giving information about the source of food (pp. 46ff). But the feeding stations were close to the observation hive. Now we enquire whether even after a long journey in the air the scent of the flowers still clings to the bee's body,[77] or whether perhaps under such circumstances the scent is communicated to the hivemates in some other manner, or possibly not at all.

The nectar secreted at the base of the flower is perfumed heavily with the specific floral fragrance. Hence the forager may carry home with the nectar in her honey stomach a sample of the scent and bring it to the attention of the surrounding bees when she distributes the drop of food. That would be another possible way of transporting the floral scent, and would keep it protected from the winds generated by flight during the aerial journey.

Hannelore Dirschedl (1960) was able to show, by adding a stain to the sugar water, that the newcomers swarming to the feeding station have previously received a sample taste from the forager in the hive. When scentless food was given, the stain was demonstrable in the stomach of 56 newcomers tested, without exception. When at the feeding station strong scent was added, 21 of 24 bees that came there contained the stain.[78]

Whether the scent brought in internally is effective we next investigated with bees whose feeding place was set up near the hive, by means of the following experiment. Single cut flowers of *Phlox paniculata* L. are stuck through wire netting into a dish filled with water, and each is supplied with a drop of 2M sugar solution (Fig. 218). After 1–2 hr the sugar water is pipetted off, collected in a weighing bottle, and kept closed up there. It has a strong odor of phlox. Now at the

FIGURE 218. Separated phlox flowers, each with a droplet of sugar water. The stems of the flowers dip through the screen into a dish of water.

---

the comb space at the rear was discovered by a bee. When she returned to the vessel she stopped for a time, everting her scent organ and fanning her wings on the side of the frame. In a little while traffic was vigorous, and the bees coming back to the food halted at various places and everted the scent organ, so that they occupied the entire route from the anterior combs to the food. Whether it happens also that under especially favorable conditions a marking out of the way is effective during flight remains a vexatious question, so long as we do not know whether the foragers fly over the route with the scent organ protruded. I believe that they do not make use of it until they are hovering over the goal.

[77] Regarding the persistence of the scents on the bees see pp. 49f.

[78] Perhaps the definite external fragrance had been enough for the three other bees and they forwent a sample taste.

feeding station (14 m from the hive) we offer the scented sugar water to the marked bees in such a way that they must suck it up from a narrow crack without their body's being exposed to the fragrance (Fig. 219). In another direction, 26 m from the hive, we have set out in the field a dish with phlox flowers and one with cyclamen flowers (as in Fig. 36). On 9 September 1944, while the foragers were being fed for an hour with phlox sugar water, 137 newcomers aroused flew to the phlox dish; 107 of them settled on the flowers and tried persistently for nectar, which they cannot reach in the deep chalice of these "butterfly flowers." To the cyclamen dish there came only 3 bees, one of which settled.[79] Hence the floral scent brought in in the honey stomach serves for identification of the species of flower visited.

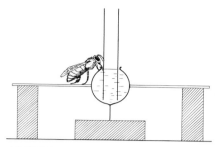

FIGURE 219. A bee is drinking sugar water scented with phlox from the narrow crack between the lip of the spherical glass bowl and the glass tube set into it. The bee's body is not exposed to the scent. The scented sugar water is replenished from above through the tube by means of a pipette.

On 12 September 1944 I performed the complementary experiment. The bees were drinking unscented sugar water from a feeding bowl wreathed with cyclamen flowers, and had to stand on the flowers as they did so (Fig. 220). Once back at the hive, they distributed unscented sugar water, but the dance followers were able to perceive the scent of cyclamen clinging externally to their body. During the hour of observation the cyclamen dish was visited by 105 newcomers (101 of them settled); at the phlox dish there appeared 7 visitors, 2 of which settled. Thus the floral scent borne externally on the body showed itself to be approximately as effective as the "internal fragrance."

In the following summer, on 4 September 1945, I repeated the experiment in the same way, only this time the feeding station was 500 m distant from the hive. Near the feeding station and somewhat to the side the flower dishes were set out. When sugar water perfumed with phlox was fed the phlox dish was visited during the hour of observation by 148 newcomers, who settled without exception. The cyclamen dish did not receive a single visit. Immediately thereafter unscented sugar water was offered and the dish was enwreathed with cyclamen. Now there came to the cyclamen dish 27 visitors, of which, however, only 6 settled. Apparently the internal fragrance had been transported home unweakened during the long flight, but the scent clinging to the exterior of the body had lost in importance.

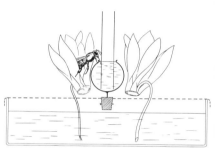

FIGURE 220. A bee is drinking unscented sugar water from the crack while her body is exposed to the smell of cyclamen flowers.

The relative effectiveness of "internal fragrance" (scent brought in in the honey stomach) and "external fragrance" (scent clinging to the body) can be tested simultaneously as follows. The drinking bees sit on cyclamen flowers (as in Fig. 220) but ingest sugar water that smells of phlox. The internal fragrance is put into competition with a different kind of external fragrance. If with such an arrangement the feeding station was close to the hive, both fragrances were effective. In the experiment on 18 August 1946 the cyclamen dish (external fragrance) received 46 visits, the phlox (internal fragrance) 89. But when the feeding station was 600 m distant from the hive (experiment on 1 September 1946) the cyclamen dish (external fragrance) had only 1 visit, the phlox dish (internal fragrance) 38. At such a distance the scent carried home

---

[79] During this and the further experiments a half-hour's preliminary observation invariably convinced me that the flower bowls are not visited when food is not offered and there is therefore also no dancing.

FIGURE 221. Dish for sugar solution, with a lucite cover through which eight capillaries are inserted. In the middle of the cover is an opening for refilling the dish. After Hannelore Dirschedl 1960.

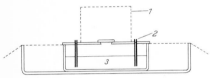

FIGURE 222. Schematic section of Dirschedl's feeding arrangement; 1, wire-screen cover; 2, glass capillary; 3, sugar water.

FIGURE 223. The feeding dish (see Fig. 221) is placed in a petri dish with a screen cover. A ring of screening keeps the bees from the hole used for refilling. After Hannelore Dirschedl 1960.

in the honey stomach, as if in a well-corked bottle, seemed definitely to have won the race (v. Frisch 1946).

When there is no competition from some different internal fragrance, even a weak external fragrance has significance, and even over a considerable flight distance. With regard to this I tested weak-smelling gentian flowers (*Gentiana asclepiadea* L.). During a 1-hr feeding near the hive on gentian flowers a vase at the side with gentian flowers received 26 visits (3 September 1947). Under similar conditions there still appeared 6 newcomers at the flower vase 600 m away (6 September 1947). It seemed desirable to test a greater variety of flowers and to vary the circumstances.

Hannelore Dirschedl (1960) devoted herself to this task. In order to separate absolutely internal and external fragrances during the feeding she made several improvements in method. The sugar water is offered in a small dish 4.6 cm in diameter and 1.5 cm high (Fig. 221). A plexiglass cover, sealed on tightly with odorless grease, has eight holes (1.5 mm in diameter) through which pass glass capillary tubes 1.5 cm in length. From these the bees drink the sugar water. With the aid of a funnel the dish is refilled with pure sugar solution or with scented sugar solution through a hole (0.5 cm in diameter) in the center of the lid. During the experiment this opening is sealed with a glass cover slip (Fig. 222); the bees are kept away from it by a ring of glass or screening (Fig. 223). The container stands in a screened petri dish. The stems of the flowers offered as a source of external fragrance can be stuck down into the water through the screen. Artificial odorants are supplied in the petri dish (Fig. 223). In order to keep newcomers away from the food, a screen cage was placed over the feeding device (this would better have been omitted; see below). The foragers soon learned to creep in and out through a yellow-rimmed hole (Fig. 224).

In the experiments five different pairs of flowers were tested in such a way that in simultaneous experiments first the one species was offered as the internal fragrance and the other as the external fragrance, and then vice versa. Both experiments were done at flight distances of 10 m and 1000 m as well.

The results were very variable. It even happened, contrary to expectation, that the newcomers were guided preferentially by the internal fragrance in the experiment nearby and by the external fragrance in the experiment at a distance. Dirschedl seeks to attribute the lack of uniformity in the results to factors that may vary from experiment to experiment: temperature, relative humidity, and wind conditions influence the persistence of the external fragrance. But above all the intensity of the scent varies within a single species of flower.

Therefore for further tests Dirschedl chose essential oils, the scent of which may more easily be offered at measured strength. She examined six different pairs of scents. With one exception there resulted a superiority of the internal fragrance at the large flight distance, but the superiority was not as marked as in my experiments. If all results with essential oils are combined, the visits to the odorants that had been offered as internal or external fragrances are, on a percentage basis, as follows:

Nearby experiment (10 m): internal fragrance 64.3 percent, external fragrance 35.7 percent;
Distant experiment (975 m): internal fragrance 83.0 percent, external fragrance 17.0 percent.

Thus in Dirschedl's experiments internal fragrance with distant sources of food did not prove to be so clearly superior as in my earlier tests. Possibly that was to be attributed to a peculiarity in her feeding arrangements. The feeding vessel and the provision of a scent had indeed been improved, but there is an objection to the screen covering (Fig. 224). The newcomers did have trouble getting to the food, but through the screen and the opening in it they could smell the scent both of the scent organ and of the flowers supplied as a source of external fragrance, and after some vain hunting about may have flown off to the freely available nearby flower bowl with this same external fragrance; if so the number of visitors to it would have been erroneously increased. Even though the screen cover was removed from time to time in order to give the newcomers an opportunity to settle down and be captured, the extent to which this occurred remains uncertain. The experiments ought to be repeated with attention to this possible source of error.

FIGURE 224. Feeding stand. A screen cover is set over the arrangement shown in Fig. 223. The foragers go in and out through the opening in the cover.

Whereas further quantitative clarification consequently must be awaited, there exists no doubt that, in addition to the fragrance of the flowers visited that clings to the bee's body, the floral odor brought back with the nectar in the honey stomach is important in communicating the species of flower, and that when the goal is far away the nectar odor transported internally and distributed on the comb gains in significance relative to the external fragrance. The observer comes to recognize this fact more strongly from the performance of the bees than from dry statistics of numbers of visits when he notes the intensity and persistence with which the bees poke about, even at great flight distances, in those very flowers whose fragrance was brought home by the foragers in their honey stomachs.

26. We Look for a Feeding Station from Directions Supplied by the Bees

That by means of their dances the bees furnish their hivemates with a description of the location of the places from which they are collecting seemed scarcely credible at first. W. H. Thorpe (1949) mentions that an experimental zoologist had told him he was "almost passionately unwilling to accept such conclusions." Thorpe himself had his doubts too. He resolved them in a congenial way by coming to Brunnwinkl to observe the experiments, working with my bees himself, and letting himself be convinced by them.

Thorpe and that experimental zoologist were not the only skeptics.

There were some in our own ranks. For when on 14 August 1946 I returned to Brunnwinkl from a trip in the mountains my daughters, at that time active assistants in the experiments, informed me that they had set up a new feeding station for the observation hive in Aich, but that they would not tell me where it was. I should enquire of the bees! So I watched their dances and then found the concealed feeding stand, though after some searching. I had inferred a distance of 340–350 m; it was 349 m. But I was mistaken in the direction by 12.5°. In a repetition of the experiment I was even 26° wrong. At that time I was not yet familiar with the light-dependent "misdirection." The dancers were able to see the blue sky. Under such conditions one may be led astray even as much as 40° (see pp. 196f).

Later we made two more of these "unwitting searches"; in these experiments the polarized light of the sky was screened off and eliminated as a source of error.

First experiment, 24 August 1949. The observation hive was in Brunnwinkl. A feeding station was set up whose location was unknown to me. I watched dances from 9 to 10 A.M. and again from 11 A.M. to 12 M. After I had derived a distance of about 330 m from the distance curve and had learned the compass direction from the curve relating the direction of dancing and the azimuth, I set out and had to look around for a few minutes at the presumed distance; then I discovered the feeding station hidden behind a bush. The direction deduced from the dances in the morning led 2° left past the goal; the line found in the afternoon ran 4° to the right of it (Fig. 225).

Second experiment, 30 August 1950. Again the observation hive was in Brunnwinkl. The feeding station, unknown to me, was established

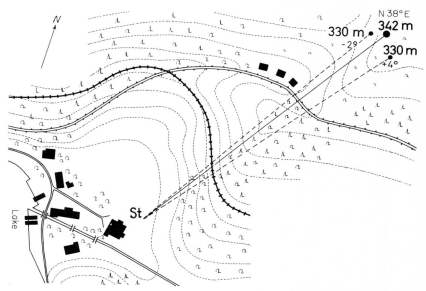

FIGURE 225. Sketch map of the experiment of 24 August 1949: *St*, observation hive. The feeding station, 342 m distant, was unknown to me. At 330 m are the locations deduced from the dances at 9:45 and 11:30 A.M.

150 m from the hive on an elevation (Schusterbergerl) to the south. From the distance curve for the individual colony, for which unfortunately I had only a few points, I estimated a distance of 190 m. The direction deduced from the dances led past the feeding stand 5° to the left, at a distance of 14 m; nevertheless I found it right away.

Later I came to suspect that the difference between the estimated and the true direction to the feeding station might be attributable to the, at that time as yet unexplained, "residual misdirection." I first tested the assumption when writing the draft of this book, and found it partly confirmed. In Table 32 are entered the expected directions of dancing for both observation periods in the first experiment and for the single observation period of the second experiment, together with the error in the direction that I deduced and the "residual misdirection" for the line of dancing in question, from Fig. 212. If this "misdirection" is taken into account, then one obtains for the three observations as differences between the deduced and true directions 0.6°, 8.7°, and 0.3°. In future experiments the "residual misdirection" could be included from the beginning in the calculations. Nevertheless, the way to the goal is never given exactly, for we infer the "residual misdirection" from the available *average* values, and individual values, particularly those near the point of reversal, may deviate markedly from these (see Table 32 and p. 215).

I should not wish to end this section without mentioning two amusing experiences that belong in the picture.

On 22 September 1951, while busy with an experiment, I noticed in the observation hive lively dances in increasing numbers among unmarked bees; these dances were pointed not to our feeding station but to a distance of about 600 m in the direction of Aich. Since natural sources of food had ceased to exist, the bees must have discovered something else. When the announcements made by the bees were entered on a map of the neighborhood, they led to a place where a dealer known to me had an apiary. I sent over an assistant who learned that he had centrifuged honey and then put the combs out in the sun so that his colonies might gather the honey left over. He was told that his honey was being harvested by my bees and that the latter had themselves revealed its origin to me. He just laughed, regarded it as a joke,

TABLE 32. Hunting for an unknown feeding station; three samples of the application by man of the bee dances that indicate direction.

Date	Mean time of observation	Direction of dancing		My error in search (deg)	"Residual misdirection"[a] (deg)	Corrected error[b] (deg)
		Determined (deg)	Expected (deg)			
24 August 1949	9:42	91 l	89 l	+2	+1.4	0.6
24 August 1949	11:35	125 l	129 l	−4	+4.7	8.7
30 August 1950	11:05	27 r	32 r	−5	−5.3	0.3

[a] From Fig. 212, p. 216.
[b] The error after the "residual misdirection" has been taken into account.

and remained convinced that his own bees had brought home their rightful possession.

A similar case occurred on 27 August 1952 in Brunnwinkl itself. Marked bees were collecting at their feeding station 200 m to the west when at first a few and then more and more unmarked bees began to point in the same direction but were dancing too rapidly. With about 10 circuits in 15 sec they were indicating a distance of 100 m. At this location there is the property of a farmer who kept a few hives of bees not far away. He too had spun down his honey and had put out the combs with the remains of the honey on the bank in front of his house. When my messenger arrived the combs had just been carried into the house because so many bees had come. That I had learned about this from my dancers did not meet with disbelief from these neighbors who had many years of familiarity with the experiments. But they were greatly surprised nevertheless that a theoretical knowledge of the language of bees might also have practical application.

Though we may be pleased that we understand the dances so well, at the same time we must be impressed with how fussily and clumsily we make use of the information given by the bees. We must possess a stopwatch in order to determine the tempo of dancing. We need a distance curve pieced together from many single observations in order to estimate the distance with some degree of reliability. We need a protractor to determine the direction of dancing, the azimuth curve for reading off the position of the sun for a given place and instant, and the circular diagram of our Fig. 212, derived from several thousand individual observations, in order to take account of the "residual misdirection." Finally we must have a compass, a surveyor's tape, and assistants, in order to apply in the open country the result we have calculated. How elegantly this task is mastered by the bees themselves, who need follow only a few dance circuits in the dark hive, then according to their innate behavior set forth in free flight, guided by the sun, and steer toward the goal. A truly splendid accomplishment!

Summary: The Indication of Direction

1. Foragers that are frequenting a good source of food quickly recruit numerous newcomers, even for distant and concealed spots. The newcomers are not led to the goal, but rather are sent to it. In addition to being informed by the tail-wagging dance of the distance to a source of food they are also told the direction to it.

2. Sometimes the bees dance on a horizontal surface, for instance on the flight board in front of the hive or (for experimental purposes) in the horizontally tilted observation hive. Then the direction of the waggling run points directly to the goal, in that the dancer maintains the same angle relative to the sun as she held previously during her flight from hive to feeding place (Fig. 114). She orients just as well relative to the polarized light of the blue sky as to the sun itself. If neither the sun nor the blue sky is visible to bees on a horizontal surface, the dances are disoriented and do not indicate any definite direc-

tion. Experimentally the sun may be replaced by a lamp. The direction of the waggling run is determined by the azimuth of the light source, not by its elevation above the horizon.

3. As a rule dancing in the dark hive takes place on the vertical comb surface. Then the dancer transposes the solar angle into the gravitational angle (Fig. 121). Well-developed sense organs for the perception of gravity make this performance possible. If these organs are eliminated by sectioning their nerves, oriented waggling runs on the vertical comb no longer are possible, while the dances on the horizontal comb, with position taken in direct relation to the sun, are undisturbed by the operation. The pattern of the cells is without significance for the direction of dancing. With nearby sources of food, diverging tail-wagging dances often are seen (Fig. 125); the bisector points in the direction of the goal.

On an oblique comb surface the bees can still make a good transposition of the visual angle to the gravitational angle if the angle of inclination of the comb is more than 15° relative to the horizontal. They become uncertain at an angle of 10° and the limit of this capacity is about at 5°. That corresponds approximately to the degree of obliquity at which other insects are still just barely able to orient to gravity. But it is a noteworthy accomplishment that, at such a small inclination, bees still are able to transpose correctly the visual angle retained in their memory.

4. As the distance to the feeding place increases, the areas of the comb where dancing takes place are shifted farther upward and backward, away from the hive entrance (Fig. 122). Foragers returning home already have in mind their future direction of dancing, for the direction in which they climb onto the comb is related clearly to the direction of the forthcoming dance, insofar as spatial conditions do not prevent this (Fig. 124).

5. Individual deviations in the indication of direction occur to a slight extent. With increasing age the indication of direction becomes more precise.

6. In the transition from round dance to tail-wagging dance an indication of direction is to be recognized when the feeding place is about 25 m distant from the hive. With increasing distance it becomes clearer and clearer and at 100 m is very evident. When alerted by round dances the newcomers search the surroundings over in all directions, up to a distance of 50–100 m. Disoriented tail-wagging dances too cause the bees aroused to swarm out in all directions, while oriented tail-wagging dances cause a directed search. Round dances as a signal of nearby sources of food and the indications of direction in the tail-wagging dances are understood and followed accordingly not only by the newcomers; after a pause in feeding old foragers also distinguish very clearly whether new stimulating dances concern the location of their own feeding place or another that has nothing to do with them.

7. Fanwise experiments reveal how precisely the newcomers swarming out follow the dancers' indications of direction. The majority of newcomers fly out in the direction of the goal or within a few degrees

to one side: 10°–15° laterally the number has declined sharply (Figs. 139–144). With an odorous food source the direction of the wind is influential. The astoundingly precise adherence to the direction indicated, regarded from the viewpoint of sensory physiology and psychology, stands as a great accomplishment in reception and evaluation of information (p. 163).

8. That the indication of direction is based exclusively on the position of the sun was demonstrated convincingly in the tropics. When the sun is in the zenith (or within 1°–2° of it) the dances on both a vertical and a horizontal comb are disoriented. They are already oriented well when the sun is 3° from the zenith.

9. The "language" of the bees has no expressions for the directions "upward" and "downward." If one feeding station is exactly above another, so that the flight path from a hive set up somewhat laterally leads obliquely upward to one and obliquely downward to the other (Fig. 145, cliff on the Schafberg; Fig. 146, Echelsbach bridge), then after an interruption in feeding the dances of only one of the two forager groups alert both of them without distinction. Newcomers too are informed by the dances only about the azimuth and not about the elevation of the goal.

If the feeding station lies directly above the hive entrance (Fig. 149, radio tower) or directly beneath the entrance (Fig. 150, Echelsbach bridge), the foragers also are unable to indicate any direction and perform round dances. Since the elevation also is not indicated, in such cases the newcomers go exploring at ground level in all directions.

If the hive is at the foot of an overhanging cliff and the feeding station on top of the wall right above the hive entrance, or the opposite (Figs. 145 and 148), when the bees fly they orient to the cliff and by correspondingly directed dances facilitate the finding of the goal by the newcomers, without giving any further indication of its elevation.

10. The waggling run indicates the direction of the outward flight to the goal. In order to learn whether the foragers take into account also in their indication of direction the impressions received during the return flight (transposed by 180°), the bees were displaced laterally after their arrival at the feeding station. They found their way home only after making long detours and had to be retrained to the direction of the return route. Under these conditions their dances indicated the direction of the bisector of the angle (Fig. 152). If one of the flight legs is designated conspicuously by markers (wooden laths, sheets of colored cardboard, and the like), or if the flight path for one of the legs is lengthened, then it acquires increased weight relative to the other leg in formulating the indication of direction.

By varying the experimental arrangements the displaced bees can be caused to find the way home at once without further training. Then with their dances they indicate unequivocally the direction of the outward flight to the goal. The differing direction of the return flight is not taken into account.

11. Bees that are traveling back and forth between hive and feeding station via a detour—for instance around a cliff (Figs. 159, 161), around

a tall building (Fig. 168), or on foot in an angular corridor (Fig. 172) —indicate in their dances the direction of the air line to the goal. That is to be seen particularly clearly and convincingly in dances on a horizontal surface: the waggling runs point through the obstacle to the goal, and thus indicate a direction that the dancers have never flown over. In doing this they are able to orient just as well by the polarized light of the blue sky as by the sun itself. The data on distance provided by the dances refer not to the length of the air line but to the detour traversed in flight. Newcomers alerted hold to the direction indicated and at first fly over the obstacle, insofar as that is possible. The (shorter) detour to the side is discovered only later.

When parts of the detour followed in flight are darkened, these stretches are not taken into account in indicating direction. That supports the view that in the detour experiment the bees derive the angle of dancing for the air line from the solar angles perceived during the different segments flown over and by integration (central nervous calculation) of their length.

Newcomers that on their first flight are led via a detour to a source of food indicate the direction of the air line in their very first dance after their return home. For scout bees, which have made their discovery via detours, this is of biological significance.

12. Bees that are exposed to a crosswind on the way to the feeding place compensate for the drift by heading obliquely across the wind. Therefore they see a different solar angle than during direct flight in calm air (Fig. 186). In spite of this the dances point straight toward the goal. Thus the bee takes into account the oblique stance of her own body and in dancing reduces the solar angle that she has seen to the angle that she would have seen in calm air. Herewith she bears evidence of the same capacity disclosed in the detour experiment (Point 11 of the Summary) and shown by dance followers who obtain information about the direction of dancing while occupying a position oblique to the dancer (cf. p. 160 and Fig. 52).

13. "Misdirections" occur in the indications of direction by bees. The solar angle indicated in the dance may deviate appreciably from the true solar angle.

(a) "Misdirections" caused by light occur in dances on the vertical comb surface when the dancing bee is able to perceive the position of the sun. For this response visibility of the blue sky (orientation according to the polarized light of the sky) also suffices. The magnitude of the "misdirection" varies; under otherwise constant conditions it shows a pronounced diurnal course (Fig. 193). "Misdirection" caused by light comes about as follows. Under normal conditions the bee dancing on the vertical comb in the dark transposes the visual to the gravitational angle. If during the dance she has a view of the blue sky she recognizes the position of the sun and her intention to align the waggling run directly with the sun's position (projected onto the vertical comb surface) comes into competition with the transposition relative to the direction of gravity. Then the dancer adopts surprisingly precisely the bisector of the angle between the directions of dancing that she would have to assume in orienting exclusively to light or to gravity, respectively (Figs. 195 and 196). If she sees the sun itself, the tendency to be guided by it alone prevails (Fig. 197).

For the newcomers alerted this performance does not imply any "misdirection." They are subject to the same influences as the dancers; hence in translating the angle of dancing back into the angle of flight relative to the sun they compensate for the deviation. Under natural conditions these things are especially significant in a hanging swarm out of doors.

This light-dependent "misdirection" can easily be eliminated as a source of experimental error by shading off the sky (the best way is to observe in the fiberboard hut).

(b) After elimination of the light-dependent "misdirection" there remains a smaller "misdirection" ("residual misdirection") that depends on gravity. It too has a systematic diurnal course. Under different conditions it follows manifold patterns that at first are confusing (Figs. 200–203). In the dances on a horizontal surface no "misdirection" occurs. It first arises in transposition of the visual angle to the sense of gravity, and in the magnitude and direction of the deviation clearly is dependent on the direction of the tail-wagging run relative to the vertical (Fig. 210). If the tail-wagging run is directed obliquely upward or obliquely downward, the run is deflected toward the vertical; if it approaches within about 30° of the horizontal, there results a (weaker) deflection toward the latter (Fig. 212). These regularities stand out in the grand average, to be sure, but show much variation in individual instances. The factors responsible for these variations are presently unknown.

Similar deviations have recently become known in the geotactic orientation of other insects. With ants and bees they occur also during training to various directions relative to gravity, that is to say, without any transposition (Figs. 215 and 217). The explanation seems to be that the geomenotactic orientation learned is influenced lastingly by the inherited fundamental orientation (positive, negative, and transverse geotaxis).

"Residual misdirection" too does not involve any error for the bees to whom information is supplied, since it is compensated for when the gravitational angle is translated back into the visual angle.

14. Foragers are able with their scent organ to attract to the proper spot those newcomers who are hunting about near the goal. The marking out of a more extended flight path (an "odor trail") by everting the scent organ is improbable and is not supported by any observations.

In addition to the floral odor that clings externally to the dancer's body, information about the species of flowers collected from is given the hivemates by the nectar brought in in the honey stomach; this nectar is perfumed heavily with the floral scent and is distributed on the comb to the dance followers. When the source of food is far away, this "internal fragrance" may become of dominant import relative to the externally clinging floral scent, which is weakened over long stretches by the wind generated in flight.

15. After receiving from the bees data relative to distance and direction a human observer too is able to find the goal indicated. Yet in order to achieve this he requires time-consuming preparations and technical aids, whereas the bee accomplishes the same task easily with her own sensory organs.

V / Dependence of the Dances on the Profitability of Foraging Activity

The pattern of the bee's dance is determined fundamentally by the distance of and direction to a source of food. But whether dancing occurs at all, and how vivaciously and how persistently, depends on many factors that significantly regulate the relation between supply and demand.

1. FACTORS DETERMINING THE RELEASE AND LIVELINESS OF THE DANCES

(*a*) *The Sweetness of the Sugar Solution*

A sugar solution must have a certain concentration if it is to be ingested. Depending on conditions, the acceptance threshold varies over a wide range from $\frac{1}{16}$–$\frac{1}{8}$M to 1–1.5M sucrose solution. A somewhat higher concentration (threshold of dancing) than is needed for bare acceptance of sugar water is always essential for dancing activity. It seemed of interest to learn whether the intensity of the sweet taste releases the dances directly or indirectly.

In order to determine the influence in this respect of individual factors, the sugar concentration of the food must be so chosen that only part of the foragers will dance. Then the effectiveness of a factor is shown by its increasing the percentage of dancers.

Distention of the honey stomach might be imagined as the key stimulus. For under otherwise identical conditions the forager ingests more sugar water the sweeter the solution. Thus, for instance, the average uptake[1] of sucrose solution in an experiment on 30 September 1928 with a group of foraging bees was: 0.5M, 42 mm^3; 1M 49 mm^3; 2M 56 mm^3.

But no connection was to be demonstrated between the degree of filling of the honey stomach and the release of dancing. Since under otherwise identical conditions more is drunk of a warmer solution, a

[1] I determined the average uptake by letting a group of marked bees drink from a known amount of sugar water in a dish. The reduction in weight of a control dish during the experimental period gave the amount lost by evaporation. The remaining decrease, divided by the number of foraging flights, corresponds to the average filling of the stomach. If the bees are allowed to drink from a graduated pipette, the amount of filling of the stomach can be determined individually (Gontarski 1935; Schifferer 1952; Schuà 1952).

greater filling of the honey stomach could be achieved, for instance, with a $\frac{3}{8}$M sucrose solution at 28° (average, 53.1 mm³) than with a 1M solution at 11° (42.1 mm³) (v. Frisch 1942:273). In spite of the less completely filled honey stomach, more bees danced for the sweeter solution (72.3 percent vs. 58.3 percent). Thus the degree of filling of the honey stomach is not decisive. Whether it is wholly uninvolved remains a question.[2]

Even more appealing is another conjecture: that a greater sugar content of the solution fed simply releases the dances via the increased fuel supply.[3] The bee does indeed give out the contents of the honey stomach again in the hive, but she takes a portion into her midgut for her own needs.

A peculiarity of the bee's sense of taste affords a test in the following manner. The sugar alcohol sorbitol, which is sweet for us, is without taste for bees (v. Frisch 1934), but according to Vogel (1931) has for them nearly twice as much food value as sucrose. That is expressed not only in its life-prolonging effect, but also directly in the increased vivacity of starving bees after they have been given sorbitol. In order that they shall drink the sorbitol solution, which has no taste for them, it must be sweetened with a certain amount of sucrose. Hence I compared the percentage of bees dancing after feeding: (1) with pure sucrose solution; (2) with sucrose solution of the same concentration plus an addition of sorbitol by which the food value was more than doubled while the sweetness remained unchanged. A clear influence of the added sorbitol was never observed, whereas even a slight increase in the sucrose content raised significantly the percentage of bees dancing.

Consequently the sweetness of a sugar solution does not act to release dancing indirectly through the degree to which the stomach is filled nor through its food value, but acts either solely or primarily directly via the sense of taste.

Once the threshold of dancing has been reached, the further sweetness of the solution has a marked effect on the liveliness of the dances. That these become more vivacious with increasing sugar content is very conspicuous to the observer, but is not to be defined quantitatively. An easily measurable quantity is their duration. Its increase with rising sweetness of the food was discussed on p. 45 and exemplified in Table 3.

Besides this Esch (1963) found a change in sound production during the waggling run with increasing concentration of the sugar solution: the frequency of acoustic bursts (frequency of vibratory episodes) rose —with a constant acceptance threshold—from 22 to 30.5 per second when the concentration of the sugar solution was increased from 0.5M to 2M (Fig. 226).[4] But it has not been shown that the rise in frequency

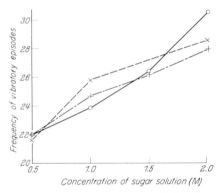

FIGURE 226. Dependence of the frequency of bursts of sound (frequency of vibratory episodes) on sugar concentration, at various distances to the feeding station. ○——○, 150 m; +- - -+, 600 m; ×- - -×, 1500 m. After Esch 1963.

[2] Uncertainty arises because of the greater viscosity of the colder sugar solution, which favors the dances (see below).
[3] In bees, in contrast with vertebrates, the blood-sugar content undergoes appreciable variation. It is influenced strongly by the food uptake (R. Beutler 1936).
[4] If the bees are warmed strongly the frequency of the vibratory episodes may increase to 60/sec, according to Esch (1963). In apparent contradiction to this Wenner (1962) found no change in frequency with rising temperature. But he measured the external temperature, whereas Esch's data refer to the body temperature of the bees.

of the vibratory episodes is comprehended by the hivemates as signaling a better quality of food. In round dances a regular production of sound is lacking,[5] while the liveliness and duration of the dances increase here too when the source becomes more profitable, and have a clear influence on the extent of arousal; thus in themselves alone they guarantee success. With the tail-wagging dance the increase in newcomers can be attributed in the same way to the greater duration and liveliness of the dances when the sugar concentration is raised. Since these expressions that are indicated by gross movements are more easily observed than the vibratory frequencies, so far as the latter are concerned there is lacking at present a foundation sufficient for judging the extent to which they depend on various conditions.

(b) The Purity of the Sweet Taste

A pure sucrose solution, near the acceptance threshold in strength, releases more dances than one slightly contaminated with salt. Thus in one experiment in which 0.5M sucrose was fed, about 80 percent of the foragers performed dances, but with 0.5M sucrose + $\frac{1}{8}$M NaCl only 50 percent (v. Frisch 1942). Eagerness to dance is inhibited also by addition of hydrochloric acid or quinine (Lindauer 1948).

(c) Ease of Obtaining the Solution

Even very sweet sugar water releases no dances if it is scanty. Thus the quantity of the solution as well as its sweetness is important. How this is "judged" is unexplained.

Probably the required expenditure of time plays a role. At the artificial feeding station the forager can fill her honey stomach in about 1 min, if there is a rich supply. With a meager supply (Figs. 19 and 20) she will often tarry there 10 to 15 min and still return home with only a small load. The same is true also of visits to natural flowers. When in the Winterhalle of the Munich Botanical Garden (p. 14) I offered locust (*Robinia*) flowers that had been screened for some hours against visitation by insects and that were rich in nectar, the bees quickly filled the honey stomach and performed dances in the hive. As the nectar grew scanty, because its production did not keep pace with the consumption, they had to visit numerous blossoms, the expenditure of time on each flight increased, and the dances became fewer and finally ceased entirely despite continuing foraging (v. Frisch 1923). The amount of work done in each flight also increases. But it is to be assumed that yet other factors are involved; see section (*e*) below.

Once the threshold for dancing is attained, easy acquisition of the food causes the dances also to become more lively.

---

[5] When a very vigorous alarm is given, waggling movements occur in the round dances too. They are scattered through the circuits irregularly and thus are no signal of the distance. According to Esch (personal communication), they involve the production of sound that is similar to that produced during the waggling run and, as is the case there, is limited to the duration of the waggling movements. Probably their only purpose is to intensify the alarm (cf. p. 45).

(d) Viscosity

An increased expenditure of energy and time need not of itself inhibit the eagerness for dancing, and under some circumstances may even augment it. On account of its high viscosity concentrated sugar solution is sucked up more slowly and with more difficulty than a thin solution. Of course it is also sweeter. But even an increase in viscosity alone may release dancing. The trisaccharide raffinose is not sweet to bees and also does not produce the least bitterness (v. Frisch 1934:49).[6] If one makes a solution that in addition to $\frac{1}{2}$M sucrose contains also $\frac{1}{4}$M raffinose, with no change in sweetness relative to $\frac{1}{2}$M sucrose, the viscosity is increased greatly. The degree of filling of the honey stomach was somewhat diminished, and the number of dances clearly augmented. With a sucrose solution of higher concentration, the viscosity of which approximately equaled that of the sucrose-raffinose mixture but which was sweeter, the dances became even more numerous (Table 33). The significance of the sweet taste in releasing dancing is confirmed thereby; at the same time an effect of viscosity in the same direction is made evident.[7] That is comprehensible because for bees under natural conditions a more viscous nectar is richer in usable sugar (v. Frisch 1942:276).

The liveliness of the dances too was increased at higher viscosity.

(e) Load

Even more surprisingly, a similar relation appears when bees are burdened with lead weights (G. Schifferer 1952; see also p. 109 and Fig. 100). The added weight of the load is accompanied by a somewhat increased flight time. Eagerness to dance was not only undiminished, but frequently was clearly increased. Thus in one experiment the control bees, not burdened, danced only exceptionally, while the experimental bees, which were able to carry their 55-mg lead weights only

---

[6] Its food value, for which in addition no dance-releasing effect could be demonstrated, is very slight for bees, about one-eighth that of sucrose (Vogel 1931:329–331).

[7] The viscosity of a sucrose solution is increased by addition of sorbitol also. That no influence of sorbitol on the eagerness for dancing was detectable in these experiments (p. 237) probably depends on its producing a slight bitterness, which would diminish the desire to dance.

TABLE 33. Effect of viscosity in increasing dancing. Solutions 1–3 are equally sweet to the bees; solution 2 is of higher viscosity than 1 and 3; solution 4 has approximately the same viscosity as 2, but is sweeter than 1–3.

Solution fed	Average uptake (mm³)	Number of visits	Fraction of these foragers that danced (percent)
1. $\frac{1}{2}$M sucrose	45.1	53	69.8
2. $\frac{1}{2}$M sucrose + $\frac{1}{4}$M raffinose	42.3	56	80.4
3. $\frac{1}{2}$M sucrose	40.9	58	63.8
4. $\frac{7}{8}$M sucrose	48.8	49	89.8

with difficulty, "showed great eagerness to dance." Like higher viscosity, under natural conditions increased weight also is connected with a greater sugar content of the nectar.

If anyone should think that the increase in eagerness for dancing on the addition of raffinose (see above) might be attributed to the increase in weight brought about thereby, and hence would not furnish evidence for an effect of viscosity, let him consider the following calculation: with $\frac{1}{4}$M raffinose and an intake of 40 mm^3 the bee's weight is increased by the small amount of 5.9 mg (in the experiment in Table 33). But if the bee drinks $\frac{1}{8}$M sucrose on one occasion and 2M sucrose on another, that means a weight difference of 40 mg in sugar. In addition there is a different degree of filling of the honey stomach (about 30 mm^3 and 60 mm^3, respectively), whence there arises a plus of 5 mg of water[8] for the 2M solution, which gives altogether an added weight of 45 mg. This agrees in order of magnitude with the weights of lead that increased the eagerness for dancing in Schifferer's experiments.

Probably the increase in load due to the pollen baskets during a period of lush production of pollen is of the same order of magnitude. According to Maurizio (1953), with many plants from which pollen is collected the mean weight for the two pollen baskets is as much as 20 mg or more, but it varies greatly and may thus be appreciably greater than this in individual cases.

(f) Nearness of the Food Source

The relations are reciprocal for feeding places at different distances. Here, under otherwise identical conditions, the greater expenditure of energy and time over longer flight paths goes hand in hand with a decrease in profitableness. If the food source is only a few meters distant from the hive, a forager can bring in about 15 loads in the course of an hour, yet only 8 from a distance of 1 km and but 4 from 3 km. The loss of time on the long flight path has the consequence that the efficiency of a nearby goal is many times greater (R. Beutler 1950).[9] Therefore it is reasonable that, given the same quality of food, a nearby feeding place should release more dancing than a more distant one, as is *de facto* the case (Boch 1956). Initially the scale for comparison was the percentage of bees dancing for a given concentration of sugar at various distances. Thus, for example, on 19 June 1953 at each of two feeding stations simultaneously, one 100 m and the other 525 m distant from the observation hive, a group of foragers was fed with 1.5M sucrose. The curves of Fig. 227 show the result (grouped for each 15-min interval). On the average at a flight range of 100 m about 60 percent of the foragers danced, at 525 m only 34 percent.

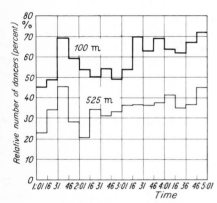

FIGURE 227. Influence of the distance to the feeding station on the dances. At two feeding stations, distant respectively 100 m and 525 m from the hive, 1.5M sucrose was fed. The curves show what percentage (ordinate) of the foragers danced. The values are combined for each 15-min interval. After Boch 1956.

---

[8] The small difference of only 5 mg in weight of water results from considering the differing volumes of sugar in the two solutions.
[9] Of course a more distant feeding place requires a greater consumption of sugar because of the greater flight demand. Yet Ruth Beutler (1950) showed that because of the loss of time over the longer flight path this increased consumption is insignificant in comparison with the decrease in sugar gathered.

By raising the sugar concentration at the distant station one can release from there approximately the same percentage of dances as from the nearby station with the lower sugar concentration. Then the curves run together. In this way Boch tested the performance of the bees both at 100 m and at greater intervals up to 6 km. In Fig. 228 the distance of the feeding station is given on the abscissa, the concentration of the sucrose fed on the ordinate. It must be kept in mind that the threshold for dancing (like the acceptance threshold) lies at different levels at different times (see (*j*), p. 243). During the period of experimentation it shifted at the flight distance of 100 m from $\frac{1}{8}$M to $\frac{3}{4}$M. The points of origin of the curves in Fig. 228 correspond to these shifts. From the further course of the curves it is evident what concentration of sugar had to be offered at the distance indicated on the abscissa in order to release the same percentage of dances as in the bees from the 100 m station. The influence of the range of flight emerges clearly. Thus, for instance, at a flight distance of 6 km one would have to offer four times the sugar concentration in order to achieve the same success in dancing as at 100 m.

Proximity of the feeding place means a smaller expenditure of energy and time. As when the food is easy to obtain (*c*), these factors seem to favor the release of dancing. With viscosity and weight (*d* and *e*), on the contrary, the greater expenditure of energy and time seems to favor dancing. In both cases the behavior is reasonable. One sees how closely the biological conditions are intertwined and how hard it is to evaluate the true releasing factors.

Hitherto we have used the duration of the dances as a measure of their liveliness. That is beyond criticism as long as the feeding place for the groups compared is at the same distance, as indeed it is as a rule. But here we are comparing bees foraging nearby with those foraging at a distance. At 100 m there are about ten and at 6 km about two circuits to the quarter minute (Fig. 63). That means that the bee foraging nearby can repeat her message ten times while the forager from a distance gives it twice. In Boch's experiments the bees foraging at a distance danced approximately as long as the foragers from nearby. But because of the larger number of waggling runs the informational value of the message was greater with the foragers from nearby. In this sense the liveliness of the dances for nearby feeding places is increased.

(*g*) *Floral Fragrance*

Lindauer (1948) succeeded in showing that the floral fragrance too might be concerned in release of dancing—although apparently it has nothing to do with the profitability of the food source. An example: Marked bees were foraging at a feeding dish that was enwreathed in phlox flowers. For comparison a control group from the same colony was collecting sugar water of the same concentration without an added scent. During the period of observation there were recorded for the phlox bees 197 foraging flights, for 104 of which (53 percent) dancing occurred. At the same time there were 163 foraging flights with 33 dances (21 percent) by the control group. Only floral scents (oil of

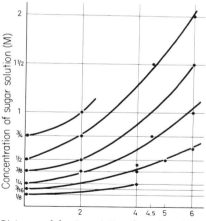

FIGURE 228. The greater the distance of the feeding station from the hive, the more highly concentrated must the sugar solution be in order to release dancing: abscissa, distance to the feeding station; ordinate, threshold concentration for dancing. After experiments by Boch 1956.

peppermint, thistle flowers, lilac, carnations, rape) have this effect, whereas the eagerness to dance is inhibited by the smell of skatole (reminiscent of carrion) or by the peculiar smell of hogweed.

Although the floral odor has nothing immediately to do with the uptake of food, it is nevertheless important because it facilitates significantly the discovery of the indicated source of food by the hivemates alerted. Under otherwise identical conditions, fragrant flowers therefore are exploited more fully than odorless flowers. They are more profitable for the colony.

In order to check the liveliness of the dances—here again well measured by their length—Kaschef (1957) fed the experimental bees with sugar water to which traces of an ethereal oil had been added. Without exception the liveliness of the dances was increased appreciably relative to that of the control group, given unscented food, by flower-like fragrances (oils of peppermint, star anise, citronella, lavender, geranium, bergamot, orange, and thyme). The positive effect begins approximately when the amount of added odorant is just perceptible to the human nose. Too great an addition has an inhibitory effect on the dances. About 1 drop of the ethereal oil to a liter of 1M or 2M sucrose solution is right.

The artificial scents bromostyrole and methylheptenone were inhibitory at all concentrations, if they had any effect whatever.

(h) *Form of the Food Container*

In feeding bees from a narrow cleft (Fig. 219) in 1947 I gained the impression that they were more eager to dance than when drinking from an open dish. That was confirmed in a study directed to the question by Irmgard Kappel (1953). She had the experimental bees drink from glass capillaries (Fig. 229) or from glass wool, while the control bees were taking up sugar water of the same concentration from an open dish. Of some 2000 bees observed after their return home, 44.4 percent of the control specimens but 58.9 percent of the bees from the capillaries danced, and the duration of dancing was greater with the latter.[10] The results with glass wool were quite similar. The duration of drinking from the capillaries was somewhat longer than from the dish (61.1 sec vs. 52.9 sec, as the average for 150 observations each). Also somewhat more sugar water was ingested from the capillaries (uptake when fed with 0.25M sucrose 69.7 mg vs. 62.9 mg for the control bees, average of 900 observations each;[10] the results of the experiments with glass wool were very similar).

That the increased interest in dancing is to be attributed to the meager increase in the weight of the load is unlikely. Probably the cause is to be sought in the fact that narrow tubes and clefts resemble more closely than a watch glass the accustomed sources of nectar in the flowers visited. Hence their greater popularity—if I may use this term for the unknown central nervous relations. This greater popularity also finds expression in the fact that the capillary bees, though they dance

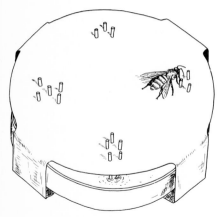

FIGURE 229. Feeding from glass capillaries. After Kappel 1953.

[10]The difference is statistically highly significant.

longer, remain a shorter time in the hive;[11] for the capillary bees the length of stay amounted to 2 min on the average, and for the control bees 2.5 min.

By taking the grooved feeding dishes (Fig. 18) we make constant use of this experience in our experiments.

(i) Uniform Flow from the Source of Food

Even with a high concentration of sugar, as a rule dancing does not occur after the day's first foraging flights. Not until the source has not been exhausted despite several flights, and thus demonstrates its productivity, do the dances set in. The same is observed when foraging is resumed after a long interruption in feeding. Only when the colony is in great need of food, or there is a moderate shortage and an unusually good find, does a forager sound the alert immediately after her first successful expedition (Lindauer 1948; Heran 1956:213, Fig. 24).

When the productivity of the food source continues, interest in dancing may increase for several hours. In 14 experiments, during which I fed a group of bees with the same concentration of sugar for a rather long time without any interruption, both the percentage of the bees dancing as well as the liveliness (duration) of the dances increased more and more, without exception, for the first 1-2 hr, on four occasions for 3 hr, and once even for 4 hr (an example is shown in Fig. 230).

(j) General Status of Nourishment in the Colony

The shortage of food just touched on above occupies an outstanding place among the factors that release dancing behavior. In the spring, at the time of abundant flowering and a good supply of food, it is often impossible even with 1-2M sucrose to induce the bees to dance. In the summer and fall, with the food supply meager, dancing can be released even with ⅛M sucrose solution. This depends not only on the natural food supply but also on the stores in the hive.

That in times of famine all possible assistance is called out for even a scanty supply of food is not to be ascribed to the foresight of the foragers. They inspect neither the supply of nectar-producing plants nor the stores in the comb before they begin to dance. As Lindauer (1948) has shown, the returning foragers, when the colony is suffering from a scarcity of food, are entreated passionately and all but attacked by the food recipients whose task it is to distribute and store up the nectar. According to Boch (1959), along with the rising state of excitement the rapidity of antennal drumming also increases. The forager's load is taken from her very quickly. When there is a superfluity of food she must look much longer for a recipient, and whereas in the first situation she will have got rid of the content of her honey stomach in perhaps 0.5 min she may now need several minutes. Being entreated and quickly freed of her burden favors the inception and liveliness of dancing (see Fig. 238).

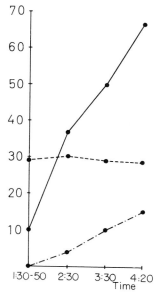

FIGURE 230. With uniform flow from the source of food, the percentage of bees dancing and the duration of dancing increase during the first few hours; experiment on 9 August 1957, 1:30-4:20 P.M.; continuous feeding with 0.5M sucrose. Ordinate numbers: —— percentage of bees dancing; —•— average duration of dancing (sec); - - - temperature (°C.).

---

[11]We are acquainted with the same phenomenon from feeding with solutions of differing sweetness. With a better quality of food less time is wasted in the hive and even the flight velocity is increased (p. 251).

(k) Improvement of the Food

Aside from the fact that an absolutely better sugar solution releases more dancing than one that is less sweet, a relative improvement as such has a dance-promoting effect. If for example one has been offering 0.25M sucrose, the percentage of dancing foragers becomes greater when a 0.5M solution is then given—under otherwise identical conditions—than when a 0.5M solution is given following feeding with 1M. That can be understood in terms of the adaptation of the sense organs of taste.[12] Through contrast with the preceding less sweet solution the 0.5M sugar water seems sweeter initially than during continuous feeding. The liveliness of the dances too is temporarily increased. The explanation of an observation by Ribbands (1955) is to be found here: a feeding station at which he first offered thin and subsequently more concentrated sugar solution was thereupon visited twice as heavily by newcomers as another station where there was continuous feeding with more concentrated solution.

The same holds, with an unchanging sugar concentration, when the sweet taste is made purer. In one of Lindauer's (1948:378) series of experiments, on feeding of long duration the bees danced no less well with 0.25M sucrose + $\frac{1}{16}$M NaCl than with 0.25M sucrose without salt. When the same sugar-salt mixture was offered immediately following a pure sugar solution of the same sugar content, the percentage of dancers declined; in contrast, it increased when still saltier sugar water had been given beforehand.

(l) Time of Day

The flight intensity of colonies of bees (number of flights from the hive) usually drops temporarily around noon. According to Lindauer (1948), when a group of bees is fed all day long with sugar water of a constant concentration there also occurs at noontime a definite decrease in the percentage of dances and in their liveliness. In his observations that was so for the period from 12 M. to 3 P.M. Boch (1956) and Bräuninger (1964) likewise noticed this noontime sluggishness. In 11 experiments in 1957 and 1958, during which a group of bees received sugar water of constant concentration for several hours or all day long, I found without exception the noontime decrease in eagerness to dance (percentage and duration of the dances) confirmed, though not invariably with the same degree of clarity (starting mostly between 11 A.M. and 12 M., ending between 2 and 3 P.M.). An example is given in Fig. 231.

In contrast with the seasonal variations, which are explained by the changing situation with respect to nourishment, neither Lindauer nor I could relate the noontime decline to an external factor, any more than Schuà (1952) could do for the noontime decrease in flight intensity. Probably both are the expression of a diurnal periodicity in activity, such as is widely known among insects and other animals. It is still discernible in bees under constant conditions (see p. 357).

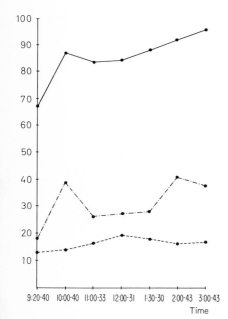

FIGURE 231. Noontime laziness; experiment on 31 August 1957; continuous feeding with 0.75M sucrose: ——— percentage of bees dancing; —·— average duration of dancing (sec); - - - temperature (°C). From 11:00 A.M. to 1:30 P.M., fewer dances and a decreased duration of dancing; then a rise despite falling temperature and increasing cloudiness.

---

[12] Such adaptation is expressed very clearly in the way in which the acceptance threshold depends on preceding feeding; see v. Frisch 1934:26ff.

(m) *Weather*

Bees are "weatherwise." Before impending rain or a thunderstorm they turn homeward in crowds to the protective hive. But their eagerness to dance may drop off strikingly long before this. That is not so strongly the case when the food source is close to the hive as when it is at a great flight distance, where a sudden storm may even threaten the lives of the foragers. I draw upon Boch (1956:150) for an interesting example. On 9 August 1952 one group of bees was being feed 100 m from the hive, another with sweeter sugar water 6 km away. From early morning on the percentage of dancers was approximately equal in the two groups, until after 3:30 P.M. the relative number of dancers declined markedly for those foraging at a distance while throughout another 3 hr it decreased scarcely at all for those foraging nearby. About 4 P.M. the first signs of an impending thunderstorm became evident to the human observer. At about 6 P.M. the bees from the distant station ceased dancing altogether, whereas those at the 100-m station kept on foraging and dancing in the hive until the storm actually broke (at about 7:15 P.M.).

Knaffl (1953:139) also noted the same sort of increase in sensitivity of the dancers to weather conditions as the feeding station was located progressively farther away.

We have now become acquainted with a large number of factors—and probably not with all—that are involved in the release of dancing. Stimuli that favor the release of dancing quite generally also increase their liveliness. Yet the acceptance threshold for sugar solution must not be linked with them. It is unaffected by the viscosity of the solution (v. Frisch 1934:14), by the addition of a floral scent (v. Frisch 1934:24), by the form of the food container (Kappel 1953:541, 542), or by the uniform flow of the food,[13] whereas the acceptance threshold is influenced similarly to the threshold for and duration of dancing by the intensity and purity of the sweet taste, by the nearness of the source of food, by a poor situation with respect to the food supply or by favorable weather. To what extent these factors, whose effectiveness has been discerned by testing them singly, may work together according to the rule of summation of stimuli (Seitz 1940) no one has as yet studied systematically. Their balanced interplay is the more remarkable because most of the influential stimuli are received at the feeding place whereas the dance takes place on the comb in the hive, at a different location and at a later time.

Delays, even very long ones, may occur with the dances. Chalifman (1950) first showed and W. Wittekindt (1955) and Lindauer (1957) confirmed that some foragers can be rearoused to dance after ending their flights in the evening or as a result of illumination of the comb at night. Then they signal the direction and distance of the feeding place frequented many hours previously (Lindauer 1955; E. and W. Wittekindt 1960). No biological reason for this reactivation of dancing by illumina-

[13] This indeed exerts its effect only some time after it has been received.

tion is evident. According to W. Wittekindt (1961), it was possible by later illumination to induce tail-wagging dances among a part of the bees that had foraged throughout the day from a feeding station at the proper distance for tail-wagging dances but that had not danced; in this event the bees still indicated the location correctly after more than 24 hr. In analogy with one's own experience one is tempted to say that the feeding place may no doubt have seemed superior in their recollection to what it actually was. Yet here we are on uncertain ground which we forsake the more gladly because further experiments can provide much clarification here.

What has been reported hitherto concerns only the way in which dances are released by the collecting of nectar and sugar water. We know almost nothing in this respect about the harvesting of pollen. Many factors that are significant for nectar foragers will be of no consequence to pollen foragers (sweetness, viscosity, form of the nectary). Shortage of pollen favors the release of dancing so strongly that in times of famine the hivemates are even aroused to garner potato flour or powdered milk (Lindauer 1948:401ff) and apparently brick dust and coal dust too, for even such useless stuff is brought in occasionally. That an easily obtained and copious supply of pollen promotes dancing particularly is probable. I recall the forager that harvested only a short time on her first visit to the gorgeous wealth of pollen of an elaborate rose bed, flew home with tiny accumulations of pollen, danced there long and stormily and only thereafter, with her miniature pollen baskets, flew back to the roses to continue her foraging (pp. 34f). Taste seems to be meaningless. The addition of a fragrance can have a dance-promoting effect (Lindauer 1948:402). That the distance of the collecting place, the time of day, and the weather are important is to be assumed; whether the weight of the pollen baskets also matters is unknown.

For bees gathering water the need of the colony at the time is probably the most important factor in releasing dances (see pp. 265ff).

In pointing out a new site for nesting the eagerness to dance is guided by an abundant and wholly different scale of values (see pp. 273ff).

Thus according to circumstances there may take part in the releasing of dances a multiplicity of perceptions, the integration of which by the bee's nervous centers remains for the time being a riddle.

2. Regulation of Supply and Demand on the Flower Market

The difficulty of understanding the harmonious interplay of the manifold dance-releasing factors is matched by the high biological significance of the process. The bees have solved the problem of extracting the best from the profuse supply of the natural flower market and of holding the risk to as low a level as is possible with the working powers they possess. Every flight out from the hive is a hazard.[14]

---

[14] Lindauer (1952) determined, for instance, that of 150 marked bees that were under continuing observation 29 (that is, almost 20 percent) had an accident on their first flight out from the hive.

In a normal colony of bees in the spring there are daily about 1000 new young bees ready to assume their foraging activity. Of these only a vanishingly few fly forth as scouts to discover new sources of food. By far the majority wait in the secure hive until they are solicited in favor of some definite goal. Scout bees, even without previous experience, are attracted by the colors and the shapes of flowers and by natural floral fragrance (v. Oettingen-Spielberg 1949; Butler 1951; Lindauer 1952; Ludwig 1956[15]). Colored marks for nectar, which are evident on flowers to the ultraviolet-sensitive eye of the bee more often than for us, and scent marks on the petals, which are hardly recognized by the human sense of smell, facilitate their finding the nectaries (v. Frisch 1943; Lex 1954; Bolwig 1954;[16] Daumer 1958; v. Aufsess 1960).

Whether they then attempt to enlist assistance by dancing does not depend solely on the quality of what they have found. What is decisive is how profitable the foraging is. This is determined by factors that have been discussed in the preceding section. They may be summarized and arranged in order as follows: (1) the quality (sugar content) of the nectar; (2) the profitability (amount and uniform flow of the nectar, the ease with which it can be collected, flight distance not too great) of the source of food; (3) factors related to quality; viscosity and weight of the load; (4) floral fragrance (facilitating discovery and communication); (5) the general nutritional state (determined primarily by the colony's requirement); (6) weather such as to keep the risk low; (7) a familiar mode of presentation, the drink being presented in a narrow cup (increased yield because of greater filling of the stomach and a shorter stay in the hive[17]).

If the sum total of these perceptions indicates that the collecting of food is worth while, it is carried forward by the discoverer, who is thereby converted from a scout bee to a forager; moreover she begins to dance and thereby transmits to her hivemates the following messages: (1) an invitation to take part in the foraging; (2) the distance to the food source; (3) what its compass direction is; (4) the floral scent as a specific sign of the goal, for it clings to her body and also to samples of the nectar that are fed to surrounding bees during short pauses in the dancing; (5) this also shows the taste, or the sweetness, of the yield; (6) the profitability of foraging, made known by the dancer through the liveliness and duration of her dances, the degree of intensity of their invitation being expressed quantitatively over the entire scale from the dull inception of a single circuit lasting scarcely a second to the stormy

---

[15] Since Luise Ludwig's teacher's thesis was not published, it may be mentioned here that she was unable to confirm a spontaneous preference for yellow by the scouting bees (v. Oettingen-Spielberg 1949); in her experiments blue was even visited more strongly primarily. Butler (1951) found blue and yellow approximately equally attractive, which is probably the true situation.

[16] Bolwig doubts the importance of scent markers. But with his own experiments he demonstrates their positive significance because small, distinct scent marks on the petals were noticed by the bees. For obviously when he makes fragrant stripes of lavender radially all over a colored plate, he cannot expect the bees to be guided to the center of it because the whole board is equally scented.

[17] The time of day fits poorly into this picture. We have interpreted the noontime sluggishness as expressing a diurnal periodicity in the bees' life.

mania, lasting for several minutes, of bees that have something especially choice to announce—effective enough by human standards, when one sees how the dance followers are carried along as the liveliness rises. A slight speeding up, not statistically significant, of the dance circuits with augmented sweetness cannot be regarded as a signal of the quality (v. Frisch and Jander 1957:256–260). It is possible but not demonstrated that the increase in number of vibratory episodes per second (Esch 1963, cf. pp. 237f) acts as an additional index of the profitability.

All this is true also of the bees that gather pollen or water, except that these bees identify themselves as such by the nature of the freight that they carry home.

Hive bees ready for work assemble on the dance floor (appetitive behavior), receive the messages, and permit themselves to be stimulated into flying forth—but in what numbers depends not solely on the urgency of the invitation but also on the content of the information. Boch (1956:139) provided quantitative evidence of this by feeding two groups of foragers 100 m and 525 m distant from the hive, respectively. At the more distant feeding station he let more bees forage, enough more that the number of foraging flights per hour was the same for both stations. In spite of that, within 4 hr there came to the distant station only 14 newcomers, but 35 to the nearer one. That this difference is to be attributed only in part to the more intense urging on behalf of the nearer feeding place (see pp. 240f) is made evident by direct observation: during our first experiment at some distance (1946) it was striking that even very lively dances that indicated a flight distance of 3 km met with less interest among the hivemates than simultaneous dances by bees foraging nearby. The comrades seemed to have little ambition to set forth on so long a flight as was urged by the former bees.

The newcomers that fly out in response to the dances search about in the direction and at the distance announced for the scent that they have perceived on the dancer, and thus reach the proper flowers. That has not only been demonstrated by means of numerous model experiments with artificial feeding stations; it has also been confirmed for the natural visiting of flowers, and for bees harvesting both pollen and nectar. With marked young bees that were watched all day long, Lindauer (1952) observed which dancers they tripped along after, whether they came home laden from their subsequent flights out, and what source they indicated with their own dances. Of 91 newcomers returning home successful, 79 brought back the same kind of harvest as the bees that had danced before them (30 nectar, 49 pollen of four different colors all of them indentical with the color of the pollen baskets of "their" previous dancers); additionally, 42 of them indicated by dancing that they had foraged in the same direction and at approximately the same distance as had those dancing before them; 37 did not dance, so that only the identical nature of their harvest but not the location of their feeding place could be determined; 12 of the 79 bees brought a cargo different from that of their dancers: 4 had foraged in another place, 2 had gathered a different kind of pollen but in the right place, 2 had harvested in the right place from the same kind of flower but had collected nectar rather than pollen; 4 supplied no information—they did not dance (Lindauer 1952:329–331 and Tables 6–9).

In their attraction for scout bees the floral colors and fragrances are of equal rank. But whereas the bees have no "word" for color, the scent they can carry home. Consequently the floral fragrance takes on a preeminent significance relative to floral color in the biology of flowers. This is widely assured by the almost ubiquitous distribution of floral fragrance among flowers pollinated by insects and by its species-specific development.

Bee-frequented flowers with odorless blossoms are rare exceptions. Of such in our flora I know of only the wild grape (*Parthenocissus quinquefolia* (L.) Planch), the bilberry (*Vaccinium myrtillus* L.), and—odorless at least frequently—the red currant (*Ribes rubrum* L.). Their blossoms are odorless for bees also (v. Frisch 1919:85ff, 1923:131ff). Perhaps some species of maple and milkweed belong here too. Common to all of them is their habit of growing in thick stands or else, with isolated trees or shrubs, that the flowers are crowded together in thousands. A heavy carpet of flowers in the fields is easier found, and here especially the attracting scent organ of the busy foragers will be of help as a signpost for the newcomers.

When the plants serving as food sources have a scattered distribution, as they often do with field flowers, the newcomers spread gradually over a rather large area, first because they do not hit the indicated direction and distance with hairline accuracy and also because they find the odor sought elsewhere in the vicinity, but secondly because blossoms not yet harvested from have more to offer than those that already have been fully exploited. But once a bee has decided upon a foraging area she remains faithful to it within surprisingly narrow limits. Where a meadow of dandelions with its glistening yellow invites visits and is exploited by innumerable bees, the individual forager comes back again and again to her own flowers and often confines herself to a range of a few square meters. If the stand of blossoms is a loose one or its yield scanty, the area may be somewhat enlarged; also it may be displaced a little or expanded on successive days. But these are unimportant variations. Even a food plant that is to be found everywhere over a domain kilometers in extent is visited by one particular bee only in a narrowly bounded region rarely more than 40 m in diameter. That is true with field flowers of scattered distribution just as it is for the profusion of plants in a field of rape (Giltay 1904; Minderhoud 1931; Butler, Jeffree, and Kalmus 1943; Butler 1945a; Kobel 1949; Ribbands 1949; Singh 1950; Bullmann 1952; Weaver 1957). Accompanying the long well-known loyalty to the kind of flower there is thus a far-reaching loyalty of the individual to a location.[18] Consequently there comes about a uniform subdivision of the whole region into little areas for the activity of the individual foragers. That is a matter of economy of labor and avoids unnecessary searching about.

---

[18]That is true of bumblebees also. Yet with them the foraging area is somewhat larger on the average and their flower constancy not so pronounced as in the honeybee (Bullmann 1952).

As soon as so many bees have been alerted that collection of the nectar proceeds more rapidly than its production, the yield becomes meager, and the dances flag and finally cease altogether despite continued foraging. To strengthen the foraging group further would be unprofitable. The reserve forces remain free for other tasks.

As a rule several or many nectar-producing plants come into bloom simultaneously. One scout bee from the hive may discover this species of flower, that one another. But because of the small number of scout bees not all the possibilities are found at one time, and after a few sources of honey have been found the hive will have its laboring force fully occupied and will not splinter them further, whereas a neighboring colony, as chance may govern its scout bees, may well exploit other sources. Thus may arise the phenomenon known to every keeper of bees, that different hives in an apiary bring in quite different harvests at the same time.

An example: in the middle of August 1962 there occurred a good yield at the Wolfgangsee; in particular, there was a profusion of "fir honey" in the woods. Some colonies of bees in the village of St. Wolfgang were flying to the stand of woods 0.5 km away. At a different apiary of the same beekeeper, right at the wood's edge, mixed honey was being collected from the meadows by almost all the colonies except for one that was specializing on fir honey (Ellmauer, verbal communication).

Where different sources of nectar present themselves at the same time and are searched out by scout bees from the same hive, the graded advertising activity takes on a useful role. One kind of flower may yield good nectar, another even better; quite apart from its variable quantity the concentration of floral nectar differs greatly. The distance of the source and whatever else determines the profitability is figured in and guides the larger stream of newcomers to the place where foraging is more worth while. Not in theory only! How competitive harvesting operates on the dances of a different foraging group can be demonstrated convincingly in experiments. Boch (1956:159) fed $^3/_8$M sucrose solution to two groups of bees each 70 m from the hive. Both groups, each of 40 marked bees, were dancing in nearly equal numbers. From 1 P.M. on one group (A) received 2M sugar solution, while the thin $^3/_8$M solution was continued at the other station (B). Although nothing had been changed at this station, after only 15 min the number of dances declined sharply and from 2 to 3:30 P.M. group B no longer danced at all. Now feeding was carried forward at both stations as it had been, thus 2M at A, $^3/_8$M at B, but at A the number of foragers was reduced from 40 to 10. Thereupon group B began to dance again (Fig. 232).

The decrease in eagerness to dance when the food supply remains constant, due only to the fact that other bees are finding something better, is to be explained by the behavior of the food recipients on the dance floor. They are drawn off by the competition. The foraging group that has something less good to offer no longer gets rid of its loads so easily. From this the releasing of the dances and their liveliness both suffer (p. 243). When—in the second part of the experiment—the bees that are bringing something better are few in number and hence are

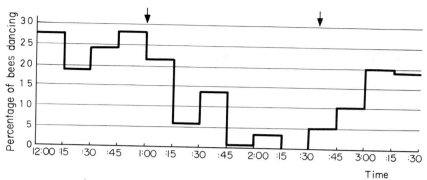

FIGURE 232. How competitive harvesting operates on the readiness to dance of a foraging group (B) of 40 bees; ordinate, percentage, combined for each 15 min, of bees dancing, from group B fed uniformly with ⅜M sucrose. Another foraging group A (not shown in the graph) of 40 bees from the same hive were fed the same concentration until 1 P.M., but were given 2M sucrose thenceforth. At 2:35 P.M. the number of foragers in group A was reduced from 40 to 10.

unable to satisfy all the recipients, the poorer quality once again gains in esteem.

Boch was able to confirm this phenomenon in many variations of these experiments (1956:159–162). By using several feeding stations it proved possible to make evident in the observation hive with surprising speed and clarity the influence of the general situation of the food sources on the eagerness of groups of bees to dance. In large normal colonies preferential recruiting for whatever source of food is most profitable at the moment will take place just as simply and reasonably, even though somewhat more slowly. That even in the natural visiting of flowers the influx of bees takes account of a changing concentration of nectar was shown in a lengthy series of observations by Butler (1945).

A better harvest is exploited preferentially not only by a greater levy of labor forces. For the better food sources the bees increase the speed of their flight, thus shortening the time spent on the trip (v. Frisch and Lindauer 1955:381), and they cut down the length of their stay in the hive between harvesting flights (p. 243); they increase the load taken on by filling the honey stomach more (v. Frisch 1934:103ff); and they prolong their working time, by starting their activity earlier in the morning and ending it later in the evening (Boch 1956:153, 154; Schricker 1965).

The daily performance of a foraging bee is sometimes large. According to Bräuninger (1964:107), for example, a bee visited a feeding station 900 m distant 60 times in a day and thus on this day alone accumulated 110 km in flight. Similar values were observed by him on several occasions and were also found by Heran (1956); see also Beutler (1954).

When a source of food is exhausted temporarily, or cannot be visited, for instance because of deterioration in the weather, the bees concerned with it are realerted as soon as an informant from the group once again finds something there. Any member of the group may fly out as a scout although they do so with quite different degrees of eagerness. Therein is a difference from the few scout bees responsible for the discovery of new sources of food. It is thoroughly reasonable too that just the successful return of working comrades, without any dancing, can provide the impetus for a revisitation of the familiar food source, whereas with new feeding places recruiting dances indicating the location of the goal are indispensable (see pp. 30, 43ff).

Production of nectar is subject to variations dependent on the time of day and environmental conditions (Beutler 1930; Kleber 1935; Beutler and Wahl 1936; Hammer 1949; Maurizio 1950; Huber 1956). In conjunction with an increasing exploitation of the flowers, this frequently has the result that not just the recruitment of further assistance but even foraging as such becomes unprofitable. Then it is customary for one part of the group to continue foraging despite the mediocre success; another part waits in the hive in case conditions improve; some members of the group let themselves be dissuaded from the deteriorated source and converted to a better one by the lively dances of another group; a few give up the failing feeding place and belatedly become scout bees—a profession that is not tied to any single age group in the colony; they set forth to spy out a more productive source of food (v. Oettingen-Spielberg 1949:483–485; Knoll 1926:500ff; and personal experiences). The degree of flower constancy differs individually and is far from frozen uniformity.

With all their hereditary ties to a racial tradition millions of years old, the bees have preserved a certain degree of individual freedom, through which alone the complexly intertwined pattern of their behavior becomes reasonable and equipped to meet the ever-changing requirements of life. Among them are not only those inconsistent individuals that do not take seriously the creed of flower constancy; not only the clever scouts, full of enterprise, that strike out on new paths while the great masses wait for a way to be shown them; there are also those "outsiders" that respond incorrectly to the instructions given: when dozens of dancers are advertising a rich source of food 0.5 km south of the hive and hundreds of newcomers are setting out for it, there are also individuals who go off searching in another direction, or in the immediate vicinity of the hive, or else kilometers beyond the indicated distance.[19] Whatever the cause of their "erroneous" performance, in

---

[19] I first encountered this at a time when I still had no idea that by their dances bees give a description of the location of the goal. I had determined that with feeding on a scented base the round dances of the foragers sent out newcomers who searched in the vicinity for the scent perceived, and I wished to know within what circumference this occurred. So on 13 September 1920 in Brunnwinkl I fed ten marked bees on a base scented with oil of basil, 16 m northwest of the observation hive. In Aich, 1 km ESE of the hive, I set out in the grass of the extensive meadows three feeding dishes on basil-scented cards. After 12 min there came the first newcomer to the feeding station itself; in the course of the 6-hr experiment 310 more followed, all of whom were captured and killed. Four hours after the experiment began there came to the dishes in Aich the first bee, looking for the fragrance with the typical searching flight, 50 min later a second one, which settled on a scent card but then flew off without having found the food, and 6 hr after the start likewise a third. The first and third bees were marked while they were drinking and were checked on their return home to the observation hive in Brunnwinkl (v. Frisch 1923:141, 142). In response to the round dances, which went on for hours, but in contradiction to their meaning, these bees had sought the perfume a kilometer away. Evidence for sporadic flights to a quite erroneous distance, in response to tail-wagging dances, will be found in the stepwise experiments in Fig. 91.

the sense of the whole they are useful eccentrics. For when, for instance, a field of rape to the south is in flower, then it is altogether worth while not only to drain it dry but to look to the chance that at the same time rape may be coming into bloom to the west and the east, nearby and far away. A few odd individuals suffice for this task, for if they are successful on their devious paths they will take care of sounding the appropriate alarm.

3. The Clocks of Bees and of Flowers

We have become acquainted with a large number of different factors that are concerned in the release of stimulating dances by foraging bees; taken in their entirety they tend to grade the mobilization of the foraging groups in accordance with the profitability of each of the several sources of food available at the time and to limit less productive foraging flights so as to conserve the working force. How simply and satisfactorily the problem of regulating supply and demand is solved in this way has been demonstrated in the preceding section.

There still remains one factor that plays a part in the frequenting of flowers by bees: their pronounced time sense.[20] Today diurnal rhythms are the subject of numerous studies on animals and plants (see Renner 1958; Bünning 1963). But I know of no other living creature that learns so easily as the bee when, according to its "internal clock," to come to the table. Stimulated by an observation of Auguste Forel's,[21] I tried many years ago to train bees to a feeding time, and was successful immediately. That gave the impetus for Ingeborg Beling's study of the temporal memory of bees (1929), which laid the foundation for further investigations. Here these will concern us only in regard to the biological relations of flowers.

Different species of plants open and close their flowers at different times of day, which are characteristic for them. According to these, Linnaeus proposed his "floral clock" (Kerner von Marilaun 1898:196). For every hour from 4 A.M. to 10 P.M. he found suitable representatives. If these were planted in proper order in a circular flowerbed, one was supposed to be able to read the time from the conditions of their blossoms—a project that failed only because of gaps in the "dial," since these flowers do not all bloom at the same time of year.

Later there was discovered another diurnal phenomenon, less conspicuous but no less important to the guests of the flowers: the amount of nectar produced, and its sugar concentration, vary in the course of the day, for each variety of flower in a manner typical for it (R. Beutler 1930; Kleber 1935). Even more drastic variations occur in the provision of pollen, which at first, at different hours depending on the species, is

---

[20] By the expression "time sense" it is not implied that a special sense organ is necessary for temporally adjusted behavior. As yet we possess only hypotheses in regard to the physiological foundations of such performance.

[21] On the veranda of Forel's country house there were candied fruits at midmorning breakfast and at afternoon tea. Once these had been discovered by the bees, the insects also appeared on succeeding days at the mealtimes mentioned, while after a few vain visits they shunned the luncheon table on which they found nothing alluring (A. Forel 1910:323ff).

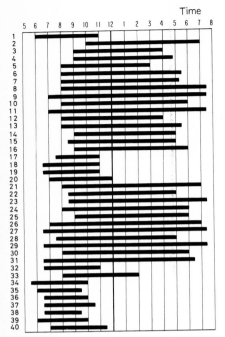

FIGURE 233. The black bars show the hours at which pollen is available from 40 different flowers frequented by bees: Nos. 1–32 according to observations by R. L. Parker (1925) in Iowa; Nos. 33–40 according to Elisabeth Kleber (1935) near Munich. The following plants are represented: 1, *Ambrosia trifida* L.; 2, *Asparagus officinalis* L.; 3, *Cercis canadensis* L.; 4, *Eupatorium rugosum* Houtt. (= *E. urticaefolium* Gray); 5, *Fragaria chiloensis* (L.) Duch.; 6, *Lonicera Morrowii* Gray; 7, *L. tatarica* L.; 8, *Lycium halimifolium* Mill. (= *L. vulgare* L.); 9, *Melilotus alba* Desr.; 10, *M. alba annua*; 11, *M. officinalis* (L.) Lam.; 12, *Prunus americana* Marsh.; 13, *P. cerasus* L.; 14, *Pirus communis* L.; 15, *P. ioensis* (Wood) Carruth; 16, *P. malus* L.; 17, *Rosa blanda* Ait.; 18, *R. rugosa* Thunb.; 19, *R. setigera* Mich.; 20, *R. xanthina* Lindl., 21, *Rubus occidentalis* L.; 22, *R. nigrobaccus* Bailey; 23, *R. strigosus* Michx.; 24, *Rudbeckia laciniata* L.; 25, *Salix fragilis* L.; 26, *Symphoricarpus orbiculatus* Moench.; 27, *S. albus* (L.) Blake (= *S. racemosus* Pursh); 28, *Taraxacum officinale* Weber; 29, *Tilia americana* L.; 30, *Trifolium hybridum* L.; 31, *Tr. repens* L.; 32, *Zea mays* L.; 33, *Convolvulus tricolor* L.; 34, *Papaver rhoeas* L.; 35, *P. somniferum* L.; 36, *Rosa arvensis* L.; 37, *R. multiflora* Thunb.; 38, *Verbascum phlomoides* L.; 39, *V. thapsiforme* Schrad.; 40, *Verbena officinalis* L.

copiously available but later in the day usually cannot be obtained (Fig. 233; R. L. Parker 1925; Kleber 1935). Thus it comes about that bees that frequent the flowers of a given species of plant do well at certain hours and yet at another time of day find the food sources less profitable or empty altogether.

These relations are easily imitated experimentally. If at an artificial feeding station one offers sugar water at a set time of day, within a day or two the visitors adjust themselves to the schedule. Thenceforth they come at the designated time, whereas before and after the hour of feeding even informational flights are almost entirely omitted. The foragers remain sitting at home, saving their strength and risking no unnecessary flights. The other situation too, which most often obtains in the actual production of nectar, is readily set up: at the feeding station there is sugar water from morning till night, but at certain hours in greater amounts or—in other experiments—at higher concentration than during the rest of the day. The artificial flower has an "optimal time." The bees quickly note this. If the visits are recorded on a "day of observation," during which the feeding dish offers nothing from morning till night, lively coming and going prevails at the optimal time. All members of the foraging group come repeatedly and search obstinately. During the other hours, when feeding was more meager or less sweet, searching is much less zealous and only part of the foragers appear (Wahl 1933). That under natural conditions bees behave in the same way relative to flowers was shown by Kleber (1935:235ff).

As a rule such pauses in foraging activity are spent at home by the bees. But they may give up the recess and visit a second food-yielding plant whose optimal time falls in the hours when the first one is unproductive. Their time sense and their adaptability are competent for this task. The foragers can be trained to visit a feeding place two or three times a day and to remain home meanwhile. Even training to five different times of day has been successful (Beling 1929; Wahl 1932). But they also learn very rapidly to seek out one feeding station at certain hours in the morning and another feeding place during afternoon hours, when at the time fixed upon food is offered now here and then there (Wahl 1932; Lindauer 1957:2). The bees even learned to appear at four different feeding places at four different times of day. But this difficult task was no longer mastered perfectly by all (Finke 1958).

Kleber was conducting experiments at an artificial bed of poppies, whose yield of pollen became exhausted every day between 9:30 and 10:00 A.M. Hence the foragers became trained to a morning time. One day the productive period was lengthened by offering at the end of the usual harvesting time poppy flowers bearing copious pollen that had been cut earlier and protected against visiting bees. With astonishment the observer noticed that individual bees, which kept on coming and carrying in the new harvest, were not mobilizing their group comrades but nevertheless recruited newcomers. It seemed as if the foragers would not respond to dances outside of the regular training time. Anyone who had seen often enough the arousing effect of the dances would hardly believe such a thing. And in fact a different, astounding solution of the riddle was found: bees trained to a definite time, once their good hours are past, withdraw to the margin or into the upper portion of the comb, or, in the observation hive—whose entrance opens on one side—very frequently to the opposite, quieter side of the comb (Fig. 234). Thus they go away from the dance floor and seek out resting places where they may doze undisturbed and where they will not be reached by a comrade dancing at the wrong hour, there to remain until the accustomed harvesting time approaches again. But then—a charming performance when the bees are marked with colors—they wander slowly from all sides toward the dancing place, until they are assembled there in a dense throng and can be alerted in a moment by the first successful informant (Ilse Körner 1939; v. Frisch 1940). In a shallow observation hive this behavior may be observed particularly easily. I. Körner made sure that the withdrawal of the foraging group from the turmoil of the dance floor and their renewed concentration there as the time of feeding draws near also take place in a normal beehive.

Thus by means of the bees' time sense their visiting of flowers is adapted to the given conditions after even short experience. The biological purpose that is found to be determinant in releasing the dances and for their liveliness apparently is the deciding factor here too: the greatest possible profitability combined with the least risk.

Summary: Dances and Profitability of the Food Source

1. The liveliness of the dances can be increased from the dull inception of a single circuit, immediately repressed, to a strong recruitment that lasts for minutes. Factors decisive for long-lasting and lively dances are: the sweetness of the sugar solution; the purity of the sweet taste; the ease of securing the food, which includes not-too-great distance to the collecting ground; a flowerlike fragrance; a flowerlike form of the food container; a uniformly continuing flow from the food source; a relative improvement in the quality; famine in the colony; favorable weather conditions. High viscosity of the sugar solution and increased weight of the load do indeed imply an increase in the work required, but in spite of this they augment the eagerness to dance. Under natural conditions both are involved when the quality of the food is better. Independently of such external factors the liveliness of the dances diminishes during the noon hours ("noontime sluggishness").

2. The harmonious interplay of the dance-releasing factors results in the wisely regulated exploitation of the sources of food in accordance with their profitability. Scout bees that have found something new arouse their hivemates only for rewarding sources. The newcomers follow the information as to distance and direction and, aided by the specific floral scent perceived on the dancer, reach the announced goal. Furthermore the foraging activity of each individual is concentrated on a relatively small area that is revisited again and again.

Different species of plants that are in flower at the same time often are very distinct as regards the copiousness of their nectar and its sugar content. Then the graded activity of the bees in recruiting regulates the

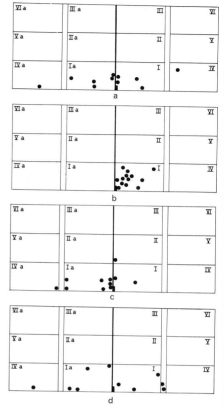

FIGURE 234. A group of marked bees from a large observation hive (see p. 7) was trained to feed between 10 A.M. and 1 P.M. The dots show how these bees—insofar as they were visible—were distributed over the combs at different times on the day of observation, a day on which no food was offered: I–VI, the honeycombs viewed from the side toward the hive entrance (black rectangle, at lower left of comb I); Ia–VIa, the same combs viewed from the other side; here there is no traffic of bees flying in and out; the bees reach this part by detouring through holes in the combs. (a) 7:25 A.M.; (b) 9:15 A.M. (the bees are beginning to accumulate on the dance floor; their concentration there lasted until 1:00 P.M.); (c) 1:20 P.M.; (d) 2:20 P.M. After Ilse Körner 1939.

size of the working groups in accordance with the supply of the individual sources of food and sends most of the assistants to the species of flower where foraging is most profitable. Very good sources of food are exploited more intensively, in that the foragers increase their flight velocity, decrease the length of their stay in the hive, take on a greater load, and prolong their working day. When a source of food becomes unprofitable the reaction of the foragers differs individually: some continue their activity even under the unfavorable conditions, many stay at home until circumstances improve, some allow themselves to be weaned away to more rewarding blossoms, and a few set forth independently on a search for something better.

3. In both quantity and quality the floral food supply is subject to regular diurnal variations. With their highly developed time sense the bees can adapt quickly to whatever conditions prevail. They avoid the risk of unnecessary flights and spend the unproductive hours at home, where they withdraw to quiet spots. As soon as the accustomed feeding time draws near they assemble from all directions on the dance floor and are ready to function again. Instead of taking a recess some bees frequent two or more sources of harvest that afford rewarding food at different hours.

VI / Guidance by Scent

1. HISTORICAL ASPECTS

About 1920 Guido Bamberger, an experienced apiarist, was my advisor in matters of beekeeping and an interested onlooker at many experiments. He had seen the dances and knew about the significance of the specific floral scent in communication. One day he told me that he had made a useful practical application of this. When he together with some other "migrant beekeepers" had transported his colonies to a neighborhood where harvesting conditions were better and had set them up there, he at once laid before them cut flowers, sprinkled with honey and sugar water, of those food plants that had occasioned the journey. These tokens were quickly discovered by a few bees. They alerted their hivemates and without delay dispatched them to the proper flowers. By this means Bamberger realized a better crop of honey than his companions. He had applied for the first time the principle of "guidance by scent" in the practice of beekeeping. But a proposal to apiarists generally that they should make use of this method (v. Frisch 1927) met with no response.

Russian scientists working with bees found a similar method for improving, by means of increased visiting by bees, the pollination and seed production in fields of red clover and other plants of economic importance. For instance, they trained the bees to red clover by feeding them in the evening or early morning in the hive with sugar water that smelled of red clover. The fragrant tincture was prepared by laying red-clover flowers for several hours in pure sugar water, which takes up the odor well. After being fed with this aromatic food the bees go hunting in the vicinity for the odor to which they have been trained and are guided to the desired blossoms.

In this way the flights to the fields were multiplied greatly (Gubin 1936:252ff, 1938; Kapustin 1938; Komarow 1939; Paparia 1940) and the production of seed increased, in part more than threefold (Sorokin 1938; Titow and Kowaljew 1939; Gubin 1939; Komisarenko 1955). Favorable results were also reported with vetch, alfalfa, sunflowers, and fruit. Control experiments where odorless sugar water was fed were not mentioned and presumably were not done. Since "stimulatory feeding" with sugar water alone results in increased flight activity, the magnitude of the effect of scent in these experiments is not clear. Only in Firson (1951) did I find mention of controls where odorless sugar water was fed; this increased the intensity of foraging, but the most prolific flights and the greatest setting of seed were obtained by feeding with the scented tincture.

With many assistants I was able in my own experiments to pursue the path opened by Bamberger: to advise the bees in their own "language" where they should fly in order best to serve the farmer or beekeeper. I reported the results in 1943 and discussed them thoroughly in 1947. Anyone who may wish to make use of this method is referred to these articles. Here only the main results will be discussed.

2. METHODS

Guidance of the bees by means of scent multiplies the flights to the harvest plant, and with this two things are accomplished: an increase in the yield of honey and a greater production of seed. No matter which of these objectives is in the foreground—the method remains the same.

If the bees are to be directed to a good food source by the dances and the scent of their comrades, the perfumed feeding must be undertaken during the daytime, about when flying out starts. With the model of the Russian scientists before one, the aromatic solution may be supplied inside the beehives just as a beekeeper does in furnishing food in autumn or when there is a shortage of nourishment.

To prepare the aromatic tincture, freshly picked flowers are put into the sugar water and this is poured off after 2–5 hr. The number of blossoms must be great enough that subsequently their fragrance is clearly detectable in the solution. But a usable aromatic tincture cannot be made from all species of flowers. It works with white clover (*Trifolium repens* L.), dandelion (*Taraxacum officinale* Weber), sunflower (*Helianthus annuus* L.), safflower (*Carthamus tinctorius* L.), thistle (*Cirsium oleraceum* Scop.), onion (*Allium cepa* L.). With other flowers the odor resembles that of the blossoms but is not identical with it: red clover (*Trifolium pratense* L.), rape (*Brassica napus* L.), turnip (*Brassica rapa* L.), wild mustard (*Sinapis arvensis* L.), mustard (*Eruca sativa* L.). But it can also be so different that there is no possibility of using the tincture for aromatic guidance: shaggy vetch (*Vicia villosa* Roth.), sainfoin (*Onobrychis sativa* Lmk.), raspberry (*Rubus idaeus* L.), apple (*Pirus malus* L.), hogweed (*Heracleum spondylium* L.), poppy (*Papaver somniferum* L.). What matters is that the odor shall exert on the bee's olfactory organ an effect comparable to that of the natural floral fragrance. This was examined in extensive experiments by Hildtraut Steinhoff and Miss Unterholzner (v. Frisch 1947:40–53).

Less troublesome than the preparation of perfumed solutions, and applicable also with blossoms from which usable aromatic tinctures cannot be obtained, is feeding the bees upon fresh fragrant flowers inside the hive.

A feeding dish with the desired amount of odorless sugar water 1:1 (1 kg of sugar to 1 lit of water) is set out in the usual way and is enwreathed with freshly picked flowers. Since the bees will try as quickly as possible to haul the flowers away as foreign bodies and throw them out of the hive, this must surely be avoided, most simply by putting the flowers into a wire basket of a mesh that admits the foragers but that makes it impossible for them to carry the flowers off (Fig. 235). The fragrance remains clinging to the bees' bodies, since they have to make their way through the blossoms and also are surrounded by them while they are drinking. When the feeding dish has been emptied the flowers are taken away. We have tested this method on red clover and thistle with excellent results.

FIGURE 235. Carniolan farmer's beehive, with a feeding bottle attached on top. The feeding dish is surrounded by a basket of wire screening for holding flowers. The hive entrance is at the lower right of the diagram.

Bees that find sugar water inside the hive alert by means of round dances. By these and with the assistance of the scent organ of the foragers some comrades are guided to the feeding saucer within. But since a feeding place inside the hive is unnatural the dances are comprehended by the majority as an invitation to fly forth. The success of such guidance by means of scent depends on this misunderstanding. Clearly this procedure is appropriate only for nearby sources of food, for in consequence of round dances the immediate vicinity of the hive is explored thoroughly.

Results are even better when the method comes one step closer to natural conditions. Comparative experiments have shown uniformly that flights to the pertinent species of flower are much more numerous after feeding on blossoms outside (in front of the hive) than after feeding within (v. Frisch 1947:65–76; v. Rhein 1957; Glushkov 1958; Pritsch 1959; Blinow 1959, 1960). The principal cause probably is that dancing is livelier and more persistent after outside feeding.

The reduction in eagerness to dance after inside feeding shows itself also in the fact that, in order to release dancing at all, a four times stronger sugar solution must be employed inside than at feeding places outside (v. Frisch 1947:72). Further advantages of feeding outside are that less sugar water is required and that there is success even with weak-smelling blossoms. Also, with little expenditure of time the feeding stand can be displaced to a distance of several hundred meters, if the field is that far away.[1] Since the dancers now indicate its correct location, it is found more rapidly than after feeding within. Of course feeding places outside may be discovered also by bees from another apiary. Where the yield of seed is concerned, it can only be advantageous if more colonies become involved in the labor of pollination. But the beekeeper does not like to promote an increased yield of honey in strange hives. In the vicinity of other apiaries one should therefore weigh whether one might not better confine oneself to using the attachment described in what follows.

Since some patience is required to attract the bees from the hive entrance to an outside feeding place, the following simplification is to be recommended in practice. In order to begin outside feeding an

---

[1] This journeying with the feeding place is necessary only on the first day; in repetitions the foragers come of their own accord to the accustomed place.

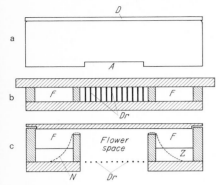

FIGURE 236. Hoehn's attachment for supplying food in front of the hive entrance: (*a*) front view; (*b*) view from above with the cover removed; (*c*) longitudinal section; dimensions, 30 × 8 × 5 cm. Traffic in and out of the hive may go on unhindered through the opening *A* in spite of the attached box. By crawling upward between the wires *Dr*, the bees pass through the flower space into the feeding areas *F*, where a wire screen *N* supplies a place for them to sit; *Z*, sugar water; *D*, cover.

attachment made of wood is fastened to the front of the hive or set into its flight opening (Fig. 236). The device was designed by beekeeper Hoehn. Its opening *A* (Fig. 236*a*) allows free traffic through the flight entrance. The bees reach the sugar water in the feeding sections *F* via the screen grill *Dr* (Figs. 236*b* and *c*). In doing so they have to pass through the flower-filled middle section and acquire the fragrance of the blossoms there. Now the bees that are alerted and come swarming out quickly find a shallow wooden feeding trough (some 30 × 40 cm in size) about 10 m in front of the apiary; the trough is full of flowers and contains about 0.5 lit of sugar solution 1:1 (for three hives). It can easily be displaced farther and carried on into the field.

Feeding during the flight period may cause piracy. After the artificial source of food has been exhausted it may happen that the foragers, now accustomed to drinking from a full cup, so to speak, will force their way into weak colonies that are little prepared to defend themselves, and steal their honey. The experienced beekeeper meets this danger by narrowing the hive entrance and by other measures. Where he brings the hives out to the natural source of food, he can prevent piracy by choosing colonies of approximately equal strength.

With properly executed guidance by means of scent three things can be achieved: (1) the flowers to which flight is desired are found more quickly; (2) more bees are active among them; (3) the bees forage more zealously. This last is not immediately obvious. But in our experiments it was often seen that bees guided by fragrance began to work earlier in the morning, continued activity later in the afternoon, and were particularly industrious.

At all events certain definite prerequisites must be met if these expectations are to be fulfilled.

1. The flowers must be fragrant. "Guidance by scent" with odorless blossoms is nonsense. A strong floral scent gives better results than a very weak one. Among red clovers there are considerable variations in this respect.

2. The flowers must yield nectar or pollen. By means of guidance by scent a good frequentation may be attained despite a small production of nectar, but where there is nothing whatever to be found the effort is as useless as an advertisement for a restaurant where there is nothing to be had. Guidance by scent should start the flights going, but they have to be sustained by the harvest supplied.

3. The work has to be done carefully and cleanly. A beekeeper who with the same hand alternately attends to his pipe and his apparatus for guiding the bees by scent will have no success.

4. The guidance by scent must be repeated daily at about the time flying begins. At first one offers 150–200 ml of sugar water 1:1; later about 100 ml suffice.

The following aspects are important when the yield of seed of fields is to be increased.

5. It is strongly recommended that the hives (two to four per hectare) be brought out to the fields and set up at their margin.

6. The most favorable time to set them up is at the start of flowering (Free 1959; Free and Jay 1960).

7. Beekeeping precautions should be taken to ensure that the colonies are well prepared for foraging (that they have many young bees).

3. Results

There is no doubt that by the methods described flight to fragrant flowers may be quickly achieved and increased, in a scientific experiment. What might be accomplished in practice still remained to be seen. This was tested in large-scale experiments of the following sort.

For each test two similar[2] fields were chosen: an experimental field and at a sufficient separation, if possible not in the same flight range, a control field. At both of them a like number of colonies of bees (that is, four per hectare) were established. At the experimental field scented feeding was carried out. The colonies at the control field received at the same times identical quantities of sugar water, but without a scent. The quantities measured were: (1) the numbers of bees visiting both fields, counted in designated control strips; (2) the weights of the guided and unguided colonies at the beginning and end of the experiment; (3) with plants of agricultural importance the yield in seed for both fields (the amount procured by threshing, in part also determination of the relative setting of seed by counting the flowers and seed of predetermined plants). With guidance by scent to wild plants for the purpose of increasing the yield of honey, the yield of seed was not of interest. In some of these experiments we did not undertake to determine the increase in numbers of flights; then experimental and control colonies could be established side by side and one ascertained whether the colonies guided by scent obtained more than the control colonies from an equal area and under the same conditions.

In the years 1942–1944, 86 experiments that could be utilized were carried out.

An increase in the yield of seed of red clover was what farmers wanted most. These flowers depend upon cross-pollination, which can be accomplished by bumblebees. But there are too few of the latter for field-scale agriculture. Honeybees can collect only part of the nectar from the long corolla tubes and mostly prefer more productive plant sources. This leads to poor harvests and harms not only the producers of red-clover seed but the entire cattle industry, because red clover is one of the most important food plants and varieties imported from abroad frequently are not sufficiently frost-resistant for the German climate.

In preliminary tests in 1942 the prerequisites were established for a quantitative comparison in our 12 red-clover experiments in 1943 and

---

[2] Complete correspondence in terms of size, type of soil, fertilization, climatic situation, processing of the harvest, and so forth will never be attainable. One can try only to keep the error as small as possible, to work *against* expectation where there are recognizable differences, and to repeat the experiments often.

1944. In all cases the fields to which bees were guided were visited more often, on the average by three or four times, than the control fields. In nine experiments it was possible to make quantitative checks on the yield of seed; on the average it was about 40 percent higher than in the corresponding control fields. Hence the application of guidance by scent is worth while despite occasional failures. With red clover a failure of the seed crop may be brought about by the weather, by very weak floral odor, or when the nectar is stolen; the latter occurs when holes are bitten into the sides of the floral tubes by bumblebee species with a short proboscis; honeybees then follow this practice (which from the viewpoint of flower biology is immoral) and steal the nectar without performing the service of pollination in return.

This bypass is not available when it is pollen that is being collected. Since bees do gather red-clover pollen too, guiding them by scent to red-clover pollen might be worth while (cf. v. Frisch 1947:91, 109). But frequently mere guidance with the floral scent will lead them to increased collection of pollen (Firsow 1951; Komisarenko 1955; Blinow 1960).

Increased yields of seed from fields to which bees were guided by scent were also achieved with Swedish clover (*Trifolium hybridum* L.), rape (*Brassica napus* L.), and turnip (*Brassica rapa* L.). With broad beans (*Vicia faba* L.) and with buckwheat (*Fagopyrum esculentum* Moench) there was some doubt as to the success.

The increase in the yield of honey among the colonies guided by scent was only moderate in the red-clover experiments; in some instances there occurred a decrease in weight relative to the control colonies, which had found better sources to harvest. Taking into account the gain in seed, it may be reckoned that on the average the work of the colonies in pollination affords the beekeeper no advantage. With the other agriculturally useful plants and with three wild plants tested the guided colonies produced between 20 and 100 percent more honey. Success was greatest with some species of clover, broad beans, turnips, mustard, onions, raspberries, and thistles. The sugar used for feeding represents no loss to the beekeeper, since it works to the advantage of the colonies and their development, and besides bears rich interest by stimulating foraging activity.

The Russian scientists initially regarded guidance by scent merely as a procedure for increasing the yield of seed. Ssacharow (1952) pointed out that in this way the yields of honey also could be increased and that by means of such methods too when food-producing plants came into bloom the bees could quickly be induced to visit them—and thus Bamberger's discovery (see p. 257) was confirmed.

4. Verification—But No Useful Application

Subsequent tests by others did not always turn out satisfactorily.

Free (1958) attempted without success to guide bees by scent to apple flowers and red clover. Conditions were unfavorable, because strong competition prevailed from other sources of food, and also most of the red-clover blossoms had been bitten into and exploited illegally. Beyond this the method of checking the results—only according to the amount of pollen collected by pollen traps—was not very dependable.[3] Stapel (1961) likewise reports negative results; he too checked only the amounts of pollen brought in. Minderhoud (1946) was able to achieve more numerous visits to red clover, but no persistent frequenting of it (presumably because of an insufficient production of nectar). With *Petunia* and *Alyssum* he (1948) had a certain amount of success, but it did not seem encouraging to him for practical application.

Such skepticism probably is unjustified. In carefully prepared field experiments lasting over several years v. Rhein (1952–53, 1954, 1957) confirmed the possibility of increasing significantly the seed yield of red clover and rape by guidance through scent.

In four scientifically performed experiments with red clover the yield of seed was increased in the experimental field relative to the control field on three occasions (by 24 percent, 46 percent, and 65 percent); on the fourth occasion it was reduced by 11 percent, this relative crop failure being caused by the appearance of "second growth" in the experimental field. In all four experiments there was a marked increase in flights to the fields to which the bees were guided. Further experiments provided practical tests in seven different farming areas and led to good results and in part to record yields.

Rape is capable of self-pollination, but cross-pollination still gives higher seed production. On three experimental fields the yield was greater by 21 percent, 23 percent, and 39 percent than in the control fields.

Gillard (1947) reports successful guidance by scent to *Allium porrum* L. Pritsch (1959) succeeded in increasing the number of flights to red-clover fields; the yield of seed was not measured. On inquiry of Dr. H. Oschmann, the Director of the Institute for Teaching and Research on Bee Culture in Tälermühle, I received the answer (1963) that in East Germany a whole group of socialist beekeeping industries have interested themselves in guidance by scent to red clover and that "inasmuch as the experiments have been so successful this procedure has since then been applied systematically. Even the Directors of the Agricultural Industries are pressing for the utilization of guidance by scent and are offering compensation for the bees used."

From the Soviet Union there is a comprehensive report by Glushkov (1958) with references to the literature. From it one may judge that the procedure is given wide practical application in the USSR. It is not to be assumed that an agricultural procedure that after all requires some expense will be retained for decades if it is unprofitable.

Experience has taught that with us too farmers and beekeepers who use guidance by means of scent in a sensible manner are satisfied with the results. Like v. Rhein more recently, in the years 1941 and 1942 beekeeper Hoehn carried out practical guidance by means of scent in cooperation with seed farmers in red-clover districts in the Rhineland

---

[3] See the critical review of these experiments by v. Rhein 1959.

and stated that the high seed yields had surpassed expectations and that the seed growers concerned had become grateful fellow workers. "There is no longer any doubt as to the efficacy of guidance by scent with red clover." Similarly, beekeepers who at that time worked with raspberries, heather, dandelion, and thistles have become convinced followers of the method and have profited thereby. Hence one must ask why, despite this, guidance by means of scent has penetrated so poorly into practice.

One cannot expect of farmers and beekeepers that they will be stimulated to activity by scientific publications or journal articles. They have to be shown. It is the task of the institutions concerned with research and instruction in bee culture to work this method out further for practical application and to make it available to farmers and beekeepers through traveling teachers and advisors on bee culture. Probably that miscarries for lack of personnel. And so, as often happens, a gain is lost for the sake of false economy.

Summary: Guidance by Scent

1. Knowledge of the way bees communicate affords the possibility of giving them directions in their own "language," to the advantage of agriculture and beekeeping, as to what kinds of flowers they should frequent.

2. By feeding sugar water that smells of certain flowers (aromatic tinctures) or with pure sugar water that is surrounded by fragrant blossoms, there are released recruiting dances that alert the hivemates and simultaneously point out the scent in question. The result is that flowers of that kind in the vicinity of the hive are found speedily, frequented strongly, and exploited zealously (guidance by scent). Feeding can be carried on in front of the hive entrance or in the open field instead of within the hive. The choice of method should be made according to the conditions prevailing and the particular objectives.

3. Through guidance by means of scent seed yields can be increased about 40 percent in red-clover fields and augmented also with other species of clover as well as with rape and turnips. With several of the honey-producing plants tested in large-scale experiments and with plants of agricultural importance the yield of honey under this procedure was increased by 20–100 percent.

4. In carrying out guidance by means of scent certain rules must be observed or else the procedure leads to failures. That hitherto the method has not found more widespread use probably is to be ascribed to such causes. Its effective introduction into practice should be the task of institutes for teaching and research in bee culture. In Russia the use of guidance by means of scent for the purpose of increasing the yield of seed has been customary for many years and has been proved to be useful.

VII / Application of the Dances to Other Objectives

There is no detectable difference between the dances of bees that are bringing in nectar or sugar water and the dances of bees gathering pollen (p. 62). That this should be so is evident just because the same bees frequently transport pollen and nectar simultaneously. But the dances may also refer to a source of water or resin. Or they may indicate a place from which nothing whatever is to be fetched, but where nevertheless there is something to be found: a dwelling for the swarming colony. In all instances the description of the location of the goal is of the same nature. What is different is the causes inducing the pertinent activity and also in great part the factors that release dancing.

1. Water

In the spring and summer months water is brought into beehives almost daily. At puddles and fountains near the apiary or in the "bee trough" provided by the beekeeper, at certain places one may see dozens of bees drinking water. In the hive they give it up on the dance floor just like nectar to awaiting hive bees who, according to circumstances, use it for two quite different purposes. If there are many larvae to be fed the brood nurses must produce large volumes of liquid food; this has a high water content, and the young bees are correspondingly thirsty. The water they require is distributed to them through the intermediation of the water recipients. Besides, this water is needed for temperature regulation. In the region of the brood nest the bee colony maintains a constant temperature of 35–36°C (W. R. Hess 1926). When there is danger of overheating, water is fetched copiously; the recipients spread it over the inner walls of the empty cells or in thin layers and shallow puddles over the surface of covered combs. Lively ventilation by fanning the wings causes rapid evaporation and effective cooling. Other bees are seen to be busy extending the proboscis. By this they spread a regurgitated drop of water out into a thin layer and remain thus for a few seconds; the process is repeated frequently and evaporative cooling is promoted in this way (Fig. 237; Lindauer 1954; Mirić 1956; Kiechle 1961).

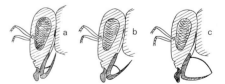

FIGURE 237. A regurgitated drop of water is spread into a thin layer by extending the proboscis, which is held in position c for a few seconds; cooling results from evaporation. After Lindauer 1954.

Differences of opinion as to whether pure, salty, or contaminated water is preferred by bees were clarified by Kiechle. On the first flight to a watering place they prefer salty or contaminated water and even urine to pure water. The odor may facilitate their finding the fluid, but the salt content is decisive for the preference.[1] Bees that already have been carrying in pure water for some time, on the contrary, customarily reject salty water, urine, or slops (Butler 1940, 1954:121, 1957:118; Piscitelli 1959; Kiechle 1961).

According to a letter from M. Lindauer, the primary preference for salty water is confirmed by the following observation. On the ground floor of the Zoological Institute of the University of Frankfurt am Main there are two aquarium rooms, one for fresh water and the other for sea water. In each room a shutter opens in summer with a slit 10 cm wide. Through this bees come to fetch water, but only from the sea-water basin; the attendant notices them here because they also often drown in it.

Dances by water foragers were first described by W. Park (1923c). Like the identical dances by bees gathering nectar and pollen, they assure an influx of helpers whenever these are needed (Lindauer 1954). But how do the water foragers assess the requirement? When on hot summer days the temperature in the brood area rises above the norm, a potentially lethal water shortage may break out in the colony with much more drastic effect than that of a famine. The water foragers are not sitting around there, nor do they come running into the brood area in order to learn whether the constancy of the temperature is endangered. As Lindauer (1954) suspected and his student Kiechle (1961) demonstrated, they find this out indirectly in a remarkable manner. When no water at all is brought in, the hive bees use for cooling—as also for satisfying the water requirement of the brood nurses—dilute nectar that they have in the honey stomach. But the use of this is not confined to the bees directly concerned, since the other bees too become involved. For a continual, extensive exchange of food goes on among the members of the colony (Nixon and Ribbands 1952; Ribbands 1952, 1953a:192; Free 1959a); the result is that among most of the hivemates the content of the honey stomach is quite similar, and when there is a shortage of water the concentration of sugar in the honey stomach soon increases. Via the perpetual exchange of food this increasing concentration of the contents of the honey stomach becomes known to the water foragers too and is their signal to fly out for water.

Once this activity has got a start, the intensity of foraging is regulated largely by the greater or lesser avidity of the hive bees that are waiting. When water is taken from the arriving foragers so stormily that they are freed of their burden in about half a minute, they dance livelily and thereby recruit additional water foragers (Fig. 238). It is the same kind of regulatory process that plays a part also in the collection of nectar (p. 243).

Nothing is known as to further dance-releasing factors with water foragers.

[1] The reason for this is unknown.

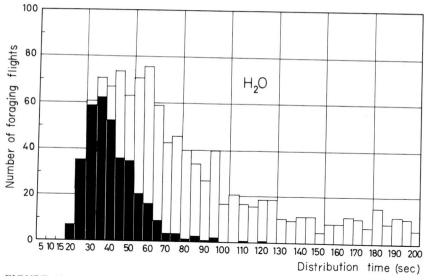

FIGURE 238. Relation between distribution time (time required for passing out the quantity brought in) and the readiness of water gatherers to dance; the ordinates are the numbers of foraging flights followed (black) and not followed (white) by dancing. Dancing occurs only when the distribution time is short. After Lindauer 1954.

Gathering water is not bound rigidly to a given age of the foraging bee. Depending on the need, the same bee may fetch alternately sugar water or pure water. If she has been doing one or the other for many hours or days a shift occurs less readily. Most specimens are fond of sugar water. Habitually bees like it better the more highly concentrated it is. In the end, however, it is the need of the colony that decides. Lindauer has studied these relations and gives the following description of the behavior of an individual bee (1954:426):

On the feeding stand there was a choice between a watch glass containing concentrated sugar solution and one of pure water. By artificial means a shortage of water had been produced in the hive. Consequently sugar water was taken from the foragers only hesitantly, but plain water avidly. Also the bee watched by Lindauer had difficulty in disposing of her sugar water and in her further flights changed to the water dish. "From time to time a small taste of the 2M solution was taken and if the reception of the fluid in the hive was delayed correspondingly the bee returned to the water dish; here at first she took only a hesitant taste but then after rinsing off the proboscis several times would drink heartily. Her individual 'personal' wishes were deferred in favor of the social requirements."

Just as with nectar or sugar water, samples of pure water are distributed to the surrounding bees before the dance and during short pauses in it. Thus these bees obtain information as to the material collected and brought in. Kiechle (1961) investigated whether with this information a "command" is given to fetch in the same substance, either sugar water or plain water. He set up, hidden in the grass, a feeding dish from which two or three marked bees collected sugar water in the presence of lavender scent. Beside it on a small stand was a plainly

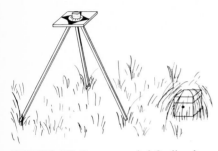

FIGURE 239. In a concealed feeding box on the ground two or three bees are fed sugar water weakly scented with lavender. At the stand in the open, marked conspicuously with lavender scent and with color, the newcomers find plain water. After Kiechle 1961.

visible dish of water (Fig. 239) that was likewise scented with lavender; this was found more easily by the newcomers. When the colony was in great need of water, often more than half of the newcomers that had been sent out by the dances of the sugar-water collectors accepted the plain water they found and became water collectors. But if there was no water requirement at the moment the newcomers declined this supply absolutely, despite the identity in odor. Thus the "command" is not a rigidly binding one. The newcomers have a certain degree of freedom of action and take into account, in addition to the material collected by the dancers, the momentary requirements of the colony.

Matters are similar with the collection of pollen. Most of the newcomers hold to what the dancers have brought, but a small part of them that by chance come upon other productive flowers seize the opportunity and take whatever they find (p. 248).

2. BEE GLUE (PROPOLIS)

Resinous exudations of various plants are collected,[2] specifically in late summer, and are brought into the hives as accumulations on the legs like pollen; there the resin is used to seal holes and cracks and also for strengthening the upper margins of the cells. Leuenberger (1954:76) watched bees collecting resin at the natural sources, namely buds on trees, and also describes the behavior on their return home. They remain sitting quietly on the comb near a place where resin is being used for building. The construction bees who are occupied in puttying come up and, as they need the material, work hard to chew off bits from the accumulations on the legs; then they carry them off in their mandibles and mold them further.[3]

According to Leuenberger (p. 77), "the resin does not stick to the very smooth bottom of the pollen basket but only to its marginal hairs; this greatly facilitates getting the resin loose. Nor does the resin adhere to the mandibles, with which the bee grasps and molds it." Rösch (1927) too saw in the observation hive how the resin collectors used to wait stolidly near a building site until they were freed of their burden. Often the load had been detached in 1 hr, but another bee had to wait 7 hr. "During this interval she did not once try to strip the resin from her legs herself."

According to Waltraud Meyer (1954), it is mainly foragers that work up the resin. The resin collectors themselves also participate in this labor.

Dances by resin collectors, identical with those of bees collecting nectar and pollen, were described but not studied more closely by W. Meyer (1954) and V. Milum (1955). The arousing effect of these dances is probable a priori and can be deduced from the following observation by W. Meyer: at various places she had set out resin; when it was discovered at one of the sites, there very quickly appeared there more

---

[2] The resin is put into the pollen baskets—in a different way from pollen—while the bees are on foot. Descriptions are given by W. Meyer (1954:186ff) and Mirić (1955).

[3] Similar observations were made long ago by François Huber (1814); see Ribbands 1953a:205ff.

bees from the same hive but none from other colonies. On another day she saw two foragers dancing with resin baskets in the vicinity of places being cemented; immediately thereafter, with the baskets untouched, they flew back again to the collecting site—a parallel to my observation on pollen collectors (pp. 34f), who on occasion flew home merely to dance and thereafter went on filling their pollen baskets.

What factors release the resin dances and affect their liveliness is unknown.

3. Dwellings

In spring, at the time when food is most copious and the deposition of brood greatest, there occurs not only an increase in the numbers of individuals in the colony but also a multiplication of the colonies by way of the act of swarming. By thousands the bees—mostly about half the colony—leave the hive in a mad whirl, after they have appropriated from the storage cells a goodly supply of honey to defray the expenses of traveling; as a hanging swarm they settle nearby, for instance on a branch, with their old queen. Timely provision for a successor in the hive has been made by building queen cells.

The hanging swarm is a well-organized structure. Its interior consists of loosely branching chains of bees that leave between them sufficient space for other bees to pass. On the outside the older bees form a compact cover; they sit close together in some three layers and at only a single place leave in the living cloak a "flight entrance" as a little opening to the world without (W. Meyer 1955).

Often dancers may be seen on the swarm after no more than 15 min; after a few hours dozens may be dancing simultaneously. These are informants, "scout bees," that were searching for a dwelling place and now are pointing out the location of a nesting site they have discovered, in the same way as a few days previously they may have been imparting the location of a feeding place. This significance of the dances was already recognized by Latham (1927). But he did not pursue these proceedings further. As a rule the beekeeper hastens to capture the swarm and roughly interrupts the normal course of events. Lindauer (1955) was the first to clarify these matters, by letting his swarms remain hanging, marking the dancers, and then following what took place until the swarm—often not until some days later—dissolved and went off to the announced nesting site.

Unlike the nectar collectors, who customarily dance for at most 1–2 min and then return to their collecting place in order to fetch a new load, these scout bees continue their dances for 15–30 min and—with interruptions—even for hours before they once again reinspect their goal, thereafter to go on dancing. They are not sacrificing any opportunity, for they have nothing to bring back except the message as to where they have found a potential home.

But it is not the case that from the start all of the scout bees announce one and the same nesting site. One bee has found something usable here, another there, a hollow tree, a hole in a stone wall, an

270 / The Dances of Bees

empty beehive, a cavity in the ground, or something of the sort, and hence dances may point simultaneously to ten or more distinct goals. The surprising thing is that in the course of hours or days agreement is reached, agreement as to the best choice of dwelling, and as soon as matters have gone that far the cluster dissolves and sets on the way.

Figure 240 gives an example of this. The swarm had settled on 26 June and remained in place until 30 June. On the first day, with increasing cloudiness, there were only two dancing scout bees, who were indicating nesting sites 1000 m north and 300 m ESE—as is shown by

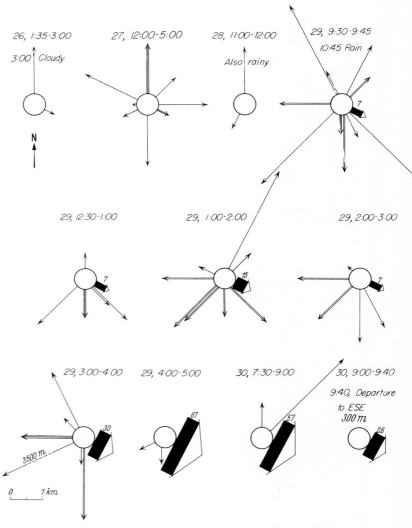

FIGURE 240. The dances of scout bees on the cluster (26–30 June) and the achievement of agreement on one of the nesting sites discovered. Throughout the experiment every scout bee was marked when she first danced. The arrows show, to scale, the distances and directions indicated by the dancers. Each arrow represents a newly marked scout bee. For the nesting site 300 m distant to the ESE, the figures show the number of scout bees marked. After Lindauer 1955.

the length and direction of the arrows in the figure. On the next day six additional nesting sites were discovered. Where several dancers indicated the same place the lines are drawn adjacent to one another or are thickened correspondingly and the number of bees is given at the side. To assist in comprehension of the figure it must be added that only the new recruiting dances are noted on each occasion; in addition those bees previously active of course continued their dances. The sequence of the observations now shows (Fig. 240) that some goals received little attention from the start, some others were designated temporarily by a rather large number of scout bees, but that finally all were given up again with exception of one 300 m to the ESE. An increasing number of bees kept advertising in favor of this site, and the dances for the competing sites gradually ceased. After reaching complete agreement the swarm occupied the home chosen: a hole in a wall 300 m away, directly ESE (Fig. 241).

As a rule the course of events was similar in the other swarms Lindauer observed. Departure occurs only when all the scouts are pointing to the same goal. Accordingly the moment when the swarm will dissolve can be foretold, and from the dances of the victorious group that is advertising it can be deduced with certainty in what direction and to what distance the swarm will fly.

Lindauer demonstrated the truth of this assertion by the fact that in some instances he went to the nesting site chosen by the bees before they did and was able to await the swarm there—a parallel with our earlier attempts at finding an unknown feeding station from the bees' description of its location (pp. 227ff). This happened, however, only three times with observations on 22 swarms.[4] One reason is that at that time it was still impossible to include the factor of "misdirection" in the calculations; further, there was the inconspicuousness of the entrances to the nesting sites: a hole in the ground beneath a patch of stinging nettles, a gap between the tiles, a little hole in a branch in a hollow tree, and the like. "The more often it was necessary to record failures in our search for nesting sites, the more our admiration grew both at the cleverness of the first scout bees that had discovered such potential nesting places and over their ability to give the other bees in the cluster such exact information about the location of these nesting sites. Only on the basis of a very precise designation of the location could the newcomers aroused find the nesting place that was announced; here there was no attraction from a conspicuous floral color or odor, which for foragers facilitate appreciably the finding of their goal" (Lindauer 1955:271). Yet it is to be assumed that scout bees, which are acquainted already with the nesting site, by means of their scent organ attract to the right spot newcomers who are hunting about in the vicinity.

Agreement on a nesting site is not always attained smoothly. For example, it may happen that two groups of scouts will advertise vehemently for two potential dwellings and that only after a rather long

FIGURE 241. Nesting site (circle) of the swarm of Fig. 240. After Lindauer 1955.

---

[4] Not included in the total of 22 are all those experiments in which the bees were offered artificial nesting sites (pp. 273ff).

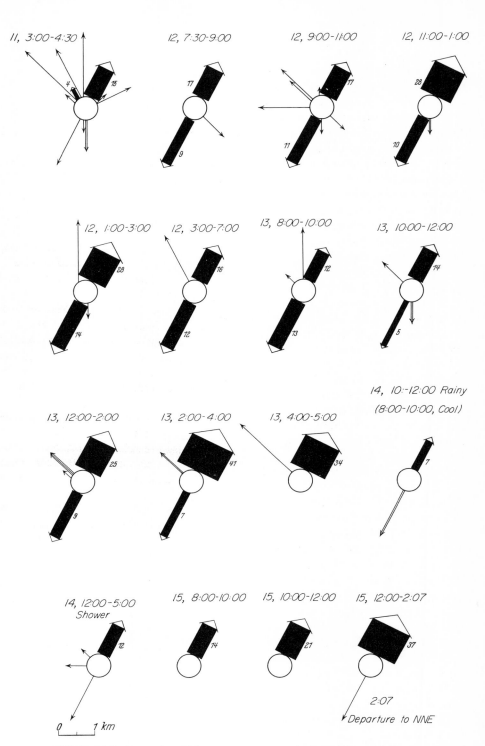

FIGURE 242. Course by which agreement was reached (11–15 June 1952) after competition had prevailed for a long time between two dancing groups of approximately equal strength (for explanation see Fig. 240). After Lindauer 1955.

contest will one of the two groups prevail. On 11 June 1952 a swarm appeared to have made its decision after only 2 hr (Fig. 242), but on the next morning competition became effective for a site in the opposite direction and the tug of war went on for two whole days before the first group finally came out on top.

It may also happen that no agreement whatever is reached (observed by Lindauer on three occasions) and that finally the swarm flies off in two parts in two different directions simultaneously. In two of the instances observed the disrupted cloud of bees returned to the former place of assembly and built a new cluster.[5] On the third occasion, after two days of disagreement there ensued a long period of bad weather and hence an acute shortage of food—the provisions for the journey had been used up. As soon as the weather improved the bees flew out after food; the swarm settled down right where it was and there built its combs (Fig. 243). In our climate that implies a catastrophe for the swarm, since a colony in the open will not survive the winter.

FIGURE 243. Comb building in the swarm cluster. After Lindauer 1955.

But these are exceptions. We return to the normal sequence of happenings and enquire: how is agreement reached? The answer is: the scout bees' advertising dances are graded in their liveliness and duration in accordance with the quality of the nesting sites discovered. Just as the nectar foragers, on the basis of many factors, communicate the profitability of a source of food, so do the scout bees "judge" in accordance with numerous quite different factors the suitability of a nesting site. Advertising by lively dances is much more energetic for good places than for less suitable ones. In this way a majority is won for the best nesting sites. Additionally, scout bees that have made a moderately good discovery become attentive to the more lively dances of some of their comrades, run after them, on the basis of the information received set out and inspect the better nesting sites, and let themselves be converted relatively easily in their favor. One is tempted to say that they have become convinced of the better quality of the other site, and now they themselves advertise in its behalf. There are some bees, too, that hold obstinately to their own goal, but their propaganda loses force gradually beside the tumultuous dancing for the favored ones, and finally agreement prevails.

That is easy to tell, but was by no means easily discovered, especially not by simple observation of the normal course of swarming.

The ordinary rate of swarming was not sufficient to provide even enough material for study. Hence Lindauer used "artificial swarms," such as are customary in the beekeeping industry. This is achieved by suspending the queen in a screen cage on a branch in a predetermined place and the colony of bees is shaken out on the ground. They gather in a cluster around the cage and behave thenceforth like a normal swarm; scout bees fly off searching and announce their findings.

The most important thing was to find a territory in which there were no natural nesting sites, but only such as were provided artificially and

---

[5] On one of these occasions the queen was lost and the swarm broke up; on the other the experiment was spoiled overnight by outside interference.

that could be shaped in whatever way might be desired. For a part of such experiments a small island in the North Sea was chosen, for another part a monotonous plain of farmed fields east of Munich. There artificial nesting sites could be provided as wanted and one could watch how the scout bees behaved toward them.

If, for example, an empty beehive was set out in a well-protected spot under a spruce sapling and covered besides with brush, while another stood open and unprotected on a tree stump (Fig. 244), dancing for the hive protected from the wind was much more intense. Evaluation by the scout bees was necessarily based on their impressions while inspecting the nesting sites. That is evident from their behavior:

"All details there were tested exhaustively; first the dwelling itself: the bees ran into the box through the hive entrance, stayed there a few minutes, came out, hovered around outside the hive and crawled over it from all directions, then flew in somewhat larger loops about the nesting place and in the vicinity of the protective brush covering, about the foliage and the groups of trees; finally they returned once more to the hive, everted their scent organ and fanned their wings at the entrance, once again inspected the interior; and so it might continue for 10 min or even for an hour. Then the scout bees flew to the cluster and in a state of the greatest excitement performed an advertising dance there or—if the preceding inspection had turned out unfavorably—remained sitting idly on the cluster" (Lindauer 1955:285).

That the protected location was really a determining factor was demonstrated when the hive that stood in the open was given a windbreak of brush. With surprising speed the same scout bees that had hitherto been so phlegmatic alerted the other bees by means of more lively dances, and the group advertising in its favor grew swiftly.

By means of this method it was possible to ascertain a series of characteristics that were heeded by the scout bees. A drop of oil of balm (*Melissa*) deprived an otherwise suitable straw container of its value as a habitation for the bees. Hence odorous characteristics are taken into account, and surely in a positive sense also when they are appropriate for the bees. What in this respect is agreeable to them and what is not is an as yet unprobed chapter in bee psychology. The interior of the dwelling must bear an appropriate relation to the size of the swarm. The distance too is given strong consideration, for the cluster is hanging in the vicinity of the parent colony. The permanent settling of different swarms in the same neighborhood would reduce the harvest for all. Preference for nesting sites at a greater distance promotes a looser distribution of the bee colonies over the territory. In an experiment with two nesting sites of equal value 30 m and 250 m distant from the swarm, the latter site released by far the more lively dances and was occupied. In exact opposition, with food sources recruiting is more intense for the nearer one (pp. 240f). Here the identical set of objective circumstances has a contrary effect on the releasing of dances, but here as there the relation to the biological significance of the goal is a meaningful one. During the flights of inspection by the scout bees attention is always paid to the conditions of wind and temperature at the nesting place. Thus in one experiment the scout bees

FIGURE 244. A choice of two artificial nesting sites: (*a*) protected; (*b*) unprotected. After Lindauer 1955.

were advertising vigorously in behalf of a well-protected empty beehive, while it was cloudy. But on the following day the hive lay in bright sunshine under a clear sky, and a thermometer inside registered over 40°C. The scout bees soon ceased dancing entirely, and not until the nesting place had been shaded with a cover of thick pine brush was their interest rearoused; it led finally to settlement of this place. In an altogether comparable manner windy days may cause the scout bees to give up an insufficiently protected nesting place. One recognizes the deep biological importance of the fact that the decision in regard to a good source of food is made very quickly, whereas for the most part a nesting site is inspected over and over again under different conditions before the verdict is reached.

One additional illuminating experiment may be mentioned. Two artificial nesting places were set up, each 75 m from the swarm; both were found by scout bees and were announced by dances. A full day later there appeared on the cluster an extremely lively dancer who had discovered a third nesting opportunity. Very quickly she won support and further tempestuous dances pointed toward the same goal. Lindauer was able to find the place from the directions given by the scout bees, even before the swarm moved in; it was a cavity in the ground, at almost the same distance (80 m) as the two artificial nesting places, beneath a decaying tree stump, protected by a heavy thicket, 0.5 m deep in the earth, of appropriate size and completely dry inside despite rain for several preceding days—an ideal habitation for bees. The scout bees that found it were unmarked, and hence they had not yet been to the other nesting sites that were up for a choice. The basis of their lively dances was not any comparison with the inferior dwellings, but obviously an innate scale of evaluation of the quality of habitations, just as the profitability of a source of food is indicated by an appropriate degree of liveliness without any comparison between it and better or worse feeding places. But here, even more than with the food sources, one has to wonder respectfully about the multitude and the complex nature of the contributing circumstances, the significance of which is weighed according to an inherited scale for evaluation.

Our setting up the artificial nesting sites in a territory that contained no bees except the observation colony confirmed an older suspicion that had not yet been proved: that a colony's scout bees become active some time before swarming occurs. On the North Sea island as much as 3 days before the swarm left the hive, scout bees, looking for a shelter, were to be detected making attentive inspection of the artificial nesting sites and other potential places for habitation. It developed that they were dancing prior to swarming in the hive just as they did later on the cluster—as though in a preview of the future home. Naturally one wonders how the hive bees know whether the dance indicates a source of food or a nesting site. The difference is not hard to discern: scout bees do not bring home any nectar and distribute nothing to their surrounding fellows; rather, during the pauses in their dancing they stretch the proboscis out toward other bees; "thus they are constantly begging for provisions for their next outward flight; foragers also require provisions for travel, but they ingest it quite according to rule just before they fly out" (Lindauer 1955:266). Additionally, at the swarming time recruiting for feeding places is almost wholly suspended, so that any danger of misunderstanding is reduced still further. That leads us to the additional question, what causes the appearance of scout bees in the first place?

The time of swarming is a period of superfluity of food, when the storage areas of the comb are full of honey and pollen and the nourishing secretion of the nurse bees is choked up in their glands because—for lack of space—there is an insufficiency of young brood. Under such conditions their burden is taken very dilatorily from the nectar foragers, and the pollen foragers too can hardly dispose of their freight. These are the very conditions that lead to swarming and to repression of food-collecting activity. Now some of the foragers who have become unemployed turn to another type of activity, more important at the time: they become scout bees. This is a temporary office, limited to the season of multiplication of the colony and bound besides to external conditions favorable for swarming and for the establishment of new settlements of bees.

We have followed proceedings in the cluster up to the reaching of agreement by the scout bees. But who finally gives the signal to depart? It is the scout bees themselves that, as soon as matters have gone thus far, gather in almost their full number on the cluster and by means of the characteristic "buzzing run" (*Schwirrlauf*) within a few minutes stimulate the entire company, sitting there stolidly, to fly off (see pp. 279f). The buzzing run is a primitive signal that contains no further information and that means no more than "Let's go!" The data relative to direction and distance have after all been given previously. Apparently it is also the scout bees, familiar with the route, that by their lively guidance further assure the finding of the goal: while the swarm cloud is proceeding gradually along one sees a few hundred bees shooting ahead through the crowd in the direction toward the nesting place, then flying slowly back at the margin of the swarm cloud, again pushing forward rapidly, and so on, until the goal is reached (Lindauer 1955:319).

When one considers from this point of view how from a hive containing 60,000 and more inhabitants about half the population is led to a suitable new home by means of a series of behavioral acts appropriate to the time, thoroughly evaluated, and adapted to the circumstances, this, more even than other processes among the bees, seems of a nature to preserve the biologist from untoward obscurity of thought and to let him discern that we are a long way from a true explanation of such powers.

Summary: Recruiting Dances for Water and Resin; Advertising Dances for Nesting Sites

1. The dances of bees that are pointing out a source discovered for nectar or resin, or the location of a nesting site, do not differ from the dances of foragers for nectar and pollen. To some extent the factors that release the dances differ.

2. Water is required by the young bees that take care of the brood

and for regulation of the hive temperature. A shortage of water leads to concentration of the contents of the honey stomach. This condition spreads rapidly among the hivemates by way of the constant interchange of food. Thus it becomes known to the water gatherers too; they are induced thereby to fly out and fetch water. The intensity of collecting and the quantity supplied—as with nectar foragers—are regulated to a great extent by the fact that when the need is great water is taken more eagerly from the foragers; that increases their zeal and their inclination to dance. Collecting water is not bound rigidly to any definite age in the life of the foragers.

3. As with the nectar brought in, the recipients in the hive are informed by sample tastes of the nature of the substance collected by the water foragers. The sample distributed can be regarded as a nonobligatory command likewise to fetch water in response to the dances—"nonobligatory" because the decision of the newcomers whether to fetch water or nectar is determined also by the momentary needs of the swarm.

4. Resin for cementing (propolis) is gathered mainly from buds on trees, is brought in like pollen in the baskets on the legs, and is used particularly for sealing up cracks and the like. Observation has shown the probability of an arousing effect of the dances by resin collectors, and the success of these dances in describing locations of resin, but this has not been studied more closely.

5. Dancing bees are seen on the swarm cluster too. These are scout bees that have discovered a nesting site and are announcing its location by means of their dances. As a rule the scout bees of a swarming colony discover several different potential dwelling places. Each bee advertises in behalf of her own finding. In the course of hours or days agreement is reached as to the best opportunity for housing. This agreement comes to pass because each scout bee advertises the more energetically for her discovery the better the habitation. Just like the profitability of a source of food, the quality of a potential nest is "judged" according to many factors (protection from wind, temperature relations, characteristics in respect of odor, roominess, distance from the present colony). Whether a feeble, a lively, or a strong dance is called for is decided after a thorough inspection of the habitation discovered, made on the basis of an inborn scale of evaluation and not consequent on examination of competing habitations. Yet bees that are advertising with feeble dances in favor of an inferior dwelling may well be induced by the strong dancing of other scout bees to suspend their dances or to take a look at the better discovery and thereafter themselves to advertise in its behalf.

6. As soon as the scout bees have attained agreement in this way and are now pointing toward a single goal, by means of buzzing runs they give the signal for the cluster to take off. Now the bees move to their new home in the direction and at the distance announced. The scout bees, who know the way, lead them and assist them in finding the goal, by flying tumultuously forward through the slowly proceeding swarm again and again, and then flying slowly back at the edge of the advancing cloud.

VIII / Other Dance Forms

Up until now we have confined our attention to the round dance (Fig. 29) and to the tail-wagging dance (Fig. 46). For the hivemates both signify a demand to fly out to an indicated goal, associated in the round dance with a vague reference to its location (near the hive) and in the tail-wagging dance with a precise description of its position.

Besides these a series of dances of other kinds have become known. Of these the jostling run, the spasmodic dance, and the sickle dance belong in the classes of round dances or tail-wagging dances as precursors or variants of them. The significance of the other dance forms has been clarified only partially.

1. Jostling Run, Spasmodic Dance, and Sickle Dance

As long ago as 1923 (pp. 30 and 40f), I mentioned that a forager who has examined the feeding dish in vain during an interruption in feeding and who returns home unsuccessful creeps placidly up into the comb and settles down; but that after a successful visit she runs upward hastily with quick, tripping steps. Upon such a lively run there follows as a rule a dance, but not when she just walks up placidly.

H. Schmid (1964) described this lively run of bees returning home successful as a "jostling run" (*Rumpellauf*). After her first distribution of sugar water the forager often continues her jostling run: for about a hand's breadth she runs linearly or in a slight arc through her densely packed comrades, as she does so pushing energetically aside all the bees that are in her way. The inception, course, and conclusion of the jostling run are so clear that its nature as a mode of expression is obvious. Its significant characteristic is not haste, but conspicuousness. Progress during this agitated run is not more rapid but slower than it may be later after the forager, having disposed of her burden, hastens back to the hive entrance. Also in the latter instance she slips artfully through gaps in the throng of bees, whereas in the jostling run she butts into her comrades as though intentionally.

For the most part dancing takes place only after several successful flights, but jostling occurs even on the first successful return home. In this way the foragers arouse the attention of bees that are sitting nearby or that get bumped into, without their learning anything except indeed that something is going on. By this means alone, without any dance, the group comrades can be induced to seek out anew the feeding place already known to them (see p. 30), especially if a specific scent is peculiar to it.

On offering food in the vicinity of the hive, Hein (1950, 1954) observed the "spasmodic dance." The bees returning home ran over the comb in varying directions, distributed food at various spots, and meanwhile made short tail-wagging movements that, in the same manner as the waggling run in the tail-wagging dance, pointed toward the goal (Fig. 245). Hein observed these direction-indicating spasmodic dances when the source of food was no more than 2 m distant. The hivemates paid little attention to this behavior, presumably because of the brevity and irregular performance of the waggles. Probably these are more the expression of a dancing mood than an effective signal. Similar behavior is seen not infrequently with far-distant feeding places, when on their first successful return home bees are in a mood to dance but are not yet in full swing. They run agitatedly about on the comb, distribute food here and there, and meanwhile perform isolated waggling runs at different places.[1] But for these runs the designation "spasmodic" would not be fitting, because they lack the brevity characteristic of these movements.

FIGURE 245. The "spasmodic dance" (*Rucktanz*). Distance to the source of food, 2m; dots, points where food was distributed. After Hein 1950.

The "sickle dance" was first described by Tschumi (1950). On p. 61 it was mentioned already as a form transitional between the round dance and the tail-wagging dance. As such it occurs with all hitherto-studied races of the honeybee except for the Carniolan bees with which most of my work has been done (cf. p. 295). The open side of the sickle-shaped curve (Fig. 246) points the direction to the goal, in the same way as the waggling run in the tail-wagging dance. That is true of dances on the vertical comb as well as on a horizontal surface.[2] Hein (1950) drew attention to short "twitching segments" before the turning points of the sickle (× in Fig. 247). Only these twitching segments of the dance before the turning points could indicate the direction to the comrades (Hein 1954). That is possible, but is not proved. Very short tail-wagging segments are hard to observe in detail. It would be necessary to make high-speed photographs of sickle dances in order to distinguish whether sickle runs without twitching segments have a direction-indicating significance or not.

FIGURE 246. "Sickle dance" (*Sicheltanz*). After Tschumi 1950a.

FIGURE 247. "Sickle dance" containing a "spasmodic dance" in some segments (×). Schematic, after Hein 1950.

2. THE BUZZING RUN

On p. 276 the buzzing run has already been mentioned as the signal by means of which scout bees, after achieving agreement as to the new nesting site, induce the swarm cluster to decamp. At an earlier point in time it is wholly corresponding buzzing runs that give the sign in the

---

[1] Wittekindt (1960) saw the same sort of performance and designated it "winding their way."

[2] In reference to the effectiveness of indications of direction in sickle dances see p. 298.

beehive for the swarm to set out. In the buzzing run "the bees concerned force their way with great excitement and nervousness through the other bees standing about; running in a random zigzag they butt them energetically aside, vibrating the abdomen violently and letting a readily perceptible buzzing of the wings be heard" (Lindauer 1955:315). According to observations by Esch the sounds produced here are similar to those made during the waggling run. P. Martin (1963) states that the buzzing run affects other bees only on close contact.

Before swarming most of the bees sit idly in the hive or in front of the flight entrance. "When now toward noon it grows warmer, all at once there is unrest and agitation in the hive and this unrest is caused by the aforementioned buzzing runs. At first there are two or three bees, then after a minute a dozen, and more and more buzzing runners go dashing about in wild haste over the combs. Like an avalanche the number of buzzing runners grows, many of them rush to the hive entrance, arousing similarly those slothful ones who had gathered together like a tuft before the flight opening, others hover briefly about the hive but return once again to continue their buzzing runs. In about 10 min the moment for the departure has arrived . . . then the bees nearest the hive entrance rush forth and in a dense stream all follow. The queen too has been aroused and if she does not follow the swarming bees out at once she is badgered without interruption by bees buzzing and running until she has found the hive entrance and hurls herself into the swarm cloud" (Lindauer 1955:316). In an analogous manner the dissolution of the swarm cluster proceeds (p. 276).

It is characteristic of the buzzing run that it has a very rapid effect, that the bees aroused pass on of their own accord the signal they have received, and that bees of all ages respond to it. Thus it comes about that all ages of bees are represented in the swarm and hence also in the newly founded colony.

On yet another occasion Lindauer saw the occurrence of buzzing runs, when during the formation of the swarm cluster an offshoot had settled to one side. Then bees buzzing and running came to drive it up again and to draw it into the cluster, by flying back there, everting their scent organ and fanning their wings. In this way the offshoot is aroused and is attracted by the scent organ to the proper place.

Mechanical disturbances alone induce neither a cluster nor an offshoot to fly away. Lindauer frequently rummaged through a cluster in search of queens, often separated a mass of bees from it and shook them to the ground, but neither the cluster nor the portion thrown down were dispersed by such a procedure. A general departure of the bees was induced only by the buzzing run of their comrades. With all its primitiveness this is therefore a specific signal.

3. Grooming Dance (Shaking Dance)

Haydak (1929, 1945) and Milum (1947, 1955) have described independently a shaking movement of the bees whereby others sitting in the vicinity are called out to clean certain parts of their body that cannot be reached by their own grooming devices. With reference to the type of movement Haydak speaks of the "shaking dance," Milum, in accordance with the objective, of the "grooming dance." In this performance the bee shakes the body rapidly to and fro, to the right and to the left, meanwhile standing somewhat unsteadily because simultaneously she raises one of the middle legs in order to comb her thoracic hairs (Milum 1955). Thereupon another bee often comes up at once (but often also not at all) as a "barber," and with her mandibles works over the hairy coat on the petiole and at the roots of the wings. While this is in progress the bee whose hair is being dressed holds quite still.

During a visit I made to the United States Milum showed me the process in his observation hive. I had not noticed it among my own bees. Lindauer tells me he has seen it on many occasions. Morgenthaler (1949) thinks it might be concerned with a reaction to an "itching stimulus" caused by external mites. Its occasional massive occurrence argues against its being a regular grooming procedure and speaks in favor of the participation of other circumstances. The matter certainly is deserving of further attention.

4. Jerking Dance (D-VAV)

It is somewhat confusing that the same expression, "shaking dance," has occasionally been used for quite a different performance that Haydak (1929) and Milum (1955) had already distinguished sharply from the grooming dance, but the significance of which is still unclarified. In it, according to Milum, a worker makes contact with another bee by brushing her with the antennae or even more often by seizing her abdomen, or even the head or thorax, with the forelegs, or by climbing altogether on top of her and then with the abdomen making a rapid vibratory movement in a dorsoventral direction (7–8 strokes); at varying intervals this is then repeated several times with other partners (Fig. 248). For this vibratory movement Milum also uses the term "dorso-ventral abdominal vibration" or the abbreviation D-VAV. The bee grasped by the "dancer" holds still and also does not respond afterward in any recognizable manner. Mostly it is a worker, but a drone or the queen also may be treated in this way (Milum 1955).

According to Allen (1959), jerking dances by the queen are to be seen very frequently before swarming, and according to Hammann (1957) very often also before the nuptial flight, so that a specific invitation to fly out might be imagined. But against this assumption is the fact that Hammann saw jerking dances even on the closed queen cells of queen bees that had not yet emerged (likewise Haydak 1945), and that before flying out the queens not only are favored with vibratory movements but also are handled and set upon in other ways.

Jerking dances toward worker bees are indeed more frequent with colonies about to fly than they are at night or in bad weather, but no rigid relation with flight activity was to be determined (Allen 1959a). With bees of all age groups Allen saw jerking dances from the third day after emergence on; they became more frequent up to about the third week of life.

Haydak interpreted the jerking dance as an expression of joy and

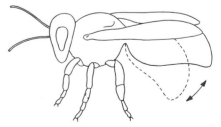

FIGURE 248. A bee in the "jerking dance" (*Rütteltanz*). The arrow indicates the dorsoventral vibration of the abdomen. Schematic, after Eleonore Hammann 1957.

contentment. But Milum points out that it is to be seen by day and by night, summer and winter, even under conditions that give no cause for satisfaction, for instance in starving colonies.

I too have occasionally watched such jerking dances in my Carniolan colonies without discovering their meaning. Butler (1958) and Wittekindt (1961) mention them without contributing to their explanation.

Schick (1953) succeeded in releasing jerking dances at will by feeding urethane-containing sucrose solution (0.3- or 0.15-percent urethane in 2M sugar solution). By feeding sugar solution with the addition of 0.025-percent Nipasol (propyl ester of p-oxybenzoic acid) he obtained jerking dances combined with "trembling dances."

Considering our present knowledge, a simple question mark is more appropriate than a long discussion of the significance of the vibratory dances.

5. Trembling Dance

At times one sees an odd performance by foragers who have returned home. "While they are running about irregularly and for the most part at a slow pace over the combs, their body, in consequence of twitching movements of the legs, continually makes quivering excursions forward and backward, to right and left. As this activity proceeds, they run about on four legs with the forepair, also quivering and twitching, held aloft about in the position a begging dog holds his forepaws" (v. Frisch 1923:90). The "trembling dance" may subside after only a few minutes, but may also continue longer, even for several hours. I think it tells the other bees nothing. They do not bother about bees doing the trembling dance and when by chance they make contact with such individuals they are not impelled thereby to any definite performance.

Additional observations by Lindauer (1948) and Schick (1953) have shown that trembling dances customarily appear under the most varied circumstances whenever the bees are exposed to a harsh disturbance or are in a poor state of health. Thus trembling dances often occur when foragers are handled roughly while they are being marked at the feeding dish, or when the surface beneath it is contaminated with sugar water thickened by evaporation, in which they get their feet stuck, or when they visit *Asclepias* flowers and make contact with their sticking organs, or when after good sugar solution has been offered at the feeding place suddenly a solution is spoiled by adding salt, quinine, or acid, or when things get too crowded at the feeding dish because the newcomers are not captured and removed, but also when the hive is being robbed. During their last days very old bees lapse into trembling dances. Gertraud Schifferer observed the phenomenon with her foragers when they were laden heavily with lead (see p. 109). Further, trembling dances may be released regularly by adding poisons to the sugar water (Schick 1953). The nature of the poison is unimportant; the most varied chemicals have the same effect provided they are harmful to the bees.

Trembling dances have been seen among foragers for nectar, pollen, and water, collecting from both near and distant sources, but occasionally with young bees 2–3 days old that neither fly out nor dance. Not infrequently foragers will endeavor to perform a recruiting dance in spite of trembling, but on their unsteady legs this turns out to be so abnormal that it is ineffectual. Whenever in an attempt at a round dance a circuit is made successfully, a few other bees will at once fall in behind; but they leave the dancer just as quickly again when she relapses into trembling.

From all this it may be deduced that the trembling dance gives the hivemates no information and they pay no attention to it. It occurs as the result of adverse circumstances and experiences and perhaps is comparable to the condition that Florey (1954) has described as a neurosis, which is seen when a situation of nervous conflict is produced artificially in bees.

F. Schneider (1949) observed the occurrence of trembling dances, which he did not recognize as such, when bees were poisoned with dinitrocresol. Rather, he interpreted these dances as a specific alarm for poison, by which flying from the hive was choked off and spread of the intoxication prevented. Schick (1953) repeated and extended these experiments and succeeded in showing that it was not a matter of special warning dances but the well-known trembling dance, and that flight was choked off simply because the poisoned foragers set out less frequently or finally stayed home altogether. Also as soon as they lapse into trembling dances they no longer arouse any newcomers.

Summary: Other Dance Forms

1. The name "jostling run" may be given to the striking motions of foragers that, returning home successful, run swiftly and trippingly forward on the comb, butting into comrades that are in the way. From it the latter learn only that something is going on. Group associates that are resting may be induced by the maneuver alone, even without any dancing, to revisit their feeding place.

2. One speaks of the "spasmodic dance" when foragers, in distributing the content of their honey stomach here and there on the comb, interpolate short, directed tail-wagging runs (Fig. 245). These are an expression of a mood for dancing, rather than an effective signal.

3. The "sickle dance" (Fig. 246) occurs in all races of bees studied hitherto with the exception of Carniolan bees, as a transition between the round dance and the tail-wagging dance. The open face of the sickle points out the direction to the goal.

4. During swarming in the hive the "buzzing run" gives the sign for departing. With a clearly audible buzzing of the wings the bees sounding the alarm run about in a random zigzag among their comrades and butt into them energetically.

5. In the "grooming dance" (shaking dance) the bee shakes the body rapidly from side to side and back and forth. In the view of many observers, other bees are induced thereby to comb with their mandibles the dancer's hair on parts of her body that are hard for her to reach.

6. In the "vibratory dance" (Fig. 248) a bee makes contact with another and vibrates the abdomen rapidly in a dorsoventral direction. The significance of such behavior is unexplained.

7. Not infrequently bees lapse for a short time, but often for hours too, into the "trembling dance." It is released by disturbances and by other adverse experiences. It gives the hivemates no information. It also receives no attention from them.

IX / Danceless Communication by Means of Sounds and Scents

1. Sounds

The buzzing of bees is of different kinds depending on their "mood." As long ago as 1810 Spitzner[1] reported that those bees that had discovered a supply of honey attracted others by a high-pitched penetrating sound. The attentive ear of the beekeeper distinguishes the "swarming tone" from the "wailing" of queenless colonies, the "stinging tone" of agitated specimens from the "hunger tone" in times of famine (v. Buttel-Reepen 1915). It is understandable that a bee language based on sound should be sought on such a basis. Yet closer study has revealed that sound waves borne by the air are not perceived by bees, and that hence they cannot hear in the customary sense. In this respect bees differ from grasshoppers, cicadas, and many other insects that, by means of drumlike structures, are able to perceive sounds.

The auditory capacity of bees has been tested in different ways. When they are flying to a good source of food their flight sound is a whole tone higher than when they are flying to a poor source (thus Spitzner's observation was correct though misinterpreted). While I was still unaware of the effect of the scent organ I suspected in the higher tone the reason why, under otherwise identical conditions, the influx of newcomers to a good feeding place is much greater than to a bad one. Yet it turned out that after elimination of the scent organ the pitch of the flight tone remains unaltered, while the influx of newcomers sinks to the value for the poor food source. The attracting effect comes only from the scent organ and not from the high flight tone (v. Frisch 1923:167–170). Kröning (1925) tried in vain to train bees to tones. Hansson (1945) conducted training experiments too, with better technique but with no better results. With the aid of a tone generator he further produced tones at the frequency occurring during scent fanning, that of the wailing tone, and so forth, and transmitted these by means of a speaker into a colony. Just as little effect was to be seen here as in other experiments in which natural tones produced by another colony

---

[1] Sect. 3, §14:71; cited from Manger 1920.

were transmitted. Wenner (1962a) corroborates the inefficacy of airborne sound.

In contrast with these negative results are positive reactions as soon as the sound waves are conveyed directly to the comb or other surface beneath the bees, and these surfaces are set into vibration. According to Hansson the effective range of frequencies extends from 100 to 1500 cycles/sec. The most striking phenomenon is that all bees on the comb interrupt their movements and hold completely still as long as the vibrations last. Even without direct contact, the reaction of holding still may be obtained if the tones are so close and are produced with such intensity that the substrate also vibrates. In training experiments Hansson (1945) found with direct transmission of the vibrations to the training box a slight preference for it relative to a nonvibrating one, but the difference was so slight that the result was not statistically significant. No distinction at all between tones of different pitch was obtained by training. That may be related also to the fact that sound-producing feeding places are a wholly unbiological affair for bees.

But from the reaction of holding still it becomes clear that sound produced on the combs can be perceived by neighboring bees and hence that for sound given out during the waggling run (pp. 57f) the prerequisites exist for perception of the sound by the dance followers. Nothing stands in the way of the assumption that by means of this "emphasis" on the waggling segments their duration is defined sharply and effectively. Here production of sound appears as a constituent of the dance.

But there is also a very striking production of sound that has no connection with dances: the "piping" and "quacking" of the queens.

When the first swarm (the "preswarm") has gone off with the old queen, in a few days a young queen emerges. Several queen cells have been made, so that after departure of a second swarm (the "afterswarm") additional queens will be available. The queen that has hatched produces at certain intervals a "piping," to which a queen ready to emerge from her cell responds with a somewhat deeper "quacking." This duet may continue for days. In the opinion of beekeepers the sounds serve for mutual communication among the rivals, so that a premature emergence is impeded; it could lead only to a life-and-death battle with the older queen who is present.

For the quacking tone Hansson's microphonic recordings gave an average frequency of 323 cy/sec, for piping by freshly emerged queens 435 cy/sec, with those 2 days old 493 cy/sec; this agrees well with data from Armbruster (1922).[2] The piping and quacking are appreciably louder than other tones audible in the hive.

With the tone generator Hansson produced quacking tones and transmitted them into a bee colony that contained a free queen and one still enclosed in her cell. If the speaker was near the free queen she replied to the artificial quacking by piping and this duet could be continued for hours. If the queen was at a considerable distance from the speaker she did not answer. Piping tones had the same effect; in spite

---

[2] After correcting his erroneous designations of the tones; cf. Hansson (1945:56).

of the different pitch there was thus no distinction made between piping and quacking. The same was true of natural piping and quacking tones that were transmitted from a different colony. Also, queens not yet emerged from their cells, but of sufficient maturity, replied to both piping and quacking tones. Hansson was unable to confirm the view that queens could be made to refrain from emerging by piping alone. Apparently olfactory stimuli from the rival still present participate. According to Hansson, the queens learn from the piping and quacking only that rivals are present in the hive, but not whether they are free or still enclosed. The rhythm of piping too seems not to have much significance. A steady tone lasting 10 sec was answered just like several pulses of sound.

According to Simpson (1964) the tones—like those in the workers' tail-wagging dance—are produced by the flight musculature and conveyed to the substrate by pressing the thorax against it.

Wenner (1962) describes experiments similar to those carried out by Hansson, without citing the latter's work. Wenner too was able to bring the queen to sing a duet with his apparatus. That she was really responding to the artificial tones, and not at random, is confirmed statistically. In Wenner's work too the tones had to be transmitted directly to the beehive. Air-borne sound was ineffectual.

In contradiction to Hansson's findings, in Wenner's study only artificial "piping" and not "quacking" was answered. Contradictory too are his data as to frequency. According to the recording reproduced the piping tone was about 1300 cy/sec, but the quacking—which should be lower—as much as 2500 cy/sec. Wenner ascribes this to a strong contribution from overtones, which were picked up by his method of recording. But it is still not clear how in this case repeated determinations by the human ear should give the values determined by Hansson. Wenner has further experiments in prospect.

The physiologic basis for the perception of tones described is given by the great sensitivity of the leg for vibrations. H. Autrum and W. Schneider (1948) determined electrophysiologically for different insects the thresholds for sinusoidal vibrations to which the sense organs situated in the extremities respond and found a high sensitivity in species with subgenual organs, to which bees belong. Frings and Little (1957) succeeded in releasing the reaction of holding still with vibrations in air, preferentially of 300 to 1000 cy/sec; these exerted an effect because the comb resonated with them.[3] With a contact microphone the reaction could be achieved by touching the tarsi of all three pairs of legs, but not by touching the antennae (Little 1959, 1962). The vibrations required a minimal amplitude of 50 mμ (cf. p. 98).

Thus the ancient concepts of a "language of sounds" among bees belong in the realm of fantasy. There is indeed communication by means of sounds, but it is of a most primitive kind.

[3] As an optimal frequency (in relation to amplitude) Autrum and Schneider give the value of 2500 cycles/sec, at which the bees can still perceive an amplitude of 13 mμ. According to Hansson and to Frings and Little, the sensitivity (in relation to the sound pressure in air) does not reach so high a frequency range. Perhaps the disagreement is to be explained by differences in frames of reference.

2. ODORS

That olfactory stimuli are of extraordinary significance for mutual communication among the bees is already evident from several examples. I recall here only the role of the floral scent in transmitting information as to the sort of flower discovered and of the scent organ for defining the location of those flowers. The odor brought home from the blossoms, as also their marking by the scent organ of the bees derive their value in connection with dances, as elements in a chain of activities and sensory stimuli that in their ordained course guide the indication of a goal to its successful termination.

The floral scent on the body of a forager returning home can alone, without any dance, arouse her group associates and induce them to undertake a renewed flight to a source of food with which they are acquainted (p. 30).

Olfactory perceptions also participate further in the colony's being constantly informed of the presence of its queen. If she is lost, so that the swarm becomes queenless, the restlessness breaks out after no more than a half hour, and after a few hours the workers institute unequivocal measures for rearing replacement queens from young larvae: they begin to build queen cells. Besides this the ovarioles of a part of the workers now develop in such a way that they themselves proceed to lay eggs, which of course results in a brood of drones and brings no help to the colony. Both the construction of queen cells and the development of the ovarioles have shown themselves to be symptoms useful in the analysis of the particular mode by which the presence of the queen is communicated among the bees. Butler (1954) discovered that a substance produced by the queen, the queen substance, is licked up by the attending bees that surround her and is distributed from bee to bee throughout the hive. The substance is produced in the queen's mandibular glands and is an acid (9-oxydecenoic acid) chemically very closely related to another agent that was demonstrated in the fluid food for queens that is important in the development of a larva into a queen. How that material exerts its effect is not known more precisely. In addition—apparently likewise by the mandibular glands—there is secreted an odorant that attracts the workers and thus promotes their contact with the queen. After extirpation of the mandibular glands, queens no longer were attractive to workers; they had no "court bees" any more. Yet oviposition remained unchanged during about a year's observation, indicating that their nutrition was sufficient (Zamarlicki and Morse 1964). Yet another odorant is given off by the queen's body and intensifies the effect of the queen substance (Butler 1956, 1957, 1961; Butler and Simpson 1958, 1965; Butler, Callow, and Johnston 1961; Gary 1961; Verheijen-Voogd 1959; Jordan 1961; *et al.*).

Doubtless odorous stimuli participate in many other ways in the harmonious cooperation of a bee colony.

If we enquire as to odorous signals that are sent out actively with the biological purpose of putting certain activities in train, I know of only a few instances to report.

Probably the nuptial flight of the queen bee belongs here. When such occurs, visual and olfactory stimuli both participate, as Ruttner (1957:587, 588) had already suggested. Initially the drones are attracted visually, though in the process they are often distracted by worker bees or by quite different kinds of insects flying past. At short distances, odorants emitted from the queen's mandibular glands are responsible for the final attraction. According to Nedel (1960:163ff), the mandibular gland is developed more strongly in the queen than in the worker, and, whereas in the latter it maintains a constant size throughout life, in the queen the apex of its development and glandular activity is reached during the period of the nuptial flight. At this time too the queen's abdominal glands (Fig. 43) described by Renner and Baumann (1964) attain their maximum development, simultaneously with the production of a scent that we too can readily perceive and that may also be of importance as a sexual attractant.[4] For the decisive observations we are indebted to Gary, who by applying a good idea brought to pass the ancient dream of having the mating of the queen bee take place before his very eyes. With a nylon thread he tied a mature queen to a captive balloon and let her hover a few meters above the ground. In this way she had sufficient freedom for flying but only a small radius of action. Several substances from the mandibular gland are concerned in the attraction; their effectiveness was confirmed also in experiments with a model, without a living queen (Gary 1962, 1963; Pain and Ruttner 1963; Butler and Fairey 1964). Since the duct of the mandibular gland is closed by a small sclerite but can be opened by muscular activity, it is to be assumed that the queen lets the attractant flow out at the proper time, just as the stingless bee *Trigona* does when depositing scent markers (see pp. 311f). The location of the gland and its closing apparatus are similar in both (see Fig. 272 and Nedel 1960:152, 153).

At many places queens flying on a nylon thread within range of observation are visited by drones only rarely or not at all, but elsewhere may be approached by them astonishingly quickly and regularly. According to F. and H. Ruttner (1965), the explanation is that the drones have assembling sites, which over a period of several years have been found with remarkable constancy in certain definite regions. Such a region has a diameter of 50 to 200 m and may be located as much as several hundred meters to 1 km distant from the nearest apiary. Apparently these places are rendezvous for the two sexes, such as are widely known among other insects. What guides the drones in the choice of these areas and how the queens find their way there is not known.

A second case of active scent signals has been mentioned already (pp. 50ff): bees use their scent glands in order to assist their hivemates, especially the as yet inexperienced younger bees, in finding their home on their return. They evert their scent organ and fan their wings in front of the hive entrance, thus driving the odorous sign toward the approaching bees (Fig. 44). In principle the procedure corresponds to

---

[4] According to Butler and Simpson (1965) these glands are not concerned in the attraction of *workers*.

FIGURE 249. A bee "alarm fanning" after being squeezed; the sting and the bulb of the stylet are visible in the opened cloaca. After Maschwitz 1964.

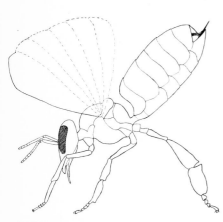

FIGURE 250. A bee "alarm fanning"; the sting and the bulb of the stylet can be seen in the opened cloaca. As when the scent organ is used (Fig. 44), the alarm odor is dispersed by whirring the wings. The alarming substance lies free in the form of oily droplets among the bristles on the bulb of the stylet. After Maschwitz 1964.

the alluring of newcomers by foragers who, after recruiting dances in the hive, have flown back to the source of food and evert the scent organ there (pp. 52ff).

The last example that I know of concerns a warning signal. The bee community has an alarm for danger. When they are direly threatened by an enemy a specific odorous signal is given that alerts the hivemates and summons them to defense. Such a thing was long suspected. Many beekeepers are convinced that the odor of the sting poison, which becomes noticeable when a bee sting occurs, has an agitating effect and calls forth further attacks. Free (1961) and Lecomte (1961) also reached this conclusion experimentally. The matter is in actuality different only in that the odor of the bee poison itself is not concerned; a special alarming odor is emitted and mobilizes the forces of defense (Ghent and Gary 1962; Maschwitz 1964).

Maschwitz (1964) succeeded in demonstrating an alarm for danger simply by holding a worker bee at the hive entrance firmly with a forceps or squeezing her. The neighboring bees got agitated at once. That there really is an alarm is shown for instance by the following experiment. Toward evening a colony of bees had ceased flying. During 24 min not a single bee left the hive; a few individuals were still sitting in front of the hive entrance. Some of these were squeezed gently with a forceps. They ran into the hive and at once other highly agitated bees came rushing out, within 16 min 140 of them altogether.

Bees that are annoyed or hurt display a typical behavior: they elevate the abdomen, open the sting chamber, and protrude the stinging apparatus (Figs. 249, 250), at the same time usually fanning the wings, like scent-fanning bees, and thus disperse a column of air behind them ("alarm fanning"). Maintaining this position, they mingle with their comrades. Occasionally there appears a poison drop at the tip of the sting.

If various isolated parts of the sting apparatus are placed on filter paper in front of the hive entrance, the contents of the poison glands are seen to be ineffective. The alarm substance consists of oily droplets sticking to the dense hairs of the bulb of the stylet (*Stap*, Fig. 251) and has a characteristic fruit-ester odor, different from that of the poison.[5] It has not yet been possible to tell which cells produce this odorant. The sting lies upon the bristles of the pad and becomes smeared with the alarm substance. Hence the alarm substance gets into the poison during the act of stinging and is transferred to the object stung.

With further experiments it was possible to show that neither the fanning nor the dashing about of the endangered bees impelled the hivemates to attack, but rather the odor of the warning material. When bees were seized whose stinging apparatus had been excised half an hour previously, they ran in agitation, fanning their wings, into the hive, but were unable to raise an alarm.

Often during "alarm fanning" the secretion of the mandibular glands

---

[5] Probably the same scent is concerned that Boch, Shearer, and Stone (1962) compare with that of banana oil. In this odorant they were able to demonstrate *iso*amyl acetate as a component that was effective in stimulating the comrades.

is emitted at the same time. This too has an alarm-producing effect, but its activity amounts to only about one-fourth that of the abdominal secretion. Participation of both the stinging apparatus and the mandibular glands in alarming for danger also occurs in many other social Hymenoptera. In the leaf-cutter ant *Atta sexdens rubropilosa* Forel the mandibular gland is developed especially powerfully in the soldiers and it alone provides the alarm odor (Butenandt, Linzen, and Lindauer 1959).

In individual instances there are appreciable differences in this respect among the social Hymenoptera. Thus, for example, what earlier was assumed for bees is the case with the genus *Vespa*: the alarm substance is produced in the poison gland; but it is not identical with the toxin. With the ant *Formica*, formic acid acts both as a poison and for spreading the alarm; but the latter effect is increased further through addition of a special alarming substance (Maschwitz 1964).

The queen bee does indeed have a sting and uses it against other queens, but she is not involved in the defense of the hive. She also does not possess any alarm substance.

The workers lack such a material only during the period immediately after emergence. As a rule it develops in the course of 3–6 days and then remains at the level reached, but is very variable with different individuals. Carniolan and Italian bees have the same alarm substance; differences specific to the different races could not be shown.

Even before there has been any stinging at all, bees can be alarmed by "alarm fanning" alone. The comrades called forth—like the assaulted individuals themselves, insofar as these are still capable—direct their attacks preferentially against moving objects in the vicinity, especially against dark-colored ones. One can also draw their attacks upon oneself by exhaling. According to Maschwitz this is not to be ascribed to the movement of the air nor apparently to the humidity of the breath, but only to its warmth. If one has already been stung, the alarm substance transmitted, by means of which the victim bears a scented label, exerts an additional attraction, and directs further attacks against him. According to a prevalent opinion, the sweaty odor of a perspiring person attracts bees that are ready to sting. Maschwitz says that is not so; the effective factor is not the sweat but the increased heat radiation.

Besides insects, the main enemies of the honeybee are mammals (badger, bear, man). When such a creature comes near, its mouth is usually the part closest to the threatened bees. Hence the warmth of the breath constitutes a dependable key stimulus for alarming and for attack.

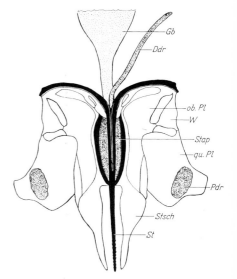

FIGURE 251. Sting apparatus of the honeybee, ventral view; *Gb*, poison sac; *Ddr*, Dufour's gland; *Stap*, bulb of the stylet; *Stsch*, sting sheath; *St*, sting; *ob.Pl*, oblong plate; *W*, triangular plate; *qu.Pl*, quadrate plate; *Pdr*, Koschewnikow's gland. From Maschwitz 1964.

Summary: Danceless Communication by Means of Sounds and Scents

1. Sounds are perceived by bees via the sense of vibration, solely after transmission through the substrate. Production of sound in the tail-wagging dance, and the piping and quacking of queens, are capable of producing an effect in this manner. No difference could be shown between piping and quacking (which differ in pitch). Artificial tones of

piping or quacking, transmitted to the comb, were responded to by queens. In this way a duet can be carried on with them.

2. Scents have manifold significance in communication within the bee colony. Moreover, odorous signals are given actively. During the mating flight the queen attracts drones by scents released from the mandibular glands. Scent-fanning bees in front of the hive (Fig. 44) waft the odor toward approaching bees and thus guide them to the way in. Bees that are threatened or set upon release an alarm signifying danger by extruding the stinging apparatus and, by fanning the wings, dispersing the odor of an alarm substance that is not identical with the bee toxin. This alarm incites the comrades to attack. The attack is directed preferentially against moving objects in the vicinity.

X / Variants of the "Language of the Bees"

1. RACIAL DIFFERENCES ("DIALECTS")

During the Ice Age the honeybee was unable to find suitable living conditions in Central Europe. But it was able to maintain itself to the southwest on the Pyrenean Peninsula, in the Balkans to the southeast, and in Italy to the south. There, obviously under the influence of geographic isolation, three races, sharply distinct today, developed and after the Ice Age colonized Central Europe; initially the Alps impeded their mixture with one another. From the southwest there came to Central Europe west and north of the Alps the dark German bee (*Apis mellifera mellifera* L.),[1] up to the Alps there came from the southeast the gray Carniolan race (*A. m. carnica* Pollm.), while in the south the Italian race with its striking yellow abdominal bands (*A. m. ligustica* Spin.) has remained the most firmly established. Today they have been exported all over the world, and have frequently been hybridized, so that colonies of dependable purity are hardly to be obtained except from rearing establishments. In addition to the three European races there are numerous others in Eastern Europe, Asia, and Africa.

I have worked with Carniolan bees, whose placid disposition makes them especially suitable for experiments. To them refers the description of the round dance and its transition to the direction-indicating tail-wagging dance (Fig. 56, *top*). In 1947 F. Baltzer (Bern) wrote me of the sickle dances that he and his student Tschumi had seen in a Swiss bee colony (Figs. 246 and 56, *bottom*); in part these dances indicated the direction when the feeding place was only a few meters distant. My suspicion that there might be racial differences, something like "dialects of the language of the bees" (v. Frisch 1948: 16n) received support from my own observation colony in Brunnwinkl in the summer of 1950. For the first time I now saw sickle dances and simultaneously it turned out that the queen of this colony, in appearance of the Carniolan race, actually was an Italian hybrid. Only part of her worker offspring were typically of the Carniolan race; many bore the yellow abdominal markings of Italian bees. The most surprising thing was that when the feeding

---

[1]The frequently used name "nigra" for this dark bee "is a designation in beekeepers' terminology for the *A. mellifera mellifera* of the Alps, and besides for a Swiss breed (that Prof. Zander cultured in his day). It is not a distinct race" (Dr. F. Ruttner *in lit.*).

place was 10 m distant the bees of Italian appearance performed predominantly sickle dances, those that looked like the Carniolan race quite predominantly round dances. A second hybrid colony, in which the queen possessed yellow rings and there were even greater numbers of yellow workers, displayed the same behavior (v. Frisch 1951), and likewise a third hybrid colony the next summer. Altogether with these three colonies I observed 321 dances for a flight distance of 10 m (on a few occasions 20 m); of these there were: with gray bees (Carniolan type) 152 round dances and 22 direction-indicating sickle dances;[2] with yellow bees (Italian type) 9 round dances and 138 direction-indicating sickle dances.

Thus the visually most striking feature of the Italian race, their yellow markings, seemed to be linked in these hybrid colonies with the behavioral characteristic of performing sickle dances for nearby sources of food.

With one of these hybrid colonies I also tested the indication of distance over a range from 100 to 500 m and found for the bees of the Italian type a clearly slower rhythm of dancing than for those of the Carniolan type (Fig. 252); the differences are statistically significant.

In 1952, with a fourth hybrid colony that had a newly introduced yellow hybrid queen, I observed, in contrast with the other three colonies, that coloration and mode of dancing did not go hand in hand to the same extent. One hundred twelve dances were observed. Of these there were: with dark bees 26 round dances, 25 sickle dances, and 5 transitional forms; with yellow bees 16 round dances, 33 sickle dances, and 7 transitional forms.

Later, in connection with the findings of Baltzer, Tschumi, and Hein, I shall return to this apparent disagreement (pp. 296f).

In the summer of 1951 I obtained a racially pure Italian colony. At a flight distance of 10 m these bees in part performed round dances, in part direction-indicating sickle dances; at a flight distance of 20 m definite direction-indicating sickle dances. This finding is in agreement with observations by Baltzer (1952) on a pure Italian colony. In contrast, with pure Carniolan colonies at short flight distances I saw, furthermore, only round dances and no direction-indicating sickle dances. Thus was confirmed the assumption suggested by the performance of the hybrid colonies: the round dance and the sickle dance for nearby sources of food are racially specific forms of behavior on the part of Carniolan or Italian bees, respectively.

In order to provide a broader basis for such statements I induced R. Boch (1957) to undertake a systematic study of the indication of direction and distance with the European races of bees and, insofar as they were obtainable, with the extra-European ones too. Besides the three Central European races there were available the bee of the Caucasus Mountains (*Apis mellifera caucasica* Gorb.), and in addition the

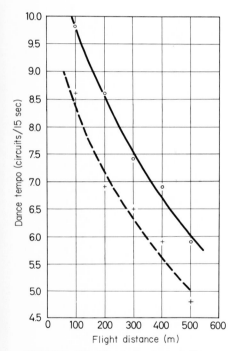

FIGURE 252. The indication of distance in a hybrid colony; ○, gray bees (Carniolan type); +, yellow bees (Italian type).

[2]Sickle dancing without an indication of direction, that is, arcs running for only one-half to three-quarters of a circle without any relation between the open side of the sickle and the location of the feeding place, also occur not infrequently with Carniolan bees, being scattered among courses that encompass one or more complete circles.

African Punic bee (*A. m. intermissa* Butt.-Reepen) and the brightly yellow-banded Egyptian bee (*A. m. fasciata* Latr.).

The result was that among these six races of bee the gray Carniolan bee (*carnica*) alone did not perform direction-indicating sickle dances. With it the transition from the round dance to the tail-wagging dance occurs as shown in Fig. 56 (*top*), with definite tail-wagging dances beginning at a flight distance of some 85 m. All other races indicate the location of the goal at still lesser distances by means of direction-indicating sickle dances, which with them constitute the transition from round dance to tail-wagging dance (Fig. 56, *bottom*). With each race direction-indicating sickle dances begin to occur at a characteristic range (Fig. 253). Thus not only the Italian bees but all the races studied display a specific performance in their mode of dancing. Sharp distinctions are not to be drawn between round dance and sickle dance, nor between sickle dances and tail-wagging dances. The transitions are gradual. In Fig. 253 the ranges of transition from the round dance (*R*) to the sickle dance (*S*) and from it to the tail-wagging dance (*T*) have been left open.[3] With feeding places near the hive entrance all races perform wholly unoriented round dances.

The Egyptian bee, which begins direction-indicating sickle dances as close to the hive as some 3 m and tail-wagging dances at about 10 m, is the smallest of the races studied (body length about nine-tenths that of the Carniolan bees). Some degree of correlation between body size, dance form, and flight distance is found also with distinct species of bees (pp. 125f).

In regard to the indication of distance, tested by Boch over the range from 100 to 500 m, the racially pure colonies confirmed the slow dance

---

[3] In order to delimit the dance forms from one another, we determined in a uniform manner a definite percentage of correctly oriented sickle runs, or circular runs and waggling runs respectively, for all the races. Further details in Boch 1957.

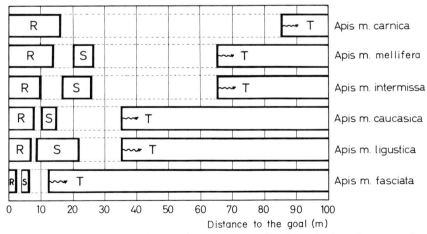

FIGURE 253. Racial differences in the indication of direction: abscissa, distance to the goal; *R*, round dances; *S*, sickle dances; *T*, tail-wagging dances; between them, regions of transition from one dance form to the other. After Boch 1957.

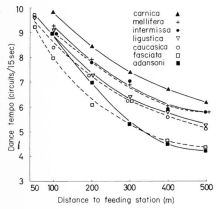

FIGURE 254. Racial differences in dance tempo for feeding stations 50 to 500 m away.

rhythm of the Italian bees relative to the more rapid Carniolan bees, the difference that had attracted attention with the gray and yellow color types in the hybrid colonies (Fig. 252). In Fig. 254 Boch's results on the six races he tested are summarized in readily visible form, supplemented in accordance with a communication of Smith (1958) on the dance rhythm of *Apis mellifera adansonii* Latr., a race of the honeybee from tropical Africa. Of all seven races the Carniolans are evidently the nimblest dancers. The German bee dances at approximately the same rhythm as the Punic bee, and the Italian race about like the Caucasian; slowest are the Egyptian *A. mellifera fasciata* and *A. m. adansonii*.

With artificial feeding stations Smith was able to entice *A. m. adansonii* only about 600 m from the hive. Thus there is confirmed with this bee too the relation between flight range and the form of the distance curve, which aims at greatest accuracy in the indication of distance over the principal flight range (pp. 125f and Lindauer 1956). Smith doubts Lindauer's conclusion and thinks that the natural flight range of *A. m. adansonii* is not much different from that of the European honeybee, because he saw his African bees, which used to come only as much as 600 m to sugar-water dishes, returning home at the same time after natural collection of pollen (from the oil palm, *Elaeis guineensis*) from some 2500 m (2–2.5 circuits/15 sec). But probably that signifies no more than that the African bee, just like our own honeybee, can be tempted to greater distances than usual by an uncommonly good food source that is not available closer by: among our bees, with a normal flight range of 3–4 km, to a distance of as much as 12 km (Knaffl 1953) or 13.5 km (Eckert 1933), with *A. m. adansonii*, in correspondence with its lesser normal flight range of ca. 600 m, to 2.5 km—in both instances about four times the normal flight range. It should be emphasized also that our own comparisons are based on experiments performed under mutually similar conditions.

It has been pointed out already (p. 84) that in the round dance no relation between the number of circuits and the flight distance is to be discerned, because in it there are run through at random now semicircles, now complete circles or several of them. In the sickle dance with its regular turns matters are otherwise. During an experiment in 1951 with an Italian colony I observed a definite reduction in the dance rhythm when the flight distance was increased stepwise from 10 to 100 m. Tschumi (1950) reports corresponding data. To what extent these announcements in regard to so small a flight range are given appreciable attention by the hivemates has not yet been studied.

We now return to the question whether, when Carniolan and Italian bees are crossbred, the behavior of performing sickle dances is linked in inheritance with the yellow body markings. The results with the first three hybrid colonies I observed suggest this assumption (pp. 293f). Baltzer (1952) reports that in Tschumi's observation colony a few specimens with yellow bands appeared subsequently, so that evidently there was an Italian strain in the stock. But the sickle dances had been seen in dark-colored bees. With my fourth hybrid colony the sickle-shaped dances predominated numerically at 10 m flight range among the yellow bees, but among the dark bees there was this time approximately the same total of round dances and sickle dances. This seemed to refute the aforementioned rule of inheritance. However, Tschumi's and Baltzer's experimental colony did not belong to the Carniolan race, but to the dark *A. m. mellifera* ("nigra"), which according to Boch and to Baltzer (1952) also performs sickle dances like the Italian bees. Here the dances first occur at a somewhat greater distance, but the regions for sickle dances overlap (Fig. 253). Thus the sickle dances of Tschumi's bees, that likewise began only at somewhat greater distances than with the Italian bees (Tschumi 1950, Baltzer 1952) need not be attributed to their weak Italian strain; they are explained simply by the bees' belonging to the race *A. m. mellifera*. The same is probably so for the Dutch colony with which Hein (1950) described sickle dances for feeding places between 8 and 30 m.[4] That my hybrid colony of 1952 had come from the crossing of the Italian queen with *A. m. mellifera* and that therefore it differed in behavior from the earlier *ligustica/carnica* hybrids is possible, because the dark race is kept at the Wolfgangsee and even is carefully purebred; and this assumption gains in likelihood because of a corresponding occurrence in 1962 (see below). The question as to the linkage of the mode of dancing with the color of the body should be reinvestigated, in fact with Italian-Carniolan hybrid colonies whose queens have been inseminated artificially, so that the nature of the cross is absolutely clear. Considering the complex inheritance of body coloration, the matter would be of particular interest.[5]

In later years in Brunnwinkl I encountered sickle dances by bees with a dark body color on only one occasion (1962). This constituted a surprise that spoiled a planned experiment, but clarified matters in another direction, particularly when the racial characteristics were analyzed subsequently.[6] What had been assumed to be a Carniolan colony proved to be a hybrid, without a demonstrable Italian strain, but between *carnica* and *mellifera*, with the *mellifera* characteristics predominating strongly.

In this experiment the intention was to test what radius of the vicinity was examined by newcomers when there were persistent round dances. The observation colony was set up at Aich and a feeding station was established 15 m WSW of the hive. On 1 September 1962 a group of ten marked bees was fed there on concentrated sugar solution with added peppermint scent. At intervals of 100, 200, 300, 400, and 500 m from the hive, distributed in various directions, two scent plates were put out at each. To our astonishment, during the 2-hr period of observation not a single bee came to any of these scent plates, whereas the feeding station itself soon received a mighty influx of newcomers; altogether during the 2 hr 229 newcomers were captured there and killed. During the final 50 min of the experimental period we set out two

---

[4] The "dark Dutch bee" (Hein) is *A. mellifera mellifera* (according to Ruttner and Mackensen 1952).
[5] According to Roberts and Mackensen (1951) and Ruttner and Mackensen (1952), the abdominal coloration is influenced by at least seven genes at different loci.
[6] I am indebted to Dr. F. Ruttner for carrying out the investigation.

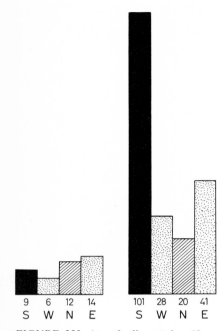

FIGURE 255. At a feeding station 10 m south of the hive, Carniolan bees or Italian bees were fed, with addition of a scent. Scent plates were set out in all four compass directions 20 m from the hive. The heights of the columns and the numbers below them show the number of visits to the odor plates: (*left*) when aroused by Carniolan bees (round dances); (*right*) when aroused by Italian bees (sickle dances); (*black*) in the direction toward the feeding station; (*hatched*) in the opposite direction; (*stippled*) at 90° to the line of flight.

additional scent plates 20 m to the NE and SE of the hive. Here there appeared two and four newcomers, respectively. Hence the bees aroused were searching approximately at the distance of the feeding station and the great majority of them in the appropriate direction. As soon as the experiment ended the dances of the marked bees were checked in the hive. These were overwhelmingly oriented sickle dances. The behavior is explained by the finding that the *mellifera* type predominated in this colony (for *mellifera* Boch records only sickle dances at 20 m). Simultaneously it became manifest that the oriented sickle dances really afford a description of the goal to the hivemates, a description of validity equal to that provided by the tail-wagging dance. This agrees with Tschumi's (1950a) observations; he records that his Swiss colony followed the indications of direction given in the sickle dances.

Boch (1957) too found, in comparative experiments, that sickle dances show hivemates the way, but that round dances do not. With sugar water in the presence of added scent he fed a group of marked bees from a Carniolan colony in a cardboard box[7] 10 m south of the hive; at 20 m from the hive scent plates were set out in the open in all four compass directions. The numbers of newcomers that flew to the four observation points are shown graphically in Fig. 255, *left*. The direction of training (black column) was not favored. The eastern station received the majority of flights, which can be attributed to the easterly direction of the wind (cf. p. 157). The reciprocal experiment with an Italian colony showed a clear preference for the direction of training (Fig. 255, *right*). In addition, here again the eastern station was visited relatively strongly (east wind).

In further experiments Boch not only succeeded once again in confirming for Italian bees an effective indication of direction at 10 m flight range, but demonstrated it also for *A. m. caucasica* and *A. m. fasciata* (in this connection cf. Fig. 253).

A glance at Fig. 254 reminds us that the dance rhythm differs racially for a given distance to the goal. Of the Central European races the Carniolan bees dance most rapidly, the Italian bees the slowest. For a flight distance of 200 m the Italian bees dance in a rhythm that in the Carniolan "dialect" points to a goal 300 m away. One can create a mixed colony by uniting in a single hive about as many bees from a Carniolan as from an Italian colony. This time it is not a matter of hybrid colonies, as on pp. 293 and 297, but rather of an artificially produced mixture of two pure races. Must not misunderstandings occur there in regard to the indication of distance, because of the differing dance rhythms?

In order to test this, Steche (1954) performed the following experiment with a mixed Carniolan-Italian colony. In a cardboard box[7] at the feeding place a marked group from the mixed colony, but exclusively Italian bees, were fed sugar water with addition of lavender scent. The feeding station was 120 m from the hive. Two lavender-scented

---

[7] By means of the box with its narrow entrance the feeding dish was concealed as much as possible from the newcomers, both optically and olfactorily.

odor plates were set out openly 20 m nearer the hive and 20 m farther away. The Carniolan bees of the mixed colony reacted to the dances of the Italian bees too, letting themselves be aroused and flying out to search, but the Carniolan bees came to the more distant odor plate relatively more frequently than the Italian bees (Fig. 256, I*a*, and Table 34, I*a*). Relative to the number of flights by the Italian bees, 3.7 times as many Carniolan bees came to the more distant odor plate than to the nearer one. The dances of the Italian bees were too slow for the Carniolan rhythm and were misinterpreted as indicating a greater distance. In the reciprocal experiment Carniolan bees flew to the feeding

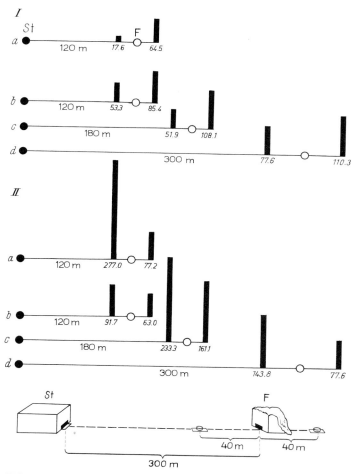

FIGURE 256. Results of the indication of distance in a mixed colony of Italian and Carniolan bees, (I) aroused by Italian bees, (II) aroused by Carniolan bees; observers and times of observation as in Table 34. The black columns and the numbers below them show how strongly the nearer or the farther feeding station, respectively, was visited in consequence of the dances, by newcomers of the *other* race of bees; the values are expressed as percentages of the number of visits by bees of the *same* race as the dancer. Bottom, diagram of the experimental layout: *St*, hive; *F*, feeding station.

TABLE 34. Effect of dances by Italian and Carniolan bees, respectively, on members of the other race in a mixed Italian-Carniolan colony.

Exp. No.	Observer	Period of observation	Distance to feeding station (m)	Distance to observation site (m)		Total number of flights		Ratio of visits (percent)		Ratio of percentages
				Near	Far	Near	Far	Near	Far	
		Carniolan bees recruited by Italian bees, slow dances								
I(a)	Steche	13 Oct. 1953, 1:50–3:20	120	100	140	20	97	17.6[a]	64.5[a]	3.7
I(b)	Boch	9 Sept. 1956, 12:40–2:30	120	100	140	230	152	53.3	85.4	1.6
I(c)	Boch	8 Sept. 1956, 12:00–2:00	180	160	200	123	77	51.9	108.1	2.1
I(d)	Boch	19 Sept. 1956, 1:15–2:30	300	260	340	136	204	77.6	110.3	1.4
		Italian bees recruited by Carniolan bees, fast dances								
II(a)	Steche	15 Oct. 1953, 12:20–2:05	120	100	140	49	37	277[b]	76.2[b]	3.6
II(b)	Boch	21 Sept. 1956, 12:00–2:15	120	100	140	92	44	91.7	63.0	1.46
II(c)	Boch	7 Sept. 1956, 12:05–2:35	180	160	200	80	47	233.3	161.1	1.45
II(d)	Boch	20 Sept. 1956, 12:30–2:30	300	260	340	39	87	143.8	77.6	1.8

[a] Carniolan in relation to Italian bees. [b] Italian in relation to Carniolan bees.

station. Their more nimble dances were misunderstood by the Italian bees as indicating a nearer goal. Now, relative to the flights by Carniolan bees, there came 3.6 times as many Italian bees to the nearer observation site as compared with the more distant one.

This single pair of experiments demanded repetition, which was carried out by Boch (1957) partly at the same flight distance and in part at a greater removal. This time the difference in the flight totals was less, but in three experiments and their reciprocals it was in the same direction as had been found by Steche (Fig. 256 and Table 34, I and II, *b, c, d*). The suspicion that the different dance rhythms of the races should lead to misunderstandings between them was thus confirmed in all experiments. That the discrepancy in the number of flights was not greater is comprehensible in view of the small difference in the flight rhythms and the relative proximity of the observation stations.

The builders of the Tower of Babel might well have been quite content if after the confusion of tongues they could have understood one another as well as did the races of bees in these experiments. Yet the racially differing response to the identical dances is clear enough to bear witness once again to the significance of the dance rhythm in the indication of distance.

2. Differences Among Species; the Indian Bees

Since the details of communication vary even among individual races of the honeybee, we were curious to know whether even more marked differences might not obtain among species of the genus *Apis* and might perhaps provide some suggestion as to the course of evolution of the "language" of our bees. This was indeed an expensive curiosity, for all the other three bee species live in India and in more distant parts of Southern and Eastern Asia. Means for conducting the "comparative linguistic studies" planned were approved by the Rockefeller Foundation for M. Lindauer. The following account is based on his results (1956).

Of the three tropical species, the Indian bee (*Apis indica* F.) is closest to our honeybee (Fig. 257). As the latter did originally, it inhabits hollow trees and builds a structure of several combs one behind the other. Of the three species it alone can be kept profitably in beehives, and in its native land replaces our honeybee as a domestic animal. Its decorative coloration is reminiscent of that of Italian bees, but it is barely half as large. To the beekeepers' sorrow its colonies are not only extraordinarily ready to swarm but, with shortage of food or in face of any disturbance to the combs, to leave en masse and settle in some other locality.

FIGURE 257. The Indian bee, *Apis indica* F.; queen with workers. After Lindauer 1956.

In the structure of their combs and in their way of living the two other species exhibit more primitive traits relative to *A. mellifera* and *A. indica*.

The dwarf honeybees (*A. florea* F.) are so tiny that they look almost like winged ants. Two brick-red abdominal segments and silvery white tomentose bands give them an elegant appearance. On a branch under the open sky they build a single comb, about the size of a small plate, that hangs vertically (Fig. 258). The storage cells for honey are two or three times as deep as the brood cells and form the upper part of the comb surrounding the branch; because of the elongation of the cells this part attains an appreciable width and forms a kind of platform for the bees sitting there (Fig. 259). The significance of this will be discussed more fully later. The lower more constricted part of the comb consists of brood cells for the larval workers, drones, and queens (Fig. 258 *W, D, Q*). The combs are never placed under a dense roof of foliage, but in thin undergrowth. Consequently they are exposed without protection to the violent monsoon rains. When there is one of these downpours the bees crowd close together on the expanded upper portion of the comb, arrange themselves one above the other like a tile roof, and thus with their living bodies constitute a porch as an umbrella for the rest of the bees and the brood cells. Their readiness to swarm and to migrate is even greater than with *A. indica*.

The giant honeybee (*Apis dorsata* F.) is about the size of a hornet and its sting is similarly feared. Like the dwarf honeybee it builds only a single, free-hanging comb, the diameter of which may, however, be more than 1 m, which is attached to the lower surface of a projecting rock (Fig. 260) or to porch roofs or to the heavy branches of tall trees (Fig. 261). According to older reports, the structure of the cells is more primitive than with the other species: there is said to be no difference between the cells of workers, drones, and queens. But Viswanathan (1950) and Thakar and Tonapi (1961) have described and demonstrated by photographs enlarged queen cells at the lower margins of the comb and drone cells that—at least by the end of the larval period—are lengthened and then protrude beyond the comb surface. The urge to

302 / The Dances of Bees

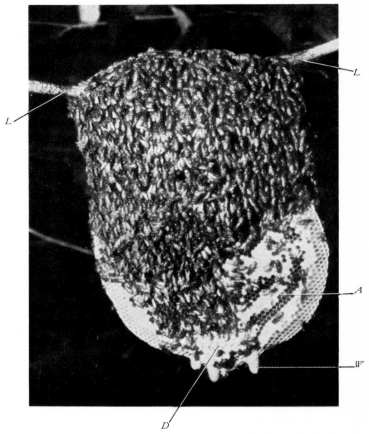

FIGURE 258. Colony of the dwarf honeybee, *Apis florea* F. The bees were brushed away from the lower part of the comb in order to show the brood nest: *A*, worker brood; *D*, drone brood; *W*, queen cells. To the right and left of the comb the bees have applied a ring of "glue" (*L*), to protect the nest against ants; it consists of a resinous substance collected from plants. After Lindauer 1956.

FIGURE 259. A normally occupied comb of *Apis florea*, viewed obliquely from above in order to show the horizontal dance floor. After Lindauer 1956.

swarm and to migrate is strongly developed. "Generally, at a given season practically every colony abandons its old nesting place, leaves the empty comb behind, and travels a long distance, 100 km or more, into a different area, then returns a year later" (this refers to Ceylon, according to Lindauer 1956:542). As with *A. florea*, the comb is always constructed in such a way that open sky is visible from it: thus it is built on the sparse foliage of trees that tower above the canopy of the primeval forest or on overhanging cliffs from which there is a free outlook in one direction.

With regard to communication by dancing among Indian bees there were only two short notes. Roepke (1930) observed among foragers of *Apis dorsata* returning home the "shaking dance of v. Frisch." Butler (1954:211) with *A. indica* saw round dances and oriented tail-wagging dances that seemed identical with those of *A. mellifera*, whereas during his brief visit he was not able to establish that such dances occurred with *A. florea* and *A. dorsata*. However, he reports (p. 141) that with *A. indica* and *A. florea* on Ceylon he was able frequently to detect the use of the scent organ during visits to flowers. Our bees will do this after they have danced in the hive on their return to profitable flowers.

According to Lindauer, all three Indian species of bees perform round dances and tail-wagging dances that have the same significance as with our honeybee. But in details the species differ from one another more markedly than do the races of our bee.

In the indication of distance there are only gradual differences. They are shown in Fig. 262, in which for comparison the dance rhythm of the Carniolan and Italian races of *A. mellifera* also are repeated once again. The giant honeybees dance with about the same rhythm as the Italian bees. *Apis indica* dances more slowly, and the dwarf honeybee the slowest. A rate of 6 circuits/15 sec, which with our Carniolan bee indicates a distance of 500 m, with *A. indica* means a distance of about 180 m to the goal, and with *A. florea* one of 100 m. The correlation between the course of the curves and the flight range of the several species was already noted on pp. 125 and 296: whereas under favorable conditions our bees may be lured as much as 12 km from the hive, and still dance for a distance of 11 km, when the same method was used *A. indica* went along for only 750 m (the last dances at 700 m) and *A. florea* only 300–400 m. The modest flight range of these small species has a good biological basis, since it is dangerous for them to be overtaken by tropical downpours. The steeper course at the beginning of the curve is appropriate for the smaller flight range, for it implies a well-defined indication of distance for short ranges to the goal, with no appreciable imprecision until domains are reached that are scarcely of practical significance. With the robust *A. dorsata* there are suggestions of a large flight range, which as yet it has been impossible to define fully.

Study of the indication of direction revealed more far-reaching and more illuminating differences. For *Apis indica* it agrees in principle with

FIGURE 260. Colony of the giant honey-bee *Apis dorsata* F. under an overhanging cliff. After Lindauer 1956.

FIGURE 261. The "bee tree" in the Botanical Garden at Paradeniya, with colonies of *Apis dorsata*. The sparse foliage is a prerequisite for colonization. After Lindauer 1956.

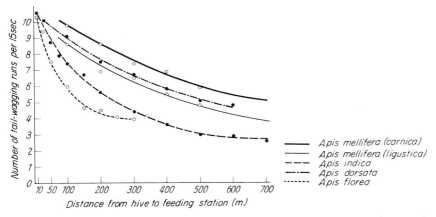

FIGURE 262. Comparison of the indication of distance by *Apis indica*, *A. florea*, and *A. dorsata* with that of *A. mellifera carnica* and *ligustica*. After Lindauer 1956.

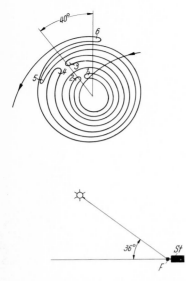

FIGURE 263. Round dance, with intermittent direction indicating, nearly closed sickles, in *Apis indica*. For clarity, the successive circles have been enlarged in the diagram; actually they lie on top of one another. Below, experimental layout: *St*, hive; *F*, feeding station 10 cm south of the hive entrance. After Lindauer 1956.

that found in our honeybee. With increasing distance, round dances shift into sickle dances and tail-wagging dances; in the dark hive the solar angle is transposed to the gravitational angle. On a horizontal surface the waggling runs point directly to the goal, and the polarized light of the blue sky as well as the sun itself is used for orientation. The record among all bees for indicating direction of nearby sources of food is held by *Apis indica*. With a distance to the goal of only 1 m it performs sickle dances, at 2 m definitely directed tail-wagging dances. At 0.5 m are seen round dances that, however, contain oriented almost-closed sickle-shaped runs interspersed among complete circles, these occurring even at a distance of as little as 10 cm (Fig. 263). In this reference to the direction of a goal that is only a few centimeters beyond the hive entrance *A. indica* stands alone.

With good reason the dwarf honeybee *A. florea* builds a comb expanded at the top. Even the slightly vaulted curvature of the upper surface is made level by a cushion of bees that sit there (Fig. 259). With their bodies they form a horizontal thatch: the colony's dance floor. Here there land the foragers returning home, and only here by means of their dances do they point out to their comrades the way to the goal, by setting themselves at the same angle to the sun or to the polarized light of the blue sky as they maintained previously during their flight to the feeding place. They are not capable of expressing the direction relative to the sun by means of the direction with respect to gravity.[8] One comprehends why they locate their nest up where it is light, at the mercy of wind and weather. Their mode of indicating direction is a stage more primitive than that of *A. mellifera* and *A. indica*, which are able to transmit their information inside their protected dark hives too. In this manner *A. florea* confirms our long-cherished suspicion that dancing on a horizontal surface, with orientation directly according to the sky, represents the more ancient form in which direction was indicated.

Lindauer made certain of this important finding by means of several kinds of experiment. If one carefully cuts off the branch on which the comb hangs and tilts it through 90° so that the former dance floor is vertical, the dances are stopped forthwith, the foragers hasten to the side that now is uppermost and, as soon as they find a horizontal surface, continue to dance with their previous liveliness. If Lindauer left the comb in its original position and laid his notebook over the colony, the foragers arriving ran agitatedly hither and yon beneath it without dancing. An hour later a few bees had moved upward over the edge of the notebook, and now the foragers would land up there, where the horizontal surface made an ideal dance floor. When a colony's view of the sun and sky was shut off by means of a linen cloth, after some hesitancy the dances were continued on the customary dance floor on top,

[8] It would be of great interest to know whether in *Apis florea* under experimental conditions a transposition of the visual to the gravitational angle occurs, as in ants, beetles, and so forth (p. 326); whether such capability is lacking altogether in this bee or whether it merely has not been developed into an unequivocal transposition in the interests of communication.

but were all disoriented. If the bees were able to see a piece of blue sky through a hole, the dances became oriented properly once again. When the foragers were forced onto the vertical comb surface, by putting a roof-shaped glass cover on top of the comb, there resulted only disoriented dances—if any at all. Even when the bees were thus compelled to dance on a vertical surface, the idea of transposing did not occur to them.

With the dwarf honeybee, *Apis florea,* round dances were limited to feeding stations closer than 5 m. At 5 m oriented tail-wagging runs commenced. In the dwarf honeybee the tail-wagging movements are not uniformly in a horizontal plane as in *A. mellifera* and *A. indica.* Now and then the abdomen also makes vertical excursions, which gives the dance a peculiar swinging appearance. Perhaps this somewhat disordered tail-wagging too is a more ancient type of behavior, from which the uniform horizontal wagging of our bees is to be derived (Lindauer 1956:538–539).

The giant honeybee *A. dorsata* performs round dances up to a distance of 3 m, and beyond that shifts rapidly to oriented tail-wagging dances. It dances on the vertical comb surface and in doing so transposes the direction to the goal into the direction relative to gravity according to the same formula as *A. indica* and *A. mellifera,* so that at first glance there seemed to be no difference. And yet there is one: like *A. florea, A. dorsata* dances only when the sun or blue sky is in sight—but does so on the *vertical* comb surface. The nests are always so constructed that sun or blue sky is visible from at least *one* comb surface, and dancing occurs there only. When with one colony Lindauer obscured the view of the sky with a cloth, the dances ceased at once. Such observations as could be made, however, were not lengthy, because he was attacked by these dreaded stingers. His stay in the tropics was nearing its end, and there was no time left for a repetition.

In *A. dorsata*'s mode of indicating direction Lindauer sees an intermediary between the behavior of *A. mellifera* or *A. indica* and that of *A. florea.* In the dark hive *A. mellifera* and *A. indica* transpose the visual angle to the gravitational angle from memory, and for their part the dance followers must take account of the gravitational angle in order, after they leave the hive, to reproduce the visual angle. During the dances on the free-hanging *dorsata* comb, dancers and dance followers have before their eyes the sun or the polarization pattern of the blue sky. "Perhaps the newcomers tripping after the dancer envisage the proper direction of flight during the very dance: if perchance the dancer transposes to gravitational coordinates: '90° left of the sun,' and the newcomer could immediately in view of the sun take aim at a right angle to the left of the sun, then indeed she would have the announced direction of flight 'before her eyes' during the dance itself, which should facilitate communication" (Lindauer 1956:546). In any case it is prerequisite to this concept that when the bee shifts into the horizontal flying position she must compensate correctly for the distortion of the solar angle due to her stance on the vertical comb surface. According to all we know of her capacities, she is to be credited with such a poten-

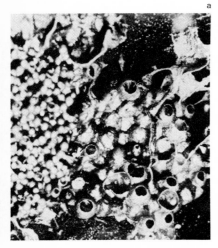

FIGURE 264. (*a*) Nest of the stingless bee *Trigona iridipennis:* (*left*) brood nest; (*right*) provision jars. (*b*) The bee, twice enlarged. From v. Frisch 1964.

tiality. One would like to know whether, when an *A. dorsata* swarm builds a new nest it retains the compass direction of its former comb, in order to avoid the further complication that would result from a differently oriented comb surface. It would certainly be rewarding if other courageous zoologists would come to closer grips with the formidable giant honeybee.

3. From Primitive to Successful Messenger Service with the Stingless Bees (Meliponini)

With the dwarf honeybee, which dances only on a horizontal surface with a view of the sun or the blue sky, the Indian bees have shown us a forerunner of the mode of dancing of our honeybee; but otherwise there are no significant differences and the hope of gaining deeper insight into the origin and phylogenetic evolution of the "language of the bees" has not been fulfilled within the genus *Apis,* that is, within the closest circle of relationship.

Systematically somewhat more distant and more primitively organized, but in many respects similar to the honeybees in their nest building and way of life, are the stingless bees (Meliponini). Especially in tropical South America they are represented by numerous genera and species.

In part they live in very populous colonies. As with the honeybee, a caste of female workers is developed and is sharply distinct from the reproductive forms. Likewise the colonies are multiplied by means of swarming, if in a more primitive manner. Primitive too is the care of the brood: it is not fed progressively, but instead is supplied once and for all with the necessary provisions, whereupon the cells are sealed. For construction material these bees use self-produced wax, to which, however—as with bumblebees—other substances are added. Among many of the *Trigona* species the lumpy comb with its irregularly arranged spherical cells also is quite reminiscent of a bumblebee comb. The honey pots and pollen containers (Fig. 264, *right*) are located apart from the brood nest (Fig. 264, *left*). Other species and more advanced genera, such as *Melipona,* construct regular, horizontal brood combs with hexagonal cells opening upward, and only the honey pots are still shaped like urns (Fig. 265). Honey may be collected in such large amounts that in South America some species are cultured, particularly by Indians, as domestic bees. However, anyone who supposes he may with impunity deprive these "stingless" bees[9] of their honey will soon learn differently from their vigorous attack and painful bites. Respecting their numerous other methods of defense, see Lindauer (1957).

Within this polymorphic assemblage there are significant differences in the level of social organization. The expectation that perhaps here there awaited discovery a scale leading from primitive to more highly perfected forms of communication was fulfilled—though in an unexpected way: on the one hand the Meliponini guide us to the roots of the "language of the bees," on the other to a higher development of it,

---

[9] The sting is not absent but is atrophied.

in which, however, the goal is attained in another fashion than with the honeybee.

The most primitive form of transmitting information about a food source that has been found was discovered by Lindauer even during his stay on Ceylon, in the little stingless bee *Trigona iridipennis* Smith, which is native there. It was possible to establish it in glass-covered boxes and observe its colonies, which resembled those of bumblebees (cf. Fig. 264). From here marked bees could be enticed to feeding stands, although the limit of the flight range seemed to be at about 120 m. Without question an arousal took place. For just as soon as a single bee or a few of them were fed, newcomers from the same nest regularly appeared at the feeding station. But information regarding the distance and the direction to the goal was lacking. For if, for instance, the feeding station was 80 m from the nest, stations with the same scent in the immediate vicinity of the nest received much more numerous visits than the feeding station itself. About as many newcomers came to stands set out at the same distance but in the opposite direction as to the feeding station. The comrades dispatched had received only a single guidepost to the route: knowledge as to the specific odor of the food source, which the foragers, just like honeybees, had borne on their body. Scent plates supplied with other volatile oils than the feeding station were not paid the slightest attention by the newcomers that swarmed out.

This modest success of the transmission of information is in agreement with the results of observations at the nest: the foragers returning home announce their find to their comrades, actively and with emphasis. They do this by running agitatedly about over the comb and seeking contact with the other bees scattered over the cells. They arouse their attention by jostling them and by means of unoriented vibratory movements of the body. The performance is somewhat reminiscent of the "jostling run" of the honeybee (p. 278). It is not a dance with rhythmic figures and it gives no information as to the direction and distance of the goal. The hivemates learn only two things: from the forager's behavior that there exists a good place for harvesting; and from the odor clinging to her what it smells like there. When then they roam over the vicinity in all directions, hunting for the odor announced, under normal conditions they will come upon the kind of flower from which the forager returned (Lindauer 1956).

Again the Rockefeller Foundation intervened helpfully and provided Lindauer with the means of undertaking a journey of investigation to South America, with its rich assemblage of meliponine species at different levels of organization. Comparative studies with them produced the following information (Lindauer and Kerr 1958).

With all ten of the species tested, when a good source of food is discovered there follows an alerting of the hivemates. But the outcome differs with the several species. If, for the sake of a better comparison, five marked bees of a colony are allowed in every experiment to collect 2M sugar solution for 1 hr at the feeding stand and the number of newcomers appearing at the feeding station during this time is recorded, then there is seen the situation shown in Fig. 266, where each new-

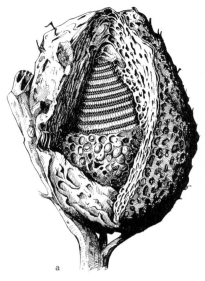

FIGURE 265. (*a*) Nest of the stingless bee *Melipona*, with protective coverings partly cut away in order to show the horizontal combs with the brood cells opening upward and the large provision jars below. One-twelfth natural size. (*b*) The bee, twice enlarged. From v. Frisch 1964.

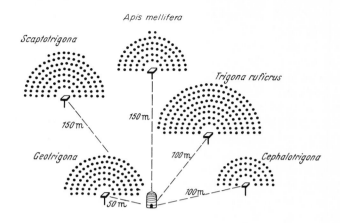

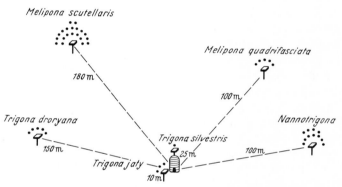

FIGURE 266. Results of alerting experiments with ten species of meliponines, in comparison with the honeybee: dots, newcomers dispatched to the feeding station in 1 hr by five foraging bees. After Lindauer and Kerr 1958.

comer is indicated by a dot. In proportion to the success in arousal, three groups can be distinguished.

1. With *Trigona jaty* D.T. and *Tr. silvestris* Friese (Fig. 266, *below, center*) only a few newcomers appear. Only near the nest can a feeding station be established; the bees will not follow to a greater distance.

2. With *Trigona droryana* Friese, *Nannotrigona,* and especially with *Melipona* species (lower diagram) the results of arousal are definitely better.

3. With *Scaptotrigona* and closely related species there is appreciably greater success yet, which may even exceed that with our honeybee, an experiment with which is shown for the sake of comparison in the upper diagram of Fig. 266.

Such widely differing results made Lindauer suspect typical variations in the behavior of the foragers. In order to be able to observe this inside the colonies he used observation hives that—in conformity with the horizontal combs of meliponine bees—were shallow boxes with glass covers. Combs arranged in layers above one another (see Fig. 265) were separated and placed side by side; this did not disturb the bees (Fig. 267).

To Lindauer's surprise, the behavior of the foragers on the combs did not reveal any significant differences, whether he was dealing with species that had little success in arousing their comrades or with others that had much. Nowhere were there rhythmic dances, as with the honeybees. Only unoriented zigzag runs by the foragers, in conjunction with energetic jostling of the hivemates, interrupted by repeated distribution of food, provide for arousal—as had already proved to be the case with *Trigona iridipennis* in Ceylon. As a further alerting signal there was evident an intermittent buzzing that was emitted by the returning foragers in shorter or longer bursts of sound (such as ··——·—···—·—·— and so forth). The pitch differs with different species. With *Melipona quadrifasciata* Lep., Kerr found 464–484 cycles/sec, with *Trigona (Scaptotrigona) postica* Latreille 391, with *Tr. jaty* 246–326. With *Tr. jaty* the buzzing is very soft, with *Melipona* loud. Lindauer succeeded in demonstrating its efficacy with a colony of *Melipona scutellaris* Latr. During a pause in feeding the foragers were distributed equally between sections *A* and *B* of the observation hive (see Fig. 267). Now the hinged flap at the hive entrance was placed so that the first forager returning could only get into *A*. Despite this she promptly alerted all her group companions in section *B* too—which came only a little later to the feeding station—evidently by sending them out in response to the sounds alone and not in response to the vigorous jostling. In the same way it was shown that newcomers too could be induced to fly out by the buzzing sounds alone.

When in control experiments the wooden floor of the box was covered with a layer of sponge rubber, the buzzing sounds had no effect. That points to the fact that the sounds—as with the honeybee—are perceived only as vibrations of the substrate and not as air-borne tones;

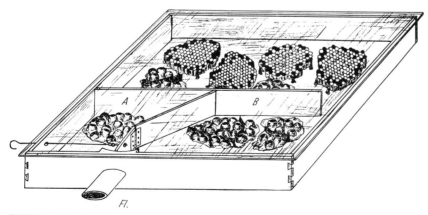

FIGURE 267. Observation hive for stingless bees; at the rear are the horizontal brood combs (light-colored cells, capped brood; dark cells, empty brood cells); in the foreground are honey and pollen containers; *Fl*, tubular flight entrance; *A, B*, front sections of the hive, of which one or the other can be cut off by the hinged flap but only *B* communicates with the brood-comb section. After Lindauer and Kerr 1958.

thus not "hearing" but rather a finely developed vibratory sense is concerned.

That was confirmed by Esch (*in lit.*), who went to Brazil to study sound production by the meliponids. Like Lindauer, he kept his *Melipona* species (*M. quadrifasciata* Lep. and *M. fasciata merillae* Cock.) in shallow observation hives. Foragers returning from good feeding places, in alerting their hivemates, emitted short, rhythmically repeated sound impulses (frequency between 300 and 600 cycles/sec). The duration of the individual bursts of sound was 0.4 sec for the distance 0 m (feeding within the hive entrance) and increased as the goal was farther removed, up to 1.5 sec for 700 m (the greatest distance tested). Thus the increase in duration of the sound signals seemed to show the same regularity as the increase in tail-wagging time and of the concomitant production of sound in the honeybee (see Fig. 117, curve o---o). Hence the principle on which the indication of distance is based is recognizable with *Melipona* too. But with the latter there is no tail-wagging run and correspondingly no indication of direction. From the following observation Esch concludes that with *Melipona* the duration of the sound signal is taken as signaling the distance. Two feeding stations were set up, at 10 m and 300 m distance, respectively. When, after an interruption in feeding, the noise for 10 m, recorded on a sound track, was transmitted to the hive by means of direct contact, the old foragers and newcomers came to the nearby feeding station; but the signal for the feeding station at 300 m had no effect in alerting the nearby foragers—nor, indeed, the foragers from the more distant 300-m station. Esch ascribes this to the absence of an indication of direction. But why did even the old foragers, familiar with this place, fail to come? Further experiments will be needed to clarify relations here.

Four species of *Trigona* studied produced similar alerting sounds, but they merely uttered an alarm, without any information about distance.

Since Lindauer found no sign of any indication of direction or distance within the colony, even in species that were most adept in arousing their hivemates, he looked outside the nest for the secret of their success. First he confirmed that the swift influx of large numbers of newcomers at the feeding station is based on an oriented guidance. Figure 268 shows the outcome of two experiments with *Trigona droryana*, a species rather poor in accomplishing arousal (see Fig. 266). Once the feeding station was 200 m south of the nest. During the observation period the foragers were joined by 17 newcomers, but 79 came to the stand with the same odor only 10 m distant from the nest where no foragers were visiting. Thus there was evidently no announcement as to distance (Fig. 268, *right*). That there is also no indication of direction is shown by Fig. 268, *left*; a test stand at the same distance to the north even received somewhat more visits than the southern feeding stand. Quite different were the results with *Scaptotrigona*, a species that is good at alerting (see Fig. 266). Feeding was conducted 150 m east of the nest. Only a very few visits were recorded at stands set out at another distance or in another direction (Fig. 269). How sharply the direction

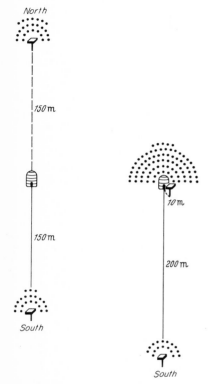

FIGURE 268. Arousal, without orientation, in the stingless bee *Trigona droryana*; the solid lines connect hive and feeding station, the dashed lines hive and observation stands; dots, newcomers. After Lindauer and Kerr 1958.

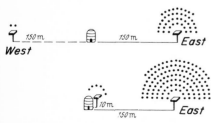

FIGURE 269. Oriented flight to the goal by *Scaptotrigona*; observation stands with the same scent as the feeding station are scarcely heeded when close to the hive (*below*) or in the opposite direction (*above*). After Lindauer and Kerr 1958.

to the feeding station was maintained is shown in Fig. 270; only a few newcomers appeared at angular separations of as little as 4° to left or right of the feeding stand.

Careful observation of the foragers outside the nest provided the solution of the riddle. Only with the most primitive species were the experiences on Ceylon with *Trigona iridipennis* confirmed, namely, that the nestmates are informed only of the existence of a source of food, whose characteristic odor is perceived on the bees returning home. With species that were adept at alerting, the newcomers were led from the nest to the goal in a surprising way, of which the following experiment with *Trigona ruficrus* Lep. affords a typical example. A marked bee was beginning to harvest at a feeding stand 35 m south of the nest. Already she had carried sugar water home 11 times and had aroused comrades in the hive. Newcomers came out and a little cloud of them swarmed about in front of the hive, but did not fly farther away. On her 12th return home the forager settled to the ground right by the feeding stand and then, as she proceeded homeward in a gentle zigzag flight, she kept landing again and again at intervals of 2–4 m, now on a blade of grass, a pebble, a lump of dirt. Each time she brushed the substrate with her outstretched mandibles, and in doing so deposited an odor mark there. After she had reached the vicinity of the nest, following some 20 such intervening landings, she flew up, mingled with the waiting crowd, and piloted a large number of them along the odor trail to the goal (Fig. 271). The odor deposited on objects is perceptible to the human nostril too and has the same definite smell as the mandibular gland, which is strikingly developed among the meliponines (Fig. 272, Nedel 1960).

That the odor trail laid down along the ground is a most important help to the newcomers in finding the goal is shown by an experiment on *Scaptotrigona*. A colony was established on the eastern shore of a pond; with a feeding station two bees from it were led in an arc along the bank to the western shore (Fig. 273). From there they at once struck out on a direct route over the water, but in several hours of going back and forth they did not bring a single newcomer. They did indeed try again and again to lay an odor trail on the water surface, but were of course unable to do so. When a rope bearing leafy branches was stretched across the water (Fig. 274), they at once put their odor marks on the twigs; and during the next half hour alone more than 30 newcomers came to the feeding station. The criticism that the optical guideline provided by the leafy branches might have been decisive does not hold. When the rope was displaced to one side on the western shore, so that from the nest it led 30° to the right of the feeding stand, the foragers flew via the direct route, the odor marks quickly evaporated, and now no newcomers flew along the leafy course any more, nor did a single one come to the feeding stand.

But merely marking the way by means of odor is not enough. The newcomers, who when aroused first come swarming out in front of the hive, must then be brought onto the trail and piloted along it to the goal. This is also to be seen from the fact that they arrive there in

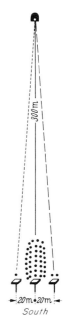

FIGURE 270. Oriented flight by *Trigona ruficrus*; observation stands displaced only as much as 4° to the right or left are scarcely heeded. After Lindauer and Kerr 1958.

312 / The Dances of Bees

batches, and in the company of marked foragers. Regarding the method of piloting there are some well-founded ideas, but these have not been thoroughly established. During the flight to the feeding place the guiding bees might emit mandibular secretion from the mouth and thus leave an odor trail behind them in the air, or they might allure the newcomers visually through their agitated way of flying. Flights following them directly were seen from time to time. As a rule what happened was that the marked guide flew off in a zigzag from the hive with a group of newcomers that were cruising about there; then she would turn back again every 3 to 5 m, gather in dawdling newcomers, and thus bring the entire crowd step by step nearer the goal (Lindauer and Kerr 1958:427). This behavior reminds one of that of the scout bees of *Apis mellifera* guiding a swarm to a nesting site (see p. 276).

On pp. 163ff it was explained that the honeybee is unable to signal the upward direction. Recall the experiment at the radio tower (Fig. 149). A group of bees foraging up above performed round dances and thereby dispatched their comrades over the meadow below where there was nothing to be collected. Lindauer repeated this experiment with *Apis mellifera* in South America at a water tower, with the same result. But *Scaptotrigona*, with her different methods of odor trailing, led her

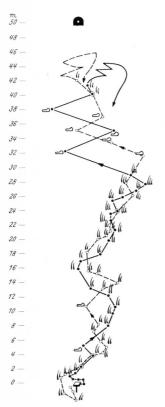

FIGURE 271. A forager's first (*solid line*) and second (*dashed line*) flights laying down an odor trail; previously she had made several foraging flights without leaving odor marks; dots and circles, sites marked with the scent; feeding station 50 m south of the nest. After Lindauer and Kerr 1958.

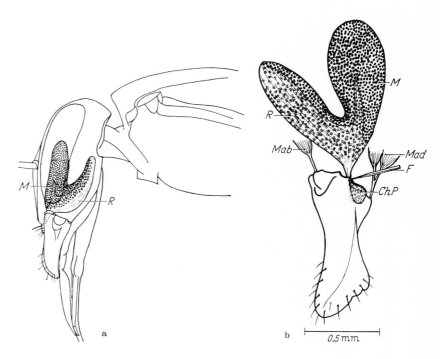

FIGURE 272. The mandibular gland of *Trigona scaptotrigona*: (*a*) side view; (*b*) right mandible with mandibular gland. The closing plate (*ChP*) permits the release of scent from the reservoir *R* as needed; *M*, mandibular gland; *Mab*, *Musculus abductor mandibulae*; *Mad*, *Musc. adductor mandibulae*; *F*, fibers running to hypopharynx; their contraction causes the opening of the closing plate and the outflow of the secretion. After Nedel 1960.

comrades up onto the water tower without the least difficulty. In the tropical forests, where in the extensive crowns of the giant trees food-giving blossoms are found far above the dense canopy throughout the year, this ability may be of no little use to them (Lindauer and Kerr 1958:432).

From Fig. 266 it was clear that species of the genus *Melipona* bring to the feeding station more newcomers than do the primitive forms, but on the other hand that they are less successful by far than *Scaptotrigona* and a few closely related species. In no instance was it possible among the representatives of the intermediate stage to demonstrate the laying down of an odor trail. Nevertheless, with *Melipona scutellaris* and *M. quadrifasciata* newcomers seem to be piloted to the goal, either optically or olfactorily or both, by the foragers in a guiding flight. Lindauer was unable to determine more exact details of the process.

Esch (*in lit.*) also reports similar observations with *Melipona*. When the foragers return to the feeding station from the hive, they make a conspicuous zigzag flight along the first portion of the route and are accompanied by hivemates that have been aroused and have been awaiting them near the flight entrance. But the latter soon lose this contact, for they return to the hive, and only after 20–30 repetitions do the newcomers reach the goal, never in company with the old forager. She has merely conducted them part of the way; then they have to follow the direction farther independently.

If the outcome of arousal with advanced meliponines is compared with that in our honeybees (see Fig. 266), it seems to be at the same level in both. But the psychological accomplishment is doubtless to be rated higher in the honeybees, since on the basis of the information received the newcomers find their goal in free flight, whereas the meliponines are led thither with the help of odor trails and guide bees. But aside from this the honeybees also achieve superior results. One becomes certain of this on recalling the circumstances in nature and thinking for instance of the flow of nectar in the temperate zone in springtime, when a multitude of different kinds of flowers simultaneously present food of unequal quality. Only the honeybees, with their graduated system of arousal, proportioned to the profitability of the food sources and combined with a precise description of the location of the various goals, are able to instruct the unoccupied waiting mass so that the labor supply maintains the proper relation to the demands of the several situations to be exploited. A corresponding accomplishment on the part of the meliponines is not known and in view of their methods is hardly to be expected. In this respect their mode of communication is at a more primitive stage. In contrast with the unique behavior of the honeybees, their way of exchanging information also is related in many particulars to the practices of other social insects.

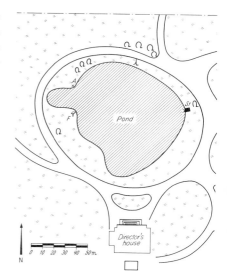

FIGURE 273. Sketch of the location of an experiment over water. The foragers flew directly across the pond from the hive *St* to the feeding station *F*. After Lindauer and Kerr 1958.

FIGURE 274. A rope bearing leafy branches is stretched across the pond and affords the foragers an opportunity of laying down odor trails. After Lindauer and Kerr 1958.

4. A Brief Glance at Other Social Insects

Among social forms that belong to the family Apidae there are besides the apines and meliponines the Bombinae (bumblebees). Their irregular combs, in which the building material is not employed economically for the construction of rationally shaped cells, their improvidence for the winter, because of which in our climate an end for every colony is foreordained in autumn, and other primitive features, led to the expectation of primitive conditions in their way of communication also. But what was found was disappointing: nothing whatever of this nature was to be made out, there was absolutely no alerting of the nestmates upon discovery of a rich source of food, let alone any communication in regard to its location. Foragers returning home successful might indeed run about agitatedly on the comb for some time, meanwhile colliding with nestmates, but never could any responsive reaction of the latter be discerned. Contact was neither striven for nor heeded. Even with activity continued all day long, the forager brings no newcomers to the feeding place (Wagner 1907:159; Brian 1954; Jacobs-Jessen 1959; Maschwitz 1964). In Brazil Esch (*in lit.*) was also unable to find any sound production in bumblebees (*Bombus atratus* Frankl.) that were fetching sugar water from artificial feeding places at various distances.

Wagner (1907:174ff) describes for several species of bumblebee a more primitive form of communication, an alarm for danger by means of warning tones. When the nest is much disturbed they are said to produce, by means of a unique whirring of the wings, loud buzzing tones, through which other bumblebees sitting nearby are induced to run about and investigate. He believes that the sounds are perceived through the sense of touch. Pertinent observations had been made by Hoffer (1889:18) even earlier. He kept a number of bumblebee nests in rearing boxes. When carefully and without causing vibrations he placed a parasitic bumblebee (*Psithyrus*) in a nest of *Bombus pomorum* Panz., there arose such a commotion in the colony that in consequence of the buzzing of these bumblebees those in neighboring rearing boxes were disquieted also. No doubt there belongs here too the ancient observation of the "trumpeter" in the bumblebee nest, who by means of loud, persistent buzzing (lasting from several minutes to half an hour) wakes the nestmates in the morning. Since the first description by the Danish artist Goedart in 1700,[10] the story of the "trumpeter" has been discussed frequently and rarely credited. Other interpretations seemed more likely. According to the most recent observations, by Haas (1961), the trumpeter is no imaginary figure, but a sentry in the bumblebee nest, who sounds the tocsin of danger when there are disturbances, and in other instances by buzzing actually routs out the nestmates as soon as it gets light in the morning. In view of the buzzing dance (*Schwirrtanz*) of honeybees (pp. 279f), their production of sound during the tail-wagging run (pp. 57f), and the alerting of meliponines by buzzing tones (p. 309), these reports merit retesting and further study. Maschwitz (1964) could not find in bumblebees an odoriferous alarm substance, such as occurs in bees (see p. 290).

[10] J. Goedart, *Metamorphosis Naturalis sive Insectorum Historia*, (Amsterdam, vol. 2, 1700); cited from Haas 1961.

In the mating flights of the bumblebee male may be seen a bridge of relationship leading to the habits of the meliponines, which guide their comrades to the feeding place by laying down an odor trail. These bumblebees use the odorous secretion of their well-developed mandibular glands[11] in order every morning to define anew a closed flight path with numerous odor marks (Fig. 275; Haas 1946, 1952; Stein 1956). Males ready for mating thus disclose their presence to the females. They proceed round and round the routes marked out, deviating from them only occasionally for flights in search of food (Fig. 276).[12] Since the odor from the scent organ is species specific and besides the different species fly over routes typical for them, one kind in the meadows, another in brush, yet others high in the crowns of trees, and so forth, and since the females are actively seeking the specific scent and the typical routes of their own males, the coming together of the sexes is guaranteed as fully as possible (Haas 1949; Krüger 1951).

Similar relations have been demonstrated for parasitic bumblebees (*Psithyrus*) and for a few solitary bees (Haas 1949a, 1960; Krüger 1951).

Among the rest of the social Hymenoptera, the wasps remind one of the bees, in their constructions and in their way of living. To summer meals in the open they often come swiftly and in numbers, as unbidden guests, so that one might well imagine a system similar to that of bees for transmitting information. But the increase in newcomers customarily does not persist. With regard to this Kalmus (1954) did comparative experiments with bees and wasps (*Vespa germanica* Fbr.): a supply of sugar water was discovered very quickly by wasps, but remained concealed for hours from a colony of bees set up in the vicinity. The number of visits by wasps proved to be fairly constant, whereas the number of bees, as soon as any came at all, grew very quickly. Wasps captured and removed were not replaced, so that their swift initial appearance and increase were to be ascribed only to zealous spying about and to their cleverness, while apparently there was no alerting of their nestmates.

But probably that is not true of all wasps. In Venezuela, Hase (1935) with *Polybia atra* Sauss., widely distributed in South America, watched for some days at a nest in the veranda of his house the comings and goings of the specimens that flew up. Generally about 50 were on the nest covering in front of the entrance, where those arriving used to land and those departing took wing. As a rule the numbers of the two were quite evenly balanced. But on occasion he saw that on their arrival the wasps performed a tail-wagging dance, meanwhile hastening through the crowd of their comrades who were sitting on the nest. Thereupon a number of specimens departed in more rapid succession than usual—as recorded in the published protocols.

FIGURE 275. A male bumblebee (*Bombus hortorum* L.) sitting on an apple leaf and laying down odor marks on it in the direction of the arrow. After Haas 1946.

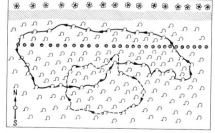

✽ Field of sunflowers ● Row of beanpoles
▨ Corn field ∩ Potato field

FIGURE 276. Flight paths of two males of *Bombus terrestris* L. in a potato field, 26 August 1947. The flights followed the odor marks, which were on the lower parts of the stems of the potato plants, 10–40 cm above the ground. After Haas 1949.

---

[11] According to Stein (1963), this odorant is identical with farnesol, a primary aliphatic alcohol that occurs in almost all floral oils. Apparently it is consumed with the food by the male bumblebee and then stored in the mandibular glands.

[12] The "flight paths" of the male bumblebee were observed even as long ago as by Charles Darwin (see Krüger 1951:62); they were rediscovered by Frank (1941), but their significance as routes marked out by odors was first recognized by Haas (1961).

In agreement with this are observations by Lindauer (1961:85) on another species of *Polybia* (*P. scutellaris* White) in Piracicaba (Brazil). Ten wasps that were visiting a feeding station 150 m from the nest brought along, on the average, 5–7 newcomers per half hour. Demonstrably these went hunting round about; they had not been informed of the direction and distance of the goal.

Zoologists who get such an opportunity in the tropics could still find out a great deal with this and other species of wasps.

Ants are able by means of odors to summon their comrades to defense, partly with their poison and partly with special alarm substances. With some species, as with *Formica polyctena* Förster, the toxic secretion serves for a primitive sort of communication about the location where food has been found. Large insects that defend themselves are sprayed with poison, the odor of which attracts other ants from the vicinity in support of the attack (Maschwitz 1964). On a higher level there is the system of communication, widespread among ants, by means of odor trails. Such methods are found also among the air-borne bees (meliponines, p. 311); with a pedestrian population, bound to the earth, this is even more likely and simpler. If, for instance, a scout from the genus *Pheidole* or *Solenopsis* or *Myrmica* has come across prey that it cannot master by itself, it returns to the nest and lays down en route an odor trail, by dragging the abdomen along the ground. Having once arrived in the nest, the informer arouses the nestmates by running about in agitation, by drumming on them with antennae and legs, or by butting against them with the head. A scent given off from the mandibular glands may also participate in arousing the nestmates (as when the alarm for danger is given; see p. 290). All of this is succeeded neither by leading them in person to the goal nor by any transmission of information as to the nature or location of the prey. The comrades learn only that there is something to be fetched. They rush out, follow the odor trail, and many of them reach the goal. Since on their return they themselves lay odor trails, they strengthen the guiding signal. On the other hand, the odor trails evaporate rapidly as soon as the source of food is exhausted and the trail is not renewed (Hingston 1928:9; Goetsch 1953, 1953a; Sudd 1960; Wilson 1962, 1963).

Extracts of these odorants have been prepared, and very effective artificial odor trails can be laid down with them. Among ants that belong to the same genus, the odor differs from species to species. Its chemical nature has not yet been defined in any instance (Wilson 1963).

Very remarkable methods of communication in some other species of ant permit one to imagine the way in which the laying of odor trails has evolved in the course of phylogeny. Where a scouting *Camponotus sericeus* Fabr. (India) finds food, she returns to the nest and leads a comrade to the feeding place. The newcomer follows immediately after the guide, and keeps touching her abdomen frequently. The leader proceeds exceptionally slowly, and if contact is lost she waits until it is reestablished (Hingston 1928:270). In Puerto Rico, Wilson (1959) saw the same behavior in another genus (*Cardiocondyla venustula* Wheeler and *C. emeryi* Forel) and called it "tandem running." Here too contact was carefully preserved, and no odor trail was laid down. The (very small) workers run ahead only a few millimeters and then wait for the ant following. Thus they proceed slowly and jerkily to the goal. This is probably the most primitive and least economical method yet known, where assistance is required.

Like *Camponotus sericeus*, *C. paria* Emery, too, conducts only a single comrade to the feeding place. But here there is a small but important advance, in that the newcomer is able to follow the leader at a distance of a few centimeters and the latter no longer has to wait for her. The leader leaves a weak odor trail. With *Camponotus compressus* Fabr., 10–20 comrades follow the leader in single file. This performance constitutes a behavioral bridge to alerting by means of intense odor trails, which is so successful (Hingston 1928; Sudd 1959).

In systematics the termites stand apart from the insects discussed hitherto. Guidance along paths is of subordinate significance to them, since for the most part they live within their source of nourishment (for example, wood) or reach their goal by way of subterranean passageways. With forms that seek their food outside the nest, pursuit of odor trails left by their nestmates has been demonstrated (Goetsch 1953).

The processes of trail laying and alerting were studied in the North American termite *Zootermopsis nevadensis* Hagen by Lüscher and Müller (1960). Stuart (1961, 1963) observed the laying of trails by the same species and by *Nasutitermes cornigera* Motschulsky. The scent glands are on the ventral side of the fifth abdominal segment. The path is marked out by touching the ground with that part of the venter where the glands open. The secretion can be extracted with ether and can be used successfully to lay down artificial trails. The odorant is effective with other colonies also, but not with other species of termites. Stimulating an individual termite makes it run in a zigzag through the nest, coming in contact with its comrades and arousing them. They may spread the alarm by similar behavior.

Certainly there are yet other ways of transmitting information among termites. But without doubt most questions in regard to these highly enigmatic social insects are still to be answered.

Thus intercommunication by laying down odor trails on the ground is quite widespread among insects, and seems to have found its highest development in a few meliponines.

Summary: Communication Among Other Races, Species, and Genera of Bees, and Among Other Social Insects

1. Distinct races of the honeybee differ only slightly but definitely in their indications of direction and distance ("dialects of the language of the bees"). The three Central European races have been studied: Carniolan bees (*Apis mellifera carnica* Pollm.), Italian bees (*A. m. ligustica* Spin.), and German bees (*A. m. mellifera* L.), as well as the bee of the Caucasus Mountains (*A. m. caucasica* Gorb.), the Egyptian bee (*A. m. fasciata* Latr.), and the Punic bee (*A. m. intermissa* Butt.-Reepen). Com-

plementary data from other sources are available for *A. m. adansonii* Latr. from tropical Africa.

2. All races indicate feeding places in the immediate vicinity of the hive entrance by means of round dances.

3. In indicating direction the Carniolan bee is distinguished from all other races by the fact that with increasing distance to the goal the round dance goes over directly into the tail-wagging dance. With the other races the transition occurs via direction-indicating sickle dances (Fig. 56). The Egyptian bee starts to perform sickle dances at a distance no greater than 3 m, and tail-wagging dances at 10 m; the other races follow in succession at greater distances, characteristic for each. At the end of the series is the Carniolan bee, which starts to indicate direction clearly only at 50–100 m (Fig. 253).

4. Indication of distance also differs specifically among the races. The Carniolan bee has the most rapid dance rhythm; of the Central European races the Italian is the slowest dancer. The African races are slower still (Fig. 254).

5. In several hybrid swarms of Carniolan and Italian bees, workers with the yellow body color of the Italian race performed sickle dances predominantly, when the feeding place was near by, while those individuals with the body color of Carniolan bees performed round dances mostly. Thus the performance of the sickle dance seems to be linked in inheritance with the yellow body color. But this question needs further study in hybrid colonies whose genetics are precisely known (artificial insemination of the queen).

6. In an artificial mixed colony composed of Carniolan and Italian bees, the faster dance rhythm of the Carniolans was misunderstood by their Italian hivemates, who went hunting for the goal at too short a distance. Conversely, the Italian foragers with their slower dance rhythm alert the Carniolan hivemates for too great a distance (Fig. 256).

7. Even greater differences in the mode of dancing occur among the different species of the honeybee. For comparison with our bees all of the three other (Indian) species of *Apis* were studied: the Indian bee (*Apis indica* F.), the dwarf honeybee (*A. florea* F.), and the giant honeybee (*A. dorsata* F.).

8. In the indication of distance there are differences only of degree (Fig. 262). The dwarf honeybee dances most slowly. The steep initial course of the distance curve is appropriate to its short flight range (Fig. 112). This implies a precise indication of distance for short intervals to the goal, and that imprecision is greater only at distances that are scarcely of practical importance.

9. In principle the mode of indicating direction by the Indian bee (*A. indica*) agrees with that of our honeybee, only the Indian bee performs oriented sickle dances when the feeding place is no more than 1 m away, and begins with oriented tail-wagging runs at 2 m. Like our bee it can tell the position of the sun from the polarized light of the blue sky too. In the dark hive it transposes the solar angle to the direction relative to gravity. The dwarf honeybee (*A. florea*) does not transpose

the visual angle. Its nest, consisting of a single comb, is always located under the open sky. The expanded upper surface of the comb serves as a horizontal dance floor. Only here, with a view of the sun or of the blue sky, is she able to point out the direction to the goal. Thus the horizontal dance, oriented directly according to the sky, such as occurs in our bees too, appears as the original form in which direction was indicated; the transposition within the dark nest cavity to the direction relative to gravity is to be considered as derived from it. The giant honeybee (*A. dorsata*) likewise builds but a single comb, from which the sky may be seen. This bee assumes an intermediate position, since she does indeed dance on the vertical surface and makes a transposition, but can indicate direction only when the sun or blue sky is visible.

10. A much more primitive method of communication is found in many species of stingless bees (meliponines), which also occupy a lower place with respect to their way of living and the construction of the nest. Thus *Trigona iridipennis* Smith (Fig. 264) alerts her nestmates by running about agitatedly on the comb, without informing them about the direction and distance to the goal. Only the smell of the food source, which clings to her body, brings the comrades greater detail: they learn the specific odor of the kind of flower that has been discovered. In such conditions may be seen the roots of the "language of the bees," from which the modes of communication among the honeybees and the meliponines have diverged. For among the numerous species of the group of stingless bees there is within the more advanced forms an effective transmission of information, but of a different sort from that of the honeybee. Direction-indicating dances are unknown. The mode of arousal is limited to an agitated running about and collision with the comrades, supported by buzzing sounds. The bees thus alerted fly out, hover expectantly about the nest, and are then led to the goal by the discoverer of the food source, along an odor trail that she has laid down on the ground (*Trigona ruficrus*, Fig. 271; *Scaptotrigona*). Intermediate between the primitive type of arousal by *Trigona iridipennis* and the highly successful process in the forms just mentioned there are *Melipona* species (Fig. 265) that do not lay down any odor trail, but pilot the newcomers in flight to the goal.

11. Meliponines like *Scaptotrigona*, which with the aid of their mandibular glands place scented markers on the ground and guide the newcomers along an odor trail to the goal, can bring them to the feeding place as swiftly and in as great numbers as can honeybees. But the latter are superior in their psychological performance, in that they do not *lead* their comrades to the goal, but rather *send* them there by means of appropriate information. Only in this way is it possible, amid a multitude of simultaneously flowering plant species, to guide the labor forces to the different work areas in a way suited to the unequal yields of the different kinds of blossoms.

12. With bumblebees neither dances nor any other form of communication in regard to food sources could be demonstrated. Nevertheless there seems to be an alarm given for danger by means of vibratory

stimuli (buzzing sounds). Like many meliponines, bumblebee males mark out with their mandibular glands scented paths on which the two sexes find one another.

13. Our native social wasps do not seem to possess any mutual communication as to sources of food. Among *Polybia* species in South America, dances and also an effective system of arousal have occasionally been observed.

14. Among ants there is transmission of information about the location of a discovery of prey. The discoverer lays down an odor trail and arouses the nestmates by running agitatedly about. The insects alerted leave the nest; many of them find the odor trail and are guided by it to the goal. There are species in which the ant that finds prey fetches from the nest only a single comrade, who lets herself be led to the location by running along behind the discoverer and maintaining antennal contact with her. Among other forms there are transitions from such behavior to the laying down of odor trails.

15. With termites too there are known an alerting of the nestmates and the laying down of odor trails that lead to the goal.

XI / Phylogeny and Symbolism of the "Language of the Bees"

To attempt a discussion of the phylogeny of the "language of the bees" sounds somewhat presumptuous. Actually we *know* nothing of this subject. The beginnings of the evolution of human language too are unknown to us, even though scarcely a million years have elapsed since that period. How then should we be informed about the roots of the "language of the bees," over which many times 10 million years have flowed? Not even the body structure of the first social insects has been preserved, let alone their way of living, which could not leave any fossil image.

All that remains to us is to attempt, from certain forms of behavior in other insects and from more primitive stages in the communication of the bees of today, to discover a way in which the "language" of honeybees *might* have developed.

In the fly *Phormia regina* Meigen, Dethier (1957) observed and described a type of behavior that—if it occurred similarly in the ancestors of the bees—might perhaps have led to the round dance as a means of communication: if one offers a fly on a horizontal surface a drop of sugar water and withdraws it before the fly has fed fully, the fly makes circling movements, alternately to the left and to the right (Fig. 277). These are *searching movements*, for the vanished sugar, and as such are meaningful. It is senseless that a fly, held in the hand and fed there, behaves in exactly the same way when afterward put down on a flat surface where there has been no sugar. Thus the searching does not take place with insight, but merely automatically. That the fly has been moved to another place after being fed transcends its horizon; under normal circumstances this simply does not happen.

There emerge certain parallels between the flies' "dance" and the round dance of the bees. With rather high sugar concentrations, or with greater hunger (decrease of the acceptance threshold), the flies dance with greater intensity and for a longer time. That too is understandable as a more lively search, following a stronger stimulus. Further, one can observe a certain degree of orientation of the "dance" in accordance with external stimuli, in that with unidirectional illumination on a horizontal surface the circles are deformed in the direction of

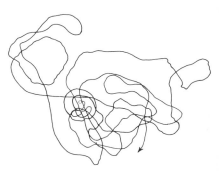

FIGURE 277. Circling course of a fly (*Phormia regina* Meigen) on a horizontal surface in daylight, after being deprived of a drop of sugar water; ○, feeding place. After Dethier 1957.

FIGURE 278. Course of the fly on a horizontal surface with unidirectional illumination. After Dethier 1957.

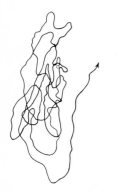

FIGURE 279. Course of the fly on a vertical surface in the dark. After Dethier 1957.

the incident light (Fig. 278), and on a vertical surface in the dark in the direction of gravity (Fig. 279). But there is no relation between the orientation of the runs and the location of the food source.

Like a forager bee returning home, when the fly enters a group of its fellows it too customarily regurgitates at least part of the sugar water it has drunk. The other flies try to lick up the sugar, grow excited, and also begin to "dance," that is to search on their own part. But no transmission of more precise information by the first fly is to be discerned.

If one puts the fly down on the surface only some time after it has taken food, the intensity of the "dance" is the less the longer the interruption, and if the latter lasts several minutes there is no "dancing" whatever. In this also Dethier sees a parallel to the bees that (he says) dance more slowly and "more diffusely" when the food source is far away and the flying time correspondingly prolonged. But this is not so. With increasing distance the bee dance loses no whit in intensity, nor does it become diffuse; only the rate of circling grows slower.[1]

There is an important difference between the "dance" of the fly and the round dance of the bee: the circlings of a fly that has lost its drop of sugar water are *searching movements*. The dancing bee has drunk her fill, and is not looking for anything. The circling runs have become a *symbol* for the hivemates, who are to do the hunting (v. Frisch 1954).

Nevertheless one will agree with Dethier that behavior similar to that shown by the fly may have been the starting point for the evolution of the bee dance. Here it is particularly significant that the "dance" of the fly also occurs after an interval, as a *delayed response* to the stimulant taste of the sugar water. If the same had been the case with the bee returning from a journey in search of food, the delayed circling on the comb would have become a signal to the hivemates. No doubt with all social insects the movements that cause arousal are more vigorous when the food source is good than if it is of lesser value. This potentiality for communicating about the quality of a discovery certainly reaches back into times before there was any information service. For even in the dance of the fly, the effort is so much the more eager the higher the quality of what has been lost.

For the bees' tail-wagging dance, too, a precursor has been found. Butterflies are no more to be thought of as potential ancestors of bees than flies are, but nevertheless it is highly remarkable that such a performance as tail-wagging occurs among other insects, without any social implication and with certain parallels to the tail-wagging dance. With numerous species of New World saturniids (a cocoon-spinning family of moths), Blest (1960) noticed, when they settled after a preceding flight, lateral vibrations of the body (about 2–4 per second); these are caused by alternate extensions of the legs of one side and the other and die away after a little while. Nothing is known as to the significance of these vibrations. With representatives of the genus *Automeris* that he let fly for various lengths of time suspended by the abdomen, there was

---

[1]This same misunderstanding is to be found in Kroeber (1952), who would relate to fatigue the slower rhythm of the dance after flights from greater distances. But no one who has seen the energetic dances of bees that are exploiting a worthwhile source of food at a distance of some kilometers would ever have any such idea.

under certain conditions a linear relation between the duration of the preceding flight—tested for intervals up to 30 min—and the duration of vibration, which increases with the time spent flying. Thus after a flying time of 2 min there were 4 waggles, after 10 min 11, after 30 min 25. The same is true of free flight. The result remains the same even when there is an interruption of 1.5 hr between the cessation of flight and the beginning of the vibratory movements. Blest notes the similarity of this phenomenon to the tail-wagging dance of bees, in which the duration of waggling increases with the duration of flight— that is, for a straight-line flight, according to the distance to the goal.

If the duration of the vibratory movements depends on the flying time, the moths must take account of that time. In order to learn how this takes place, Blest subjected his experimental animals to all kinds of demands. Before the flight he cut off the abdomen; he also removed the wings, the legs, or the antennae, sectioned the flight muscles, or extirpated the wing bases, all without influencing the relation between the duration of (attempted) flight and of the succeeding vibratory movements. From this he concludes that no sensory perceptions are concerned in determining the duration of the vibratory movements; rather the flying time must be recorded by the central nervous system itself. What is decisive is the length of time during which the central nervous mechanism mediating flight was active.

Blest thought that the same thing is true with respect to the announcement of distance in the bees' tail-wagging dance. That is possible, but is not proved. Against a simple transferral of these results to bees is the fact that with the latter the relation between distance flown and duration of waggling is nonlinear (v. Frisch and Kratky 1962). Nevertheless it is plausible that in the ancestors of the bees a performance similar to that among the moths might have furnished a starting point for the evolution of the tail-wagging dance. Whether in actuality exploratory circular runs, such as are seen in flies, and waggling movements, as in moths, were available to the ancestral bees as elements from which methods of communication could be developed, we know nothing whatever.

Not until we make a comparative evaluation of the social insects do we reach firmer ground. The simplest and probably the oldest form of giving information about discovered food is the agitated behavior of the finder and the transmission of this excitement to other members of the community. Even the bumblebees display a striking degree of excitement on their return home after they have found good food. But this is confined to an increased activity on the part of the discoverer herself (p. 314). With primitive meliponines and with ants the excitement is transmitted to nestmates, via contacts that are sought with them, and they are activated thereby. Frequently this arousal is bolstered by the emission of sounds, which are perceived with the vibratory sense. With stingless bees these are buzzing tones, such as occur during the production of sound in the course of the waggling run of honeybees.

In some instances, as among the seed-gathering ants of the genus *Messor,* the nestmates informed swarm forth in all directions without any point of reference, searching in circles and spirals, and it is owing only to their great numbers that a few reach the goal (Goetsch 1953a: 86ff). Among other species the discoverer during her return home lays down an odor trail that serves her searching comrades as a guide. But apparently it does not happen with ants that they receive from the one who finds prey information about the goal, as for instance by transmission of its specific smell. Yet there is something of this kind among the stingless bees, which take heed what sort of odor is clinging to the body of the agitated discoverer. Though they do not in this way learn where they should go hunting, they find out what they should look for.

Thus with primitive meliponines the nestmates are alerted, but they receive no information other than that foraging is worth while[2] and how the goal smells. In such relations no doubt are to be seen the beginnings of the "language of the bees." From here on the pathway divides: among the more advanced meliponines there evolves conduction to the goal by means of guiding flights, which are perfected further by laying down an odor trail on the ground (p. 310ff). Within the related genus *Apis* there arose, presumably from similar beginnings, the supplying of the comrades with information about the location of the goal, by means of rhythmic dances that enabled them to seek for it and find it in free flight without direct guidance.

With all apines the round dance signifies to the nestmates a demand to fly out and go searching around nearby. Whether they do this with circling flights or whether they proceed successively in different directions and then return is not known, but could be determined with a completely numbered observation colony. Probably the round dance should not be interpreted as an "intention movement," for the forager has no tendency to fly in circles. On her next excursion she returns in a straight line to the goal she knows. The round dance is a kind of jostling run (*Rumpellauf*) performed at a single location. That it is performed only within a throng of other bees and not at empty places on the comb emphasizes its nature as an active transmission of information to the hivemates. Interpreting it as a symbolic demand to go searching round about the hive leaves open the question of its origin. In this respect Dethier's hypothesis may be recalled (pp. 321f).

The tail-wagging dance probably can be interpreted as derived from "intention movements."[3] If we think of the dwarf honeybee, which dances on a horizontal surface exclusively and only in view of the sun or blue sky, we can interpret the tail-wagging runs as repeated starts to take off, with each run oriented toward the goal. That might be compared with the intention movements of communally living birds, among which before they depart from the feeding grounds to the nocturnal sleeping places one or another initially expresses its being in a mood to fly away by just starting to take off, smoothing its feathers, and stretching its neck toward the goal; this serves to transmit the mood to its comrades and leads in the end to mutually coordinated action.

[2] Alerting takes place only when food is good. Just what detailed factors are determinative here, even where communication is very primitive, remains to be investigated.
[3] Haldane (1954:267ff) expresses himself to this purpose with respect to the round dance as well as to the tail-wagging dance.

But in its content of concrete data the dance of the bee is superior to these simple intention movements. The gestures of a bird, appropriate to its mood, are something different from the precise transmission of the direction to a strange goal, such as the dance follower receives from the dancing bee, and nothing comparable to the description of a location by combining the indication of its direction with data as to its distance is known anywhere else in the whole animal kingdom. The greater the distance and hence the longer the flight path, the more the dancer prolongs the tail-wagging period, which we regard as the definitive signal of the distance (p. 104). The expenditure of time or energy in the waggling stretch is by no means identical with that for the flight course. But in its relative scale the tail-wagging period serves as a signal for the distance of the goal. The production of sound during the waggling run gives this phase of the dance pattern an additional special emphasis. And since the sound evidently is produced by the activity of the flight musculature—for it comes from the thorax and its frequency of about 250 cycles/sec accords with the frequency of wingbeat (Esch 1961)—one is tempted to regard this emission of the flight noise as a symbolic urge to take off.

This thought was expressed in conversation by Esch and was developed further on the basis of his most recent experiments with stingless bees (*in lit.*). The buzzing after the return home may have arisen originally as an accompaniment of continued contractions of the flight muscles that had the purpose of keeping the muscles warm between flights. Thereafter the sound was taken by the hivemates as being connected with the food and in this way became an alerting signal in the primitive *Trigona* species. On a higher level, with *Melipona*, the excitation that leads to warming up and to the emission of sound is subject to a rhythmic variation, with dependence on the distance of the source of food. The differing durations of the bursts of sound are comprehensible to the comrades. With the honeybee the excitation and the production of sound, now linked with the tail-wagging movement, occur whenever the forager moves in a direction toward the feeding place (also when this direction is transposed on the vertical comb). The indication of distance has been combined with that for direction. These are attractive notions. Whether they are correct we do not know. As a point of cleavage between the meliponines and the honeybees there remains the fact that the former give the direction to the goal by means of guiding flights and by laying odor trails, whereas the honeybees send their comrades on the right course with the symbolic gesture of the tail-wagging run.

The tail-wagging run, by thus combining a precise description of the location of the goal with the alerting process, surpasses all other forms of social arousal. Unfortunately in the genus *Apis* there have been preserved no intermediate stages that might tell us how the transmission of information about the food source has actually evolved.

Only one noteworthy advance is to be seen among the apines of today: the transition from the horizontal tail-wagging dances of *Apis florea,* with their orientation directly to the sun, to the dances of *Apis mellifera,* which have become independent of the sight of the sky and

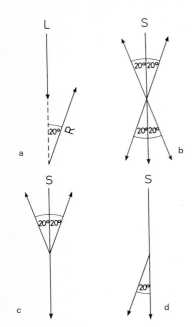

FIGURE 280. (*a*) A straight course *R* is maintained on a horizontal surface by keeping a constant angle—here 20° to the right—relative to the light from a source *L*. (*b*) Ant, (*c*) ladybird, and (*d*) dung beetle exhibit different ways of transposing angles in the dark with reference to the direction of gravity *S* after the supporting surface is tilted to the vertical.

the sun and that express the solar angle via the gravitational angle. This capability for transposing (p. 137) was prerequisite to living in protected, dark cavities, and hence for the occupation of regions in which colonies that build in the open perish in the winter.[4]

For a long time this evolutionary step from the direct to an indirect method of indicating direction seemed enigmatic. But with discoveries in other insects it has now come closer to being comprehensible.

Ants (*Myrmica ruginodis* Nyl., *Lasius niger* L.) often use the sun as a compass. When placed on a level surface in a dark room they accept a lamp in lieu of the sun and maintain a straight course by holding a constant angle relative to the light, for example 20° to the right of it (Fig. 280*a*). If now the light is turned off and simultaneously the surface on which they are running is tilted to the vertical, they set out on the same angle relative to the vertical that previously they had followed relative to the light. They too transpose the angle of orientation from light to gravity, though to be sure somewhat generously: if for instance they had been running 20° to the right of the light source, now they run 20° to either the right or the left of the vertical, and either upward or downward (Fig. 280*b*; Vowles 1954).[5] In a comparable experiment, a ladybird (*Coccinella septempunctata* L.) comes closer to the performance of the bees, by equating the direction toward the light with direction upward, though indifferently to right or left (Fig. 280*c*; Tenckhoff-Eikmanns 1959). Dung beetles (*Geotrupes silvaticus* Panz.) make a proper lateral transposition but—just opposite to the bees—identify the direction toward the light as the direction downward (Fig. 280*d*; Birukow 1954). As these randomly chosen examples show, it happens with many different insects that—for no obvious biological reason—when compelled to shift their orientation from light to gravity they preserve the original angle; in doing so not all species keep to the former direction laterally, and some equate the course toward the light with the upward, others with the downward line. Only the size of the angle is very generally preserved (approximately) when the regulation of movement passes from the visual to the gravitational sense, or vice versa. Apparently this is effected by higher central-nervous centers that are responsible for steering the movements and that regulate orientation in accordance with a definite point of reference, whether this is provided by sense organs for light or for gravity (Birukow 1954; Jander 1957:220ff).

Birukow (1956a) endeavored to interpret from the biological point of view the differing ways in which various groups of insects respond. Even though assumptions, some more, some less firmly established, are involved, nevertheless we know that there is a connection between ori-

---

[4] It may happen that a swarm of *Apis mellifera*, unable to reach agreement on a nesting site, will build in the open (p. 273). In the vicinity of Lawrence, Kansas, a large structure of this sort lasted for 3 years (Byers 1959). But for places with a cold winter that is a rare exception.

[5] This finding should not be interpreted as indicating that the ants are unable to distinguish directions to the right and left of the vertical. Under other experimental arrangements they make this distinction with great certainty (see pp. 219ff, and Markl 1964:563).

entation to light and to gravity, which occurs under natural conditions and is biologically meaningful. Bumblebees, preparing to fly out into the open from their subterranean nests, are positively phototactic and seek the light. Jacobs-Jessen (1959) displaced them, on their way into the open, onto a vertical surface; there they proved to be negatively geotactic and sought to go upward—hence in the direction that would lead them from their dark nest in the ground to the outdoors. On the flight home matters were precisely the reverse; negative phototaxis was combined with positive geotaxis. Honeybees behaved similarly. There were to be sure many aberrant responses, but Lindauer and Nedel (1959:347) showed that those bees that on a horizontal surface in the dark aim with precision toward a point source of light also maintain a well-oriented course upward on a vertical surface after the light is turned off. A similar explanation had already been suggested by Birukow (1956a).

Thus there is to be recognized a way that the bees might have hit upon for expressing in their "language" the direction toward the sun by the direction upward. Perhaps it might be thought that, if such a connection existed from the beginning, the dwarf honeybees too should have accomplished the transposition. Yet there is a great difference whether a bee in a positively phototactic mood runs upward in darkness or whether later, when dancing in the dark hive or just as well in the sunlight on the freely suspended honeycomb, she reproduces from memory, in such a way that her comrades then find the goal, the visual angle she maintained during a foraging flight. When contrasted with the variable and never so completely exact transposition of other insects, this remains an accomplishment the level of which the dwarf honeybees have not attained.

The bee's life is so short that one never would expect her to have to learn the art of dancing. That this capacity is innate was demonstrated convincingly by Lindauer (1952:331), who put together a small colony of newly emerged bees and took care that no older bees from elsewhere could join it. After only a week a few of these workers, reared in isolation, came home with food they had collected and danced normally, vivaciously, and with success. The reception and understanding of the information, however, need some practice and maturation of the instinctive behavior. Initially young bees are somewhat clumsy as dance followers. Especially at the moment when at the end of the tail-wagging run the dancer hastens in a swift semicircle back to the starting point, they lose contact easily and let themselves be shoved aside by others in the retinue, whereas after a little experience they share unhesitatingly in the turns (Lindauer 1952:327).

Summary: Phylogeny and Symbolism of the "Language of the Bees"

1. When deprived of a drop of sugar water, flies perform circling movements, similar to the bees' round dance. Some moths make lateral waggling motions of the body when they settle after a flight. With flies and moths these forms of movement have no social significance. But

comparable behavior among the ancestors of the bees may have become a starting point for their mutual communication.

2. Among social insects the most primitive form of transmitting information about a discovery of food is the communication of the excitement from the discoverer to her nestmates. Ants thus alerted are able to find the goal by following an odor trail laid down by the discoverer. Primitive species of stingless bees learn from the scent that clings to the excited discoverer the specific smell of the flowers that have been visited, and find them independently. Buzzing sounds, perceived by means of the sense for vibration, are significant signals in the alerting process. From such beginnings higher forms have evolved two types of mutual communication: some stingless bees accompany the newcomers in flight to the goal, and others improve this method further by previously marking out the flight path with scent. The honeybees have managed to inform their comrades in the hive so well about the location of the goal that the latter find it unescorted in free flight.

3. For all apines the round dance is a symbolic demand to go out searching on all sides around the home. The tail-wagging dance points out the direction to the goal with respect to the sun, and can be conceived as intention movements that stimulate the comrades to fly out in this direction; the duration of tail-wagging is a symbolic indication of the distance of the goal.

4. With honeybees only a single intermediate stage in the evolution of their "language" has been preserved to us: the dwarf honeybee (*Apis florea*) can indicate the direction only on a horizontal surface in view of the sun or of the blue sky. She cannot transpose the solar angle to the gravitational angle.

5. The phylogenetic step from the direct indication of direction to its transposition was facilitated by the fact that this capability is a primary possession of many insects, without having any biological significance for them. When, while they are running on a horizontal surface, they are maintaining a specific angle relative to a source of light and the surface is tilted to the vertical, with simultaneous darkening, then they retain approximately the same angle in reference to gravity as they held previously toward the light. Higher central-nervous centers direct their course at a definite angle to the source of stimulation, be this light or gravity. In the precise way in which transpositon is made this process displays many variations. The biologically important linking of positive phototaxis and negative geotaxis may have contributed to the unequivocal nature of the information symbolized in the bees' dancing.

# Part Two / The Orientation of Bees on the Way to the Goal

XII / Orientation on Long-Distance Flights

In their studies of the homing ability of birds Watson and Lashley (1915) distinguished between orientation from a distance, in which the goal cannot be perceived directly by any known sense organ, and orientation from nearby, in which there is an effect of stimuli that emanate from the goal itself. This same division was adopted by Tinbergen (1932) in his fine studies on the digger wasp *Philanthus triangulum* Fabr., the "bee wolf." Baerends (1941) did not consider this terminology very fruitful, because the wasp was oriented according to optical cues to the same degree on extensive journeys and in the vicinity of the nest; the signs merely took on smaller dimensions, little by little. At that time, of course, he did not suspect that birds and insects might use methods in their navigation similar to those that man has used for ages. Just as a sailor on a long journey relies on landmarks only where such things are visible and for the most part steers according to the stars or the compass, so these animals have a compass at their disposal during extended flights. Only when they are within view of the goal does the compass become for them as insignificant as it is for the ship's captain entering a harbor. Since there are two different methods that occupy equally prominent positions, separate discussions of orientation from a distance and from near at hand seem thoroughly justified.

A. LANDMARKS

Even on flights over a rather extended range, visually conspicuous markers are no doubt significant in orientation. Especially during their first excursions I frequently saw bees on their way to the goal deviate to the right or left of the air line, because they headed to one side toward conspicuous points, like a thicket or a house, as an intermediate goal. E. Wolf (1926) publishes sketches of flight routes that were directed toward a ridgepole as an intermediate goal. He also noted (1927) that bees that were displaced from their feeding station to other spots in a level waste, devoid of bushes and trees, took a long time to find their way home, whereas they were able to do so rapidly if the hive was set up in the richly broken landscape of the Botanical Garden.

When in establishing an artificial feeding station one journeys with a group of bees to a considerable distance from the hive, one soon learns that it is necessary to pass visually conspicuous points (hedges, buildings, road crossings, and the like) as quickly as possible, because otherwise the entire crowd becomes so accustomed to flying to these markers that they can be conducted farther only with an appreciable waste of time.

Near the goal small objects often take on decisive importance. In setting up in Brunnwinkl a feeding station intended to stand 200 m west of the hive we had not reached the end of the journey on the first day; at its end the feeding stand remained for some half an hour 1 m to the side of a sign for protection of the meadow. We noticed that the bees were first heading toward the sign and then flying down over it to the stand. When on the next morning the feeding place was shifted some 10 m (Fig. 281), for a time thereafter they still aimed for the sign and flew about it before they approached the stand from there.

Yet another situation, demonstrating a visual intermediate goal, is today as clear before my eyes as it was 14 years ago. In an experiment on the Schafberg (1951) we wanted to learn whether the "language of the bees" contains an expression for elevation. From 15 to 19 August the observation hive stood at the base of a cliff (Fig. 145a). When the experiment was over we brought the colony back to the valley. After the bees had been flying for a week at Brunnwinkl, they were again taken to the Schafberg, but this time they were set up on top of the cliff. When the hive entrance was opened, I was below the cliff preparing at the former site of the hive the feeding station that was intended to go here now. Almost at once there came some eight or ten

FIGURE 281. The signboard in the picture served the bees as a final intermediate goal in flying to the feeding stand (to the left of the observer).

bees that apparently in their rush to fly off had not paid good attention and, without having become oriented to the new position of the hive, were now looking for it in the place below the cliff with which they were acquainted from previous experience. But now there was nothing there but the inconspicuous board, supported by two laths, on which the hive had stood. Not a flight to this exact spot was to be seen. Rather the bees hovered obstinately about 2 m to one side, over a prominent rock some 4 m high, in fact over the same side of it that faced the place where the hive had sat previously (Fig. 282). Since their final goal, the hive, had disappeared, they were left hanging hopelessly at the last intermediate goal.

Indirect evidence for the attention that is paid to landmarks on the flight to a feeding place can be drawn from Otto's experiments on the relative significance of outward and return flights in the indication of direction (see p. 171). He showed that in this respect a flight path that is well marked visually is given greater weight than a poorly marked stretch.

These examples no doubt suffice to show that visually conspicuous objects play an important part in the maintaining of a course by bees.

Baerends (1941) in his studies on the orientation of *Ammophila* came to the same conclusion. On the heath where these digger wasps hunted their prey and then dragged it with assurance into their inconspicuous nests in the ground, pine trees that projected above the low vegetation were prominent landmarks for them. Baerends produced elegant proof of this by planting little pines in transportable containers; when these were moved about he was able to change the route of the wasps at will.

The importance of landmarks relative to that of the celestial compass will be discussed later (pp. 339ff). But one point should be emphasized here: bees use landmarks for their *own* orientation. Landmarks are of no account in the orientation of the colony, in the transmission of information to the hivemates, because the "language" of bees has no "words" for them.

FIGURE 282. Location of the observation hive at the foot of the cliff on the Schafberg (Fig. 145) 10 days before the same colony was set up on top of the cliff. Bees that had gone astray and were hunting for the hive down below did not hover about its former location, but instead flew about over the nearer surface of the conspicuous rock 2 m distant.

B. THE SUN AS A COMPASS

Insects are able to use the sun as a compass to ensure a straight-line course (pp. 132ff). On the basis of this "light-compass movement" bees have evolved the capacity of making known to their hivemates the location of a goal with respect to the sun's position (pp. 129ff). In discussing the indication of direction we at first left out of account one difficulty: since the sun's position in the sky changes in the course of the day, either the sun can be useful as a compass only for short intervals or in determining direction its diurnal course and the time of day must be taken into consideration. Bees know the time of day, but it struck me as fantastic to presuppose that the shifting position of the sun was known and allowed for. Consequently I kept deferring an envisaged "displacement experiment" that was designed to throw light on the matter. When at last we carried it out, there opened before us a view of unsuspected new territory.

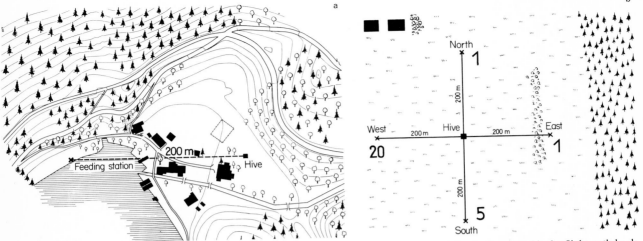

FIGURE 283. Displacement experiment. (*a*) Training in Brunnwinkl; the flight path leads between houses and trees and across a bay to the feeding station 200 m west of the hive (see Fig. 285*a*). (*b*) The beehive was moved to an environment unfamiliar to the bees; the numbers beside the feeding stands, 200 m to the W, N, E, and S, respectively, show how many visits there were from bees trained in Brunnwinkl to the westward direction (see Fig. 285*b*).

1. DISPLACEMENT EXPERIMENTS

A group of marked bees from an observation hive in Brunnwinkl had been fed for several days, with scent of anise added, 200 m west of the hive (Fig. 283*a*). On the morning of 24 September 1949 before flight began we took the colony into a region unknown to the bees and established it about 5 km distant from Brunnwinkl (Fig. 284), in the middle of a broad, level field. Four feeding stands of the same sort used at the accustomed feeding station were set up to the W, E, N, and S, each 200 m from the hive (Fig. 283*b*), with a sugar-water dish and 2 drops of oil of anise on the paper beneath. At 8:25 A.M. the hive was opened. Orientation flights, which began at once, indicated that the bees were taking note of the changed location. At each feeding stand there was an observer. His task was to catch and kill all bees that came to the sugar water, so that there could be no chance of communication with the hivemates. Any bee that found a stand would have come there spontaneously.

I had taken in charge the western station, the direction of which was that of the training place. Twenty minutes after the hive entrance was opened the first bee of our marked group appeared here. In astonishment I saw others succeed her, each coming individually. After an hour there had been 10, after another 18, and during the third (final) hour of observation there appeared 2 additional stragglers. Of the 29 marked bees that on the preceding afternoon had visited the feeding station to the west in Brunnwinkl, 27 of them had come to a dish in this unfamiliar neighborhood, 5 of them to the south, 1 each to east and north, but 20 to the west. The sun must have guided them. Yet it now stood to the southeast of the hive, whereas on the day before it

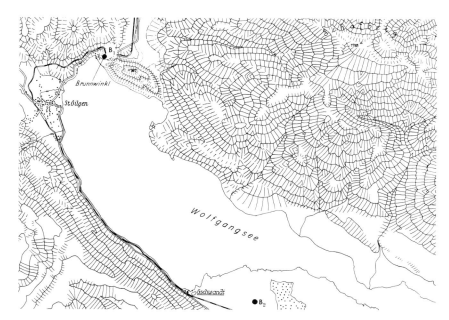

FIGURE 284. Map of the displacement experiment at the Wolfgangsee: B_1, location of the observation hive in Brunnwinkl; B_2, location near Gschwandt after the hive was shifted, 5 km from B_1. To the northeast is the Schafberg.

had been approaching the western horizon during the last foraging flights (v. Frisch 1950).

That after the hive had been displaced the foragers continued to maintain the accustomed westerly route, although it led now across a level field instead of between the tops of trees and along wooded slopes that bordered the flight path in Brunnwinkl (Fig. 285) gives evidence of a surprising degree of reliance on the celestial compass. One might ask how the bees could be so obstinate as, when in a different neighborhood, to seek their feeding station in the earlier direction. But one has no right to expect that they should grasp correctly the consequences of so unbiological a type of interference as is represented by displacement of the home hive. Judgments of that sort do not conform with the mental capacity of insects.

The slight preference for the southern station over those to north and east probably should not be regarded as fortuitous. First, the place was favored by a south wind, which bore the scent toward the searching bees. Additionally there is the second factor that young bees must first learn the diurnal course of the sun (see pp. 364f). Initially they are not capable of computing the sun's course. In a displacement experiment they then fly to the actual angle, that is, they maintain the same angle relative to the sun as prevailed during their last foraging flights. If our group contained a few of these still young bees, flights toward the south are comprehensible. For during their last flights in Brunnwinkl the feeding station, located to the west, was about 40° to the right of the sun's position. On the day of the experiment all flights to

FIGURE 285. Displacement experiment at the Wolfgangsee. (a) View to the east from the feeding stand F, located to the west. The flight path goes between houses and trees. The large linden tree in the middle of the picture is halfway along the route; (b) after the hive was shifted: view in the same direction, from the western feeding stand F toward the hive, which stood in the open field behind the two figures visible at the right of the picture against the dark woods.

the southern station took place between 9:14 and 10:13 A.M., with the sun in the southeast. Bees that held constantly to an angle about 40° to the right of the sun arrived at the southern station. In five additional displacement experiments I noticed a small secondary maximum. It was always in the direction to be expected if the bees had oriented to the actual angle.[1]

In our first experiment the hive entrance was directed eastward, both in Brunnwinkl and in the new location. The criticism was possible that the bees had been oriented by the position of the hive and had gone searching for the feeding station to the west, in a direction opposite to that of their taking off. On this account, in two repetitions and likewise in subsequent displacement experiments, we set up the hive in the new location in such a way that the direction of takeoff was changed by 90°. That had no effect on the result. By far the majority of the visits were recorded at the feeding station in the accustomed compass direction.

There were two conceivable explanations for the performance of the bees. Either the foragers had noted for each time of day what solar angle they had maintained during their flight to the feeding station, or they were so thoroughly familiar with the sun's diurnal course that even without a previous foraging flight they could at any time strike out in the compass direction in which they had found food at some other time of day. In order to decide this question, newcomers were trained on the afternoon of 1 October 1951 to the western feeding station in Brunnwinkl and were displaced to a strange location on the following morning (as in Fig. 283). The result was positive, but on account of autumnal cold and violent wind was numerically unsatisfying. Most of the bees stayed at home. We had to be patient.

On 29 June 1952 we took an observation hive from Munich into a region 17 km to the east near the Grub Estate, and this time set it up in an area unknown to the bees even before training them to a feeding direction. At 11:30 A.M. the hive entrance was opened and a feeding station set up 180 m from the hive in a direction 30° N of W. The site was reached at 3:15 P.M. Marked bees were fed there, with oil of anise added, until 8:00 P.M. (Fig. 286a; arrangements for the scent as in Fig. 83). Then we closed the hive entrance. When the colony flew out again on the next morning, it was once more in a strange environment, of a yet different sort. The site was on the Dachau Moor, 23 km by air line from the previous day's location. The position of the four feeding stands set out there, with anise-scented odor plates (as in Fig. 84), may be seen from Fig. 286b. Figures 287a and b afford a comparison of the two landscapes. Of 30 bees that had been marked on the preceding afternoon, 15 appeared during the observation period (7:15–11:15 A.M.)

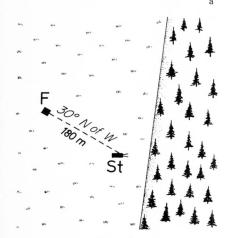
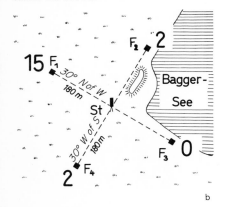

FIGURE 286. Displacement experiment on 29–30 June 1952: (a) first day, at the Grub Estate; St, observation hive; F, feeding station; (b) second day, on the Dachau Moor; the numbers of visits to each of four feeding stands are shown.

[1] With reference to the technique of these experiments it should be noted that success on the day of experiment depends heavily on there having been food of the best quality given on the preceding day. It would have been tempting to supply dilute sugar solutions so that not too many newcomers would come flying in during the course of training and have to be killed. But with such a procedure the incentive to hunt spontaneously for food the next day suffers; and only a few of the marked bees appear at the feeding stands.

FIGURE 287. Displacement experiment on 29–30 June 1952: (a) the observation hive at the Grub Estate (flight funnel facing east); view toward the NE; training on the afternoon of 29 June to a direction 30° N of W; (b) the observation hive (left foreground) on 30 June after it had been moved to the Dachau Moor (flight funnel facing south); view toward the north.

in the direction of training and were captured. Altogether 4 came to the three other stands (Fig. 286b).

Thus it was made clear that bees can at any chosen time find a given compass direction according to the sun's position, by taking account of the time of day and the sun's diurnal course. The sun is a perfect compass for them.

With somewhat different methods Gustav Kramer (1950, 1952, 1953) obtained corresponding results with birds. Our experiments were conducted simultaneously, but without our knowing of one another. Afterward each of us found pleasure in the unexpected discoveries of the other.

Several repetitions of such displacements substantiated the finding. Yet it must be mentioned that among a total of 12 experiments 2 turned out negatively: at the feeding stand in strange territory there appeared only two and four bees, respectively, from the group trained on the previous day, and the direction of training was not preferred above the others. Only on these two occasions was the day of experiment interfered with by a closed cloud cover, whereas the other displacement experiments were favored with blue sky and sunshine. But it cannot be maintained with certainty that there was a connection with the heavy cloudiness, since on both those days the weather was unfavorable in other respects too. Considering the importance of blue sky in orientation (pp. 380ff) and the capacity of bees for perceiving the sun behind clouds, it would be interesting to learn whether a displacement experiment may also turn out positively when there is a complete cloud cover.

A variant of the experiment was designed to show whether the bees are able to learn to disregard the diurnal course of the sun and to maintain the same direction relative to the sun at every hour: experiments with a traveling feeding station and their technical performance have already been discussed in the section on "misdirection" (pp. 207ff). In the present experiment (actually done somewhat earlier) the question investigated is different: will foragers who have consistently found their feeding stand in the direction of the sun from noon until evening also search in the direction of the sun the next morning in a strange landscape? The experiment (done in cooperation with Lindauer) took the following course.

On 22 July 1956 we set up an observation hive in a broad, level sheep pasture within the eastern limits of the city of Munich, near the Botanic Garden. At 10:00 A.M. the hive entrance was opened and a feeding station was established in the direction of the sun. When the planned distance (180 m) from the hive had been reached, between 12:18 and 2:24 P.M. the old foragers were weeded out and simultaneously 40 newcomers were marked. Feeding was continued until 7:30 P.M. All the foragers had consistently found their goal in the direction toward the sun (Fig. 288, open circles).

In the morning of 23 July we took the colony to another level sheep pasture northeast of Munich, beside the highway to Nürnberg. Feeding stands were set out in four directions (as in Fig. 286b). From 6:53 A.M. on, when the hive was opened, a fifth stand followed the course of the

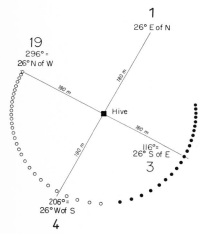

FIGURE 288. An attempt to train bees, by means of a traveling feeding station, to the direction toward the sun. Open circles, positions of training on the afternoon of 22 July 1956; the feeding station was moved every 5 min so that it remained in the direction toward the sun; each point in the figure represents three positions. Solid circles, positions of the feeding station during the experiment on 23 July from 6:53 to 11:53 A.M., after the hive had been shifted to another area; the station went unnoticed. The numbers show the visits to the four stationary feeding stands.

sun. All five feeding stands looked exactly like the training stand from the day before, and like it were supplied with orange-blossom perfume. The experiment ran till 11:53 A.M. on (the traveling stand is represented by the solid circles in Fig. 288). Of the 40 foragers from the previous day 27 came during this time: not one of them to the station traveling in the direction of the sun, but 19 to the stand in the WNW, where the last feeding of the evening had been given; one, three, and four visits were recorded at the other stands, respectively.

Thus the bees did not grasp the fact that the feeding station was being moved. In the life of a bee the sun journeys, but the goal stays put. The bees are thoroughly impressed with these facts, and hence in the experiments on misdirection, with a traveling feeding station, the dancers pointed to the left of the traveling stand, to the place where actually it should have been, lagging behind the advancing sun.

We planned to do a repetition of the experiment, in which training to the direction of the sun should be carried out not just in the afternoon but from morning to night. A change in the weather thwarted this intention, and later we did not get back to it. Probably the result would have remained the same. But perhaps it will be different when the behavior of young bees, reared without experience with the sun, is tested (see p. 209).

2. COMPETITION BETWEEN THE CELESTIAL COMPASS AND LANDMARKS

The objective of additional displacement experiments was to clarify the relative significance of landmarks and the sun's position as concerns finding the correct direction again (v. Frisch and Lindauer 1954). Hitherto, displaced bees had been deprived of any opportunity of making use of known landmarks, because there were none such in the strange territory. But what will happen when, after the bees are displaced, they are offered landmarks of the accustomed appearance but in a different direction, and in this way earthly and celestial modes of orientation are brought into conflict?

In the neighborhood of Munich, near both Ödenstockach (Putzbrunn District) and Möschenfeld (Zorneding District), there is a broad meadowland that is enclosed on three sides by a tall stand of woods, but in one the open side is toward the east, in the other toward the south. Early in the afternoon of 6 July 1953 we set up a bee colony near Ödenstockach 5 m from the wooded margin, which runs from north to south, and fed a group of marked bees between 1:15 and 6:10 P.M. at a stand 180 m from the hive in a direction 2° east of south (Fig. 289a). Hence the flight course ran parallel to the edge of the woods, 5 m away. In the morning of 7 July we established this same colony on the Möschenfeld Estate, 5 m away from the edge of the woods, which here ran from east to west (Fig. 289b). Three feeding stands, scented with oil of anise, which had served as the training odor, were in the locations shown in Fig. 289b. Comparison of Figs. 290a and b shows the similarity of the two wooded margins. Predominantly the bees followed the edge of the woods. Of 35 foragers marked on the preceding day, 22 appeared at the

340 / The Orientation of Bees

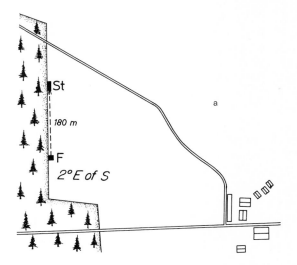

FIGURE 289. Displacement experiment, 6–7 July 1953. (a) Training on afternoon of 6 July, Ödenstockach; F, feeding stand south of the hive St; distance of the line of flight from the woods, 5 m. (b) The experiment on the morning of 7 July, Möschenfeld; F_1–F_3, feeding stands; the numbers show the number of visits. Most of the bees, trained to fly south, held to the forest margin and came to the western station.

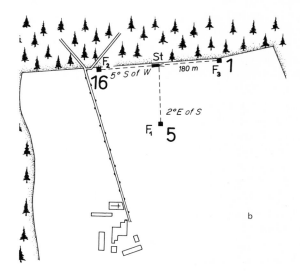

stands, 16 of them at the western station, 5 at the southern station, 1 at the eastern station. As a guideline, the margin of the woods surpassed the celestial compass.

In two further experiments with other colonies we tested at what distance the woods lose their directing effect. Whereas the arrangement remained unchanged in every other way, on the first repetition we

FIGURE 290. (*a*) Ödenstockach, 6 July 1953; view from the observation hive *St* toward the southern feeding stand *F* (see Fig. 289*a*). (*b*) Möschenfeld, 7 July; view from the hive toward the western feeding stand F_2 (see Fig. 289*b*); a measuring stand *M*, with compass, is stuck into the ground beside the hive *St*.

increased the separation of the bee colony from the margin of the woods to 60 m. Once again the majority of the bees, trained to go south, came to the western station (Figs. 291*a*, *b*, and Fig. 292). But when the distance from the edge of the woods was increased to 210 m, the trees were appreciably less important than the celestial compass as a guideline (Fig. 293*a*, *b*). From this interval of 210 m the woods subtended an angle of only 4° (from the ground to the treetops, Fig. 294); from 60 m away the angle was 13–14°. During the period of training in Ödenstockach the values had been similar (ca. 3° and 10–11°, respectively).

Just as by increasing their distance from the flight path we had reduced the apparent height of the determinative landmarks, so in other experiments we allowed them to contract in length, and in the extreme case replaced the woods by a single tree. On 5 August 1953 an observation hive was set up south of Dachau in a broad meadowland in which a single large willow was conspicuous (Figs. 295*a* and 296*a*). During the afternoon 40 marked bees were foraging at a feeding station 180 m to the south, where their flight path went close by the willow. After the hive had been displaced it stood in another broad meadow-

342 / The Orientation of Bees

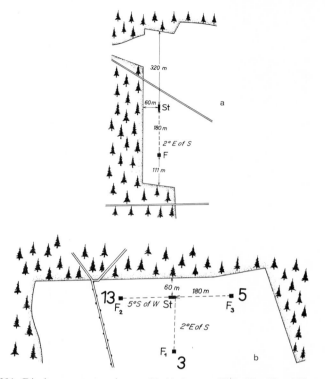

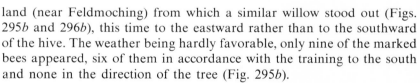

FIGURE 291. Displacement experiment, 10–11 August 1953; like Fig. 289, except that the line of flight was 60 m from the edge of the woods: (*a*) training, Ödenstockach; (*b*) experiment, Möschenfeld. Most of the bees again flew along the forest margin in the wrong compass direction; see the numbers beside F_1–F_3.

FIGURE 292. (*a*) Ödenstockach, 10 August 1953; view from the observation hive westward to the forest margin 60 m distant (see Fig. 291*a*). (*b*) Möschenfeld, 11 August 1953; view from the observation hive northward to the forest margin 60 m distant (see Fig. 291*b*).

land (near Feldmoching) from which a similar willow stood out (Figs. 295*b* and 296*b*), this time to the eastward rather than to the southward of the hive. The weather being hardly favorable, only nine of the marked bees appeared, six of them in accordance with the training to the south and none in the direction of the tree (Fig. 295*b*).

A somewhat larger group of trees also was without decisive influence. On the afternoon of 15 September 1953, the flight path of a colony near Wagenried (between Ingolstadt and Dachau) led past an alder thicket that extended to a depth of 40 m perpendicular to the direction of flight and about 25 m along it (Figs. 297*a* and 298*a*). On the following morning the colony was set out near Stetten (between Dachau and Augsburg), again in an open field. A similar stand of alders, mixed with a few ash trees, was visible from the hive, to the SW (Figs. 297*b* and 298*b*), whereas on the previous day the clump of trees had been in the SE. Of the 40 marked bees, 16 reappeared, 9 of them at the southeast station (the compass direction to which they had been trained), 3 at the southwest station (the "path of deception"), and 2 at each of the remaining test stands.

Despite their visual conspicuousness, the individual trees or mere groups of trees were not able to compete with the sun, but the woods, as a continuous guideline, had indeed been decisive.

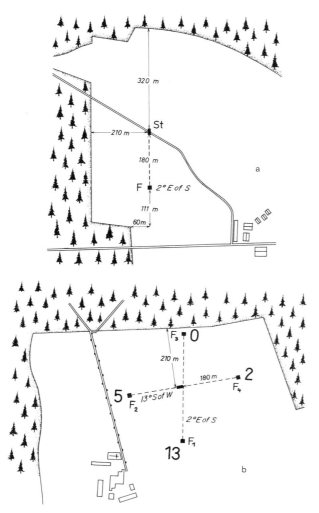

FIGURE 293. Displacement experiment, 15–16 August 1953; like Fig. 289, but with the hive and line of flight 210 m from the forest margin: (*a*) training, Ödenstockach; (*b*) experiment, Möschenfeld. Most of the bees now followed the solar compass and flew to the southern station.

The great importance of continuous landmarks along the flight path was confirmed in two further experiments. Bees that had been established on the north shore of the Speichersee near Munich were flying along the margin to a feeding station 187 m distant 9° S of W (Figs. 299*a* and 300*a*). On the next day they were brought to the shore of a dredged lake near Dachau that ran from north to south (Figs. 299*b* and 300*b*). As Fig. 299*b* shows, after the displacement 14 bees followed the shoreline, and only 1 held to the compass direction of her training. In another experiment the flight path went along a straight road, 4 m wide (from Salmdorf to Ottendichl, east of Munich), in a westward direction. There were open fields on all sides (Figs. 301*a* and 302*a*). On the following morning deception was provided by an equally wide

FIGURE 294. Möschenfeld, 16 August 1953; view from the observation hive northward to the forest margin 210 m distant (see Fig. 293*b*).

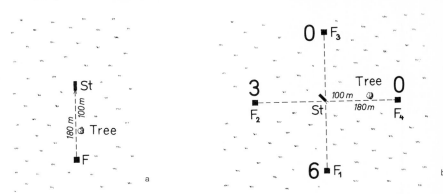

FIGURE 295. Displacement experiment, 5–7 August 1953 (on 6 August the colony was in the laboratory cellar, because of rain): (a) training at Dachau Moor on the afternoon of 5 August to the feeding station F due south of the hive St; (b) experiment at Feldmoching on 7 August 1953; the number of visits is shown beside the feeding stands F_1–F_4. The tree had no influence as a landmark (see Fig. 296).

FIGURE 296. Displacement experiment of Fig. 295: (a) Dachau Moor, 5 August 1953; view from the feeding station F northward to the hive St (see Fig. 295a); (b) at Feldmoching on 7 August 1953; view from the feeding stand F_4 westward to the observation hive St (see Fig. 295b).

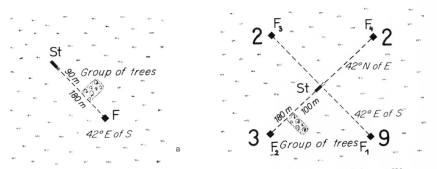

FIGURE 297. Displacement experiment, 15–16 September 1953: (a) training at Wagenried on the afternoon of 15 September; St, observation hive; F, feeding station; the flight path leads past a group of trees (see Fig. 298a); (b) experiment at Stetten on the morning of 16 September 1953. The group of trees (see Fig. 298b) was unable to lead the bees astray.

Orientation on Long-Distance Flights / 345

FIGURE 298. Displacement experiment of Fig. 297: (*a*) Wagenried, 15 September 1953; view from the observation hive *St* southeastward to the feeding station *F* (see Fig. 297*a*); (*b*) At Stetten on 16 September 1953; view from the observation hive southwestward toward the feeding stand F_2 (see Fig. 297*b*).

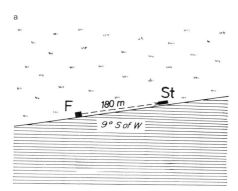
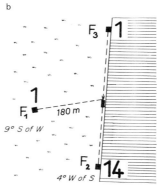

FIGURE 299. Displacement experiment, 20–21 July 1953: (*a*) training on the afternoon of 20 July along the shore of the Speichersee near Munich; *St*, hive; *F*, feeding station; (*b*) experiment in the forenoon of 21 July on the Baggersee near Dachau; F_1–F_3, feeding stands. Fourteen bees followed the shoreline, only one the compass direction to which they had been trained.

FIGURE 300. Displacement experiment in Fig. 299: (*a*) Speichersee, 20 July 1953; view from the observation hive *St* westward to the feeding station *F* (see Fig. 299*a*); (*b*) 21 July 1953 at the Baggersee; view from the hive *St* to the southern feeding stand F_2 (see Fig. 299*b*).

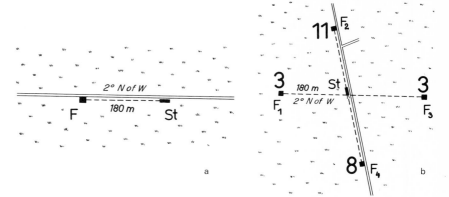

FIGURE 301. Displacement experiment, 8–9 August 1953: (*a*) training at Salmdorf on the afternoon of 8 August westward along the country road; (*b*) experiment at Grub on 9 August on a road with a different course; F_1–F_4, feeding stands, with numbers of visits adjacent. Most of the bees used the highway as a guideline, without a clear decision between N and S.

road (Grub-Parsdorf), running approximately on a north-south line at a site 4.5 km distant from the earlier location of the colony (Figs. 301*b* and 302*b*). Of 40 marked bees from the previous day, 25 came to the feeding stands: only 3 to the westward direction of training, 11 to the north along the line of deception, and—to our surprise—8 to the south. These last bees, too, had flown along the road, at right angles to the compass direction they had learned, but a clear point of reference for deciding between north and south was lacking. In the experiment on the lake shore the water on one side probably had served as a guide for the bees (see Figs. 299 and 300), but now open fields were on both sides and when the bees went searching around above the road the situation was ambiguous to them (see v. Frisch and Lindauer 1954).

On the whole then continuous guidelines from the hive to the goal, whether the edge of woods, a road, or a shoreline, proved to be such significant landmarks that they surpassed the celestial compass in impor-

FIGURE 302. Displacement experiment in Fig. 301: (*a*) Salmdorf, 8 August; view from the hive *St* to the westward feeding station *F* (see Fig. 301*a*); (*b*) Grub, 9 August; view from the hive *St* to the northern feeding stand F_2 (see Fig. 301*b*).

tance. What matters here is not the finer structure of the terrain from the visual point of view, but the broad general picture. For the wood near Möschendorf was not exactly like the wood near Ödenstockach (see Figs. 292a, b), differing from it in the distribution of taller and shorter trees and in many other details. The two lake shores also were by no means identical visually, and the Speichersee was much larger than the dredged lake.

Though such guidelines were dominant, yet the celestial compass did not go unheeded. In several of our experiments an observer at the hive recorded the direction of takeoff for all marked bees, insofar as he could distinguish them. Initially only about a fifth of them followed the path of deception, two-fifths held to the celestial compass, and two-fifths flew off in other directions. In this there is expressed clearly the conflict between landmarks and the celestial compass; for most of the bees it was only during further search for the feeding stands that the deceptive course set by the landmarks became decisive.

Where only a single tree or group of trees was in the middle of the course the behavior was different. After displacement, in the experiments with the willow tree and the alders, the oriented takeoffs of 24 marked bees were observed: 19 of them struck out in the compass direction to which they had been trained, and only 2 on the path of deception. The landmarks halfway along the course had no attraction.

We had intended to continue with such experiments. We were attracted by the prospect of choosing other landmarks in such a way as finally to bring them into equilibrium with the celestial compass, and in this way gaining insight into their relative importance. But other problems intruded and thus to this day what we had hoped to develop into a detailed picture remains a mere sketch.

3. The Contribution of the Time Sense to Orientation, and Knowledge of the Sun's Course

In the section on "the bees' clock and the flowers' clock" (p. 253), we spoke of the ability of bees to take note of feeding times favorable for them and then to make their appearance at the proper hours when, according to experience, the sources of nourishment were flowing profitably. Even in the dark hive they know the time. For when the hour for feeding approaches they move out from their resting places to the dance floor (p. 255 and Fig. 234). Thus they do not need to see the sun in order to know the time of day. Just the opposite is the case: by means of their sense of time and their familiarity with the course of the sun, they know at every hour what its compass direction should be, and thus they are capable of using it as a compass. The displacement experiments showed us this. There was not long to wait for confirmation of this astounding finding. Kramer's similar and simultaneous discovery in birds, and many subsequent observations, disclosed, in fact, that such a capability is very widespread in the animal kingdom (see pp. 441ff). Additional data obtained with bees confirmed the result and led to a better comprehension of their abilities.

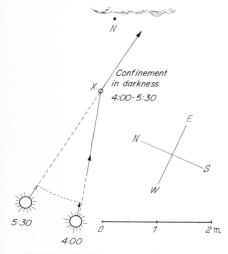

FIGURE 303. Solar orientation of ants: *N*, ant nest (*Lasius*); solid line, route of an ant returning home. After confinement in the dark for 1.5 hr at the indicated spot, she shifts her direction by 23.5°; meanwhile the sun's position had advanced 22.5°. After Brun 1914.

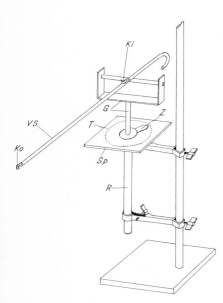

FIGURE 304. Direction finder, for measuring the angle of takeoff; description in text. After Meder 1958.

Orientation according to the sun after imprisonment in the dark. As long ago as 1914 Brun came upon the good notion of testing the solar orientation of ants by confining them for a few hours. Set free once again, some of them (genus *Lasius*) ran in an incorrect direction, deviating by the angle by which the sun had changed position meanwhile (Fig. 303, and see p. 448). In a level area without conspicuous landmarks, bees that had been kept in darkness at the feeding station for an hour sought their hive in the wrong direction, corresponding to the shift in the sun's position, according to E. Wolf (1927). But Wolf did not actually see this; rather he deduced it indirectly, from the length of time required for the flights. Moreover, since too little is to be learned from his publication as to how the work was performed, E. Meder (1958) rechecked Wolf's experiments thoroughly and extended them. Confirmation of Wolf's statements would have meant that the movement of the sun is not allowed for by bees imprisoned in the dark. But with a precise repetition of Wolf's experiments, Meder reached a different conclusion.[2] From the flight times recorded by Meder it was to be deduced that even when the foragers had been held captive at the feeding station for a rather long time they flew home to their hive without any detours once they had been set free. Meder became convinced of this by means of direct observations, by watching with a direction finder the course taken by the prisoners and then comparing it with the flight course of control bees.

The apparatus for measuring the direction of takeoff (Fig. 304) consisted of a vertical tube *R* supporting a horizontal wooden plate *Sp* on which a circle *T* was inscribed. Turning in the tube *R* was the forked axle *G*, carrying the aiming rod *VS* (length 110 cm), which was equipped with sights (*Ki* and *Ko*) for precise determination of the direction of flight. This rod could be rotated both horizontally and vertically. The pointer *Z* made it possible to read the azimuth angle on the graduated circle on *Sp*. The observer lay on the ground and was able to follow for long distances the flight of the bees as dark points against the brighter horizon.

With respect to technique the nature of the box used for confinement is also important. It had to be lighttight but also sufficiently ventilated, and had to contain a feeding dish so that after their period of custody the bees would fly off fully fed and undisturbed (see Meder 1958:616, Fig. 5).

As an example, Fig. 305 shows an experimental result of a kind that proved to be general. The takeoff of unmolested control bees (*above*) and of experimental bees that had spent an hour in darkness (*below*) was always measured at the same time, so that the wind and other external factors would exert the same effect. The direction from the feeding station to the home hive, 188 m away, is designated as 0°. Each dot indicates a measurement; deviations to right or left of the line toward the hive were read from the scale. In contrast with the control bees, it was customary for the bees that had been imprisoned to make a

---

[2] For a detailed critique of Wolf's work the reader is referred to the publication by Meder, who seems to have considered all imaginable possibilities.

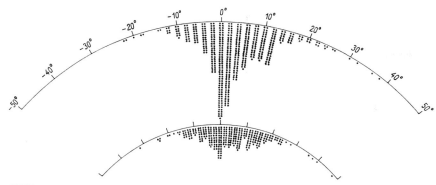

FIGURE 305. Direction of takeoff from the feeding station for the hive: (*above*) control bees; (*below*) bees after confinement in the dark for 1 hr at the feeding place. Each point shows one flight; 0° is the direction to the hive. After Meder 1958.

few orientation passes over the feeding station. But then, like the former, they would strike out directly for the hive.[3] They had included in their computations the duration of their confinement and the corresponding course of the sun.

The criticism that during the return flight they had oriented by landmarks is refuted in two ways: in the first place, landmarks of such an extent that they could have offered competition to the celestial compass (see pp. 340f) simply were not present. In the second place, bees that were displaced to one side of the feeding station actually did orient according to the sun and thus flew off in a direction that paralleled the accustomed flight course. They got home via detours and on the average took 13 times the normal amount of time in doing so.

Even after 2 hr of darkness, the bees headed home directly when released, although meanwhile the position of the sun had shifted by more than 30°. On the basis of experiments with a shorter time of confinement Meder suspects that bees that were shut in even for only 7–11 min (corresponding to a change of 3–4° in the sun's azimuth) had made a temporally accurate estimate of the sun's movement. The number of observations was too small to be statistically significant. But further understanding was provided by data of quite a different kind.

Spontaneous dances in accordance with the sun's position, without any view of the sun. In the year 1949 three independent and geographically widely separated observers noticed that in the beehive there occurred spontaneous dances, even without any immediately preceding foraging flight. At 3 P.M. on 18 July 1949 in St. Wolfgang I went to fetch an observation hive that had sat in a half-darkened hallway since morning. When I opened the protective cover a bee began to dance uninterruptedly on the comb. The waggling runs were pointed upward, the rhythm of dancing corresponded to a goal 1500 m distant (note in the protocol of 18 July). On 22 July 1949 Lindauer (1954a) noted with a free-flying colony in Graz that between 12:36 and 2:00 P.M. a bee started to dance

[3] Statistical evaluation did not yield a conclusive difference between the directions of flight ($P = 0.54$).

over and over again, without any intervening flights, and in doing so shifted the direction of dancing 33° counterclockwise. During this interval the change in the sun's azimuth amounted to 34.5°. And on 2 October 1949 the Russian biologist Chalifman (1950), after removing the protective cover and illuminating the honeycomb, saw several bees perform short but definite dances.

From further observations by Lindauer (1954a, 1955) it emerged that scout bees that go out seeking a new home for a swarming colony (see p. 275)—frequently even before the act of swarming and hence directly from the hive—often for hours at a time make known to their hivemates, without further exploratory flights, the location of a nesting site they have discovered ("marathon dancers," or *Dauertänzerinnen*). Unlike foragers, they do not need to bring in additional burdens from time to time. Indeed, there is nothing to be fetched from the nesting site. Wittekindt (1955) observed that, after they had stopped flying, food foragers could be stimulated to dance anew by illuminating the comb. No doubt this sort of thing was at the basis of the observations that Chalifman and I made. According to Wittekindt, supplying sugar water inside the hive favors the release of belated dances. Then some bees will perform round dances, as is usual with feeding inside the hive; with others their recollection of the external source of nectar is awakened, and they dance in favor of it.

We are indebted in many respects to such "marathon dancers" for valuable information. On 30 August 1953 Haidl, who was occupied in our laboratory with experiments of another sort, chanced to notice in his observation hive a numbered bee that even after darkness had fallen went on dancing for a goal 7.5 km away. At 7:05 next morning the same bee resumed her dancing, even though the hive entrance remained closed and the hive was standing in diffuse light in an inside room, the window of which was covered with cardboard. Up until 10:46 A.M. Haidl determined with the protractor the direction of dancing in repeated dances. He was not aware of the sun's position. Subsequently the results of his observations were compared with the corresponding solar azimuthal values (Table 35). During the period of observation the azimuth of the sun had shifted 54.5° clockwise, the bee's direction of dancing 53.5° in the opposite sense. She was maintaining the proper compass direction without being able to learn from the sky the position of the sun (see v. Frisch 1956:368–369). Closer examination of the table shows that in her "blind" indication of direction she took account of the sun's progress whenever two dances succeeded one another by as much as 6 min (corresponding to an azimuthal difference of 1.5–2°) or by a greater interval (an exception at 9:34 A.M.). When the difference in time was 4 min (about 1° difference in azimuth) the direction of dancing was unchanged on one occasion, increased once, and decreased once. Meder (1958) had reached the conclusion that after being imprisoned in darkness for as little as 7–11 min bees apparently took the changed position of the sun into their calculations. According to Haidl's series of measurements they were able to read correctly from their "clock" the time of day down to an interval of 5 min. More observational data on this question would be desirable.

TABLE 35. Indication of direction by a "marathon dancer" on 30 August 1953, without any view of the sky (after Haidl).

Time of observation	Time since preceding dance (min)	Change in solar azimuth since preceding dance (deg)	Change in angle of dancing since preceding dance (deg)
7:05	—	—	—
8:09	64	+12.5	−14.5
8:13	4	+ 1	+ 2
8:17	4	+ 1	− 3
8:20	3	+ 0.5	− 0.5
8:31	11	+ 2.5	− 2.5
8:44	13	+ 3	− 7
8:57	13	+ 3	− 7
9:20	23	+ 5.5	− 10
9:34	14	+ 3.5	+ 4
9:46	12	+ 3.5	− 4
10:02	16	+ 4.5	− 3
10:06	4	+ 1	0
10:23	17	+ 5.5	− 2
10:29	6	+ 2	− 3
10:35	6	+ 2	− 2
10:46	11	+ 3.5	− 1

Haidl's "marathon dancer" pointed correctly toward the goal on the next day, without any more recent flight. But the bees' memory is capable of even greater accomplishments. Between 24 October and 1 November 1957 Bräuninger (reported by Lindauer 1960:373) had been feeding marked bees 385 m distant from the hive. Then it got cold and the hive was transferred to a warm room. At 10:15 A.M. on 8 December, 1.5 months after the last feeding, sugar water was supplied inside the hive. Two bees began to dance on the vertical comb and (without any view of the sky) signaled the compass direction to their last feeding place.

From such experiences it becomes comprehensible that free-flying bees have sometimes performed directed dances on the vertical comb when it would have been impossible for them to have seen the sun on their last flights. During an experiment on 8 September 1958 the sun disappeared at 3:30 P.M. behind storm clouds that were so heavy after 4 P.M. that certainly the bees could no longer see the sun (see p. 367). Also there was no longer any blue sky anywhere. Nevertheless dancing continued for 20 min and the direction to the feeding place was indicated correctly. New reports that many of his bees were able to indicate the direction even when the sun was in the zenith (see p. 162). In both instances the bees apparently kept on dancing in accordance with their "clock." One comes almost to marvel that they are not always able to do so. Perhaps under normal conditions they are more responsive to the sun and more readily distracted by its failing than when they are dozing in the hive secluded from the outside world.

Nocturnal dances. Lindauer (1954a) notes 20 "marathon dancers" that signaled fairly correctly the direction to the goal on the day after their last excursion. He was surprised again when at night (3:22 A.M.) a bee performed a definite tail-wagging dance and (with 12.5° error) pointed out the azimuth of the sun, which she could never have seen at that time of night. Later (1957a) he obtained more nocturnal dances[4] from foragers stimulated by diffuse illumination and which he fed sugar water in the hive. Uniformly they pointed out, though not so precisely as by day, the direction to the feeding place that had been frequented the previous afternoon (example: Table 36; an additional 16 comparable observations from August and September also are available). I shall return (p. 365) to the question how the bees know about the nocturnal position of the sun.

Astounding accomplishments provoke ever-increasing demands. Lindauer trained his group of bees to two feeding stations: food was offered to the south for an hour following sunrise, to the east for an hour before sunset. We know already that bees learn to seek out the proper place at the right time on one occasion, and then the other (p. 254). But which place would the nocturnal dancers indicate? Two dances in the evening (8:36 and 9:31 P.M.) pointed to the afternoon site, three dances in the early morning (3:54, 4:11, 4:18 A.M.) to the morning site; seven dances took place in the period around midnight (11:10 P.M.–1:14 A.M.); of these five were incorrect or disoriented, one signaled approximately the resultant, and one the morning site with a strong error (Table 37). Consequently the nocturnal indication of direction referred, with a 2-hr transitional period of uncertainty, to the temporally closer feeding place.

The internal clock. Without doubt the time sense of bees functions inside the dark hive too, and even at night. This argues that bees read

[4] Fewer than Wittekindt, however, for in order not to distort the indication of direction only diffuse illumination could be used. Thus Lindauer kept watch through many a night without being able to record a single dance.

TABLE 36. Indication of direction by nocturnal dancers; feeding station 102° E of N; last feeding time 1 hr before sunset (after Lindauer 1957a).

| Day (1956) | Time of dancing | Bee No. | Solar azimuth (deg) | Direction to right of sun | | Error (deg) |
				Feeding station (deg)	Angle of dancing (deg)	
12 June	9:06 P.M.	106	317.5	144.5	126	− 18.5
15 July	9:10	61	316.5	145.5	138	− 7.5
15 July	9:35	21	321.4	140.6	156	+ 15.4
17 July	9:16	202	317.2	144.8	148	+ 3.2
17 July	9:23	14	318.7	143.3	142	− 1.3
20 July	9:51	67	324.4	137.6	134	− 3.6
20 July	9:56	67	325.5	136.5	138	+ 1.5
29 July	1:24 A.M.	26	16.3	85.7	71	− 14.7
30 July	11:02 P.M.	51	340.2	121.8	97	− 24.8
30 July	11:20	56	344.6	117.4	86	− 31.4
31 July	10:29	67	332	130	101	− 29

Orientation on Long-Distance Flights / 353

TABLE 37. Indication of direction by nocturnal dancers that had been fed in the morning at a southern feeding station (10° W of S), in the afternoon at an eastern feeding station (12° S of E).[a]

Day	Time of dancing	Bee No.	Angle of dancing (deg)	Solar azimuth (deg)	Direction of dancing[b]	Station indicated	Error (deg)
22 August	8:36 P.M.	116	154 r	304		Afternoon	− 4
11 August	9:31 P.M.	105	156 r	317.5		Afternoon	+ 11.5
10 August	11:10 P.M.	160	14 l	342		Wrong	—
10 August	11:21 P.M.	131	14 r	344		Wrong	—
19 Sept.	12:40 A.M.	109	157 l	10		Morning with large error	+ 23
15 August	12:48 A.M.	122	134 r	8		Resultant	+ 40 − 48
19 Sept.	12:58 A.M.	150	—	16	Disoriented		
19 Sept.	1:02 A.M.	168	42 l	17		Wrong	—
24 Sept.	1:14 A.M.	111	—	21	Disoriented		
13 August	3:54 A.M.	141	132 r	53		Morning	− 5
14 August	4:11 A.M.	156	116 r	56.5		Morning	− 17.5
14 August	4:18 A.M.	171	121 r	57.5		Morning	− 11.5

[a]The first row may be explained as follows: at 8:36 P.M. bee No. 116 danced at an angle of 154° to the right. At this time the sun's azimuth was 304°. Thus the eastern (afternoon) feeding station (azimuth 102°) was 158° to the right of the sun, the southern (morning) feeding station (azimuth 190°) 114° to the left of the sun's position (dashed line in row 1 of the table). The direction of dancing (arrow) deviates by −4° from the expected direction of dancing for the eastern station.
[b]Arrow, actual; dashed line, expected; E, afternoon station; S, morning station.

the time of day not from external factors but from an "internal clock," in other words, that their sense of time is not exogenous but endogenous. The question whether this really is true has been the subject of discussion since the beginning, as a short résumé will indicate.

The basis of all pertinent experiments is training the bees to a definite time, such as was first done systematically by I. Beling (1929). A group of marked bees, which had, for example, received sugar water at a feeding stand only between 4 and 6 P.M., was observed thereafter for a whole day without food, and the number of each bee that appeared at the feeding station was recorded. Figure 306 gives an example from the protocols. Combined for half-hour intervals, each visit is noted in a square with the number of the bee. In this instance the visits were limited almost entirely to the accustomed feeding time. But

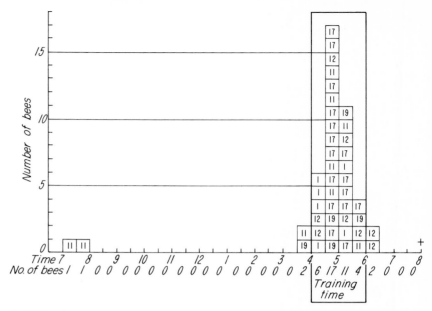

FIGURE 306. Experiment with bees that had been trained to a feeding time from 4 to 6 P.M. (enclosed area): on the day of the test (20 July 1927) five marked bees were foraging; each square (with the bee's number) denotes a visit to the empty food dish during the half hour in question. After Ingeborg Beling 1929.

for the most part the foragers actually come a little too early (Figs. 307a, b), as is comprehensible biologically. Better be early than late at the feeding place, in a world full of hungry creatures. The phenomenon should not be interpreted as indicating any imprecision in the time sense.

Beling had already discovered that training to a definite time is successful in a closed room with constant illumination, as well as training to a time during the night, and that the result is independent of periodical variations of temperature, humidity, and the electrical conductivity of the atmosphere. In a deep rocksalt mine cosmic radiation also was eliminated without interfering with the time sense of bees (Wahl 1932). The sense seemed to be governed endogenously. Just how it is regulated remained unexplained and still is unknown. It could not be founded on a "hunger rhythm" merely because bees that find nourishment at the feeding dish only at certain hours can meanwhile sate themselves in the hive at any time, while pollen collectors too can be trained to a time for collecting (Beling, Wahl). Seeking the foundations of the sense of time in rhythmic metabolic variations runs into the difficulty that drugs that increase or interfere with metabolism do not affect training to a time (see Werner 1953, Renner 1957, *et al.*). Only after having been cooled, in a way that never occurs normally in the life of a bee, did the trained bees arrive too late (Fig. 308). But this observation did not settle the question, for when the bees were stiff with cold not only was their metabolism in all probability depressed, but also their responsiveness to stimulation, whence various

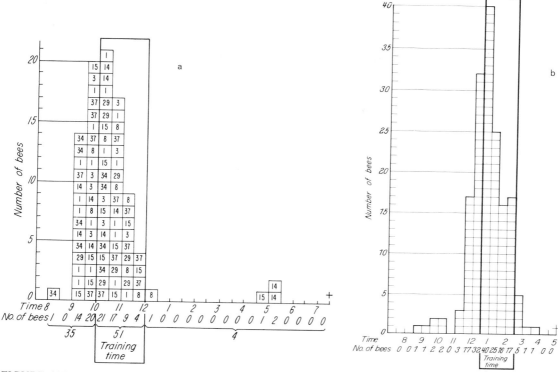

FIGURE 307. (a) Experiment on 16 June 1927: training time, 10 A.M.–12 M.; seven marked bees were foraging (see legend to Fig. 306). (b) Experiment on 21 September 1926: training time, 12:30–2:30 P.M.; 21 marked bees were foraging. After Ingeborg Beling 1929.

diurnally periodic external factors, present though unknown to us, could have escaped detection.

The suspicion that the bees' clock is regulated by some unknown diurnally periodic external factor was kept alive by an additional observation: it proved to be impossible to train bees to other than a 24-hr rhythm. They would indeed learn, for example, to appear at the feeding station at three different times of day, but each feeding time was repeated after exactly 24 hr. They would not learn a 19-hr rhythm. I. Beling fed a group of bees in the darkroom with constant illumination for a period of 3.5 hr every 19 hr; in doing so the feeding time of course was shifted from day to day. After 16 days (23 feedings), immediately following the last feeding, the empty food dish was observed continuously for 25.5 hr. As Fig. 309 shows, the 19-hr training was unsuccessful, but 24 hr after the last feeding the number of visits was increased. A 48-hr rhythm too was not learned; after such training an augmented frequency of visits was seen every 24 hr (Beling 1929, Wahl 1932). A 19-hr training schedule also failed with bees that had been held even during their larval period under an artificial 19-hr day (9.5 hr light alternating with 9.5 hr darkness; Paschke 1956).

A clear decision as to whether the time sense functions without

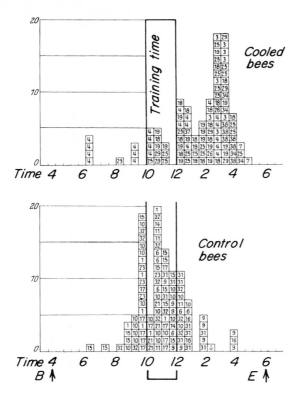

FIGURE 308. Experiment on 11 September 1954; results of training to a time (10 A.M.–12 M.) bees that had been held at 4–5°C for 5.75 hr previously (*above*) and for control bees (*below*); *B*, *E*, beginning and end of experiment. Eleven cooled bees and 13 control bees were foraging. After Renner 1957.

periodic diurnal external factors was to be expected from a displacement experiment, which, to be sure, had to assume dimensions different from those already discussed. The plan existed some decades ago (v. Frisch 1937:17–18). Preparations for carrying it out with an ocean liner failed for unrelated reasons. Eventually, with better means of transportation, it was performed by Renner (1957). In a bee room (as in Fig. 11, but demountable; see Fig. 310), with constant illumination and temperature, he trained a group of bees for several days to an afternoon feeding time, in Paris. Immediately following the last feeding, on 13 June 1955 he took the colony by air to New York, where the

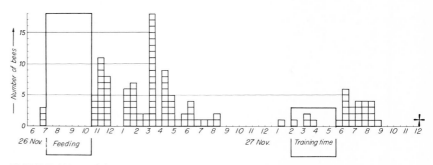

FIGURE 309. Training to a 19-hr rhythm; results of observation after 16 days (23 training feedings). An increase in searching occurred about 24 hr after the final feeding. After Ingeborg Beling 1929.

FIGURE 310. The two demountable bee rooms for the transoceanic experiment; trial assembly in a classroom at the Munich Zoological Institute, before the bee rooms were sent to Paris and New York.

hive was put into a second, identical bee room. If the sense of time was regulated by an external factor, unknown to us, that depended on the earth's rotation and hence had diurnal periodicity, the bees should now appear according to New York local time, 5 hr later than Paris time. But they came according to Paris time (Fig. 311); and they did it, somewhat less, on the two following days also, without renewed feeding. This observation showed that bees are capable of being guided by an internal clock. The result was confirmed by the reciprocal experiment (training in New York; testing after the return flight, in Paris; see Fig. 312).

How does it happen that bees do not learn an interval that deviates from 24 hr? Probably the explanation is that in the final analysis their sense of time has in fact been forged on the basis of the diurnal rhythm. No doubt the most important "index of time" (Zeitgeber[5]) in this process was the alternation of light in the day-night cycle. The rhythm has now become so imprinted that it cannot be changed experimentally. It is to be recognized even when a colony, in continuous light and under constant temperature and humidity, regularly finds sugar water at its feeding place for weeks at a time in a closed bee room (Fig. 11). In these conditions the bees will indeed collect food both by day and by night, but a diurnal rhythm of 24-hr duration is suggested in the activity of the foragers (Bennett and Renner 1963).

Periodic diurnal processes of a simple sort have long been known with other animals and plants, some much longer than in bees. Recently

[5]This good designation for external diurnal factors that regulate rhythmic life processes comes from Aschoff (1954).

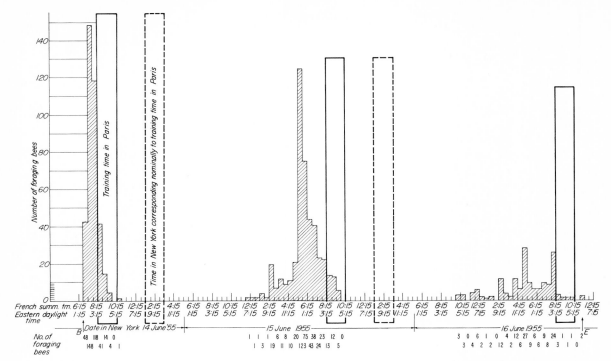

FIGURE 311. Transoceanic experiment; visits to the feeding stand (without food) on the three days of experimentation in New York, 14–16 June 1955, by the bees trained in Paris: abscissa, French Summer Time and (below) Eastern Daylight Time; time difference, 5 hr; solid frame, training time in Paris; broken frame, nominally corresponding time (local time) in New York; *B, E,* beginning and end of experiment. The bees were not guided by local factors corresponding to the sun's position, but came in accordance with their internal clock, at intervals of about 24 hr after the last training feeding in Paris. After Renner 1957.

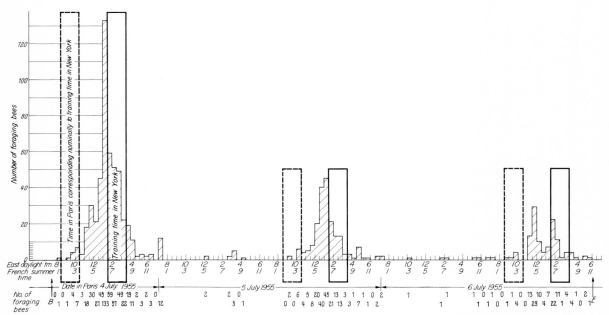

FIGURE 312. The reciprocal of the experiment of Fig. 311, 4–6 July 1955: training in New York, testing in Paris.

they have been investigated on a very broad basis. Whether flowers open and close at certain hours, whether the position of leaves changes in the morning and in the afternoon, or whether mice display more lively activity at certain times of day than at others—such life processes are repeated with remarkable punctuality according to the time of day. In many instances they continue in accordance with an internal clock when external factors have been excluded. That is more easily demonstrated for the regular diurnal repetition of some process than for the behavior of bees, which has so great a capacity for pertinent temporal adjustment. Thus, for instance, even under constant conditions mice maintain their activity rhythm, but after some time there appear small variations from a precise 24-hr cycle, and since these frequently differ somewhat in different individuals they cannot be regulated by some (unknown) external factor (Aschoff 1955). The principle on which the internal clock operates is as little known here as it is with bees. There is much to suggest that its basis is of a similar nature in animals and plants (Bünning 1963).

The sense of time is put in conflict with the sun's position. In Renner's transoceanic experiment the bees in their shut-off rooms were compelled to depend on the internal clock. Thus it was made evident that the latter does indeed function. But what will happen if the experiment is done out of doors and puts the sense of time into conflict with the sun's position?

An additional displacement experiment gives information in regard to this question (Renner 1959). The bees were trained on the east coast of North America in Saint James on Long Island (near New York City), in a level region without conspicuous landmarks. The colony belonged to the Italian race. According to local time there, the training period was from 1:02 to 2:32 P.M. Following the final training session, Renner flew overnight with the beehive to the west coast, and set it up in a similar experimental field at Davis, not far from San Francisco. The difference from Saint James in local time was 3.25 hr. On the three succeeding days the performance of the bees, not fed further, was observed under a clear sky. Figure 313 shows the result. On 3 September 1958 the bees began to search vivaciously, 23 hr after their last feeding and thus 1 hr too early according to their internal clock. Such prematurity is a frequently recurring phenomenon, and had also happened during a preliminary experiment, without displacement, on the east coast. But thereupon, diverging from the norm, a drop in frequency was followed by a second rise with a peak lying 1.5 hr after the first. At the local time that at the western station should have corresponded nominally to the training time, not one bee appeared (Fig. 313). On the two following days, both peaks were repeated, but were shifted nearer to the local time that would have corresponded nominally to the training hour in the east. Probably the correct interpretation is that at the first maximum the bees were following their internal clock, while the second depended on local external factors that grew yet more influential on the two succeeding days. On the west coast it got light or dark 3¼ hr later than on the east coast. Probably this displacement of the light time of day

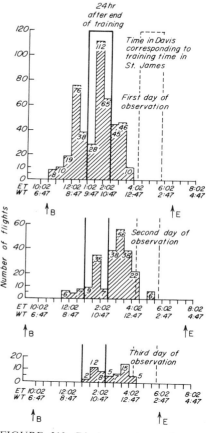

FIGURE 313. Displacement experiment, after preliminary training in the open to a feeding time. Bees were trained on the North American east coast (near New York) to the time from 1:02 to 2:32 P.M., and were then flown to the west coast and tested at Davis (near San Francisco). Shown here is the number of searching flights on three successive days of observation, 3–5 September 1958, without feeding: *ET*, local time at the eastern site (training); *WT*, local time at the western site (experiment); solid frame, training time; broken frame, local western time corresponding to the time of training, *B*, *E*, beginning and end of observation. After Renner 1959.

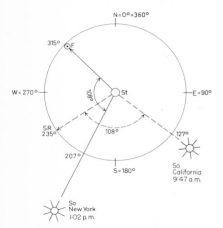

FIGURE 314. Additional information relative to the displacement experiment, after preliminary training in the open to a feeding time: *St*, hive; *F*, feeding station; *So*, sun's position; *Sr*, expected direction of searching. Further explanation in text. After Renner 1959.

relative to the night is responsible for the second peak and for the gradual shift in the time of visiting. This latter behavior showed that the internal clock could be reset, (see pp. 363, 445f, and 453).

This long-distance field experiment afforded an opportunity of combining with the training of the bees to a certain time their training to a compass direction, and of testing, following their displacement, what compass direction they would pursue in the altered geographic longitude. The meaning of this experiment is made clear in Fig. 314. During the training procedure on the east coast (Saint James) Renner had established the feeding station *F* 140 m from the hive exactly to the northwest, at 315° (the azimuthal angles are reckoned continuously from N = 0° through E, S, W, and back to N again). At the beginning of the training time (1:02 P.M.) on the last day of training, the sun there was in the SSW (azimuth 207°). At this point the feeding station was 108° to the right of the sun for the bees (Fig. 314). On the next day in California (Davis) 24 hr later, after the displacement westward, the sun was only in the SE, at 127°; time locally was 9:47 A.M. When the bees, obeying their internal clock, flew out some 24 hr after the beginning of the last training period—this they did first—and if they were to seek the feeding place at the accustomed solar angle, 108° to the right of the sun, then they would have to fly to the NW instead of to the SW, in the direction of the dashed arrow (azimuth 235°). In the experiment they did in fact fulfill this expectation quite exactly.

In order to check the direction to which they flew, Renner had developed the counting device described earlier on pp. 20ff. Even during their training the bees collected sugar water from inside a little aluminum box (Fig. 22), into which they crawled through a small opening. In the experiment, exactly similar devices, without sugar water of course, were set out in a circle, each 140 m from the hive, at angular separations of 45°. When a bee crept through the opening her visit was recorded photoelectrically by interruption of a light beam. This automatic registration not only excluded any influence of an observer, but was the only way to obtain data in eight directions. The observers needed for doing this would not have been available. By sinking the boxes in the ground (Fig. 22) they were made so inconspicuous that they could not be seen by the bees at any great distance.

In maintaining the solar angle in accordance with the internal clock, but at incorrect local time, the bees demonstrated once again that they find the proper direction by means of the sun's azimuth. It was possible to draw a series of further conclusions beyond this result, and to test the role of other possible means of orientation. One can calculate the compass direction in which the bees would have had to search in California had they been guided by the lines of force of the earth's magnetic field. In Saint James the magnetic lines of force point 11° W of N (azimuth 349°), in Davis 17° E of N (azimuth 17°). During their training the bees had to head 34° to the left with respect to the magnetic lines of force; after they had been displaced, in maintaining the same course relative to the magnetic field they would have had to search at an azimuth of 343°. The straight line (curve 1) in Fig. 315 corresponds

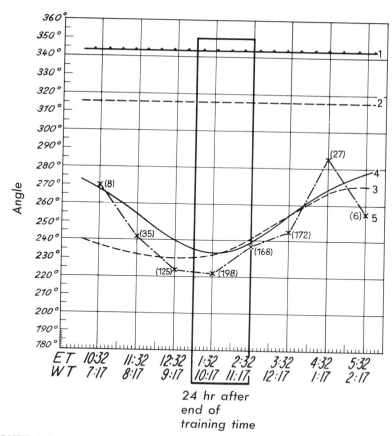

FIGURE 315. Theoretical and actual direction of searching by bees trained to a definite time after they were transported from an eastern to a western site: curve 1, direction of search expected if the bees were oriented by terrestrial magnetism; curve 2, if they maintained precisely the compass direction to which they had been trained; curve 3, if they maintained the solar angle learned at the eastern site, without taking into account the changed azimuthal angular velocity; curve 4, if the varying azimuthal angular velocity was considered; curve 5, the values actually observed at the hours indicated, shown by crosses, each of which represents the mean direction of searching for all three days of observation. After Renner 1959.

to this direction. Curve 2 shows the line that would be appropriate in case the bees, after having been transported to the west, were to find once again, in accordance with some unknown capacity, the compass direction (azimuth 315°) they had learned in the east. But the directions in which they actually searched (curve 5) in the course of the day, after their displacement, show that in their flights the bees were striving at all times for the solar angle that they had maintained at the eastern station.

Now, Renner had hoped at the same time to obtain an answer to another question, for the comprehension of which it must be recalled that the (apparent) movement of the sun occurs with an average angular velocity of 15°/hr, making 360° in 24 hr in correspondence with the full circle. But since the arc followed by the sun rises steeply in the sky

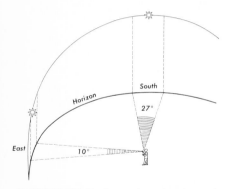

FIGURE 316. During a given time interval the sun's azimuth changes less in the morning or the afternoon, when the sun is following a steep course, than at noontime, when the course is flatter. Schematic.

in the morning, falls off steeply in the afternoon, and is flat during the noonday hours, the sun's azimuth changes less during the morning and afternoon hours than during the corresponding interval of time around noon; the azimuthal angular velocity of the sun is greatest during midday; for instance, in Saint James on 2 September from 6:30 to 7:30 A.M. it is 10°/hr and from 11:30 A.M. to 12:30 P.M. it is 27°/hr (Fig. 316). Whether in their orientation according to the solar compass the bees take account of these diurnal differences or whether it is based on an average value for the angular velocity is still an open question. Now, the changes in the azimuthal angular velocity in California, in consonance with the local time (later by 3.25 hr), occur less early than on the east coast. When the rapid midday course is already in progress there, in the west the sun is still following the slower morning tempo; thus the changes do not occur conformably. Bees that when in the west are still maintaining the solar angle in adherence to an internal clock set in accordance with eastern conditions would consequently have to change their direction somewhat during the course of an experiment lasting several hours, and their manner of doing so would vary depending on whether or not they were taking account of the different azimuthal angular velocity of the sun. If not, then on the basis of calculation the expected direction of search would be that shown by curve 3 in Fig. 315. If the bees do take account of the changing azimuthal angular velocity, then the direction of search should be that shown by curve 4. Curve 5 gives the experimentally observed direction of search; each point on the curve stands for the mean of the individual observations, and the adjacent numbers give the number of visits. The similarity of the course of this curve to that of curve 4 leads one to suspect that the changing azimuthal angular velocity of the sun is included in the bees' calculations. However, a definite decision is not possible on the basis of this single experiment (for further details see Renner 1959).

In the tropics, conditions for a clarification of the problem are much more favorable, since such contrasts increase when the sun is higher in the sky. If the sun passes through the zenith, the azimuth even remains unchanged all morning long, then suddenly shifts by 180°.

Other long-distance displacement experiments. Renner transported the bees from east to west. Lindauer took advantage of a stay in tropical Ceylon to make a south-to-north displacement of his zenith across the ecliptic. At Kandy, Ceylon, marked bees were trained for a whole week at noontime to a feeding station 150 m north of the hive. On 24 April, when the course of the sun was midway between Kandy and Poona, India (Fig. 317), the hive was taken to Poona overnight and established in an open field. Feeding stations were set up at the accustomed flight distance, to the north, east, south, and west. At noon on this day the sun stood in the sky 5° 35′ to the north of the zenith in Kandy, 5° 35′ to the south in Poona. When the flight entrance was opened at 11:45 A.M., the sun was still so far to the east that for bees the accustomed northerly direction (left of the sun in the morning) must still have been recognizable. By 12:02 P.M. there had been three visits to the northern station. The sun crossed the meridian at 12:30 P.M. Between 12:13 and

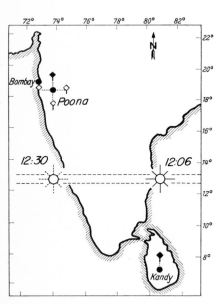

FIGURE 317. Displacement experiment from Kandy, Ceylon (where the sun is in the north) to Poona, India (where it is in the south). Bees were trained to the north at Kandy at noon and were tested at Poona on 24 May at noon. After Lindauer 1957.

12:24 P.M. three more bees visited the test stand, this time to the south. Here, as in Kandy, they were maintaining the direction toward the sun at noon and had confused north with south. Unfortunately, it was much hotter in Poona than in Ceylon, and most of the bees that started to fly out from the hive turned back at the entrance and stayed inside. But even only three bees carry conviction when one recalls how rarely in our many other displacement experiments flights have occurred in a direction diametrically opposed to the direction of training. And so this result confirms the fact that bees make use of no other compass than the sun.

On his return to Munich Lindauer conducted displacement experiments with two Singhalese colonies of bees under our sky, which was strange to them. They had not only been transported far to the west—the local time was shifted 4 hr 36 min relative to that in Ceylon—but also the sun, which in Ceylon they had seen in the northern sky at noon, now stood in the south at midday, and moved clockwise instead of counterclockwise. After the bees had been trained on the day of their arrival in Munich to a given compass direction—and on several later occasions—the colonies were displaced to another region and their direction of searching was tested. The trained bees did indeed continue to fly out, but initially only a small number of them, without displaying any preference for the direction of training, found their way to the test stands. After 4 days it was no different. But when an experiment was tried again 1.5 months after their arrival in Munich, the training to a direction was successful and the displaced bees maintained the proper compass direction. They had reoriented themselves to the new course of the sun (Lindauer 1957a).

Data of Kalmus (1954, 1956) seemed contradictory to these findings. The offspring of a California queen, mated there but then shipped to Brazil in the Southern Hemisphere, performed in a displacement experiment as though trying to compute the sun's movement as clockwise—which would have been correct in their original home. Even a hybrid colony, whose queen had been imported from California 8 years previously but had been mated with Brazilian drones, had not yet been able to orient properly to the southern sun. From this Kalmus concluded that the direction of the sun's movement to be included in their calculations apparently was innate.

Lindauer (1959) showed convincingly that Kalmus' displacement experiments, for causes inherent in his methods, led to this result and that the conclusions were not justified. I do not need to go into details (see Lindauer 1959:45–51), for more convincing than any discussion is the repetition of the experiment at the very place, with due attention to the sources of error revealed by years of experience. Lindauer proceeded in this way and came to the following results.

An Italian queen bee from North America, which had been imported into Brazil 14 weeks earlier and—as with Kalmus—had been placed with native black bees, had subsequently crowded out the black bees in the colony with her own (yellow) offspring. When trained on 16 May 1957 from 2:00 to 5:45 P.M. to a feeding station in the NW and tested

364 / The Orientation of Bees

the next morning in a strange territory, the foragers responded correctly to the compass direction: from 8:00 to 11:00 A.M. 17 of them flew in to the NW station, 1 to the SE, none to the NE and SW. Although their mother had lived in the Northern Hemisphere up to a few weeks before, her offspring were familiar with the course of the sun in the Southern Hemisphere. A repetition of the experiment with a hybrid colony of the sort used by Kalmus led to the same result.

Thus the internal clock evidently is not coupled rigidly to the position of the sun. That agrees too with Lindauer's experiences with colonies that he transported from Ceylon to Munich. In Renner's long-distance east-west displacement, adaptation to the changed conditions was noticeable even by the second and third day of observation. That brings us to the question whether in their native haunts also bees have to learn by experience the sun's path and its temporal course, or whether the knowledge of local conditions is at least roughly innate with them.

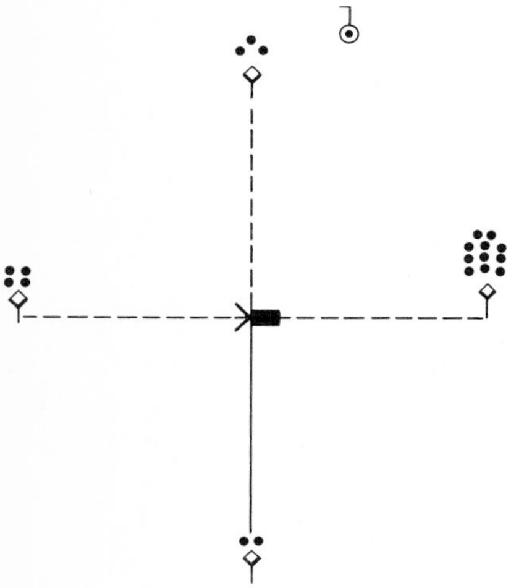

Bees learn how to use the solar compass. Lindauer let bees emerge in an incubator and from them formed a colony that for several weeks was kept in a cellar without a view of the sky. If he then trained these bees on their first day of free flight to a given compass direction, on the day following they were incapable of finding the direction again. But the experiment did succeed after a week of free flight.

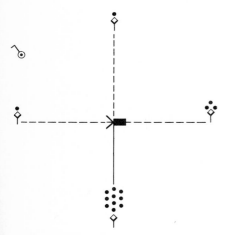

FIGURE 318. Results of a displacement experiment with bees reared in a cellar in the dark and then trained to the south for three afternoons: solid line, direction of training; the dots above the feeding stands show the number of visits in the morning. After Lindauer 1959.

FIGURE 319. After five afternoons of training, the bees included the sun's course in their calculations and, after being displaced, maintained the proper compass direction in the morning. After Lindauer 1959.

In order to acquire a more precise picture of the learning process, cellar-reared bees, without any experience of the sun, after being trained for 3 days to a southern feeding station, were tested under the open sky. They were trained only during the afternoon; in the mornings they were always put back in the cellar again. Now the displacement experiment had a positive outcome, yet it was not the southern direction to which they had been trained that received the most visits, but the eastern feeding stand (Fig. 318). The foragers still had not grasped the motion of the sun. All they had noted during the training afternoons was that the feeding place was to the left of the sun. After the displacement the majority of them continued to go to the left of the sun during their searching flights in the morning, and that brought them to the eastern station. Their flight was correct as regards the angle, but not in reference to the compass direction. Only in an additional experiment, with five afternoons of training, had their solar compass begun to function properly. After these bees were displaced into strange territory, flights to the southern station (the direction of training) were then in the majority (Fig. 319).

The last-mentioned experiment was intended simultaneously to provide information about another question: is what matters for appreciating the course of the sun how long the hive has stood out of doors, or how often the bees have flown in the open? This thought was in mind at the very beginning of Lindauer's experiment. Hence on the fifth afternoon of training he enlarged the foraging group with 20 newcomers. The latter surely had not yet been to the southern feeding station, for every newcomer that appeared there had been killed. Like the others, these new bees had had 4 days for observing the sun's position, yet very probably they had not flown out regularly, but only occasionally in preliminary orientation flights around the hive, for instance. Only on the fifth day did they visit the feeding station. Now they were taking part in the experiment simultaneously with the old foragers—and they performed differently: the majority still came at the correct angle and hence to the eastern station (Fig. 320). Thus the response to the correct compass direction seems to be learned from the experience gained in several hundred flights. There remains much to be clarified in this matter, but the first worthwhile insight has been achieved. Though this has shown that the bees master the solar compass only after a certain amount of experience in flight, and thus that not merely a maturation of their instincts but a true process of learning is involved, yet there certainly are bound up with this method of orientation innate components also, such that the bees pay attention to the sun's location even on their very first excursions, or that in their first indication of direction they point out the location of the goal with respect to the sun, or that the dance followers find the goal on the basis of such announcements.[6]

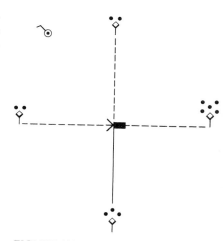

FIGURE 320. The behavior of bees reared in a cellar, which were allowed to fly out freely for four afternoons, and on the fifth were trained to the south. The majority performed disoriented flights or maintained the actual solar angle. After Lindauer 1959.

Under natural conditions the young foragers in their first open-air flights to known sources of food certainly will not fly, using only the proper angle, in the wrong direction. For they are not in a strange territory and they see familiar landmarks. One might imagine that the latter even accelerate the process of learning the sun's course.

Particularly remarkable is a further capacity inherited by the bees: from a fraction of the sun's path they are able to fill out its whole diurnal course. It was intentional that Lindauer, in the experiments discussed and even in earlier ones (1959, 1957), trained the "cellar bees" in the open only during the afternoon and kept them in the dark basement every night and every morning. Despite this they sought their goal in the correct compass direction when they saw the morning sun for the first time. Now, this gift for completing the sun's entire course on the basis of knowledge of a portion of it helps one to understand also the behavior of the marathon dancers during the night. They have never seen the sun's location at night, but they have extended the daytime section of the sun's course to a complete circle. Observations by Lindauer suggest that the shifting azimuthal angular velocity is not included in their calculation of the sun's nocturnal path.

The biological aspect. After having let the facts speak for themselves, a word as to their biological interpretation is appropriate. A rigidly determined hereditary knowledge of the sun's course was not to be anticipated, for the reason that bees range about quite freely when they swarm and may in the course of years disperse to such an extent that the sun's path may no longer agree with that at their earlier dwelling place. In the annual migratory flights of the giant honeybee extensive shifts of location are even quite frequent. Additionally, between

---

[6] Lindauer (1959:60) mentions the precise indication of direction by a pollen collector 15 min after the hive entrance of a colony reared in the cellar was opened for the first time under the open sky, and likewise the immediate correct understanding of these announcements by the hive bees raised in the absence of sunlight.

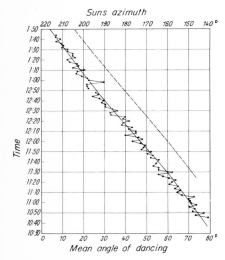

FIGURE 321. Indication of direction by bees after foraging flights to a feeding station 6 km distant; 17 and 18 August 1954. Each point on the zigzag curve denotes the average value of the angle of dancing measured during a 2-min interval; solid line, sun's azimuth at the time of the dances; dashed line, sun's azimuth at the time of the preceding outward flights of the dancers in question. After Boch 1956.

the Tropics of Cancer and of Capricorn the direction of the sun's motion changes twice a year, from clockwise to counterclockwise and vice versa, and thus makes necessary an internal reversal even in a single location.

The capacity of finding at any time a compass direction with the aid of the solar compass—apart from the question of its being learned or inherited—is of value to bees in many respects. When, in search of new sources of food, they undertake extended flights, their solar compass guides them home despite the advancing sun. Also foraging flights not rarely last for more than an hour (Grout 1949:46, Ribbands 1949), exceptionally for even as much as 3 hr (Singh 1950). The solitary bee *Xylocopa* may even remain abroad for 4–9 hr before she returns to her nest (Rau 1933). By means of persistent dances scout bees are able to announce to their hivemates the compass direction of the dwelling they have discovered, over and over again without the risk of renewed excursions. In their dances the foragers indicate the direction of the goal that they perceived on their outward flight, corrected, however, to the time when dancing occurs. This difference becomes important for long flights (Fig. 321). In visiting flowers the foragers quickly note the hours when the quality of food is best; they regulate their flights accordingly (pp. 253f), and set out in the correct compass direction at all times of day. Thus orientation by the solar compass is very much involved in the daily life of a bee colony and—once recognized—can hardly be imagined as nonexistent.

4. Perception of the Sun Through a Cloud Cover

Even with a complete cloud cover the bees can indicate the solar angle of the feeding place correctly in their dances on the vertical comb surface. It was imaginable that from their earlier flights they know at every hour where the sun stands in the familiar landscape. But are they perhaps capable also of perceiving the sun, when it is invisible to us, through the clouds? In order to exclude the first possibility, on the morning of 23 July 1946 with a completely closed cloud cover we displaced to Aich an observation hive that had been sitting in St. Gilgen, and conducted newly marked bees to a feeding station 10° E of S. The territory was strange to them and looked quite different from their last location. The sky remained covered, the sun invisible to us. By 8:30 A.M. the feeding station had reached a distance 100 m from the hive. We observed 20 dances; the indication of direction was right. After moving on to 220 m, another 20 dances were measured; they likewise indicated the appropriate solar angle, within the customary limits of scatter. Since soon thereafter the sun became recognizable for us, we concluded the experiment. It had shown that bees are able to perceive the sun when it is invisible to us behind a cloud cover. And in the unfamiliar landscape they could not know the location of their hive from memory (v. Frisch 1948:18).

The same result was obtained in an experiment at Moosach (near Munich) that was undertaken for a different reason on 25 June 1952 with an Italian colony. The colony was in this territory for the first time. With addition of oil of anise we set up a feeding station 76 m west of the hive. Dancing was proper despite a closed cloud cover. For reasons not important here, scent plates with oil of anise were set out 66 m from the hive in three compass directions. From 11:18 A.M. to 12:18 P.M. the southern and eastern stations received two and three visits, respectively, but the western station—in the direction of the feeding place—177, with a never-ending stream of newcomers. Thus the latter too are able to orient in the strange territory on the basis of the information they had received, according to the sun which was invisible to us.

The ability of bees to perceive the sun behind clouds can be demonstrated more simply than by means of the time-consuming displacement of the colony and the establishment of a new feeding place, in the following manner. The observation hive is laid horizontal and the dancers are given a free view of that portion of the sky where at the time the sun is behind the clouds. We know that bees are able to point out the direction to the goal on the horizontal comb only when they can see the sun or the blue sky (see pp. 131ff). In these experiments there was no blue sky and the sun was not visible to our eyes through the clouds. Nevertheless the dances were oriented correctly; hence the bees must have perceived the sun through the clouds.[7]

This does not hold, however, for every form of cloudiness. With increasing density of the clouds the indication of direction becomes imprecise, and the scatter increases. Behind heavy rain clouds the sun becomes invisible to the bees too, and the dances are then disoriented.[8] The experiments gave positive results when the sky was covered approximately evenly by stratus, stratocumulus, or altostratus clouds.

With this sort of cloud cover the bees orient by the sun even if it has long since disappeared for our eyes. For the bees are disoriented at once if their view of the sun's position is screened by a sheet of paper or cardboard, a board, or objects of that sort. All the rest of the cloud-covered sky will not enable them to perform oriented dances if the view of the sun is shut off. On the other hand, I have still obtained well-oriented dances in the canvas tent with the "sun pipe" (Figs. 118 and 119), when in the direction of the sun the bees could see from the comb surface only a circular spot of the cloudy sky with a diameter of about 15°. As a rule I allowed them a free view of a larger section of the clouded sky, to the extent of about 50° × 40° around the position of the sun. Whether it is done in this way or whether the view is open in all directions is of no consequence.

Physicists react to such information with the suspicion that infrared radiation from the sun is responsible for the bees' perceiving it through the clouds. But it is not. For the dances on a horizontal comb are dis-

---

[7] In such experiments I removed the second pane of glass, which is put over the combs to preserve warmth. I have noticed repeatedly that with a cloud-covered sky orientation is worse under two layers of glass.

[8] As a rule such conditions do not prevail on the days when bees are flying; hence they are of subordinate significance to the biologist.

oriented at once if one interposes a Wratten 87 filter (Eastman Kodak), which absorbs the wavelengths that are visible to bees but lets through the infrared. That is equally so for clear sky (Heran 1952) and when there is a cloud cover (my own experiments). Beyond this Heran (1952:200ff) found that bees could not be trained to infrared radiation (Wratten 87).[9] This agrees fully with our knowledge of their color sense (see pp. 471ff).

It was natural to suspect that near the sun the brightness was greater than elsewhere. But the dances frequently proved to be oriented according to the sun's position when, with the cloud cover thinned out in spots, other portions of the sky were clearly brighter. I examined this with a photocell mounted on a tube, set up beside the observation hive, that could be rotated and tilted so that the photocell could be aimed quickly at any chosen azimuth and elevation in the sky.[10]

An example: on 4 September 1953 there was a complete cloud cover over Brunnwinkl. Marked bees were visiting a feeding station 200 m to the west. The observation hive was laid horizontal, the roof and roof supports were removed, and thus the bees were given a free view of the entire clouded sky. From 8:52 to 8:56 A.M. seven bees were watched; their dances were well oriented toward the west. At this time at the position of the sun (azimuth 118°, elevation 32.5°) a brightness of 98 lux[11] was measured; measurements in 10° steps in azimuth at the same elevation above the horizon showed no significant difference in brightness for the neighboring region, but for azimuths of 340° and 350° a brightness of 142 lux. Since the elevation of the sun is irrelevant in the indication of direction (see p. 136), there remained to be tested whether the brightness might not be still greater in the sun's meridian at some other elevation. That was not the case. All the values were less than the brightness that had been measured W of N. To us too the cloud cover seemed clearly brighter there than at the position of the sun. Despite this, the bees had oriented according to the location of the sun. If the clouds are so heavy that the sun is no longer perceptible to the bees, then indeed they are able to take guidance from other, conspicuously bright regions of the sky. That is comprehensible. If there is no possibility of finding a point of reference for indicating direction in the waggling run, the dancer acts as though she were in the darkened hut: she adopts the brightest spot as a replacement for the sun (in the dark hut, the lamp: lamp effect; see pp. 135f).

In order to learn whether all regions of the spectrum of light visible to the bees were equally effective in the perception of the sun through clouds, on the hive, laid horizontal, I tested the dances beneath glass filters (Schott and Co.; size of the glasses 12 × 12 cm, thickness 1 mm).

[9] As is well known, training experiments to wavelengths that the bees can perceive succeed very easily.

[10] This little device was constructed in the workshops of the Physical Institute of the Technische Hochschule, Munich. For it I am obligated to Prof. Joos.

[11] The lux is a unit not of the brightness of the sky itself, but of the strength of illumination exerted from this direction on the photocell. The photocell (SAF Type 2501) was calibrated in the Physical Institute, and after the experiments the calibration curve was retested and found to be unchanged.

To my astonishment this showed that, with a clouded sky and a free view toward the sun's location, the dances are well oriented only beneath filters that pass, completely or to a great extent, the ultraviolet range from 300 to 400 mμ, even when the light visible to us is absorbed. On the other hand, the dances beneath filters that held back the ultraviolet light were disoriented, even when they passed undiminished the light visible to us and the longer wavelengths out to more than 3000 mμ (v. Frisch 1954a, 1956a). The transmittances of the filters used (according to data furnished by Schott and Co.) are shown in Fig. 322. In Table 38 is recorded how many dances well oriented toward the goal, imprecise, or disoriented ones were observed beneath each filter. The experiments extended over 19 days with a clouded sky, in 5 years (1949, 1953 to 1956). For comparison, on these experimental days there were observed altogether 302 dances without any filter; 266 of these dances were well oriented, 18 were imprecise, and 18 were disoriented; in no instance was the sun perceptible to us.

In Table 36b are assembled the filters beneath which the dances were predominantly or fully disoriented. Despite all other differences in their absorption spectra, these filters have in common the fact that ultraviolet, at least in the short-wavelength region from 300 to 350 mμ, is mostly or completely absorbed.

Table 36a shows that the dances beneath all filters that pass the majority of radiation from 300 to 400 mμ are well oriented, however different the rest of the absorption spectrum may look. With

TABLE 38. Number of well-oriented, imprecisely oriented, and disoriented dances beneath filters whose transmittances are shown in Fig. 322 (experiments with a complete cloud cover).

	Fig. 322(a)				Fig. 322(b)		
	Number of dances observed				Number of dances observed		
Filter	Well-oriented	Imprecise	Disoriented	Filter	Well-oriented	Imprecise	Disoriented
WG_8	8	0	0	BG_{17}	5	3	14
WG_7	3	0	0	BG_{34}	0	0	12
WG_6	6	0	0	BG_{18}	1	1	28
WG_5	7	0	0	BG_{13}	1	0	7
WG_4	10	2	1	WG_1	5	3	24
WG_3	19	2	9	GG_{18}	0	0	13
WG_2	5	4	2	NG_5	1	3	33
BG_3	36	0	0	NG_4	0	0	10
BG_{26}	11	3	0	GG_4	0	3	6
BG_{21}	5	0	2	GG_{13}	1	0	22
BG_{25}	5	1	5	VG_{10}	0	0	16
BG_{23}	10	0	0	GG_9	0	0	20
BG_{20}	3	3	1	VG_5	0	0	10
BG_{12}	35	3	6	OG_4	0	0	7
BG_{14}	16	1	0	RG_1	1	1	24
UG_1	24	1	1	RG_7	0	0	12
UG_2	—	—	—	UG_6	0	0	15

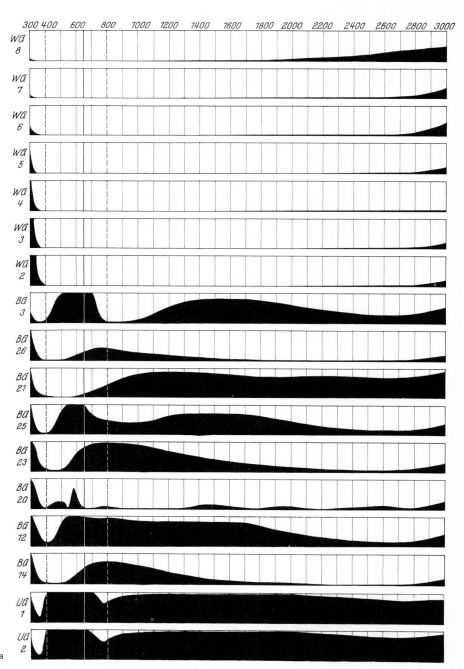

FIGURE 322. The transmittances of the Schott and Co. glass filters used, in the spectral region from 300 to 3000 mμ. The range of human vision lies between the two dashed vertical lines, that for the bees between the left-hand margin and the solid vertical line (300–650 mμ). (*a*) Filters beneath which the dances were well oriented; (*b*) filters beneath which the dances were predominantly or completely disoriented. See Table 38.

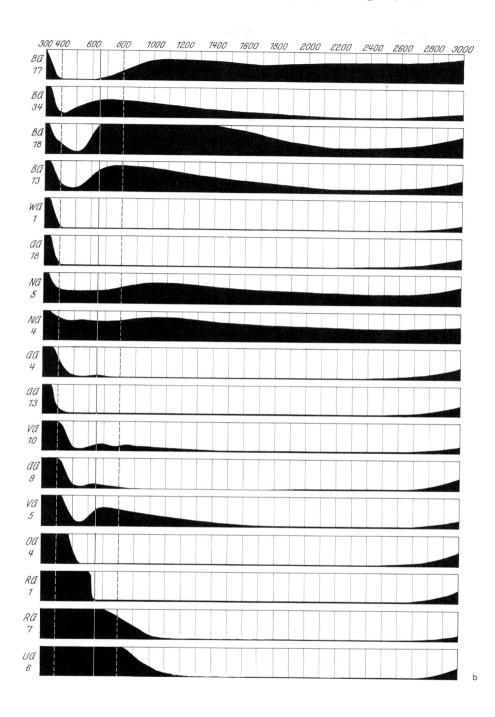

the WG filter series it may be seen particularly clearly that orientation gets worse with increasing absorption of the ultraviolet. The same thing appears, somewhat less definitely, in the series of the BG filters. In such a comparison a rigid relation according to physical criteria can of course not be expected, because the cloud density, which was not identical in all experiments, also influences the goodness of the orientation.

With respect to technique it should be mentioned that, after removing the outer glass pane of the observation hive, I put the filters on the inner pane immediately above the dancer. Glass squares that were not sufficiently transparent for us were lifted slightly on the side turned away from the sun, so that we could see the bees (Fig. 323). In our doing this the bees got no view of any part of the sky except through the filter. Numerous controls with filters transparent for us showed that the dances were not affected by this procedure.

Particularly impressive is the behavior of the bees with alternate interpolation of the GG_{13} filter, which absorbs ultraviolet but is fully transparent for us (disoriented dances), and the UG_1 filter, which lets ultraviolet through but that we find opaque (dances exactly in proper orientation).

From the data it is to be deduced that the sun, behind a cloud cover that already makes it invisible to us, remains visible to the bees' eye in the region of the ultraviolet, through a greater density of clouds. This presupposes that more ultraviolet penetrates the clouds from the position of the sun than from the surrounding region.

Efforts to demonstrate the anticipated increase in brightness by means of photographs and physical measurements were at first without result, until we once again studied the problem in cooperation with the Second Physical Institute (Prof. Rollwagen) and the Munich Observatory (Prof. Schmeidler and M. Eckstein; see v. Frisch, Lindauer, and Schmeidler 1960).

Since the sun is not perceptible to bees under every kind of cloudiness, we combined the physical measurements with observations of bees. In the summer of 1959 an observation hive was set up on the flat roof of the Munich Observatory, and a feeding station was established at a distance of 275 m from the hive in a direction 24° E of N. There marked bees received food almost daily at various hours, so that they were always available if a suitable weather conformation occurred. In such a case the hive was laid horizontal. As soon as a marked bee from our feeding station began to dance, the biologist informed the man at the telescope only a few meters away and recorded the direction of the waggling runs while a photograph of the sky was made simultaneously.

A plate camera was used to photograph the sky light in the ultraviolet region. In place of the objective a quartz lens with a 10-cm focal length was used in combination with a UG_2 filter.[12] The camera was attached to a parallactic mounting; the instrument was aimed at the sun, and by means of the clockwork was driven so as to follow the apparent rotation of the heavens.

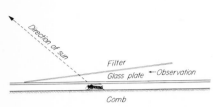

FIGURE 323. Observing the dances beneath an opaque filter.

---

[12]The transmittance is almost identical with that of the UG_1 filter (see Fig. 322a).

For photometric analysis the lower portion of each plate was illuminated through a graduated filter, so that after developing there were 12 different degrees of blackening of known intensity, and the individual blackening curves of the plate could be determined from them.

After it had been found impossible in preliminary experiments in the summer of 1958 with plates of the Perutz-Astro type to demonstrate any increase in brightness at the position of the cloud-covered sun, in the summer of 1959 we used Agfa Blue-Ultrahard spectroscopic plates, which have a particularly high contrast. Thanks to the extremely fine gradation of the material of the plates an accuracy of about 1 percent could be attained. The plates were measured with the Hartmann microphotometer in the x-coordinate (direction of declination on the sky) and in the y-coordinate (direction of right ascension on the sky). The two coordinate axes intersected at the position of the sun.

Not until the experiments had been concluded were the observations on dancers and the results of the photographs of the sky compared. Of the 26 plates, 2 were discarded, because at the time of these pictures the dances were not oriented clearly. In both instances a very slight increase in brightness could be seen at the position of the sun. All the rest of the photographs were divided into two groups. The first group comprises 11 plates; according to Lindauer's data, the dances were well oriented when these pictures were taken. In the grand average of all these plates there was an increase in brightness of 5 percent at the position of the sun (Fig. 324). Let it be emphasized that in each of these 11 exposures an increase in brightness at the position of the sun[13] or near it,[14] was to be seen. The second group comprises 13 plates, at the time of whose exposure the bees' dances were disoriented. In these instances no increase in brightness at the sun's position is to be discerned (Fig. 325).

The figures show a uniform change in the brightness of the sky over the entire field of the plates. The reason is that the general brightness of the sky light depends on the elevation and azimuth of the spot observed. The direction of the coordinates of the sections measured in right ascension and declination always forms a certain angle with the lines of equal elevation or azimuth, respectively; only when the exposures are made precisely at noon, as in practice never happened, do the lines of equal declination run parallel to the horizon. Hence it is understandable that in both coordinates a variation in intensity runs right across the field of the photograph.

A single plate on which the increase in brightness at the position of the sun is especially clear was measured, near the center of the plate, at intervals of 1 mm in both coordinates. The result, in the form of isophots, is shown in Fig. 326. The increase of 20 percent in brightness at the center is obvious. The deviation of the 20-percent isophot from the circular form is readily explained by the general variation in brightness over the surface of the plate and by the effect of the grain size

---

[13] In three instances it was slight (1.5, 2.5, or 3.5 percent), once it was as much as 4 percent; otherwise it was always more.

[14] More than half of the plates show the increase in brightness at the position of the sun; in a few instances the maximum is 0.5–1.0° away, and as an extreme 2°.

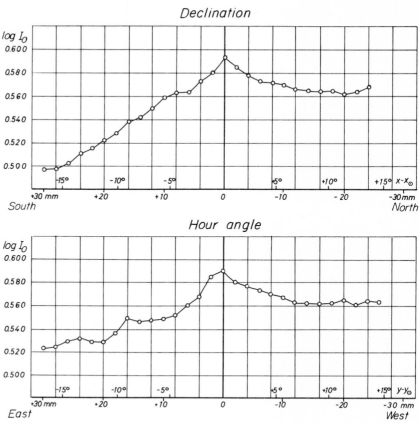

FIGURE 324. The variation in light intensity with declination and hour angle under a complete cloud cover; mean of 11 exposures during which the dances were well-oriented: abscissa, angular deviation from the position of the sun (0°), and distance (mm) on the photographic plate; ordinate, logarithm of the intensity. After v. Frisch, Lindauer, and Schmeidler 1960.

in the emulsion. That the sun's disk was not perceptible to our eyes in spite of the 20-percent greater brightness in the center probably is to be attributed to the gradual decrease of brightness in the surrounding field.

The bees did not let themselves be deceived by brighter spots that appeared, scattered randomly in the cloud cover. Presumably they have learned to distinguish the sun's image from these thinner areas because of its regular shape, just as we too do not confuse the disk of the sun shining through thin layers of clouds with small light spots in the general cover.[15]

It seems astonishing that bees can indeed perceive so slight an increase in brightness at the sun's position as was determined in the measurements (on the average 5 percent). In order to obtain a basis for judging their capacity in this respect, Daumer (1963) conducted experiments on the contrast sensitivity of the bees' eyes. With an ar-

---

[15] Whenever the sun is not perceptible to the bees they can adopt another bright area in replacement for it and be guided accordingly (see p. 368).

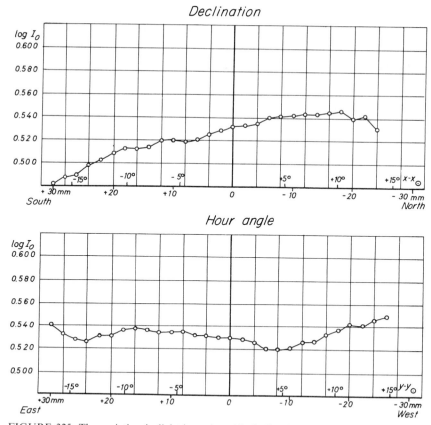

FIGURE 325. The variation in light intensity with declination and hour angle; mean of 13 exposures during which the dances were disoriented. After v. Frisch, Lindauer, and Schmeidler 1960.

rangement that had a common basis with the customary training to food and color but that afforded a quantitative test of spectral colors and color mixtures (including the ultraviolet), and using white lights of equal ultraviolet content—hence in studying the pure contrast in brightness for the bees' eyes—he found a difference in brightness of 15 percent necessary for perception. If more ultraviolet compared with light of different wavelengths should penetrate the cloud cover in the vicinity of the sun than in the surrounding area, then, on account of the sensitivity of the bees' eyes to ultraviolet, a color contrast would exist for them. Under such conditions Daumer found a difference of 5 percent in ultraviolet content sufficient to make the distinction perceptible to the bees.

However, a preferential transmission of ultraviolet in the direction toward the sun is in contradiction with the teachings of physics. Besides, it has also been learned that the sun can be perceived through the clouds even in pure ultraviolet light. Hence it is probable that the contrast of brightness in the region of the ultraviolet should be sufficient.[16]

[16] Theoretically, however, a weak color contrast even within the ultraviolet is imaginable, since the sensitivity curves for the blue and green receptors extend into the ultraviolet (see p. 477).

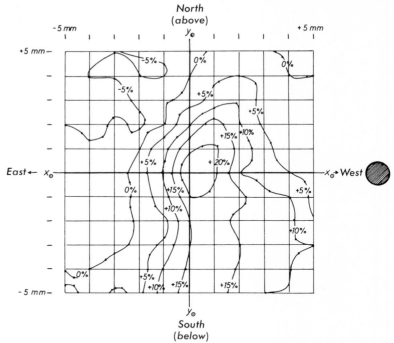

FIGURE 326. Isophots from the central portion of a plate that shows a particularly clear effect: surface measured, 1 cm² ($\cong 5° \times 5°$); isophots at 5-percent differences in intensity. At the right is shown the apparent size of the sun's disk on the same scale. After v. Frisch, Lindauer, and Schmeidler 1960.

Daumer's experiments provide no information as to the sensitivity for contrast in the ultraviolet. There is the further consideration that in his experimental arrangement the downward-directed regions of the bees' eyes were tested, whereas it is the upper part of their eyes that is called on for perception of the sun through the clouds. Among drones the ultraviolet receptors seem to predominate in the dorsal portions of the eye.

As yet nothing about this is known with regard to worker bees. Their sensory cells are much smaller and less readily accessible to experimentation (Autrum and v. Zwehl 1962, 1963, 1964). It may be, and would be logical biologically, that with them too the parts of the eye directed toward the sky possess relatively numerous ultraviolet receptors. An increase in sensitivity to contrast in the ultraviolet could be the result, and would make possible the perception of the sun through a cloud cover in this very region of the spectrum. Daumer has planned additional experiments in this direction.

In the summer of 1962 photographs of the cloudy sky were made once again by the Munich Observatory (Prof. Wellmann and H. Guckelsberger) in cooperation with the Zoological Institute (Prof. Lindauer, H. Markl); this time the region of the sun was photographed not only in the ultraviolet but simultaneously (or almost simultaneously) through both blue and red filters. Unfortunately, during that summer and fall the kind of weather necessary for such experiments occurred very seldom. Hence there were finally a number of exposures made during which the cloud cover was appreciably less uniform than during Eckstein's photographs, and when it also failed to match the conditions that had prevailed during my observations. At the time of seven of the exposures the bees' dances were disoriented, but with six they were oriented. Evaluation of the plates in the ultraviolet range at the sun's position during the time of the oriented dances did not yield the increase in brightness found by Eckstein, or only in a minor degree. Guckelsberger reaches the conclusion that oriented dances under a cloudy sky do not necessarily require an increase in brightness exceeding 2 percent at the position of the sun. But this does not refute Eckstein's findings, which were obtained under a uniform and not very heavy cloud cover.

The oriented dances during Guckelsberger's exposures are limited to 21 August 1962, 3:27–4:25 P.M., and 19 October 1962, 1:58–2:04 P.M. On 21 August 1962 there was at the indicated time an irregular cloud cover (cumulus clouds, sporadic showers). Since the photographs show no clear increase in brightness at the position of the sun, it is to be assumed that the sun's disk was not visible to the bees. We know that under such conditions the bees can adopt instead of the sun the brightest area visible to them in the sky ("lamp effect", see pp. 135f). On 21 August Guckelsberger's photographs in all four cases of oriented dances show a very strong maximum in the ultraviolet,[17] about 10° above the sun; since the elevation of the sun is not heeded by the bees (see p. 136), this can account for the oriented attitude of the dances. On 19 October, as is seen from the data, at the time of the photographs there was to the southwest, near the position of the sun, a conspicuous increase in brightness relative to other regions of the sky. That too would necessarily lead to oriented dances if the bees, the sun being invisible to them, were guided by the brightest spot in the sky.

Additional experiments, with photographs taken in several regions of the spectrum, would be desirable. But they will have to be made when the cloud cover is uniform.

Summary: Orientation According to Landmarks; the Sun as a Compass

1. En route from the home hive to a distant goal conspicuous landmarks are paid much attention and are used in orientation. Figures 281 and 282, as well as other examples, are discussed.

2. In addition to landmarks the sun serves in orientation. Bees use it as a compass and are able at any time to find again a given compass direction, since they know the time of day and the diurnal course of the sun. This was discovered by means of displacement experiments (for example, Fig. 286).

3. A "traveling feeding station" that is kept always in the direction toward the sun is unnatural for the bees and they did not understand

---

[17] In the blue and the red this maximum was less well developed or even completely absent.

the situation. After they had been transported to an unfamiliar neighborhood, they searched for the station the next morning in the westerly direction in which they had last been fed on the preceding day.

4. The relative significance of landmarks and the celestial compass may be tested by transporting bees, after they have been trained to a flight path along which there are conspicuous landmarks, to a territory in which corresponding landmarks run in a different compass direction. A forest margin paralleling the flight path from the hive to the goal (Fig. 290) is of greater importance than the celestial compass, so long as it is not too far to the side (Fig. 294). Of equal importance are other continuous landmarks along the flight path (shoreline, Fig. 300; highway, Fig. 302). On the other hand, individual trees or small groups of trees are not able to compete with the celestial compass.

5. For orienting according to the sun's position, which is linked with the time, the bees' sense of time is an essential prerequisite. It functions in darkness also. In a territory in which there were no landmarks usable for determining the flight path, bees at their feeding station were placed in darkness for 1–2 hr. After they had been set free they took account of the altered position of the sun and flew off in the direction toward their hive. Scout bees that in the dark hive are indicating the location of a nesting site may dance for hours ("marathon dancers," *Dauertänzerinnen*); in doing this they compensate for the advancing position of the sun by changing their angle of dancing correspondingly. Even on the following day, without having glimpsed the sun meanwhile, they can, with spontaneous dances, point out the proper direction to the goal.

6. With diffuse illumination bees can be stimulated to dance in the hive at night. Then they point out the direction to the goal—though less precisely—in accordance with the nocturnal position of the sun, that is, at a time when they could never have seen the sun. After they had been trained to two feeding places, one of which was in operation in the morning and the other in the afternoon, the nocturnal dances indicated the afternoon station initially and then, after a 2-hr period of uncertainty (from about 11 P.M. to 1 A.M.), the morning station.

7. Prerequisite for these performances is the bees' internal clock. Their sense of time continues to function after exclusion of external factors that have diurnal periodicity. Bees kept under constant conditions may be trained to any time of the day or night. But they will not learn to come to the feeding place in accordance with a time cycle that deviates appreciably from the usual diurnal rhythm (thus not on a 19-hr cycle nor on a 48-hr one). This is comprehensible on the assumption that their sense of time was imprinted originally on the basis of the diurnal rhythm.

8. Their appearing at feeding places at the proper time cannot be attributed to diurnally periodic external factors that are unknown to us. Such factors would necessarily be changed appreciably when the bees were displaced a long way over many degrees of longitude. Bees in Paris were trained to a given time of day in a closed room with constant illumination and temperature, were then taken by air to New York, and after this displacement from east to west were tested there in an

identical closed chamber. They came to be fed in accordance with Paris time. The reciprocal experiment, with transport from west to east, had a corresponding outcome.

9. By conducting such an experiment in the open, one brings the bees' internal clock into conflict with the position of the sun. When bees that had been trained to a given time on the east coast of North America were displaced to California, near the west coast, they came initially to feed according to their internal clock. At a time when this frequency was already declining, a second maximum appeared; evidently this was released in response to local time. During the two following days it became even more prominent—without further training.

10. In the long-distance displacement experiment in the open, training to a time was combined with training to a compass direction. By this means it was confirmed that the solar angle is decisive for the direction of search. The lines of force of the earth's magnetic field had no influence. This experiment also demonstrated that the variation with time of day in the azimuthal angular velocity of the sun probably is taken account of by bees in calculating the sun's course (see Fig. 315).

11. A long-distance displacement from north to south in the tropics provided further evidence that the bee has no compass other than the sun. At the time of the experiment the noonday sun stood in the north at Kandy, Ceylon, and in the south at Poona, India. Bees trained to the northern direction in Ceylon searched for their feeding place in the south in Poona at noon.

12. With bees that were transported from Ceylon to Munich the course of the sun at home had been counterclockwise, while it was clockwise in Munich. In addition, the local time differed by 4.5 hr, because of the displacement from east to west. At first the bees were incapable of locating according to the solar compass a compass direction to which they were trained. After some weeks they learned to do so. This ability to adjust was confirmed in experiments in which the bees were displaced from North to South America.

13. Even in her place of birth the bee is not endowed from the start with the ability to employ the solar compass. This capacity is acquired during the life of the individual, in the course of a few days of flight experience. During the process, when bees are displaced experimentally into a strange territory without familiar landmarks, they orient initially at the correct angle to the sun; only with additional experience do they learn to include the diurnal course of the sun in their computations and to compensate for it.

14. Young bees, allowed to fly out only during the afternoon and hence able to become acquainted by experience with only a fraction of the sun's diurnal path, can complete the entire diurnal course of the sun from this fraction. They find the proper compass direction in the morning, on the basis of the sun's position. This makes comprehensible the performance of bees that point out the correct direction in reference to the sun's azimuth in the nighttime. They have expanded to a complete circle the daytime sector of the sun's course.

15. The capacity of finding a compass direction in accordance with

the solar compass is of biological significance in many respects. However, it is biologically meaningful also that knowledge of the sun's course is not rigidly determined by inheritance.

16. Bees are able to perceive the sun also through a complete cloud cover, long after the sun has ceased to be visible to us. That is shown most readily in dances on a horizontal comb. The indication of direction is appropriate when the dancers have a free view of the position of the sun, which is invisible to us. Dense cumulus clouds and rain clouds render the sun invisible to bees also.

17. Perception by bees of the sun through a cloud cover is possible only in the ultraviolet region (300–400 mμ). They perceive the position of the sun even through a dark glass that is transparent to ultraviolet, yet not through glasses that seem fully transparent to us but that do not pass any ultraviolet.

18. Since the sun is not perceptible to bees under all cloud covers, observation of their dances was combined with photographing the cloudy sky at the locus of the sun and in its vicinity. The photographs were made in the ultraviolet region; high-grade contrast-sensitive plates were used. They were evaluated photometrically. In those photographs that were made at the same time as well-oriented dances it was possible to determine an increase in brightness at the sun's location averaging 5 percent. As soon as—with thicker cloud cover—the dances became disoriented, the exposures ceased to reveal any increase in brightness in the ultraviolet range at the sun's locus (Figs. 324 and 325).

19. Regarding the sensitivity of the bees' eyes to contrast in the ultraviolet range, particularly in the dorsal parts of the eye, nothing is known. As to other points also these problems need further study.

C. ORIENTATION BY POLARIZED LIGHT

On a horizontal surface the dancers point by means of the direction of the tail-wagging run directly to the goal, by maintaining the same angle relative to the sun as prevailed during their preceding flight to the goal (see p. 131). In the closed fiberboard hut horizontal dances are disoriented. They turn immediately into dances directed toward the goal when a point of reference, the sun, or even just a piece of blue sky, is made visible to the bees. This does not have to be a broad blue area of the sky. An opening 10 cm wide in the fiberboard hut is sufficient. In other experiments a stovepipe, 40 cm long and 15 cm in diameter, was set into the north wall of the hut, so that from the dance floor the bees could see a circular piece of blue sky whose diameter subtended a visual angle of 10–15° (Fig. 327). That was enough for them to orient by; they pointed toward the west, where the feeding station was. If I set a mirror in front of the stovepipe, so that on the comb the reflection of a portion of the southern sky was seen instead of the northern, the dances were inhibited to a noticeable degree but were oriented uniformly toward the opposite direction, the east (see p. 383). If with a second mirror doubly reflected northern sky was seen

FIGURE 327. Schematic sketch of the observation hive laid horizontal in the closed fiberboard hut. For the illustration the panels have been removed on two sides. The dancers can see blue sky to the north through a stovepipe.

through the tube, the dances again pointed correctly toward the west.

From this I concluded "that the bees perceive in the segment of blue sky some feature influenced by the sun and according to which they are able to orient in relation to the sun's position. Perhaps this is the polarization of the light of the sky, which does have very definite relations to the sun's position . . . Initially it is not apparent whether and to what extent the hypothetical perception of the polarization of of the light of the sky participates in free flight and in orientation in the countryside. The rigidly positioned compound eye with its facets aligned in all directions could . . . provide an essentially ideal device for surveying the intensity and direction of polarization simultaneously over the entire expanse of the sky and for keeping this continuously in view. But it would be premature to weave the reader a sweet-smelling wreath from imaginary though attractive flowers. One must wait until the coarser foliage of facts has matured further" (v. Frisch 1948:41, 42).

During the years that have elapsed since then an abundance of facts has developed. Before we consider them, let us glance at the polarization of the vault of the sky.

5. The Polarized Light of the Sky

Natural light as it comes from the sun may be regarded as a vibratory phenomenon in which the direction of vibration of the transverse light waves changes rapidly and randomly. With linearly polarized light, the direction of vibration lies in a single plane (Fig. 328). Polarized light arises frequently in nature, for instance during the reflection of the sun's rays from a water surface.

The light that comes from the blue sky also is partly polarized, with both the direction of vibration and the intensity of polarization displaying definite relations to the position of the sun. Figure 329 presents a picture of relations when the sun is on the horizon at S. In addition to the horizon, the parallels of altitude are shown at 15° intervals.

Direct sunlight is not polarized. Moreover, there are three neutral points that have unpolarized light. They lie on the vertical circle of the sun, that is, the celestial meridian that runs across the zenith from the sun to its antipodal point, and are: Babinet's point B, about 20° above the sun, Arago's point A, 20° above the antipodal point, and Brewster's point, about 20° below the sun. With increasing distance from the sun or from its antipodal point the intensity of polarization increases and reaches its maximum (about 70 percent polarized light, depending on atmospheric conditions) at a distance of 90° from the sun.

The double-pointed arrows indicate the direction of vibration of the polarized light. To a rough approximation this follows concentric circles around the sun or its antipodal point, with strong deviations in the regions of the neutral points. As a general rule the direction of vibration at a celestial point looked at is perpendicular to the plane defined by this point, the sun, and the eye of the observer. When the sun is on

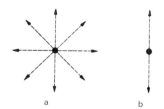

FIGURE 328. Diagram to explain the difference between (*a*) natural light and (*b*) polarized light: point, ray of light coming toward the observer; arrows, planes of vibration.

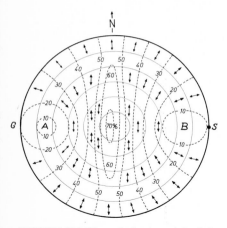

FIGURE 329. The polarization of sky light with the sun on the horizon: *S*, sun; *B*, Babinet's point (neutral point above the sun); *A*, Arago's point (neutral point above the antipodal point *G* of the sun); arrows, planes of vibration; dashed lines, loci of equal degrees of polarization; numbers, polarization (percent). Schematic, based on Müller-Pouillet, *Lehrbuch der Physik* (11th ed., 1928), V/1, p. 234.

the horizon, the half of the vault of the sky toward the sun is the mirror image of the opposite half, with respect to the intensity and direction of vibration of the polarization (Fig. 329). That changes as the sun climbs. Intensity and direction of vibration of the polarized light retain approximately their location in relation to the sun's position, so that Arago's point vanishes below the horizon while Brewster's point comes into view above the latter (below the sun; Fig. 330). In contrast to the ambiguous polarization pattern seen when the sun is on the horizon, with the sun in a higher position each area of the sky is characterized uniquely in its position relative to the sun. At different elevations of the sun above the horizon, the direction of vibration remains about the same in a given meridian.

The polarization of the sky is affected by a number of factors, but they have less influence on the direction of vibration than on the intensity of polarization. The degree of polarization declines with increasing atmospheric turbidity. Light from clouds is not polarized. According to a verbal communication from Professor Sekera (April 1957), although the intensity of polarization is less in the ultraviolet than in the blue, the direction of vibration is here most precisely oriented in relation to the sun's position. Disturbances of polarization by atmospheric dust also are least in the ultraviolet.[18] (For further details and references see Sekera 1951, 1955, 1957, Stockhammer 1959.)

6. Demonstration of Orientation by Polarized Light

Many different experiments were required before I myself would believe that an animal is capable of perceiving the direction of vibration of polarized light and of orienting itself accordingly. In order to convince others too, I planned an exhaustive publication of the extensive protocols. Today that is doubtless no longer necessary. For meanwhile the finding has not only been confirmed as regards bees; beyond this it has emerged that this seemingly incredible capacity is widespread in the animal kingdom (see pp. 439–444). Consequently a résumé of the course of the investigation and of its nature will suffice.

(*a*) We start with the fact that under certain conditions dances on the horizontal comb in the fiberboard hut become oriented correctly at once when the bees can see a spot of the sky, laterad of the sun, with a

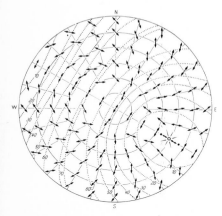

FIGURE 330. The polarization of sky light with the sun at an elevation of 30° and azimuth of 120°. Arago's neutral point has already vanished; Brewster's neutral point (black dot below the sun) has come into view above the horizon. Further explanation as in Fig. 329. After Stockhammer 1959.

[18] In reply to my question whether recent investigations had brought any changes in this, Dr. Sekera wrote me on 21 February 1964:

"The statement about the skylight polarization in ultraviolet is still valid. The polarization in ultraviolet is smaller than that in the blue or green (where it usually reaches maximum); however, in this region the polarization is unaffected by the great variability of atmospheric dust or other forms of atmospheric turbidity.

"New measurements have confirmed fully this statement; unfortunately, I have not published any new paper on this subject, nor can this be found in any other paper on skylight polarization. In our applications we are more interested in the spectral region, where the effect of atmospheric turbidity is largest, and therefore there is little mentioned about ultraviolet, where daily variations are smallest."

Further he wrote on 21 September 1964 that from available measurements it was to be concluded that the irregular variations in polarization that depend on local weather conditions are greatest (up to as much as 10–20 percent) in the range of yellow and red, but least (only 1–2 percent) in the ultraviolet.

visual angle of 10–15°. The conditions under which this holds already point to the decisive importance of polarization. The piece of the sky must be blue; with increasing cloudiness orientation grows worse, and if clouds alone are to be seen the dances become disoriented or are oriented incorrectly. The same is observed with a clear sky if the tube is aimed at areas near the sun. In both instances the bees are given light with very little polarization or without any.

In the experimental arrangement initially employed—a closed fiberboard hut with a stovepipe insert (Fig. 327)—the dances are not always disoriented when there is a view of a cloudy sky, but frequently are oriented incorrectly, because the dancers in the dark hut adopt the bright spot on the sky instead of the sun—just as one can direct the dances in a completely closed hut with a lamp (see p. 135). For brevity I call this the *lamp effect*. In order to exclude such an influence, in later experiments a canvas tent (Fig. 118) was used instead of the fiberboard hut. The difference in brightness between the piece of the sky and the canvas tent is so slight that the lamp effect is not manifest.

The latter enters by no means only into the experiments with the stovepipe. Even when there is a free view in all directions, the dancers—when the sun itself is obscured by dense clouds and no blue sky is to be seen—may be oriented according to the brightest spot in the sky. This connection becomes obvious if attention is paid to the bees that are about to leave the hive. In their positively phototactic mood for flying out, instead of pressing toward the entrance, in the horizontally placed hive, they make their way precisely in the direction that determines the orientation of the dances, and thus they themselves tell us that this is the brightest spot for them.

When there is a scudding overcast, so that first blue sky and then clouds are to be seen through the pipe,[19] oriented and disoriented dances succeed one another promptly, in impressive fashion.

(*b*) Bees in the closed hut that see blue northern sky through a tube dance correctly. It has been mentioned already (p. 380) that they would dance incorrectly, in the reverse direction, if a mirror were placed in front of the tube in such a way that they saw the reflected southern sky instead of the northern sky. These experiments were conducted in 1947 at midday, between 12:30 and 1:45 P.M. In order to find an adequate degree of polarization in the southern sky, before 12 M. the mirror was aimed somewhat west of south, after 12 M. correspondingly east of south. In both instances the plane of vibration in the south was approximately parallel to the meridian of the sky, but for the northern sky approximately parallel to the horizon. When they were en route to the (western) feeding station the bees would see the northern pattern on their right, the southern pattern on their left. Since during the experiment they could see only a piece of sky that was in the north but that showed the southern pattern, they quite properly turned so that the left eye was toward this pattern and thus danced erroneously, in the mirror image toward the east. Here again it is evident that the plane of vibra-

---

[19] With the help of a mirror the observer in the hut can quickly find out what the bees are seeing through the pipe at the moment.

tion of the sky light is decisive for the direction of dancing. Striking, and emphasized repeatedly in the protocols, was the strong degree to which the dances were inhibited beneath the reflected light of the sky. As the cause I suspect the slight change in plane of vibration due to reflection,[20] as a consequence of which the field displayed did not correspond precisely in pattern of polarization with any spot on the sky. But this question was not pursued further.

(*c*) On the vertical comb surface the dancers transpose the solar angle perceived during flight to the direction relative to gravity. But as soon as they catch sight of blue sky they are diverted from this direction ("misdirection," pp. 197ff). On pp. 198–202 it was explained that under such conditions the tail-wagging run corresponds precisely to the bisector between the two directions the dancers would have to maintain, one in transposing the solar angle to the direction relative to gravity, the other in adopting a position in immediate relation to the sun's position. Since the bees could not see the sun from the comb, they must have deduced the sun's position from the direction of vibration of the polarized sky light.

(*d*) Orientation according to the polarized light of the sky can be demonstrated by altering its plane of vibration artificially. At that time we did not have available polarizing sheets, which polarize the transmitted light linearly in a definite direction. I received some, 15 × 30 cm in size, through August Krogh, who was just then in the United States. There they were being sold as visual aids for motorists. I put this long-wished-for polarizer on the glass window of the horizontal observation hive over the dancing bees and eliminated the sky light from three sides by a three-part screen, which was placed on top of the comb; the view upward was restricted by the roof over the observation hive. From the comb the bees could see outward in only one direction, for instance to the north; an area of blue sky some 40° broad and 30° high, with its center about 45° above the horizon, was visible to them. If the polarizer was so placed that the plane of vibration of the transmitted light was the same[21] as that of the polarized light of the portion of the sky in question, the bees danced correctly. But if I rotated the polarizer, for example 30° clockwise, then I was dumfounded to see that the bees at once shifted their direction of dancing in the same sense and by about the same amount; when it was rotated in the opposite direction their indication of the line was in error by a corresponding amount toward the other side. With these observations all doubt was removed that the plane of vibration of the polarized light of the sky determines the direction of dancing (v. Frisch 1949).

(*e*) When the polarizer is rotated, the direction of dancing shifts in the same sense, but not necessarily to the identical degree. That is

---

[20] As mentioned, the plane of vibration in the southern sky was only approximately parallel to the celestial meridian. Reflection reversed the existing deviation, which thereby became unnatural.

[21] The "agreement" refers to the *average* plane of vibration. Agreement is not complete, because the set of polarizers polarizes the light in a single direction, whereas even on this limited area of the sky differences in the plane of polarization are noticeable.

comprehensible, because when the azimuthal angle is changed the plane of vibration of the light does not necessarily shift to the very same extent. In order to examine the relations in greater detail and to clarify for myself how rotating the polarizer affects the bees' eyes, I constructed a simple model of an ommatidium. It is based on the concept developed by Autrum for the analysis of polarized light by the insect eye (see p. 416). According to it the radially positioned sense cells of each ommatidium (Fig. 331) themselves polarize the transmitted light, each in a different plane according to its position, and hence like a radial Nicol prism would allow determination of the plane of vibration of incident polarized light. How this is envisaged may be understood at once by considering the model.

We first glance briefly at Fig. 332 in order to recall a well-known phenomenon. Strips were cut from a polarizing sheet in such a way that the plane of vibration of the transmitted light was parallel to the long sides of the rectangles. If two sheets are superimposed so that the planes of vibration coincide (Fig. 332 *left*), the light transmitted by the first sheet will also pass through the second without hindrance. The covered area merely looks somewhat darker, because the sheets are lightly tinted and a double layer absorbs more light. But if the sheets are rotated relatively, the covered area gets darker and darker, until when they are at right angles the light transmitted by the first sheet is completely absorbed by the second.

For the model, eight equilateral triangles are cut from a polarizing sheet and are joined into a star in such a way that after light has traversed it the plane of polarization of the light is parallel to the base of each triangle (Fig. 333). Whether matters are so arranged that the plane of vibration is parallel to the base of the triangles or everywhere perpendicular to it of course has no effect on the result. These triangular sheets in the model correspond to the eight rhabdomeres in the cross

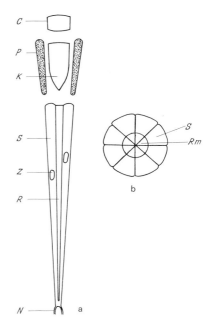

FIGURE 331. An individual eye (ommatidium or omma) from an insect's compound eye: (*a*) longitudinal section; (*b*) cross section, at higher magnification, through the sensory cells; *S*, sensory cells; *Z*, cell nucleus; *N*, nerve fiber; *C*, corneal lens; *K*, crystalline cone; *P*, pigment. On its internal face each sense cell forms a visual rod (rhabdomere, *Rm*); the rhabdomeres may unite to form a rhabdom, *R*.

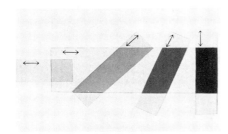

FIGURE 332. Polarizing sheets are placed over one another at different angles, with increasing extinction of the light. The plane of vibration of the light transmitted by the sheets is indicated by the double-headed arrows.

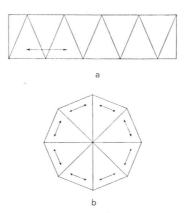

FIGURE 333. (*a*) Pattern for cutting a polarizing sheet to make a star-shaped model. (*b*) The double-headed arrows indicate the plane of vibration of the polarized light.

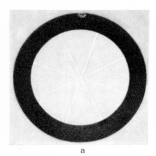

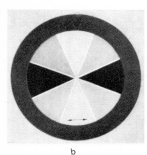

FIGURE 334. Views through the star-shaped model: (*a*) toward a bright surface that is emitting natural light; (*b*) toward a bright surface from which polarized light is coming; its plane of vibration is indicated by the double-headed arrow.

section shown in Fig. 331*b*, that is, to the light-sensitive inner portions of the sense cells. If one looks through the star-shaped model at a bright surface that is emitting natural light, all the triangles appear equally bright (Fig. 334*a*); we do not recognize, indeed, that the light has been polarized by the sheets. But if we look at a surface from which polarized light is coming, then there appears a brightness pattern (Fig. 334*b*), since depending on its plane of polarization the incident light is either transmitted fully, weakened, or cut off (see Fig. 332). Thus with such a model, which is based on the same scheme as an insect ommatidium with eight sense cells, the plane of vibration of incident light can be analyzed. I shall return later to the special relations peculiar to the eye of the bee (pp. 423ff). The correctness of the basic idea has been confirmed experimentally by Autrum himself, and by others since.

The star-shaped model was mounted on a frame so it could be rotated and tilted. Its position can be read from the appropriate graduated arcs (Fig. 335). This device has the advantage that with regard to any part of the sky that is of experimental interest it provides information of an accuracy adequate for our purposes both about the plane of vibration (through the brightness pattern) and about the intensity of polarization (through the contrast between the several fields). Figure 336 shows how each azimuth of the celestial circle is characterized by its polarization pattern and the contrast of the fields. At each point of the sky the patterns change, of course, with the time of day, since they depend on the position of the sun. Photographs of the sky through the star-shaped model, taken both at different times of day and at angular separations of 20° (from N to SE), are shown in Fig. 375.

FIGURE 335. The star-shaped model, mounted; azimuth and inclination may be read from two graduated arcs; in the center of the model is a small mirror for monitoring the vertical direction of sighting.

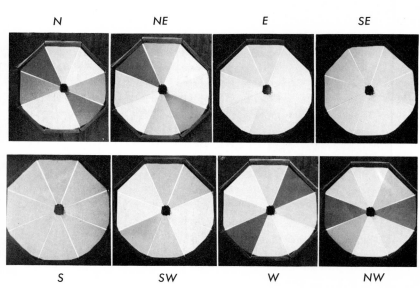

FIGURE 336. Views through the star-shaped model in eight directions; elevation above the horizon 45°; photographs taken at 10 A.M., 25 September 1949. Close to the sun the degree of polarization is slight, and hence the pattern contrast is poor. For corresponding exposures at a different time of day and at smaller angular intervals, see Fig. 375.

The purpose of the experiments that follow was, by rotating the sheet above the bees, to alter for them in a controlled fashion the plane of vibration of the sky light and simultaneously to test whether the shift in direction of dancing obtained was in harmony with the pattern in the sky.

Arrangements were now improved in comparison with the first experiments. To avoid lateral illumination the observation hive, lying horizontal, was covered with a sheet of plywood in which a square window, 20 cm on a side, was cut above the bees' dance floor. Over this window was placed a large polarizing sheet,[22] mounted in a rotatable circular wooden frame (Fig. 337); the diameter of the transparent area was 38.5 cm. The plane of vibration of the transmitted light was known, and its position in space could be read from a graduated arc. A three-piece screen (not erected in Fig. 337) around the sheet, and the protective roof above the hive, limited the bees' visual field to a patch of blue sky of some 40° in breadth and 30° in height, with its center about 45° above the horizon. In order to be able to read off the direction of the tail-wagging runs without interpolating an additional transparent pane, a net of white threads was stretched across the window at 30° angles (see Fig. 27).

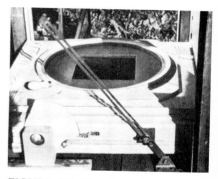

FIGURE 337. The observation hive, laid horizontal; on the hive is a board containing a square window, over which is the rotatable polarizing sheet.

To show what effect the sheet placed over the hive ("cover sheet") might have on the bees' eyes, a corresponding cover sheet in a circular rotatable frame was placed over the star-shaped model and adjusted in the same way as the sheet above the hive. The model stood in the open field beside the observation hive. The feeding station for the marked bees was 200 m west of the hive.

First, an example will clarify the principle of these experiments. For the sake of simplicity we choose an instance in which the dancers were given a view of the open sky to the west, in the same direction as that of the feeding station. The sun was not in the visual field, but only blue sky. Through the star-shaped model was seen the pattern M_1 (Fig. 338a). No other patch of sky afforded the same picture. En route to the feeding station the bees had this pattern before them. Simultaneously, of course, they saw other patterns in the rest of the sky. While dancing they were able to see only the western sky, and they oriented their tail-wagging runs in the direction toward this pattern, which they had had before them in their flight also. They danced correctly, with great precision, when the polarizing sheet was placed over the comb in such a way that the plane of vibration (and with it the polarization pattern) in the west remained unchanged.

---

[22] I wish to thank E. H. Land, president of the Polaroid Corporation, Cambridge, Mass., for generously supplying polarizing sheets. For many experiments it is important to know that these sheets are almost opaque for wavelengths between 300 and 400 mμ. According to measurements for which I am grateful to Prof. G. Joos (Physical Institute of the Technische Hochschule, Munich), transmission in this range amounts to only about 2 percent. On the other hand, the polarizing effect is still good, since two crossed sheets transmit at most 1 percent of the light that penetrates in this range through sheets placed in parallel. Later experiments with sheets that transmit ultraviolet (see p. 399), for which I am indebted to the optical shops of E. Käsemann (Oberaudorf am Inn), showed that in both cases the fundamental results, discussed below, were the same.

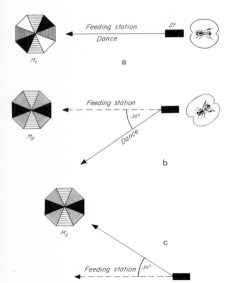

FIGURE 338. Experiment of 5 September 1949: in flying from the hive St to the westward feeding station the bees see before them the pattern M_1; when dancing on the horizontal comb they have a clear view only to the west. (*a*) From the comb they see the pattern M_1 in the west, and properly point in this direction. (*b*) The polarizing sheet is rotated above the comb, so that pattern M_1 is changed to M_2; thereupon the direction of dancing is shifted 35° to the left. (*c*) The pattern shown is found, by means of the star-shaped model, 34° N of W in the sky. In their dancing the bees have maintained the same angle (with an error of 1°) to this pattern as on their free flight to the feeding station.

Now the sheet above the beehive was rotated 30° counterclockwise. At once the direction of the bees' dancing changed, and pointed 35° S of W. When I moved the cover sheet over the star-shaped model into the same position as the sheet over the beehive, the pattern M_2 appeared in the west (Fig. 338*b*). Upon searching the sky with the star-shaped model (now uncovered) I found that this pattern occurred only at a region 34° N of W (Fig. 338*c*). On a free flight to the feeding station the bees would have had to keep to a line 34° to the left of this pattern in the sky. Now that it was their only point of reference for dancing, I found their dances directed 35° to the left of it—whether the imprecision of 1° was the bees' fault or my own is an open question. In artificially displacing the polarization pattern in the sky, the direction of dancing had been shifted to the very same extent. In pursuance of this principle, in the summer of 1949 I did many experiments that may be summarized best by combining them in groups.

Group 1. The plane of vibration of the polarized light is allowed to remain unchanged by the sheet, in the region of the sky visible to the dancers. If the star-shaped model is aimed toward this part of the sky and a polarizing cover sheet placed in front of it in the same orientation as over the honeycomb, the polarization pattern is unchanged, except that it shows more contrast because the light transmitted by the sheet is completely polarized, whereas that from the sky is only partly so (Fig. 339). In 14 experiments of this kind (10 measured dances each) the bees, as they did without any polarizing cover sheet, indicated the direction of the feeding station with an average error of 5.8°.[23] In the experiments the bees were given a view, alternately in all compass directions to N, E, S, or W.

[23] This error of course includes the observer's reading errors and his imprecision in positioning the polarizing sheets and in evaluating the pattern. In comparison with the drastic effects observed when the positions of the sheets are changed, errors of a few degrees in these experiments are of no consequence.

FIGURE 339. Experiment of 7 September 1949, 2 P.M.: (*left*) eastern sky, seen through the star-shaped model; (*right*) the same, with a polarizing cover sheet in front of it, so placed that its plane of polarization is the same as that of the polarized light from the sky. Because the sky light is only partially polarized, the pattern seen through the added sheet has stronger contrast; otherwise it is unchanged. The dark spot in the lower triangle is a tree top.

Group 2. By means of the polarizing cover sheet, the plane of vibration of the polarized light is changed in the portion of the sky visible to the dancers, in such a way that here a polarization pattern can be seen that actually is present in some other part of the sky at the time. An example of this was discussed on pp. 387f and has been further explained in Fig. 338. The bees' direction of dancing was shifted in the same sense and to the same extent as the polarization pattern in the sky was dislocated for them by the interpolated cover sheet. In 48 such experiments (with ten measured dances each) all four compass directions were chosen, partly in the morning, partly in the afternoon. The deviation of the observed direction of dancing from that expected theoretically amounted on the average to 8.3°.

Group 3. By means of the polarizing cover sheet, the plane of vibration of the polarized light is changed in the portion of the sky visible to the dancers into a pattern that cannot be recognized anywhere in the sky at that time. Under such circumstances the bees' dances were completely disoriented. In a total of 83 experiments this occurred 16 times. On 5 occasions the pattern produced with the cover sheet was not to be found anywhere in the sky, 10 times such a pattern could be made out, but it was unusually faint and invariably even poorer in contrast than the picture reproduced in the lower half of Fig. 340.

Figure 340 supplies the photographic evidence from an experiment in which an indication of direction was still just barely possible. The bees had a view to the south, through the polarizing sheet. The upper picture shows how the star-shaped model, directed southward, looked when the cover sheet placed before it was oriented like the polarizing sheet above the dancers. A version of the pattern, but with very poor contrast, was to be seen in the sky 60° W of N (lower picture), otherwise nowhere. The bees pointed to the SE, deviating on the average 15.5° from the direction expected theoretically. Their uncertainty was

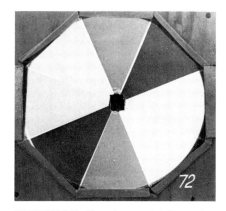
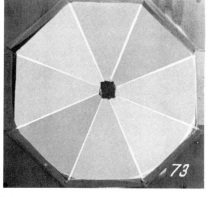

FIGURE 340. Experiment of 6 September 1949, 1:27–1:36 P.M.: (*left*) the southern sky, seen through the star-shaped model with a cover sheet in front of it; (*right*) without a cover sheet, the same pattern, poor in contrast, is seen in the sky, 60° W of N. When the cover sheet is placed as in the left-hand picture the dancers indicate 45.5° E of S, hesitantly and uncertainly. Deviation from the theoretically expected indication, 15.5°.

expressed also in the fact that, in contrast to their behavior under other conditions, they frequently ran in circles, as though searching, before, with a sudden "decision," they struck out on a definite course. Besides, in addition to 10 dances recorded as indicating a direction there were 4 that were completely disoriented. Similar behavior was seen in 14 experiments altogether, all of them with sky patterns that were poor in contrast. The evident relation between clarity in the pattern of the sky and in the indication of direction was particularly impressive.

Perhaps in these borderline cases the dances beneath the polarizing sheets would have remained oriented longer, as contrast was diminishing, if the sheets had transmitted ultraviolet (see note 22, p. 387).

In one experiment the dances were disoriented, although the pattern shown was clear at one spot in the sky. But in this case the pattern that the dancers saw through the polarizing sheet was poor in contrast and unusually dark (see v. Frisch 1950:217 and Fig. 17).

Group 4. Through the polarization sheet (or in its absence) the bees see a polarization pattern that occurs at this time in two different regions of the sky. Then the dances, like the pattern, are ambiguous and point alternately in two different directions. An example is shown in Fig. 341.

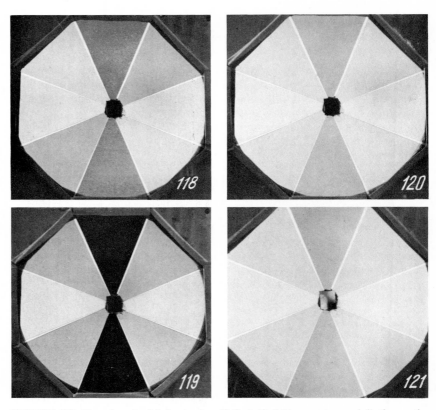

FIGURE 341. Experiment of 7 September 1949, 5:12–5:21 P.M.: (*upper left*) the northern sky, seen through the star-shaped model; (*lower left*) by means of the cover sheet placed in front of the model the pattern is altered a little and made bilaterally symmetrical. With the model this pattern was found at two places in the sky: 20° W of N (*upper right*) and 16° W of S (*lower right*). Dances: indication of direction in mirror images, to west and to east.

On 7 September at 5 P.M. the northern sky, viewed through the star-shaped model, gave the picture shown in Fig. 341, *upper left*. With the cover sheet it was altered slightly, so that it was bilaterally symmetrical (Fig. 341, *lower left*). The polarizing sheet was placed in the same orientation over the hive. Thus the bees saw this pattern in the northern sky. It was present in the sky in the same form but with poorer contrast 20° W of N and 16° W of S. The bees danced partly toward the west, where their feeding station was, and partly toward the east. Some shifted the direction of the tail-wagging run in regular alternation between west and east in the same dance. From within the hive they could not decide whether the pattern shown was the one from the northern sky or from the south, since they could not see any landmarks; in free flight the latter would have enabled an immediate decision.

Among the 83 experiments such ambiguities in the polarization pattern of the sky occurred five times, and in all five instances the bees' dances indicated two directions.

(*f*) In the experiments discussed hitherto the bees had a view of a laterally situated patch of sky, with its center about 45° above the horizon. If by means of a tube set into the roof of a closed fiberboard hut I made visible to them a blue spot (some 10° in diameter) in the zenith, then in their dances they pointed quite at random to the west (the direction of the feeding station) and to the false mirror image of this direction in the east. Not infrequently they also turned about, searching, in circles. Taken alone, the pattern of the zenith is ambiguous. The dancers see the same pattern with the body in two symmetrically opposite positions.

During the noon hours, whenever the sun was high in the sky, the pattern at the zenith became of such poor contrast that the dances—in good agreement with the other experiments already discussed—grew more and more uncertain and eventually became disoriented, until, with increasing distance of the sun from the zenith, the earlier situation was restored (v. Frisch 1951 and unpublished experiments).

From the experiments it emerges that the plane of vibration of the polarized light (which is expressed in the pattern shown by the star-shaped model) determines the position taken by the bees. The degree of polarization (on which the intensity of contrast in the star-shaped model depends) is comparatively without importance, or plays only a subordinate role. For while the bees respond immediately to every change in the pattern with a change in the direction of dancing, there is no noticeable effect on the dances when the degree of polarization shifts from the greatest intensity of contrast to near the limits of perceptibility. That is appropriate biologically, for under natural conditions the degree of polarization is subject to strong variations because of atmospheric turbidity and dust, whereas the plane of vibration remains relatively constant, especially in the ultraviolet (see p. 382).

When the star-shaped model is directed at the sky in aiming for a certain pattern, finding the latter is facilitated by slightly turning to left and right, going beyond the pattern sought, much as in fine-focusing a microscope or a camera one approaches the sharpest picture by screwing the lens back and forth. Frequently bees flying from the hive to a feeding station are seen to progress toward the goal on a line that weaves from right to left. This oscillating flight too might have the significance of being a constant retesting of the position taken according to the polarization pattern by means of slight excursions to right and left. If such is the case, this manner of flying should occur particularly when orientation is predominantly to the pattern of the sky, and not so much when guidance is taken over by landmarks, fine-focusing on which is less difficult. I have not pursued this question further. It should be easy to test.

7. The Connection Between the Polarization Pattern and the Position of the Sun. Experiments in the Shadow of a Mountain

From the experiments discussed hitherto it is evident that bees perceive polarized light as such and that they are able to orient their dances by its plane of vibration. Just as in view of the sun itself, when they can see the blue sky they point out by means of their tail-wagging runs the direction to the goal with respect to the sun's position. Thus the sun remains the reference point. But the dances on the horizontal comb do not yet prove that the bee is able to draw a parallel between blue sky and the sun's position. Her performance can also be explained more simply: in the horizontal dance she seeks the position of the body in which she again observes with the same parts of the eye that specific polarization pattern, retained in memory, that she saw during flight. Then she has adopted the proper stance relative to the goal, even without any reference to the sun's position. The following experiments will show that her accomplishments surpass this.

On the afternoon of 18 August 1953 we established an observation hive on the southwest shore of the Wolfgangsee near the Hotel Lueg and, starting at the flight funnel, at once set up a feeding station with which we journeyed along the lake shore. At 4 P.M. we were 10 m from the hive; at 5:10 P.M. the final site, 175 m from the hive, 38° E of S, was reached; and the marked bees were fed there until 6:30 P.M. (Figs. 342 and 343). During the entire time, hive, flight course, and feeding station lay in the shadow of the high mountain ridge that follows the lake shore to the southwest. Even when the hive entrance was first opened, the boundary of the shadow already lay so far out over the lake and so high above the hive[24] that even during their orientation flights the bees did not get into the sunshine.

On the following morning the colony was set up in the level fields near Aich, which were unfamiliar to it. The result of this displacement experiment was that during the time of observation (7:54–11:54 A.M.), of the 30 marked bees 9 came to the dish in the compass direction to which they had been trained, 0, 1, and 1 in the three other directions (Fig. 344a, b). Considering the brevity of the training time, that was good success. It shows that the bees are just as well acquainted with

FIGURE 342. Experiment in mountain shadow, on the shore of the Wolfgangsee: *St*, observation hive; *F*, feeding station, 175 m distant from the hive; behind the flight path is the steep slope of the Zwölferhorn; right, Hotel Lueg.

FIGURE 343. Experiment in mountain shadow: observation hive and feeding station (*F*), during training.

[24]Ascertained by measuring the distance from the shadow boundary to the shore, and upward by releasing children's balloons.

Orientation on Long-Distance Flights / 393

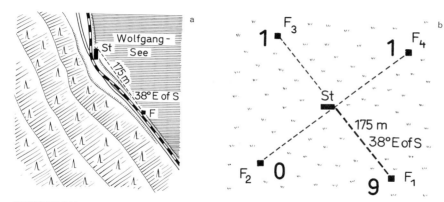

FIGURE 344. (a) Sketch map of the experiment in mountain shadow: *St*, observation hive; *F*, feeding station 175 m from the hive. (b) After being trained in the mountain shadow on 18 August 1953 from 4:35 to 6:30 P.M. (Fig. 343) the colony was moved to Aich on 19 August; of 30 marked bees, 11 flew to the feeding stands between 7:54 and 11:54 A.M., 9 of them in the direction of training.

the diurnal course of the polarization pattern as with the diurnal course of the sun. For in the mountain shadow during the late afternoon they had seen only the blue sky, and on the morning of the next day they found the accustomed compass direction in spite of the different distribution of the pattern over the vault of the sky.

On 31 August we repeated the experiment in the same manner with a second colony, only on the following morning the moor at Strobl was chosen as the site for the displacement. Again by far the majority came in the direction to which they had been trained (Figs. 345, 346; v. Frisch and Lindauer 1954).

From other experiments (pp. 339ff) we know that landmarks, such as here were provided for the bees during their training by the lake shore and the woods, are superior to the celestial compass in a competitive experiment. If the course deviates, they are the deciding factor. It is noteworthy that their absence did not lead the bees astray. That agrees with our experience in other displacement experiments. Only in those other experiments a lake shore and a forest margin were never realized simultaneously as landmarks all the way from the hive to the goal to the extent that they were here. The presence of just one of these marks, a shore or a wood, is enough to mislead the bees into a wrong compass direction. But the disappearance of both did not distract them from steering their flight in the accustomed direction according to the celestial compass.

When dancing on the horizontal comb the bees, if they see the blue sky, could direct their tail-wagging run toward the goal by adopting such a position that the polarization pattern of the visible patch of sky coincided with the area of the eye that had viewed the same pattern during the flight to the feeding station. During the experiment in the shadow of the mountain the hive was vertical. According to the well-known code, the bees transposed the solar angle into the direction relative to gravity. On their flight to the feeding station they had seen only blue sky and no sun. When dancing they saw neither blue sky nor sun.

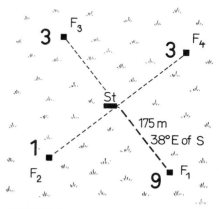

FIGURE 345. Results of an experiment with another colony: after being trained in the shadow of the mountain (Fig. 344a) on 31 August 1953 and moved on 1 September onto the moor at Strobl (Fig. 346), of 40 marked bees 16 flew, with the distribution indicated, to the four feeding stands. Period of observation, 7:54–11:24 A.M.

FIGURE 346. Experiment in mountain shadow: St, the observation hive after it was moved onto the moor at Strobl; F_1, the feeding stand in the direction of training. View in the same compass direction as in Fig. 342.

Since nevertheless they indicated on the vertical comb the solar angle, it follows that for them the polarization pattern is linked with the position of the sun.

In one of the two experiments, newcomers found the way to the feeding station in large numbers. They had been informed of the position of the sun, but reached the goal in accordance with the blue sky alone, without seeing the sun. Thus the position of the sun and the polarization of the sky light can substitute mutually for one another. When both are visible at once, the entire vault of the sky provides the all-inclusive field of the compound eye with a powerfully impressive compass.

A surprising observation during the experiments in mountain shadow bears evidence of another sort for the familiarity of bees with the diurnally changing course of the polarization pattern. On 18 August I recorded between 4:48 and 6:28 P.M. on the vertical comb 29 dances of completely normal character. During the second experiment on 31 August the time of observation was from 3:52 to 6:59 P.M. Up until 6:17 P.M. I had seen 36 normal dances. From 6:18 P.M. onward there occurred, in addition to normal dances, also some with erroneous tail-wagging runs, mirror images of the correct course. Of a total of 25 dances during this period, 16 were correct, and 9 were wrong but mirror images of the others. Among the incorrect dances all the tail-wagging runs, of which there were often many, pointed toward the wrong side. Yet more: it was always the same individuals that gave the alarm for the wrong direction. With bees Nos. 1, 20, 34, and 40, nine dances altogether were observed after 6:18 P.M.; all of these were incorrect mirror images, while simultaneously 16 normal dances by 15 other individuals were seen.[25] At 6:18 P.M. on 31 August the sun was only 4° 37′ above the horizon. That meant that in the eastern sky, visible to the bees in their flight, there was the mirror image of the identical polarization pattern that occurred in the west, where it was screened from them by the mountain ridge. Four of our foragers oriented their dances on the vertical comb according to the polarization of the western sky, which they had not seen. Evidently they were familiar with the symmetric polarization pattern of the afternoon sky, and in orienting their dances in accordance with the image remembered from their preceding flight they had interchanged the two directions. I see no other explanation for their performance. In the previous experiment on 18 August, observations had been concluded at 6:28 P.M., with the sun 6° 49′ above the horizon. Presumably indications in the direction of the mirror image would have taken place then too if cool weather had not dictated earlier cessation of the training.

Unfortunately, after the colony had been displaced on 1 September to the Strobl moor, of the four bees that had danced according to the mirror image of the proper direction, only one (No. 34) appeared at a feeding stand—as we had anticipated, at the northwest station. She

[25] After I had noticed the occurrence of dances in the mirror image of the proper direction, I tried insofar as possible to observe at least one dance by each bee of our group; with six individuals I did not succeed.

stayed loyal to her (symmetrically) mistaken course even when flying in a strange landscape, whereas during the preceding afternoon she had been able to steer correctly to the feeding station in accordance with the landmarks. If, in view of the fact that the indication of direction was given in the form of its mirror image, we evaluate as correct a flight to the incorrect station (because it is the mirror image of the proper one), the result of the second displacement experiment (Fig. 345) is improved, to ten correct choices and on the average two flights to each of the three incorrect compass directions.

With a complete cloud cover, an indication of direction from the position of the sun should be impossible in the shadow of the mountain. Unfortunately, I have only one experiment under such conditions. It confirmed expectations. On 25 September 1959 a complete cover of high fog lay over the Wolfgangsee all day long. We took an observation hive from Brunnwinkl to Lueg, to the same place as before, and opened it for flight at 2:22 P.M. The bees showed little interest in flying off for collection. By 4 P.M. we were only 10 m from the hive with the feeding station, at which oil of lavender had been added. At 4:45 P.M. we reached 80 m. After 5:10 P.M. there were no more dances, and soon flying ceased altogether.

It was not possible in the dances on a vertical comb to measure a definite direction, because it varied greatly from one tail-wagging run to another. I had to confine myself to sketching according to visual estimation the direction of the individual runs. Often the dancers merely turned searchingly round and round. According to the sun's position during the period of observation, an indication of 110–113° to the left would have been correct. But not one of the 29 tail-wagging runs recorded was oriented even approximately in this direction. Most of them pointed to a direction between 40° and 120° to the right. The preference for this direction is understandable on the basis of other experiences. In the absence of a view of the sun and without blue sky, the bees select the brightest spot in the visual field as a replacement for the sun, be it a lamp or a bright patch of sky (*Lampeneffekt*, p. 383). In the present instance there was the dark wooded slope in the SW, and to the NE a uniform broad bright area of sky. The bees seemed, with considerable scatter, to have oriented to about the center of the bright area of sky, and in allusion to this inadequate point of reference they announced: "Feeding station to the right of the light" (see Fig. 344a).

I regret being unable to support this interpretation with more copious data. But there can be no doubt of the significant result, that in this experiment the indication of direction according to the sun's position failed.

8. THE USE OF ARTIFICIAL POLARIZATION PATTERNS WHEN THE SKY IS CLOUD COVERED

In September 1949 it was the perception of the sun through clouds that I was studying. In a control experiment with a complete cloud

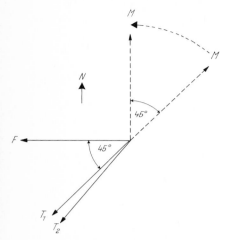

FIGURE 347. Example from the protocol of the experiment of 21 September 1949, 3 p.m.; looking down on the horizontal comb; F, feeding station. From the dance floor, the bees have a view to the north and see through a polarizing sheet the pattern M that with blue sky would have been visible 45° E of N. The shift in pattern would correspond to a change of 45° in the direction of dancing T_1; observed direction of dancing, T_2; difference from expectation, 5°.

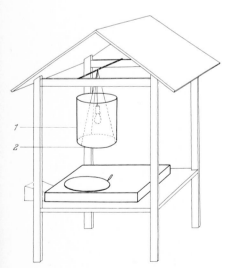

FIGURE 348. Observation hive, placed horizontal in the closed fiberboard hut (here shown open), with a rotatable polarizing sheet above the dance floor; 1, black paper screen; 2, round wooden frame with a pane of ground glass.

cover the bees looked out from the horizontal comb at a patch of northern sky far to one side of the sun's position. As expected, they performed disoriented dances. Now I laid a polarizing sheet over the comb. To my surprise the dances at once began to indicate a definite direction. In a number of tests I placed the sheet in differing orientations and recorded the direction of dancing. When on one of the following days I used the star-shaped model to examine the sky, now blue again, the following facts became evident. If the bees that had been foraging at the western feeding station under a complete cloud cover were shown the beclouded sky to the north through the polarizer and if the latter was adjusted so that the plane of vibration agreed with the one that would have been seen when the northern sky was blue at this time, then the bees danced correctly and pointed to the west. If the sheet was put in a different position so that the polarization pattern was what would have been visible with blue sky in some other part of the heavens at this time, then the tail-wagging runs maintained the angle with the northerly direction that corresponded to the angle between the true celestial location of the pattern shown as being in the north and the westerly direction to the feeding station (Fig. 347). In other words, the direction of dancing was shifted in the same sense and by approximately the angle through which the polarization pattern of the sky had been moved for the bees. But if there came into view through the sheet a pattern that at the time of the experiment was not realized anywhere in the sky, then the bees performed disoriented dances (v. Frisch 1950:219). All of this fits with the results that we had obtained with the blue sky by interposing the polarizing sheet (pp. 387ff). The difference is merely that this time, in consequence of the cloud cover, the bees had not perceived any polarization in the sky during their foraging flights. They must have had the location of the patterns and their diurnal course in their head.

During these experiments, and likewise in repetitions in the summer of 1950, I sometimes had the impression that new members of the foraging group, who had not yet frequented the feeding place when skies were blue, danced beneath the polarizer with orientation inferior to that of the old visitors to the feeding station. But I was unable to obtain clear evidence on this point. That was partly because denser clouds too will interfere with the certainty of orientation, when on account of the decrease in brightness the contrast of the polarization pattern falls too low. In order to be independent of this, in September 1957 under a beclouded sky I supplied in lieu of artificially polarized light from the clouds the polarized light of a lamp that was suspended in the closed fiberboard hut over the horizontal hive.

The setup may be seen from Fig. 348. The observer is in annex A of the closed fiberboard hut (see Fig. 5), which in Fig. 348 would be in front of the side of the stand toward the reader. Above the bees' dance floor there hangs a 200-watt bulb, beneath which there is a ground glass in a circular wooden frame (2). A black paper cylinder (1) shields the inner walls of the fiberboard hut from light emitted sidewise. The rotatable circular polarizing sheet is placed beneath the lamp on the window of the observation hive. The lamp is turned off between measurements in order not to disturb too greatly, in consequence of the illumination from above, traffic in and out of the hive entrance. The direction of dancing is measured in degrees and dictated to an assistant, or sketched from visual estimates, since errors of a few degrees are not important to the problem.

In contrast with the earlier experiments, this time the dancers saw the polarized light in the zenith. Precisely like the view of the blue sky through a stovepipe aimed upward, the pattern of the zenith is ambiguous. Correspondingly the dances, insofar as they were oriented, always pointed in two directions, one the mirror image of the other (see p. 391). Often the tail-wagging runs signaled the two directions in regular alternation, but often, too, several runs in one direction would succeed one another, then to be replaced by runs in the opposed sense. The experiments on 11 September 1957 lasted from 11:00 A.M. to 3:35 P.M. At times the sun was weakly visible through the closed cloud cover, but no blue sky was to be seen. The plane of vibration of the artificially produced polarized light was adjusted alternately to be from north to south and from east to west. During the period of observation the polarizer was rotated 90° ten times. The feeding station was 200 m west of the hive. On the preceding day the foragers had traversed the course under blue skies. But between 2:00 and 2:15 P.M. there were incorporated into the group three newcomers, who consequently were acquainted with the flight path only under cloud.

Twenty-six dances were recorded for bees that had earlier passed over the flight path under blue skies. One dance was disoriented; the other 25 agreed with the direction to be expected from the adjustment of the polarizer.

When for instance at 2:30 P.M. the sun was in the SW (azimuth 225°), the foragers, with blue sky in the zenith, had seen a plane of vibration running from NW to SE (S_1, Fig. 349) and had maintained an angle of 45° with it. When illuminated now from above with a west-east plane of vibration (S_2), they positioned themselves accordingly and danced alternately to the SW and NE (T). When the polarizer was rotated 90° the dances at once shifted uniformly into the directions NW and SE. As the sun's position advanced the directions changed correspondingly. In general, the indication of direction was less precise, the scatter greater, than when orientation was according to the blue sky.

With the three bees that were acquainted with the flight path only under a cloudy sky, during this same time 20 dances were observed. Of these 17 were oriented correctly, or predominantly so; 3 were disoriented.

Because at that time there were but few newcomers available, I repeated the experiment on 2 September 1963 under a complete cover of high cloud. We were able to incorporate into the foraging group seven newcomers that had not yet visited the feeding station under blue skies.

This time the evaluation was not for entire dances, but the directions of the individual tail-wagging runs were recorded. With one of the seven bees only a single very short dance was seen, with a second a

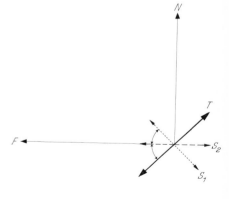

FIGURE 349. Experiment with a complete cloud cover: observation hive in the closed fiberboard hut, illuminated from the zenith through a polarizing sheet (see Fig. 348); F, feeding station; S, sun. With a blue sky, the bees on their flight to the feeding station would have seen the plane of vibration S_1 in the zenith; in the experiment the dancers see through the polarizer above them the plane of vibration S_2. The direction of dancing observed (in opposite senses) is T.

somewhat longer one; altogether these two bees made four tail-wagging runs that accorded with expectations and six that did not. The five remaining newcomers danced vigorously. During the time of observation, from 12:44 to 1:31 P.M., I saw them perform 25 dances; these were well oriented in the direction to be expected from the placement of the polarizer. With the adjustment of the polarizer used at the beginning of the experiment, there were 42 tail-wagging runs directed according to expectation and 3 that deviated. After the polarizer had been rotated 90°, 62 tail-wagging runs corresponded to the direction now to be expected; 7 (distributed through different dances and interspersed among correct runs) were in other directions.

Thus it was shown that not only foragers that had already flown over a route under blue sky are aware beneath a cloud-covered sky of the polarization patterns that would appear at various locations if the cloud cover were dispersed; newcomers, too, that have joined in only under a cloud-covered sky and that in their flights have not been able to see any celestial polarization but only the sun (through clouds) have mastered this capacity. They provide further evidence for the fact that in the nerve centers of bees the polarization pattern of the sky is linked with the position of the sun.

9. The Relative Significance of the Sun and Polarization of the Sky

If the entire celestial compass is put into competition with landmarks, it is the nature and course of the latter that determine whether they or the signals from the sky are to regulate the direction of the bees' flight (see pp. 339ff). Now we play the two signposts of the heavens against one another and enquire whether the sun or the polarized light of the sky is the more potent. Here too the details prove to be of importance.

As long ago as 1949 and 1952 I already did experiments according to which the laurels of victory clearly rested with the sun. At that time the dancers on the horizontally situated comb had a view approximately to the south, and regarded a patch of blue sky about 40° wide and 30° high with its midpoint about 40° above the horizon. When the dancers saw only blue sky, they indicated a westerly direction (toward the feeding station) and could be shifted by rotating a polarizing sheet, as has been described on pp. 387ff. But as soon as I expanded the view to such an extent that the sun shone on the bees through the sheet, the change in the direction of dancing was eliminated at once, and the tail-wagging runs pointed to the west. The altered plane of vibration of the polarized light was powerless against the guiding of the dances by the sun itself.

The polarizing sheets used at that time are almost opaque to ultraviolet light (see p. 387n). When the significance of ultraviolet for the orientation of bees became apparent, doubts arose whether those earlier findings were indeed valid under natural conditions. Therefore I repeated the experiments in the summer of 1956 with polarizing sheets that have considerable (about 50 percent) transmittance in the range from 400 to 300 mμ (Fig. 350).[26] For the sake of better comparison, in this work I used alternately the old, ultraviolet-absorbing, and the new, ultraviolet-transmitting, sheets. To begin with, arrangements remained otherwise entirely unchanged. If the bees saw only blue sky and if the plane of vibration was altered by rotating the sheet, they reacted in the same way with and without ultraviolet light. But when the sun shone through the sheets the response was different in the two instances; when the ultraviolet-absorbing filter was used the dances performed under influence of the sun pointed westward (toward the goal), without being diverted by the polarized light. Beneath the ultraviolet-transmitting filter there were indeed also dances oriented to the west, but more often there were dances appropriate to the altered plane of polarization or that suited some intermediate direction. Thus, when the ultraviolet range was participating, in addition to the sun itself the polarized light had clearly gained in influence on the direction of dancing, and there arose a competition of uncertain outcome.

As yet this was no honest competition. For opposed to the full power of the sun there stood only a little segment of sky, the plane of polarization of which did not accord with the position of the sun. In addition, the ultraviolet radiation was still weakened. I wanted to pit the full extent of the natural polarization of the sky light against the sun. Since one cannot mimic the complete expression of the pattern over the entire sky, the only alternative was to replace the sun.

Thus the following experimental arrangement was conceived. The roof of the observation hive and its entire superstructure were removed. From the horizontally situated comb the bees had a view over the whole of the blue celestial vault. The sun was cut off by a small screen and its mirror image[27] was thrown onto the dance floor from a direction shifted by 90° without changing the elevation above the horizon. After each observation a dark cloth was put over the hive while the mirror was being reset in accordance with the position of the sun. Polarizing sheets were not involved. The idea was that the natural polarization of the sky should show what it was capable of in competition with the reflection of the sun. Of course the latter alone shifted the dances by 90°, in a northerly direction. By means of the polarized light alone the direction of dancing was directed to the west (Fig. 351a), as control experiments regularly confirmed. When the bees glimpsed simultaneously both the blue expanse of the sky and the reflected sun, they adopted a compromise direction. Their direction of dancing was no

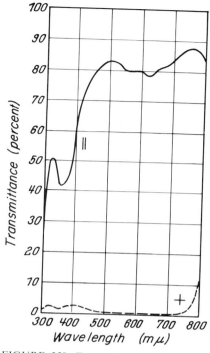

FIGURE 350. Transmittance of a Käsemann ultraviolet-polarizing sheet (unprotected by glass) of light polarized by passing through a similar sheet. After E. Käsemann.

[26] The E. Käsemann Company, Oberaudorf am Inn, makes such polarizing sheets. I am indebted to them not only for supplying these without cost, but also for preparing such sheets in the sizes necessary for my work. The sheets are very thin. Hence they have to be enclosed between thin plates of lucite (which will transmit ultraviolet). During this summer the glass windows of the observation hive were replaced with ultraviolet-transmitting WG$_1$-plates (Schott).

[27] Polarization of the sunlight by reflection is not a factor as a possible source of error. For it proved to be of no consequence whatever to the position toward the sun adopted by the bees, no matter whether or in what position a polarization sheet was interposed. In the experiments described on pp. 396f (for setup see Fig. 348), conditions were different, because a ground glass intervened.

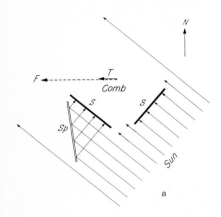

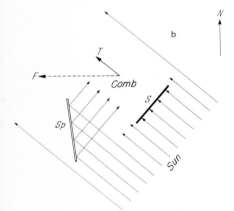

FIGURE 351. View looking down on the horizontally placed comb of the observation hive; the bees have a free outlook over the entire expanse of the blue sky; *F*, direction to the feeding station; *Sp*, mirror. (*a*) Direct and reflected sunlight both cut off by screens *S*; direction of dancing *T* according to the polarized light of the sky, correctly toward the west. (*b*) Direct sunlight cut off by a screen; the dancers see the sun displaced 90° in the mirror, which should shift the direction of dancing to the north; the observed direction of dancing *T* is a compromise between orientation according to the sun and that according to the blue sky.

longer so variable and impossible to reckon as with the earlier experimental arrangement. In spite of the marked scatter among the individual runs, a definite indication of direction stood out clearly. But it changed with the time of day: in four series of experiments, on 5, 9, 23, and 24 September 1956, in the morning between 9 and 10 A.M., the compromise direction always lay between about 20° and 30° N of W, between 10 and 11 A.M. it shifted to about 40°–60° N of W, and between 11 A.M. and 12 M., with increasing scatter, it was about 60°–70° N of W, with even a few individual tail-wagging runs aimed directly N. Thus the influence of the sun relative to the polarized light increased with increasing elevation, and declined again in the afternoon as the position of the sun fell.[28]

I believe that this change in the relative effectiveness of sun and polarization of the sky light is determined only indirectly by the sun's elevation. For in other experiments it has been evident that the direction of dancing does not respond to changes in the sun's elevation but only to alteration of the azimuth. Another relation might exist: when in our experiment the bee is guided by the sun and makes the tail-wagging run in a northward direction she comes into conflict with the polarization pattern, which pulls the run westward. As the sun climbs on its way toward the south, the pattern changes, so it seems to me, in such a way that toward noon it conflicts less with a more northerly directed dance than it does during the morning hours. In order to judge this surely one would have to possess precise polarization charts for the entire sky during the times in question.

Another interpretation, pointed out to me by K. O. Kiepenheuer, is simpler: as the sun rises its brightness in the ultraviolet increases extraordinarily, whereas the sky light in the zenith remains at a constant intensity. On a clear summer day the intensity in the region around 400 mμ is often about four times as great at 10 A.M. as at 8 A.M., and for shorter wavelengths it is even significantly more, whereas in the green there is scarcely any increase in brightness. In view of the great stimulatory action of ultraviolet on the eye of the bee, this increase in brightness must have a strong effect. In some other experiments the dances had been subjected to the competitive influence of sun and gravity. When the light of the sun was gradually weakened by the increasing density of a cloud cover, the sun's directing influence diminished (see p. 202). Hence it may be assumed, on the other hand, that in competition with the polarization it gains in potency as brightness is increased.

Whether now one of these interpretations is right, or whether both or neither are correct, the experiments leave no doubt that the natural polarization of the entire expanse of the sky has a clear and often a predominant influence when competing with the sun. A small segment of the sky did not have this effect. In this may be seen a confirmation of our assumption that in their orientation the bees pay attention to the polarization of the whole celestial vault.

[28] This last I can substantiate with only eight dances, from 9 September between 2:53 and 3:06 P.M.; on the average these pointed, without much scatter, 20° N of W.

10. What Color Range is Effective in the Perception of Polarization?

In the perception of the sun through a cloud cover only the ultraviolet is important (pp. 368f). The same is not true of the perception of the plane of vibration of polarized light. For this plane was recognized beneath polarization sheets that transmitted almost no ultraviolet. In order to learn more about the effective ranges of radiation, I tested the bees beneath the same wavelength filters (Schott and Genossen) that had been used in the experiments with a cloud-covered sky.

Matters were so arranged that from the horizontally situated comb the dancers saw only a patch of blue sky to the north, at a mean level of some 40°–50° above the horizon. They oriented themselves according to the polarization and pointed correctly to the west, where the feeding station was located 200 m away. Then the indication of direction was examined beneath various colored filters, with frequently interpolated controls without a filter. Two series of experiments (1953 and 1955) were conducted under the canvas tent with the "sun pipe" (Fig. 118); here the filter was placed in front of the pipe. In two other series of experiments (1949 and 1955) the bees were given a clear view of a rather large patch of blue sky in the north and their outlook was screened only on three sides and upward; the filters were laid above the dancers on the window of the observation hive. In 1955 the hive's outer pane of glass was removed during the experiments and the inner window was replaced by a WG_7-filter (Schott) transparent to ultraviolet.

The results of all series of experiments were in agreement and may be summarized as follows. Ultraviolet is adequate by itself for orienting by polarized light. Under the UG_1-filter, which looks black to us, the dances were always directed clearly toward the goal. But without ultraviolet the dances are well oriented too, as long as sufficient blue is present. Beneath filter GG_4 an indication of direction was still possible, but it was obviously distorted, while dances under VG_{10} were completely disoriented—with the green and yellow regions scarcely weakened. Hence the plane of vibration of the polarized light in the green and yellow is not effective for orientation. In regard to the transmittance of the filters used see Fig. 322.

In detail, the following were the results: the dances were oriented well beneath filters BG_3, BG_{12}, BG_{13}, BG_{14}, BG_{17}, BG_{18}, BG_{20}, BG_{21}, BG_{23}, BG_{25}, BG_{26}, BG_{34}, GG_{13}, GG_{18}, NG_4, NG_5, UG_1, WG_1, WG_2, WG_3, WG_4, WG_5, WG_6; they were partially disoriented beneath GG_4; they were completely disoriented beneath filters GG_9, OG_4, RG_1, RG_7, UG_6, VG_5, and VG_{10}.

Since the polarization of the light of the sky includes the region of long-wavelength radiation too, the reason for the ineffectiveness of green and yellow must be sought in the bees' eye itself or in the central nervous system.[29]

The dorsal portion of the bees' eye is not by any means blind to yellow light. When in the canvas tent I let them see the sun through

---

[29] See footnote, p. 417.

the pipe with filters, the dances were disoriented only under RG_7 and UG_6. These filters pass no light, but only heat radiation. Beneath filter RG_1 the dances were directed precisely toward the feeding station, although this filter transmits only a narrow band of radiation at the long-wavelength end of the spectrum visible to bees. In this test there were involved exactly the same regions of the eye that had failed to analyze the polarized light in the long-wavelength range.

The objection might be raised that in green and yellow light the bees do not orient themselves in accordance with the plane of vibration merely because in this range of color the brightness beneath the filters is too low. But this assumption is opposed by the following experiments.

The hive was laid horizontal beneath the canvas tent. Through the pipe the dancers saw instead of blue sky a lamp in front of which was interposed the color filter and—to replace the polarization of the sky light—a polarizing sheet. In this way the degree of polarization was increased almost to 100 percent. The area visible through the pipe was much brighter than the patch of sky visible in the earlier experiments. The greatest brightness was attained with a ground glass and combined source (mercury vapor + filaments), the luminous intensity of which was about 50 percent more than with an ordinary 200-watt filament bulb. That in the yellow-green range (filters GG_4, VG_5, and OG_4) the bees now perceived the opening of the pipe as a very bright spot—very bright in comparison with the canvas tent, which screened off the rest of the sky for them—they themselves demonstrated, by orienting to the azimuth of this spot of light as though the sun itself stood there (lamp effect, p. 383). Thus the dances were clearly oriented, but not according to the plane of vibration of the light. If the bees' stance had been taken in accordance with the polarization, the dances would have had to point to an entirely different direction. Now the intensity of light was decreased stepwise, by placing in front of the pipe instead of the combined source a 200-watt bulb, then 100-watt and 40-watt, and finally a sheet of white paper illuminated from the side. In no instance was there orientation of the dances by the plane of vibration of the light. The response remained one of the lamp effect, that is, the bees oriented to the source of light as though it were the sun. Even with the 40-watt bulb there were still some dances directed in this manner, whereas others became disoriented. With the weakest illumination (white paper in front of the pipe) only disoriented dances occurred. Thus as the intensity of light decreased there was an immediate transition from orientation toward the light (lamp effect) to disoriented dances, without any suggestion of orientation by polarization. In view of the broad range of brightnesses that caused orientation toward the light, the absence of any orientation relative to the plane of vibration cannot be ascribed to insufficient light intensity. In comparative experiments with a blue filter (BG_3) orientation to the goal in accordance with the plane of vibration of the polarized light was unequivocally clear with all the lamps, from the mixed source to the 40-watt bulb.[30]

[30] This is important because interposition of a polarizing sheet in the line of sight to the sun has no effect (see p. 399, note 27). But in this respect the lamps were not to be equated with the sun, whether because they were less bright or because of the much larger illuminated surface.

Why only the blue and ultraviolet regions are decisive for the orientation of bees by the polarized light of the sky, while the polarization of the rays of longer wavelength remains ineffective, is unexplained. But that this is so must be of the greatest significance for the proper functioning of this method of orientation. For it is in the ultraviolet that the plane of vibration is the most precisely directed relative to the position of the sun, whereas in the long-wavelength region atmospheric distortions are very large (see p. 382 and the pertinent footnote).

11. What Degree of Polarization Is Needed for Orientation?

Close to the sun the polarization of sky light is so weak that no pattern shows in observations through the star-shaped model. Not until one reaches a lateral angular separation of 30°–40° from the sun does any pattern become manifest. In agreement with this is the fact that the bees' dances on the horizontal comb were disoriented within the canvas tent when looking through the pipe (Fig. 118), they saw a patch of blue sky that was less than 30° removed from the sun.

In order to learn more precisely what degree of polarization was necessary for orientation, I combined observations of dances with measurements on those areas of the sky that were displayed to the bees.

For measuring the degree of polarization an illumination meter (photocell with an iris diaphragm and a rotatable polarizing sheet[31]) was mounted at the inner end of the "sun pipe" (Fig. 352). The little device[32] could be swung quickly out of the way to one side, so that observations on the dances and measurements of the polarization could follow one another in rapid alternation. The photocell was connected to a galvanometer (mirror galvanometer from Hartmann and Braun, Frankfurt). Before the measurements, after selecting the appropriate sensitivity range and balancing the galvanometer, we positioned the polarization sheet so as to obtain the biggest galvanometer excursion. Then the iris diaphragm was opened up until the spot stood at 100 on the scale, and thereupon the minimal excursion was determined by rotating the sheet. From the value observed during the experiment the degree of polarization could be read from a table.

With areas of the sky near the sun, a suitably positioned screen prevented the inside of the pipe from being illuminated unilaterally by the sun—which might have distorted the orientation of the bees.

In the experiments (summer of 1960), I observed the orientation of the dances in the canvas tent on a horizontal comb, while the bees could see through the pipe a patch of blue sky 15° in diameter.[33] The degree of polarization at that spot in the sky was measured immediately afterward.

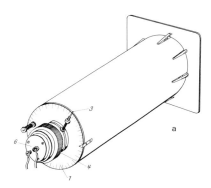
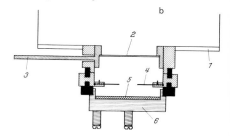

FIGURE 352. Apparatus for measuring the degree of polarization: (a) over-all view; (b) section through the lower end. At the lower end of the "sun pipe" is fastened a plate 1 that can be swung aside, with an attached polarizating sheet 2 that can be rotated by the handle 3; 4, iris diaphragm; 5, photocell with light-sensitive layer; 6, housing and output leads.

[31] In a part of the experiments an ultraviolet-transmitting sheet was used (see p. 399), but that had no appreciable effect on the measurements.
[32] The device was constructed in the workshops of the Second Physical Institute of the University of Munich; for this I am indebted to Professor Rollwagen.
[33] Control experiments showed that the glass pane (window glass) over the bees did not interfere with the degree of polarization. In other control experiments the "sun pipe" was closed off with a canvas cover, whereupon the dances always became disoriented. From this it was evident that local differences of brightness on the canvas tent did not provide the bees with points of reference for orienting.

TABLE 39. Orientation of the dances with different degrees of polarization of light from the blue sky. Each number is based on observation of about 6 or 7 dances; naturally the number of tail-wagging runs was many times greater.

Dances	Percent polarized light in individual experiments
Well-oriented	12, 15, 17, 17, 19, 20, 21, 23, 24, 28, 28, 30, 31, 32, 35, 37, 39, 40, 43, 51, 51, 58
Still clearly oriented	10, 12, 13, 14, 15
Imperfectly oriented	10, 11, 11, 13, 13, 14
Not completely disoriented	7, 8, 8, 9, 9, 11
Completely disoriented	2, 3, 3, 3, 4, 4, 4, 4, 5, 5, 5, 7, 8, 10

The results of the total of 53 experiments show that light polarized about 10 percent is needed for perception of the plane of vibration. With less than 7 percent the dances were always completely disoriented, and with more than 15 percent well oriented. Between 7 and 15 percent there lay a transitional region of recognizably but imperfectly oriented dances (Table 39).

The polarization of sky light extends over all ranges of wavelength visible to bees, but is perceived by them only in the ultraviolet and blue. This had been deduced from the performance of dancers beneath light filters (p. 401). I rechecked this finding, monitoring simultaneously the degree of polarization. In the experiments, on each occasion a light filter (Schott) was attached outside in front of the pipe opening; and the course of dancing, and immediately thereafter the degree of polarization, was tested in the filtered light.

It was confirmed that ultraviolet alone (UG_1[34]) would suffice for orientation, even with as little as 11-percent polarized light (Table 40); and that on the other hand ultraviolet is not essential for orientation, for even beneath the GG_{13} filter, which transmits no ultraviolet, the dances were aimed well for the goal, and were still partially oriented in two experiments with 16-percent and 14-percent polarized light, respectively. Even beneath WG_1, which transmits only a part of the longer-wavelength ultraviolet, orientation was still quite good. On the other hand, beneath filters GG_4,[35] VG_5, and OG_4 there were only completely disoriented dances, even at a high degree of polarization.

12. On the Function of the Bees' Ocelli

On the frontal area of the head between the compound eyes the bee bears three visual organs of simple structure (ocelli). A variety of functions have been ascribed to them. A possibility to be considered was

---

[34] For the transmittances see Fig. 322.

[35] Beneath GG_4, which does indeed weaken the blue region but not as much as VG_5, in two earlier experiments I had obtained dances that still were partly oriented ones (p. 401). That may have been because in one instance the bees saw through the pipe deep blue sky at a great distance from the sun, thus presumably with a rather high degree of polarization, and on the other occasion they had a view of a rather broad expanse of sky.

TABLE 40. Orientation of the dances in colored light with a regulated degree of polarization. The transmittances of the filters are given in Fig. 322. (For further explanation see Table 39.)

Dances	Filter	Color	Degree of polarization (percent)
Well-oriented	UG_1	Ultraviolet	36, 33, 24
	WG_1	Colorless, with long-wavelength UV	37, 36, 33
	BG_3	Blue with UV	38, 33
	BG_{14}	Light blue, some UV	33, 23
	BG_{18}	Blue-green, little UV	33, 28, 22
	GG_{13}	Almost colorless, no UV	60
Still clearly oriented	UG_1	Ultraviolet	11
Imperfectly oriented	GG_{13}	Almost colorless, no UV	16, 14
Disoriented	GG_4	Pale green-yellow, no UV	39, 33, 33
	VG_5	Bright yellow-green, no UV	60, 51, 39, 33, 30
	OG_4	Orange-yellow, no UV	39, 33, 23, 22

that they and not the compound eyes mediate orientation to polarized light. But the ocelli are neither necessary for analysis of the plane of vibration nor capable of it.

In order to eliminate their ocelli, bees were captured at the feeding dish and anesthetized in a vial with a two-hole stopper by letting CO_2 flow in. As soon as they had been immobilized (after about 5 sec), I fastened them with insect pins in the proper position in a wax-bottomed dish and under the dissecting microscope covered the ocelli with opaque caps. For this purpose I used an alcohol solution of shellac with which a large amount of fine lampblack had been mixed. The bees did not awaken from their narcosis for 0.25–0.5 min—time enough to apply the opaque eye covers. The bees recovered quickly thereafter and went on foraging. When enough dances had been watched, the insects were renarcotized so that the caps could be inspected. The results were used only when the caps were found to fit perfectly.

One may easily be deceived as to the opacity of a black eyecap if one merely examines against a light background the covering material that is to be used. I put the mixture of lampblack and shellac onto glass in a thickness corresponding to that of the caps, and tested this against the sun.

The bees thus treated were given a view of a limited patch of northern sky in clear weather, from the horizontal comb. On 5, 14, and 16 September 1952 I succeeded in observing 18 dances by 9 different bees with covered ocelli; several of these dances were of long duration, with numerous tail-wagging runs. Not one of these dances was disoriented. On 16 September the observations were made on 3 bees whose eye covers had been put on on the 14th and were still holding perfectly. Hence the plane of vibration of polarized light is perceived by the compound eyes.

The question whether the ocelli too are capable of this is to be answered in the negative. In other experiments the dances were abnormal or disoriented after the compound eyes had been partially covered (pp. 411f), even when the ocelli had an unimpeded outlook. Stockhammer (1959:44) too reports that bees no longer responded to polarized light after the compound eyes had been eliminated by means of black covers while the ocelli were left free. According to him (*in lit.*), these experiments involved the spontaneous orientation of bees in an arena to the plane of vibration (see p. 407).

Wellington (1953) states that the fly *Sarcophaga aldrichi* Parker orients beneath a polarizing sheet to the plane of vibration not only by means of its compound eyes (after elimination of the ocelli), but also with the ocelli (after the compound eyes have been eliminated). A retesting of these findings would be desirable.

In the bee the ocelli do not serve for perception of polarized light. Despite this they certainly are not to be regarded as functionless organs. Their action does not seem to be identical in all insects.

Disregarding numerous hypotheses that are not well enough founded, the significance of the ocelli is to be suspected of lying in two directions. Many investigators see them as stimulatory organs, by the illumination of which the activity of the animals is increased. Recently Bayramoglu-Ergene (1964, 1965) succeeded in confirming this for grasshoppers (*Schistocerca gregaria* Forskål and *Anacridium aegyptium* L.). Their flight speed is increased by illuminating the ocelli. Electrophysiological studies with ocelli have provided yet another angle on the problem. The frequency of their excitatory potentials is greatest in the dark and decreases with increasing intensity of illumination. When illuminated at low intensity the ocelli are especially well suited for registering slight changes in brightness (Autrum and Metschl 1961; Metschl 1963; literature review in v. Buddenbrock 1952; Autrum and Metschl 1963; Goldsmith 1964; Schricker 1965).

For bees we are indebted to Lindauer and Schricker (1963) and Schricker (1965) for informative experiments. No stimulatory effect could be demonstrated. Readiness to forage, rhythm of dancing, and speed of flight and of running were uninfluenced by opaque covers over the ocelli. On the other hand, it was possible to show experimentally the importance of these organs for perception of the degree of dim lighting. The time for beginning foraging activity in the morning as well as for ending it in the evening is regulated by the intensity of the dim light. Bees whose ocelli are covered with black hoods begin their morning activity later than their normal hivemates, in fact not until the brightness has increased some four or five times. In the evening the foraging flights cease correspondingly earlier.

Of especial interest is the fact that the starting and conclusion of the foraging flights does not occur according to a rigidly determined light intensity. Here too the bees bear witness to their capacity for paying heed to the length of the prospective flight.[36] In the evening they choose for their last excursion the degree of brightness that, in

---

[36] Compare their taking along a supply of honey, the amount of which is proportioned to the flight range (p. 29, n.1), or their earlier cessation of flight activity at distant foraging places when a storm threatens (p. 245).

consideration of the duration of the flight, will still allow them to reach home again in sufficient light despite the growing darkness. This becomes evident if one sets up for two groups of bees from the same hive two feeding stations, one close to the hive, the other some kilometers distant. The nearby group keeps on foraging until it gets too dark for them to orient. The distant group forages until—taking into account the time required for flying there and back, the length of their stay at the feeding station, and the decrease in light intensity during this time—it would be too dark on their return.

It was possible to demonstrate also, by means of experiments on phototaxis, that the ocelli of bees serve for perception of slight differences in brightness (Schricker 1965). Positively phototactic bees were given a choice between two dim light sources of somewhat different brightness. Normal bees chose the brighter lamp when its luminous intensity was twice that of the other. After one, two, or all three ocelli had been eliminated, the light intensity had to be increased four, six, or eight times (0.8 lux vs. 0.1 lux) for a correct choice.

13. Spontaneous Orientation Relative to the Plane of Vibration of Polarized Light

When with the help of polarized light bees find once again a familiar flight path or fly to a new goal in the direction of which they have been guided by their comrades' dances, or when they orient their own dances in accordance with the polarization of sky light, the stance assumed relative to the plane of vibration—based on previous experience—may assume any desired angle. All the observations discussed hitherto belong in this category of menotactic orientation (Kühn 1919). Their biological significance is obvious.

But there exists also a spontaneous positioning of the body, independent of individual experience and comparable to positive and negative phototaxis, at certain preferred angles to the plane of vibration (fundamental orientation or *Grundorientierung*, Jander 1957). This sort of response was first seen in the dung beetle (Birukow 1953a), in lower crustaceans (Eckert 1953; Baylor and Smith 1953), and in various other arthropods (see p. 439 and Table 41). When the animals were illuminated from above with polarized light, they moved predominantly at certain angles to the plane of vibration, in fact either parallel or perpendicular to it or at 45° to the left or right. Thus there were four different known preferred angular orientations to the plane of polarized light—not considering the several mirror images as distinct orientations.

Jacobs-Jessen (1959) found such spontaneous responses in bees too, among which can be demonstrated besides menotactic orientation the—certainly more primitive—fundamental type of orientation to polarized light. This is true also of bumblebees and ants, among others.

As apparatus Jacobs-Jessen used an arena, like the one described on p. 13 and in Fig. 11. The glass plate was replaced by a circular polarizing sheet (40 cm in diameter); 90 cm above the center of the arrangement there hung a 200-watt lamp. Bees captured in the open were liber-

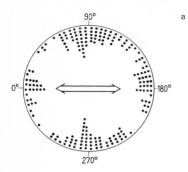

Apis mellifera

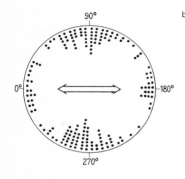

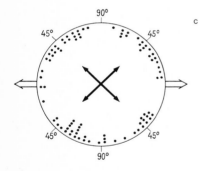

Andrena

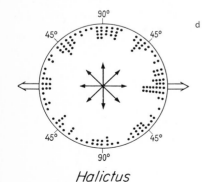

Halictus

ated into the arena from below through the central opening, and their direction of running toward the periphery was recorded. Each dot in the circular diagram (Fig. 353a) marks the direction taken by a single bee. Most of them ran approximately perpendicular to the plane of vibration, which is indicated by the double-headed arrow inside the circle. Secondary maxima stand out clearly parallel to this plane. When the polarizing sheet was taken away, the courses pursued were scattered at random in all directions.

These tendencies toward definite directions are innate. Bees from a colony that Renner kept in a closed chamber for more than 3 years and that had no experience out of doors behaved identically (Fig. 353b).

In Jacobs-Jessen's experiments bumblebees showed the same behavior as honeybees, but solitary bees of the genus *Andrena* oriented diagonally to the plane of vibration (Fig. 353c) and other solitary bees (*Halictus*, Fig. 353d) combined the two tendencies and had four recognizable preferred directions.[37]

These noteworthy differences seem to have been caused, at least in part, by differences in experimental technique. Jander (1963) obtained the same results as Jacobs-Jessen when he let bees captured out of doors into the arena through its central opening. But if he first gave them sugar water in the middle of the arena, so that their agitation disappeared, they strove afterward in all four preferred directions toward the periphery (Fig. 354). The same was the case with bumblebees and other Hymenoptera, whose behavior had differed in Jacobs-Jessen's experiments. Jander suspects that only their great agitation, from being handled when captured and put into the arena, suppressed their orientation to two of the four fundamental directions.

Jander (1963:420ff) has tried to provide a physiological explanation for their orienting to the preferred directions. At present this is hypothetical, especially for the bee with its divergent arrangement of rhabdomeres (see pp. 423ff). In this connection, it would be of importance to know the position, over the visual field involved, of the rosettes formed by the sensory cells. The outward shape of the bee's eye complicates these relations.

In the fundamental orientation to polarized light no biological significance is evident, unless one chooses to suppose that the animals

[37] According to Rensing (1962), the water strider *Velia* has six preferred directions. Beneath a polarizing sheet, maxima were to be seen at 30° intervals. However, the numerical differences are slight. In comparison with the other experiments discussed here, his methods were less exact. The individual determinations were not independent. A precise statistical analysis is lacking. In some unpublished tests of his own, Jander (*in lit.*) obtained other preferred directions. A careful reinvestigation is needed.

FIGURE 353. Spontaneous orientation of bees to the plane of vibration of polarized light: (a) honeybees *Apis mellifera* captured in the open; (b) honeybees *A. mellifera* reared under artificial light, without experience of the sky; (c) the solitary bee *Andrena*; (d) the solitary bee *Halictus*; white double-headed arrow, plane of vibration. Each dot indicates the direction in which a bee liberated at the center runs to the periphery. After Jacobs-Jessen 1959.

tend to maintain a given line and thus are kept from making undirected circle movements. Regarded phylogenetically, fundamental orientation is no doubt the basis on which menotactic orientation could be erected with the participation of higher nervous centers (Jander and Waterman 1960:154; Jander 1963).

14. Is Perception of Polarization Direct or Indirect?

Demonstration that the plane of vibration of the light is perceived by the compound eyes does not suffice to show that it is perceived by the sense cells according to the principle of the star-shaped model (pp. 385f). Under certain conditions orientation to polarized light is possible in other ways also. If, for instance, a shiny black surface is illuminated with linearly polarized light, the latter is reflected more strongly when its plane of polarization coincides with that of the reflecting surface than when it is perpendicular to it (Fig. 355). In consequence there is visible on the background a brightness pattern that is oriented in accordance with the plane of vibration and that rotates when the latter is shifted.

Kalmus (1958) illuminated with polarized light from above flies, ants, and moths on a shiny black background. They responded to rotation of a polarizing sheet, following the brightness pattern by changing the position of the body (optomotor response). According to Smith and Baylor (1960), with such an experimental arrangement the direction of running of bees, too, can be influenced by the brightness pattern. These observations demonstrate that insects on a black, reflecting background are able to respond to differences in brightness produced by polarized light. But that bees in free flight or dancers on the comb, when they glimpse blue sky, might be able to orient to brightness patterns of the background is a generalization for which the laboratory experiments mentioned do not provide any foundation. Nevertheless, this conception has found acceptance elsewhere too (De Vries and Kuiper 1958).

In considering this one should not overlook the fact that bees dancing on the comb or flying under the blue sky are exposed to wholly different conditions than the insects in the laboratory experiments done by Kalmus and by Smith and Baylor. Instead of orienting to completely polarized light, the bees in our experiments are guided by a slight degree of polarization, presented by a small patch of blue sky. Instead of a black shiny surface they see beneath them during flight meadows, fields, brush, or trees, that is to say, certainly no substrate that is able to reflect as a definite brightness pattern the varied polarization of the sky; and if there were such a pattern on the horizontal comb the dancers could not perceive it, because they are moving about in the midst of a dense throng of other bees. In more recent publications even Kalmus (1959) and Kennedy and Baylor (1961) are no longer so opposed to the possibility that polarized light is analyzed by the sense cells. New experiments, planned with direct reference to this moot problem, should resolve the last doubts (v. Frisch, Lindauer, and Daumer 1960).

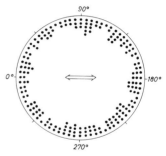

FIGURE 354. Spontaneous orientation to the plane of vibration of polarized light by honeybees that were calmed by being fed sugar water before they started from the center. Each dot summarizes seven measured values. The eight maxima are statistically significant. After Jander 1963 (somewhat altered for better comparison with Fig. 353).

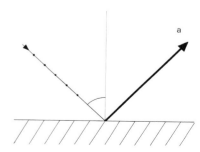
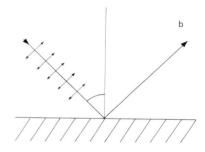

FIGURE 355. Linearly polarized light is reflected more strongly from a shiny black surface (*a*) when its plane of vibration is parallel to the surface than (*b*) when it is perpendicular to it. This may result in patterns of brightness. Schematic.

(a) Partial Elimination of the Eyes

We wish to learn whether bees read the pattern of polarized light from the blue sky above them or from a reflected pattern beneath. Theoretically that can be decided by covering with black caps the upper half of the two eyes on one occasion, the lower halves on another (for technique see p. 405), and then testing whether in dancing on the horizontal comb the bees are or are not able to orient according to the polarized light of the sky. In practice the experiment in this form was almost a complete failure, because bees thus treated, if they find their way home at all, soon go hopelessly astray.

Of 16 bees with the dorsal parts of the two eyes covered, only two came home. At first both performed disoriented dances; later they danced correctly in part, but with these two individuals the eyecovers had cracks, so that they could see portions of the sky. That was probably why they succeeded in returning home.

With 22 bees the lower parts of both eyes were covered. Of them 11 found their way home, and 6 back to the feeding station again. They had trouble landing. With only one of them could dances be seen, one dance with 11 tail-wagging runs and one with 6, all of which were oriented exactly toward the goal. On her next return to the feeding dish this bee was captured and killed so the eyes could be examined. The covers proved to be opaque (Fig. 356).

Thus, orientation by the polarized light of the sky was clearly deranged in two bees that had small, imperfect covers on the upper parts of their eyes, but was perfectly correct in one bee that had larger, opaque covers on the lower parts of the eyes.

An incidental result worth recording is that the upper portions of the eyes, directed toward the sky, are more important than the lower parts for orienting in the countryside. For of the foragers with covers on the lower part of the eyes half found the way home, but of those

 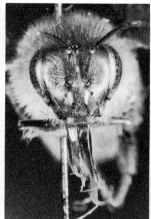 

FIGURE 356. Bee No. 4 from 28 August 1959: (*right* and *left*) lower parts of the eyes covered with lampblack in shellac; she danced absolutely correctly; (*middle*) a normal bee for comparison.

with caps on the upper parts—insofar as the covers were not defective—none.

In order to obtain more copious data in regard to the principal question, in further experiments only the right eye was covered, either above or below, on the basis of the following consideration. The dancers have a view of a patch of blue sky to the north, and see it with the right eye during their tail-wagging runs, which are directed to the west toward the feeding station. If they read the polarization directly from the blue sky, the maintenance of the westerly direction should be made difficult for them by eliminating the upper half of the right eye; although it might not be rendered impossible, because in the turns of the dance the bee might be able to recognize her incorrect orientation by means of the left eye. Eliminating the lower half of the right eye would not be expected to cause any disturbance. But if orientation were in accordance with a pattern of reflection, the opposite should hold true; elimination of the lower half should interfere with the dances.

Our hope that by treating the eyes on a single side the loss of experimental bees would be reduced was fulfilled. Of nine bees with eyecaps on the *upper* right, six found the way home; of nine bees with eyecaps on the *lower* right, all returned. This confirmed the secondary result of these experiments, that the celestial compass is of predominant importance for orienting in the countryside.

For the six bees with a cap on the upper part of the right eye that continued to frequent the feeding station, 46 dances were observed. Not one had a normal course, although the eyecovers in part were rather small (Fig. 357). The most striking feature of the behavior of the dancers was that over and over again they turned searchingly in circles to both sides. That is comprehensible if they are reading the polarization directly

FIGURE 357. The right eye with a lampblack cap on the upper portion: (*left*) bee No. 41 from 8 September; (*right*) bee No. 64 from 12 September 1959; the dances were considerably disoriented; (*middle*) a normal bee for comparison.

from the sky. For with the cap on the upper right they lose the view of the sky at the very moment when they are turning into the position for starting the tail-wagging run. Almost half of the tail-wagging runs had an improper direction. There occurred also some round dances, a behavior of the bees "for getting out of some difficulty" that we have already mentioned in a comparable situation (pp. 135, 167).

In clear contradiction to this, with the bees whose right eye was covered below, the 37 dances observed were normal, although some of the eyecaps were appreciably larger than with the other group (Fig. 358). Of a total of 229 tail-wagging runs, 225 pointed westward, and only 4 in other directions.

In conclusion, we have established the fact that the plane of vibration of polarized light is perceived with the parts of the eye that are directed upward, that is, it is read immediately from the blue sky.

(b) Alteration of the Reflected Pattern

It was possible to demonstrate the same thing in another manner also. Bees were trained in an arena to maintain a certain compass direction during journeys afoot. If they were unable to see anything except blue sky and the floor of the arena, a course aimed for the goal was to be expected, provided orientation was indirect, only when the floor showed a reflection pattern. Both Kalmus (1958) and Smith and Baylor (1960) had emphasized previously that a polarization pattern is clear only on a shiny black surface, while the light is largely depolarized by a light background, and then even in laboratory experiments there is only a very indefinite response, or none at all, to a background pattern.

In choosing suitable papers for the background, numerous samples were illuminated from above on a table top in the darkroom and were

FIGURE 358. The right eye with a lampblack cap on the lower portion: (*left*) bee No. 37; (*right*) bee No. 39, both from 8 September 1959; their dances were correct; (*middle*) a normal bee for comparison.

observed from the side at various angles. When a polarizing sheet placed above a black, highly calendered paper was rotated, there appeared very striking differences in brightness, which were measured with the help of a selenium cell. They amounted at a maximum to close to 50 percent. On the other hand, with a white, highly calendered paper and a white mat sheet of cardboard, we could not detect with certainty changes in brightness when the polarizer was turned. Measurement yielded differences of 3-5 percent at most.

Since the differences in brightness were so slight even under the most favorable laboratory conditions when the white background was illuminated with linearly completely polarized light of various planes of vibration, orientation beneath the open sky over white paper would surely be impossible, insofar as it was guided by the reflected pattern.

This question was tested with an experimental arrangement like that worked out by Jacobs-Jessen (1959) following Jander (1957). The bees can leave their hive (*St*, Fig. 359) only through a vertical wooden tube, and emerge at the center of a circular horizontal rotatable plate 70 cm in diameter; except for the sky above them they have here no possibility for orienting. A pane of glass 1 cm above the floor of the arena prevents the bees from flying off. They are able to leave the space only on foot through an opening *F* situated to the south and can fly off from *A*.

In the experiment a marked group is collecting sugar water from a feeding station located to the south. Every 10 min the plate is rotated and another one of the total of eight flight openings is brought into the southern position and uncovered while the other seven are kept shut. Before each experiment the floor of the arena is screened off with a fresh sheet of cardboard. In this way the bees are prevented from finding their way by means of odor trails on the substrate. The direction to the exit, and on return the direction to the central opening, have to be guided by the sun, and, when the sun is covered, by the polarized light of the sky.

On their way to flying off the bees pressed toward the south even on a white background, and those returning set out on a northerly course in the arena. Whether the background consisted of mat white paper or of highly calendered white or black paper had no influence on the assurance of orientation. From this it is apparent that the bees had oriented directly according to the polarized light of the sky.

Warm weather on the days of experimentation had the result that often hive bees would stay sitting before the exit in the middle of the rotary plate and relieve the foragers of food here, just as is to be seen occasionally with normal hives on the flight board in front of the entrance. Consequently dancing occurred in the arena. During all observations the sun was so screened that the entire rotary plate lay in shadow. We succeeded in observing 13 dances, with a total of 122 tailwagging runs, on a white background; without exception these were oriented correctly and pointed exactly south.

In order to test whether a reflection pattern such as occurs under the most favorable conditions (darkroom, completely polarized light,

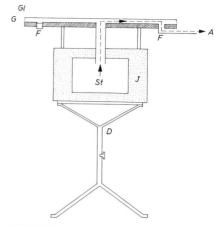
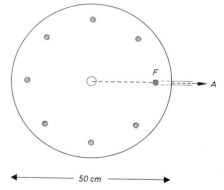

FIGURE 359. Arrangement for training bees to walk in a given compass direction: (*above*) *St*, observation hive; *F*, hive entrances; *A*, site of departure in flight; *G*, narrow-meshed gate; *Gl*, glass pane; *J*, wadding for keeping the hive warm; *D*, rotating stand; (*below*) top view. After Jacobs-Jessen 1959.

black highly calendered paper) can ever be detected under the conditions of an experiment out of doors, the distribution of brightness was measured with the selenium cell in the arena in the open. Then it appeared that because the arena was shaded during the bees' dances, because of the constantly shifting position of the observer, and because of the coming and going of individual cumulus clouds there resulted a continual variation in the distribution of brightness, a variation that in every case surpassed by far and at random any reflection pattern, dependent on the polarized light, that might have been present. On this account, derangements of the orientation would have had to appear in the experiments, but they did not take place. From this it is evident again that, in their directed excursions on foot and in the dances in the arena, the bees were orienting to the sky and not to a reflection pattern beneath them.

(c) Indirect and Direct Orientation in Other Animals

Responses to polarized light were first discovered in bees, but have now been demonstrated in many other arthropods and in mollusks (see pp. 439ff). Is the capacity of directly perceiving the plane of vibration perhaps limited to the highly differentiated insects, while in other instances there is an indirect orientation to reflection patterns?

Under laboratory conditions there may occur a directed orientation to brightness patterns produced by polarized light. Bainbridge and Waterman (1958), by illuminating the crustacean *Mysidium* from above with polarized light, found that it spontaneously took a position primarily perpendicular to the plane of vibration. In turbid water this was more definite than in clear water. Since turbidity favors a brightness pattern that depends on the plane of vibration, it was tempting to suspect the existence of phototactic orientation to the brightness. But with improved technique Waterman (1960) succeeded in showing that mysids and *Daphnia* also orient in clear water. Jander and Waterman (1960) demonstrated convincingly, in several ways, with *Mysidium, Daphnia,* and the water beetle *Bidessus,* that they orient directly according to polarized light, independently of a brightness pattern in the environment. Schöne (1963) showed the same for a representative of the higher Crustacea, the mangrove crab *Goniopsis,* and Görner (1962) for the funnel spider *Agelena.*

When marine snails (*Littorina*) are illuminated from above they orient to polarized light and respond clearly to a change in the plane of vibration (Burdon-Jones and Charles 1958, 1959). In experiments by Baylor (1959, *Nassa*) and Charles (1961, 1961a, *Littorina*) the snails oriented to a brightness pattern. But *Littorina* is capable also of perceiving the plane of vibration directly (Charles 1961b).

Octopuses (*Octopus*) can be trained to a specific plane of vibration of polarized light (see pp. 440f). Rowell and Wells (1961), Moody and Parriss (1961), and Moody (1962) furnished proof that the plane of vibration is perceived directly and not indirectly from a reflection pattern; Jander, Daumer, and Waterman (1963) showed the same for the spontaneous orientation of squid in polarized light.

Wherever hitherto an indirect orientation to polarized light by means of an environmental brightness pattern has been indicated as probable or has been demonstrated, this has happened under laboratory conditions extremely favorable to the production of such brightness patterns. To my knowledge no one has as yet observed in the natural surroundings of animals orientation to such reflection patterns, nor even their existence. On the other hand, it has been shown that crustaceans, insects, spiders, and mollusks are able to orient directly to polarized light, independently of environmental patterns.

15. THE ANALYZER FOR POLARIZED LIGHT

In order to detect the plane of vibration of polarized light there is required—in the organism just as in a physical experiment—an analyzer, which itself produces polarized light of a definite plane of vibration. If the incident light has the same plane of vibration, then it can exert a maximum effect. If its plane of vibration is different, the light is weakened progressively to a minimum at an angle of 90° (see Fig. 332). How this principle can be realized in the compound eye by means of radially positioned sense cells was stated earlier (pp. 385f); our star-shaped model is based on this concept, but it is not the only possible one. The analyzer might also be provided by the refractive components that are situated in front of the receptor cells.

(a) Is the Analyzer in the Dioptric System?

With studies using polarization optics on the compound eyes of Diptera and Hymenoptera, Stockhammer (1956) showed that neither the cornea nor the crystalline cone, nor both together, have the properties required for an analyzer. They exert no polarizing effect on light rays that fall approximately in the optic axis and that might reach the sense cells. This was also determined for the eye of the back swimmer *Notonecta* (order Rhynchota) by Selzer (1955) and Lüdtke (1957:337), and for ants by Vowles (1954a).

When light strikes the cornea obliquely, polarization may occur through refraction and reflection. Such processes apparently caused Waterman's findings (1953, 1954, 1954a) on the *Limulus* eye. Using electrophysiological recording from a single ommatidium, when he stimulated with polarized light and rotated an interposed polarizer he obtained maxima and minima at angular separations of 90°, but only with oblique lighting. As the obliqueness was reduced the effect grew weaker and finally vanished when the light fell in the direction of the optic axis. Possibly this functioning as an analyzer of oblique light is to be ascribed to the monstrous development of the cornea-cone complex of the *Limulus* eye. Further details about it are not known. Also, hitherto no orientation to polarized light has been described in the behavior of *Limulus*. From the findings with the *Limulus* eye general conclusions cannot be drawn because of its anatomical peculiarities. Waterman has directed attention to this also.

Refraction phenomena with light rays that fall on the eye at an angle to the optic axis are responsible with water mites and Cladocera, according to Baylor and Smith (1953), and with the fruit fly (*Drosophila*), according to Stephens, Fingerman, and Brown (1953), for their spontaneous orientation perpendicular or parallel to the plane of vibration of polarized light. These assumptions are based on theoretical considerations. Studies employing polarization optics were not done. Stephens, Fingerman, and Brown emphasize that with bees the responses to polarized light are too complex for so simple an explanation.

In contrast, Berger and Segal (1952), on the basis of findings with polarization optics, assume for the honeybee too a mechanism of analysis by means of refraction in the dioptric apparatus. At all events, the "normal" ommatidia, in which the corneal lens and crystalline cone form a common linear axis, are said to be incapable of perceiving the plane of vibration, but only those ommatidia situated at the margin of the eye, in which the crystalline cone is oblique to the cornea. Additionally, in their opinion, sense organs that lie beneath the strongly birefringent marginal ridge of the eye could serve for analyzing the plane of vibration.

These hypotheses cannot be maintained. According to the investigations of Stockhammer (1956) in Diptera and Hymenoptera, even with the oblique marginal ommatidia the dioptric apparatus is unable to analyze polarized light, either with normal or with oblique illumination.

Besides, in the summer of 1959 I coated with black shellac the marginal ommatidia in both eyes of a few bees. The method was the same as that used for applying half eyecovers (see p. 410). With two bees the marginal masks were so wide and came out so well that orientation would have had to fail if analysis of the plane of vibration had been a specific property of the marginal ommatidia. But these bees had perfect orientation to a patch of blue sky and danced completely normally.

(b) *The Radial Analyzer in the Insects' Ommatidia*

In July 1949 Autrum wrote me that he suspected that the visual cells in the ommatidium are the analyzer for polarized light, since they themselves polarize the incident light in different directions, depending on their radial position. How such an analyzer performs has already (pp. 385f) been discussed in reference to the star-shaped model. By electrophysiological experiments Autrum was able to establish the probability that the premise for his hypothesis is correct and that in passing through the sense cells light actually is polarized in different planes of vibration. Since then this has been demonstrated directly by various workers.

Electrophysiological experiments. First, Autrum and Stumpf (1950) found in compound eyes of flies and bees that, when an individual ommatidium was excited by a flash of light, rotating a polarizing sheet in front of the eye had no effect on the amplitude of the potential resulting from illumination. From this it follows that the analyzer is not situated in the dioptric apparatus in front of the receptor cells, thus not in the cornea or the crystalline cone—as Stockhammer confirmed in another way (p. 415). But this conclusion does not apply to the rosette of sensory cells. For in the event that they polarize the light in different planes (Figs. 331b, 333b), when the polarizer in front of them is rotated excitation is decreased in one and increased in another; this cancels out, because the several receptor cells of the ommatidium are recorded from simultaneously in the experiment. Nevertheless, it was quite apparent that polarized light was more effective than ordinary light of the same intensity. That is comprehensible only if light, in passing through the sense cells, is polarized in different planes. For then polarized light incident on a receptor cell that itself tends to polarize light in the same plane must produce an effect of greater brightness than unpolarized light of the same intensity. That the other receptor cells with differing orientation show smaller potentials is not observed, because they are masked in the large potential.

What was thus deduced indirectly received direct confirmation when it became possible to insert into the receptor cells very fine microelectrodes, with a tip diameter of about 0.1 μ. The insertion of the electrode into the receptor cell cannot be checked visually; but it is assured by the nature of the excitatory potentials.

Success that had not been achieved with simultaneous recording from all sensory cells of an ommatidium now appeared in all clarity with the testing of single cells. Rotating a polarizer placed before the eye yielded maxima and minima of the excitatory potential for every shift of 90° (Kuwabara and Naka 1959, in the fly *Lucilia caesar* L.; Burkhardt and Wendler 1960, as well as Autrum and v. Zwehl 1962a, in the fly *Calliphora erythrocephala* Meig.; Burkhardt, verbal communication on the basis of new experiments with *Calliphora*). This work provided direct proof that in flies the individual visual cells respond with excitatory potentials of different amplitudes to polarized light in different planes of vibration.

In passing through the visual cells light is not completely polarized, as it is with our polarizing sheets; some notion of the partial degree of polarization is derived from the observation that rotating a polarizer placed before the eye from the maximal to the minimal position has the same effect as a decrease in intensity by $\frac{1}{2}$–$\frac{1}{4}$ has with ordinary light.[38]

When different receptor cells are tested, the maximally effective plane of vibration changes (Burkhardt and Wendler 1960). According to Autrum and v. Zwehl (1962a:2, 3), in a given area of the *Calliphora* eye, the planes of vibration that have the maximal effect in different sensory cells are at definite angles to one another. Though with some receptor cells the largest excitatory potentials occur with the rotating polarizer in the 0° position, with other cells they are at 90° or in the diagonal positions at +45° or −45° (each ±20°; the filter was rotated at 20° intervals). That agrees with what might be expected from the placement of the receptor cells in the fly's eye (see Fig. 361).

---

[38] The effect on the fly's eye (*Calliphora*) of rotating the polarizer is the same with illumination by short-wavelength light (400–450 mμ) and long-wavelength light (over 570 mμ) (verbal communication from D. Burkhardt). Whether this is true also for the bee's eye is unknown. If it is, the fact that the bees' orientation to polarized light fails at wavelengths in excess of 450 mμ is not to be ascribed to the analyzing mechanism.

Autrum and v. Zwehl obtained the described variation in excitatory potentials from only about half the tested cells when the polarizer in front of them was rotated. But, according to a verbal communication from D. Burkhardt, it has now been possible, with the use of point-source light stimulation and improved methods, to demonstrate the effect of rotation of the polarizer in all cells (over 200) pierced by the electrode.

From all of this it follows that the principal features of the star-shaped model are realized in the fly's eye.

Objections to this idea have been raised in various quarters.

Kennedy and Baylor (1961) consider it possible that in recording the potentials from individual receptor cells the glass capillary used as a microelectrode polarizes the stimulating light, by means of refraction and reflection, and thus simulates a polarization by the sensory cell itself. This was refuted by Autrum and v. Zwehl (1962a) in new experiments on the *Calliphora* eye. If the objection were indeed well founded, the plane of maximal effectiveness should depend on the position of the capillary and not of the sensory cell. But it turned out that, with a constant position of the microelectrode, the plane of maximal effectiveness of the polarized light differs in different sense cells.

A further objection by Kennedy and Baylor is based on the authors' own experiments. After adaptation to bright polarized light, the sense cells with a differing plane of vibration would be expected to respond better to weak light stimuli than the (more fatigued) sense cells with the same plane of vibration. Experimentally no difference was to be found. But Autrum and v. Zwehl point out that the light is only partially polarized by the sense cells and that, with the intensities used for adaptation, dark adaptation proceeds with almost equal rapidity, so that a negative outcome of the experiments was to be expected.

H. De Vries and Kuiper (1958) point out that insects in the center of a round plate respond with optomotor reactions to the rotation of a peripheral pattern of black and white stripes. They found that such reactions were absent when they substituted for the black-and-white pattern strips of polarizing sheet with alternately horizontal and vertical planes of vibration. But in this instance one would not expect a reaction, for there is no general polarizer in front of the insect's eye, as there is in front of our own eye when we look at the pattern of strips through a polarizing sheet (Fig. 360). Taking the star-shaped model as a basis, and considering the microscopic dimensions of the radially arranged sense cells in the insect eye, it is clear that when the polaroid strips are moved past it the brightness pattern at the margins of the strips would each time shift suddenly by 90°, without any direction of movement becoming detectable (Fig. 360b).

With the same reasoning Baylor and Kennedy (1958) expected without justification an electrophysiological effect when they suddenly shifted by 90° the plane of vibration of the stimulating light. All that changes is the pattern of excitation over the receptor cells, and this cannot be recognized with the system of summed recording potentials. Thus the objections are not valid.

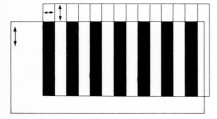

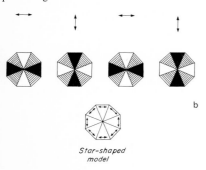

FIGURE 360. Pattern of strips from a polarizing sheet, with the planes of vibration of transmitted light (double-headed arrows) perpendicular to one another in adjacent strips: (*a*) seen through a polarizing sheet; (*b*) seen through a star-shaped polarizing model.

Morphological findings and studies with polarization optics on insects. Now we compare the results of electrophysiological examination of individual receptor cells with the findings in morphological studies and in those using polarization optics. With all three methods of investigation the fly's eye is more favorable than the bee's eye, because in the ommatidium of the fly the sensory cells are more loosely arranged and the visual rods (rhabdomeres) are separated clearly from one another. Being the sites where light exerts its effect, these latter deserve our particular interest.

Figure 361*a* shows a longitudinal section of an ommatidium from the right eye of a fly; Fig. 361*b, c* are cross sections from the upper and lower halves of the ommatidium respectively. The cross sections cut the eye in the region of its receptor cells and show that the latter—as in general with flies (Dietrich 1909)—are arranged differently in the dorsal and ventral parts of the eye, as mirror images of one another (cf. Fig. 362). Here *Tr* is a trachea accompanying the sensory cells. The visual rods *Rh* on the inner margin of the sensory cells protrude inwardly. In each ommatidium one finds in flies seven well-developed receptor cells; an eighth is rudimentary. The seventh sensory cell lies outside the wreath formed by the others and sticks out into the space surrounded by them. Through it one may draw a plane of symmetry, designated in Fig. 361*b, c* as the 0° direction. This enables one to define the relative position of the rosettes over the entire eye. Since in all parts the 0° direction runs parallel to the meridian of the eye, the

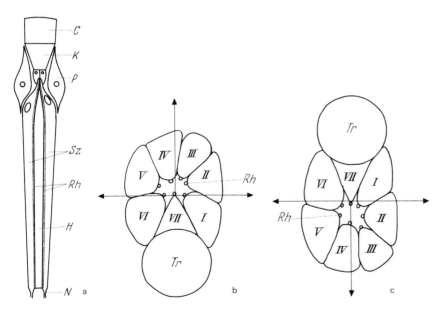

FIGURE 361. (*a*) Longitudinal section through the ommatidium of a fly, schematic: *C*, corneal lens; *K*, crystalline cone; *P*, surrounding pigment cells; *Sz*, sensory cells; *Rh*, their rhabdomeres (visual rods); *H*, space between the sensory cells; *N*, nerve fiber. (*b*) Cross section, at the level of the sensory cells, through the ommatidium of a fly; from the dorsal region of the right eye; (*c*) same, from the ventral region of the eye; I–VII, the seven sensory cells; *Rh*, rhabdomeres; *Tr*, trachea; (*b*) and (*c*) after Stockhammer 1956.

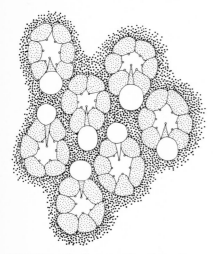

FIGURE 362. Cross section through several ommatidia of the fly *Lucilia caesar* L., at the boundary between dorsal and ventral halves of the eye, showing the reversal of structure to the mirror-image arrangement. After Stockhammer 1956.

FIGURE 363. Cross section of the eye of a fly (horsefly, *Tabanus bovinus* Loew), in the region of the sensory cells, viewed under crossed polarizers. In each ommatidium only a few of the seven rhabdomeres appear bright, these being always the ones in corresponding locations; others become bright when the section is rotated. After Stockhammer 1956.

orientation of the visual rods proves to be extraordinarily exact and constant. In adjacent regions of the eye their arrangement is precisely identical (Dietrich 1909; Stockhammer 1956).

Light coming from the crystalline cone and entering the visual rods remains confined there. Since the analyzer is not in the dioptric portion, it must be localized in the rhabdomeres (Stockhammer 1956).

The visual rods are birefringent in the direction of their long axes (Stockhammer 1956). When transilluminated they appear in the microscope as bright dots, of which only a few grow brighter under crossed polarizers while the rest are extinguished (Fig. 363). When the stage is rotated they grow bright or dark, respectively, at 90° intervals. This is so both with fresh preparations of living material and with those suitably fixed. In the individual rhabdomere the planes of vibration are approximately parallel or perpendicular, respectively, to the surface of its junction with the visual cell to which it belongs.[39] The planes of vibration of low index of refraction are approximately radial in the entire ommatidium.

That the measured values[40] fit only approximately with a radial arrangement may be a result of the preceding preparation of the eyes. In this procedure it is easy to shift the location of the rhabdomeres somewhat. "It is indeed quite possible that in the undamaged ommatidium there exists an equiangular radial symmetry, in eight directions, of the planes of vibration" (Stockhammer 1956:71).

The radial pattern of the planes of vibration in the individual ommatidium is in harmony with the star-shaped model and with the measurements of Autrum and v. Zwehl as well as with those of Burkhardt, who determined the planes of vibration electrophysiologically (see p. 417). Unfortunately, with this method one does not know which sense cell one is measuring from, so that one determines merely that there are directions of preference without being able to assign them to particular visual rods. In this respect the method of investigation with polarizing optics is superior.

As yet it has not been possible to determine experimentally the nature of the mechanism in the visual rods that makes one of the two planes of vibration exert a preferred effect; it is of course prerequisite to the functioning of an analyzer that this should occur.

Information from the electron microscope. The concept that the rosette of visual rods serves as a radial analyzer for polarized light re-

[39] In earlier measurements by Stockhammer, he did not find the planes of vibration perpendicular to one another. This gave rise to a hypothesis regarding the perception of polarization (Menzer and Stockhammer 1951) that was given up again when the abnormal finding proved to be an artifact. It was observable only in glycerin-gelatin (Stockhammer 1956:67).

[40] Stockhammer has summarized the results of his measurements in a diagram (1956:72, Fig. 24). It shows clearly that in principle the vibration planes of low refractive index in the rhabdomeres of the ommatidia are radially situated. That agrees with the theory of a radial analyzer. But in this figure I do not understand in detail the way in which the planes of vibration are related to the individual visual cells. Thus, for example, with rhabdomere 7 the plane of vibration of low refractive index is not radial but tangential, in contradiction to Stockhammer's express statements on p. 68 (see also his Fig. 19, p. 58) and in his later summary (1959:50).

ceived further support from investigations made with the electron microscope. This advance to a range of dimensions that previously had been concealed from our eyes discloses in the visual rods of flies, as in the bee and other insects too, a radially arranged pattern of very tiny tubes, which with the highest precision are shaped alike and situated in parallel. In adjacent ommatidia they are repeated in identical orientation. Their discoverer, Fernandez-Moran (1956), himself saw in these structures the basis for the analysis of polarized light—a conjecture that has been strengthened by the findings of subsequent investigators.

We shall next consider the relations in the fly's eye. The observations with different genera are in agreement (housefly *Musca domestica* L., Fernandez-Moran 1956, 1958; fruit fly *Drosophila*, Wolken, Capenos, and Turano 1957; Danneel and Zeutschel 1957; Yasamuzi and Deguchi 1958; Nolte 1961; flesh fly *Sarcophaga*, Goldsmith and Philpott 1957). Figure 364 shows schematically the structure of a visual rod, after Goldsmith and Philpott. Perpendicular to its long axis, and hence perpendicular also to the direction of transmission of light, run very fine tubelets, packed tightly together. It is assumed that they arise as outgrowths from the visual cells (microvilli); that, slender as they are, they are enclosed by a membrane some 10 mμ thick; and that their interior (diameter about 30 mμ) communicates with the cytoplasm of the sensory cell. Figure 365 gives a picture, likewise schematic, of their radial arrangement in the receptor cells of an ommatidium. For the sake of clarity the seven visual rods are drawn larger than actual size and the sensory cells that belong to them are merely indicated by a ring. How the structure actually looks in the electron microscope is shown in Fig. 366. Depending on the plane of sectioning, a parallel

FIGURE 364. Small segment of a rhabdomere of a fly's eye, schematic; as determined by electron microscopy, the tiny tubelets are perpendicular to the direction of transmission of light. After Goldsmith and Philpott 1957.

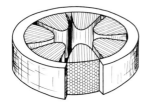

FIGURE 365. Cross-sectional slice cut from an ommatidium of a fly's eye, showing the radial arrangement of the fine structure in the rhabdomeres. For the sake of clarity, the rhabdomeres are shown as overlarge, and the sensory cells belonging to them are merely indicated as a ring. Schematic, after sketches by Wolken, Capenos, and Turano 1957.

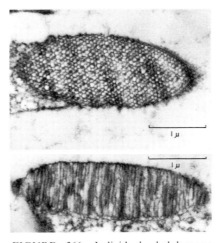

FIGURE 366. Individual rhabdomeres from the eye of *Drosophila*; electron micrograph. The tubelets have been cut (*a*) in cross section, (*b*) longitudinally (see Fig. 365). After Wolken, Capenos, and Turano 1957.

structure or a honeycomb pattern is discernible (see also Figs. 364 and 365).

With such fine tubelets the direction of sectioning is rarely so exact that they are cut precisely lengthwise or crosswise. Oblique sections result in combinations of the striped and honeycomb patterns, such as are seen in Fig. 367. This photograph shows the visual rods in the ommatidium of a housefly (*Musca domestica*). Only the innermost parts of the receptor cells to which they belong are in the picture. The rhabdomere labeled *D* corresponds to the one for visual cell VII in Figs. 361*b*, *c*. For the six remaining visual rods Fernandez-Moran (1958) remarks that they can be grouped into three pairs lying opposite one another (A_1 and A_2, B_1 and B_2, C_1 and C_2), whose radial structures are nearly identically arranged. This may be seen more clearly in Fig. 368, which has been sketched from an electron micrograph by Goldsmith and Philpott (1957) and shows the cross section of an ommatidium of the flesh fly *Sarcophaga bullata* Parker. During the year 1957, so successful for the electron microscopists, Danneel and Zeutschel studied the fly's eye (with *Drosophila*) and made some measurements on the course of the fine striation; Fig. 369 shows their results. They emphasize that the directions are inexact, since it may be seen just from the outline of the ommatidia that the rhabdomeres have undergone a certain degree of displacement during fixation and sectioning. Nevertheless, the basic radial plan emerges clearly. That there exists a connection between the radial structural arrangement and the planes of vibration is highly probable.

In most insects the rhabdomeres, in contrast with their clear separation in the fly's eye, are fused together in a rhabdom (see Fig. 331). In a cross section the rhabdom lies centrally, enwreathed by the visual cells. Often it is so compact that its construction from individual visual rods can be discerned only through the electron microscope.

With fused rhabdomeres there was also found a radial arrangement of the visual rods, with opposite pairs corresponding: in the tropical moth *Erebus*,[41] in a tropical butterfly from the family Hesperiidae, and in grasshoppers (*Dissosteira, Schistocerca*); in these forms superposition eyes are concerned (Fernandez-Moran 1958). The fine structure in the eye of an apterygote (*Lepisma*, Brandenburg 1960) likewise fits into this scheme.

A somewhat different picture is provided by the rhabdom in cockroaches, an ancient and primitive group of insects. According to the electron micrographs of Wolken and Gupta (1961) with *Periplaneta americana* L. and *Blaberus giganteus* L., the fine structures of two neighboring, closely adjacent cells are identically arranged. The three pairs are oriented in three different directions. Whether the seventh rhabdomere represents a fourth direction or fits with one of the pairs

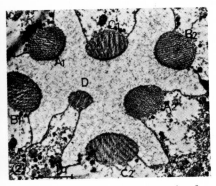

FIGURE 367. Electron micrograph of a slightly oblique section through an ommatidium of a housefly; only the inner portion of the sensory cells, with the rhabdomeres, is reproduced; *D*, the protruding rhabdomere of the seventh sensory cell; rhabdomeres opposite each other are marked with the same letter. Magnification 25,000 ×. After Fernandez-Moran 1958.

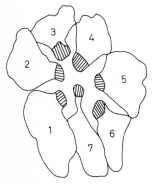

FIGURE 368. Schematic cross section through an ommatidium of the flesh fly *Sarcophaga bullata* Parker, showing the arrangement of the fine structure. From an electron micrograph by Goldsmith and Philpott 1957.

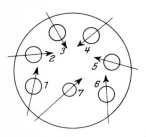

FIGURE 369. Schematic cross section of an ommatidium in *Drosophila*; the arrows indicate the approximate direction of the fine striation of the rhabdomeres. After Danneel and Zeutschel 1957.

[41] T. H. Goldsmith (1964:453, 454) draws attention to the fact that in the V-shaped rhabdomere of *Erebus* the structures are not parallel in the two arms, but that they are so in the contiguous arms of adjacent rhabdomeres. More detailed morphological and physiological studies of such eyes would be of great interest, for at present there is no clear understanding of their afferent innervation.

cannot be judged with certainty from the available pictures (Fig. 370). For the compact rhabdom in the apposition eye of the dragonfly *Anax junius* Drury, Goldsmith and Philpott (1957) also indicate three directions in the course of the fine structures (Fig. 371a). Apparently here too each third of a rhabdom consists of two rhabdomeres with identically directed tubelets; but these relations have not been studied in detail. According to Naka (1961), damsel flies (Zygoptera, for example, *Agriocnemis*) should have three rhabdomeres, which are formed by four receptor cells (Fig. 369b).

A surprise was found in the investigation of the bee's eye with the electron microscope (Goldsmith 1962). In the worker bee the rhabdom is so compact and thin that its study by polarization optics is very difficult. In accordance with the regular wreath formed by the surrounding receptor cells, we have assumed hitherto eight radially arranged rhabdomeres (see the scheme on p. 385, Fig. 331). But analysis of the cross sections with the electron microscope shows clearly that here too the visual rods of each two adjacent visual cells are fused into a unit with identically directed fine structures. Since there are eight equally well-developed retinula cells with their visual rods, there results a four-branched star-shaped figure, the opposing sectors of which are similarly oriented (Fig. 372). Cross sections at various levels show that the picture remains the same over the entire length of the rhabdoms. According to Goldsmith (*in lit.*), it is not known to which portion of the eye the sections belong. This question is important, however, because only the dorsal half of the eye is involved in the analysis of the polarized light of the sky.

Dr. F. W. Schlote (Göttingen) was kind enough to prepare electron-microscope sections in the region of the receptor cells for ommatidia

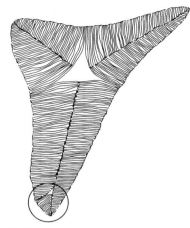

FIGURE 370. Cross section through the rhabdom of a cockroach: each rhabdom has seven rhabdomeres, six of which lie adjacent to one another in pairs with their fine structure aligned similarly; in the circle is the asymmetric seventh rhabdomere. Sketch from an electron micrograph by Wolken and Gupta 1961.

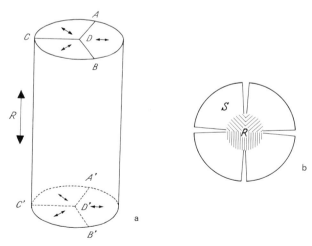

FIGURE 371. (a) Schematic sketch of part of the rhabdom from the eye of a dragonfly, *Anax junius* Drury: AD, BD, CD, boundaries of rhabdomeres; the small double-headed arrows indicate the lengthwise direction of the fine tubelets (microvilli); R, long axis of the rhabdom. After Goldsmith and Philpott 1957. (b) Schematic cross section of an ommatidium of *Agriocnemis* (Zygoptera, damsel flies): S, visual cell; R, rhabdom. After Naka 1961.

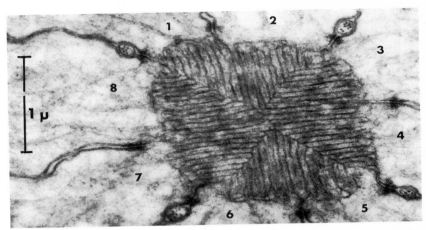

FIGURE 372. Cross section through a typical rhabdom from the eye of a honeybee worker; the microtubules arise from the surrounding eight (numbered) sensory cells and run in the same direction in pairs of adjacent rhabdomeres. Electron micrograph, magnification 29,000 ×. After Goldsmith 1962.

that are directed upward when the bee's body is in a normal position (Fig. 373). Comparison with Fig. 372 (from Goldsmith) shows fundamentally the same relations, only the cross section is more elongate. But there were also (less frequently) ommatidia that were identical with Goldsmith's figure in this respect. On the other hand, Goldsmith too mentions variants from the cylindrical form.

It is noteworthy that in addition to ommatidia with eight sensory cells Schlote found, less frequently but also not rarely, some with nine sense cells.[42] With these, variants from the four-armed structure can be seen. Although the possibility has to be considered that the arrangement has been disturbed during preparation, yet it is evident in the ommatidium shown in Fig. 374 that in at least one rhabdomere (that of cell No. 5) the microvilli run obliquely to the others. The directions are indicated by the black lines. Other sections make it even more probable that a six-branched arrangement of the fine structure occurs. Yet at the present stage of investigation final judgment as to these relations cannot be made. If it is found that a multibranched star-shaped figure is realized in many of the upward-looking optical units, then the model of an eight-branched (or a six-branched) star-shaped polarizer for the bee's eye might be justified. But if the four-branched construction predominates in the dorsal region of the eye, then the model of the star-shaped polarizer requires modification.

Photographs of the sky through a four-branched polarizer (Fig. 375, *lower*) show a less differentiated pattern in comparison with the old model (Fig. 375, *upper*, and Fig. 336). Should the analysis of the plane of vibration take place in accordance with this scheme, that might mean a simplification in the central nervous processing of the polarization pattern. Now instead of a more richly structured star-shaped pat-

[42] The occasional occurrence of nine receptor cells in the bee's ommatidium was mentioned long ago by E. F. Phillips (1905).

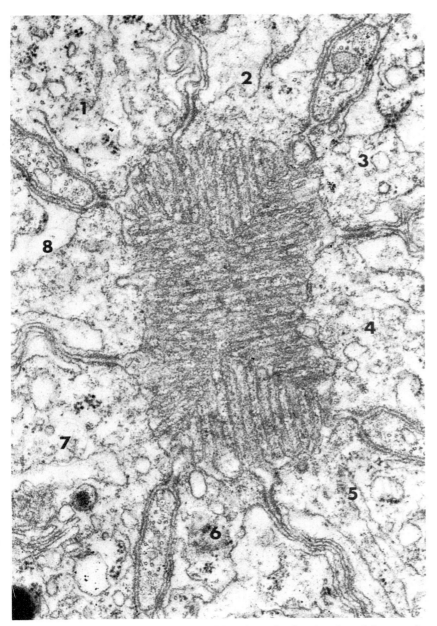

FIGURE 373. Cross section through a rhabdom from the dorsal region of the eye of a worker bee. Electron micrograph, magnification 50,000 ×; preparation by F. W. Schlote, Göttingen, 1964.

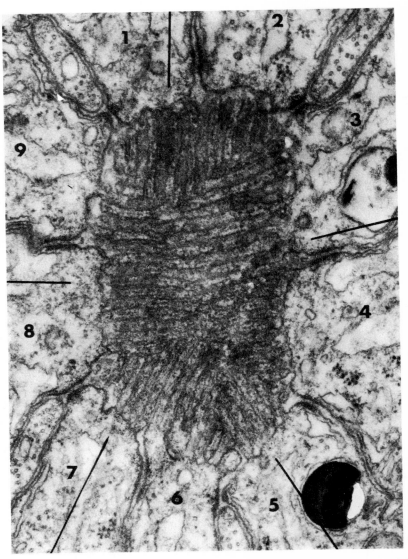

FIGURE 374. Cross section through another rhabdom from the dorsal region of the eye, showing nine sensory cells; the black lines indicate the course of the fine structure. Electron micrograph, magnification 50,000 ×; preparation by F. W. Schlote, Göttingen, 1964.

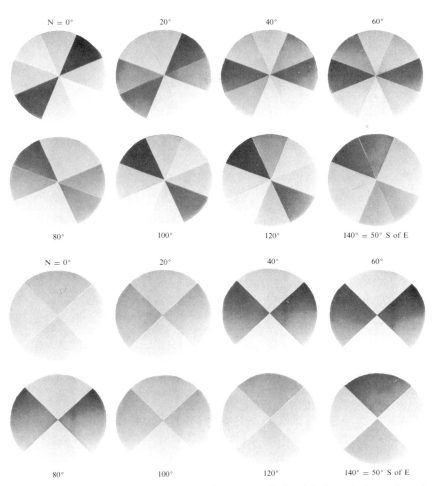

FIGURE 375. Photographs of blue sky: (*above*) through the eight-branched star-shaped polarizing model, 3:03–3:11 P.M.; (*below*) through a four-branched model, 3:26–3:32 P.M., 11 September 1964, near Munich; pictures at azimuthal separations of 20°, from N to 50° S of E; elevation above horizon, 45°. Photographs by M. Renner.

tern each ommatidium would produce only two different intensities of light. The relative brightness of adjacent fields of view is determined by the plane of vibration in the portion of the sky looked at. But the relation is no longer unequivocal. When the plane of vibration is altered, as in moving across the expanse of the sky, the same pattern is found at another place, whereas with the eight-branched star-shaped model any confusion would be excluded by the difference of the intervening visual fields (compare 0° with 120° or 40° with 80° in Fig. 375, *upper* and *lower*, respectively).

On first thought it seems that this would reduce the bee's capacity. But she does not observe the sky with a single ommatidium. Thousands of ommatidia are aimed at the heavens, and even a great many adjacent ones on one small patch of blue sky. The arrangement of their fine structure agrees in the minutest detail, so they provide exactly comparable patterns. And if the azimuth is shifted by even a few degrees, the plane of vibration changes (see Fig. 375, *upper*). This change follows a definite course; and this, when we refer once again to the like patterns of Fig. 375 (*lower*), is precisely opposite in the two instances. This is in fact what is expressed, with the eight-branched model, in the difference of the intervening visual fields. Thus as soon as the bee gives heed to the pattern of adjacent ommatidia, the analysis of the plane of vibration becomes as unique as it is with the eight-branched model, and is perhaps even simplified for its processing by the central nervous system.

At this point it should be emphasized once again (see pp. 419f) that the extraordinarily precise arrangement of the sensory elements in the ommatidia over the broad visual field is of fundamental importance for analysis of the polarization of the sky. The orderly patterns, an esthetic pleasure for the human observer, are for the insects prerequisite to their orientation in accordance with the plane of vibration of the light (see Figs. 362 and 363).

In various experiments regarding orientation by polarized light, the bees saw blue sky through a pipe at a visual angle of 15° (or sometimes 10°). Under such conditions the dances clearly were still directed, if at times less precisely than when a larger area of blue sky was viewed. We have explained the unequivocal analysis of the plane of vibration by visual rods in a four-branched arrangement on the basis of cooperative activity of groups of adjacent ommatidia, which permit recognition of the changing pattern of polarization over the sky. This assumption presupposes that in the experiments with the pipe a change in azimuth of 10°–15° was sufficient to render the alteration in plane of vibration apparent to the bees. With the eight-branched star-shaped model I tested this repeatedly and, at an elevation of 45° above the horizon (as in the experiments), detected the change in pattern on moving laterally 6°–13°; of course the clearness of the sky and the degree of polarization had some effect. Thus this prerequisite is fulfilled. For experience has shown that the bees are able to orient in accordance with polarization patterns in the sky as soon as these become just perceptible to our eye with the eight-branched star-shaped model.

A comparable but bilaterally symmetric four-branched arrangement is present in the ommatidia of water bugs and water striders. In the back swimmer (*Notonecta*), Lüdtke (1957), with electrophysiological recording from small groups of ommatidia stimulated by polarized light, obtained potentials of clearly different heights when he changed the plane of vibration of the stimulating source; here the maxima and minima were separated by angles of 90°. In the bee's eye he found no such effect, in agreement with Autrum and Stumpf. He ascribes the difference to the asymmetric arrangement of the sensory cells in the back swimmer, which is similar in water striders too; with the latter (*Velia, Gerris, Hydrometra*), six rhabdomeres in cross section form a rectangle that encloses two rather small visual rods (Fig. 376).[43]

FIGURE 376. Cross section through the rhabdom of the water strider *Velia*, with six large and two small rhabdomeres. Two of the large ones are separate; the other four are joined together closely in two pairs, the indentations indicating the boundaries. After Rensing 1962.

[43] According to Rensing (1962), with the water strider (*Velia*), when a polarizer is rotated 90°, the two opposed pairs of rhabdomeres, or the remaining two outer ones together with the two inner rhabdomeres, appear alternately light or dark, respectively. This was not, however, a demonstration of a natural dichroic absorption in the visual rods, but was an artificial dichroism resulting from gold-chloride staining.

16. Structure of the Visual Rods and Perception of Polarized Light in Other Groups of Animals

The fundamentally similar fine structure of the visual rods from the lowest to the highest insects (Apterygota, cockroaches, grasshoppers, bugs, dragonflies, butterflies, flies, bees) justifies the expectation that this is a common possession of all insects. The ability to analyze polarized light has been demonstrated in so many insect groups (see pp. 439ff) that its ubiquitous distribution also may be considered probable. The assumption of a causal connection between the two phenomena is strengthened by the fact that recently the occurrence of that fine structure and the perception of polarized light have become known far beyond the class of the insects.

Spiders do not have compound eyes, but only simple ones (ocelli). Among the species of highly organized araneids of interest here they are grouped in four pairs; three pairs (in part with large lenses) have a broad field of vision (accessory eyes, 2–4 in Fig. 377) but only one pair, the anterior median eyes (principal eyes, 1 in Fig. 377), are so constructed that they mediate a modest degree of form perception and with their mobile retinas are able to fixate the prey. These principal eyes also perceive the plane of vibration of polarized light, while the accessory eyes cannot do so or play a wholly subordinate role (Görner 1958, 1962 in *Agelena*; Magni, Papi, Savely, and Tongiorgi 1964 in *Arctosa*).

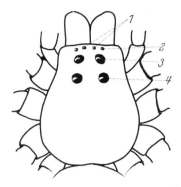

FIGURE 377. Cephalothorax of a spider, schematic, for designating the ocelli: 1, anterior median eyes (principal eyes); 2, anterior lateral eyes; 3, posterior median eyes; 4, posterior lateral eyes: 2, 3, and 4 are designated accessory eyes.

Miller (1957:224) found microvilli in the typical arrangement in the visual rods of the principal eyes of an undetermined species of spider. Baccetti and Bedini (1964) give a comparative description of the principal eyes and accessory eyes of *Arctosa*. A section perpendicular to the optical axis in the principal eye shows on each sensory cell a wreath of rhabdomeres, usually five (Fig. 378*a*). In the electron micrographs one sees clearly microvilli that are grouped radially around the axis of the sensory cell, as is pictured schematically in Fig. 378*b*. In the accessory eyes the radial arrangement is lacking; here the tubelets run in a single direction. The sensory cells lie in folded rows, and only where these change course does a different direction occur (Fig. 379). Baccetti and Bedini (p. 120) think that the attractive theory of an analysis of the plane of vibration via the structure of the rhabdomeres finds no confirmation here, because it is just in the principal eyes, which are capable of perceiving polarized light, that no preferred direction is to be seen.

In actuality there does seem to be here a difficulty in the way of our view of the mechanism of analysis: the rhabdomeres—and only those of the principal eyes—do indeed exhibit a radial grouping that is strikingly reminiscent of the rhabdomere rosette of an insect retinula (see Fig. 331). But whereas in the latter the differently directed tubelets belong to different, separately derived cells, in the spider's eye they are developed in a single sensory cell and cannot function separately.

In an exchange of letters about this question with P. Görner and U. Thurm, the latter pointed out that closely packed pentagons cannot be equiangular pentagons, because when these are placed together they have to leave intervening spaces. As a matter of fact, Baccetti and Bedini (p. 102) emphasize their irregular shape, and the picture of the

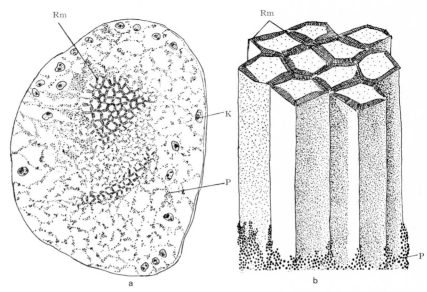

FIGURE 378. (*a*) Cross section through a principal eye of *Arctosa variana* Koch: two groups of visual cells are sectioned in the region of their rhabdomeres, *Rm; K*, nucleus; *P*, process of a pigment cell. (*b*) Schematic picture of the distal parts of a group of visual cells from a principal eye. After Baccetti and Bedini 1964.

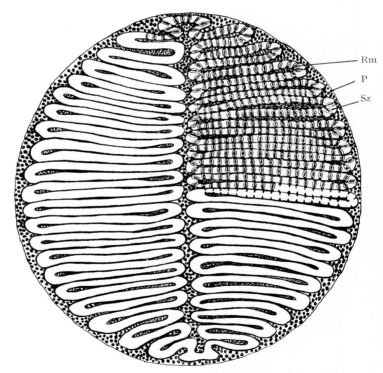

FIGURE 379. Schematic cross section through an accessory eye of the spider *Arctosa variana*, in the region of the rhabdomeres, *Rm; Sz*, sensory cells; *P*, pigment cell. After Baccetti and Bedini 1964.

cross section (Fig. 378a) shows such an irregularity clearly. That seems to us to be the salient point.

Görner (*in lit.*) did a simple experiment. He cut strips from a polarizing sheet and made a model similar to the star-shaped model but with an arrangement corresponding to that of the rhabdomeres in a spider's eye (Fig. 380a). The strips were joined in an irregular pentagon, such as may occur with contiguous pentagons and as is realized in the principal eye shown in the cross section of Fig. 378a. For comparison a model of an equiangular pentagon was used (Fig. 380b). The middle portions were screened with black paper, the artificial "rhabdomeres" were projected by means of a lamp onto a photocell, and the total photoelectric current was read from a microammeter. A polarizing sheet placed in front of the model was rotated in 10° steps and the strength of the photoelectric current noted; its maximal strength was set at 100 microamperes. With the equiangular "sensory cell" the brightness showed a maximal fluctuation of 4 percent,[44] but with the nonequiangular one 48 percent, with maxima and minima occurring at 90° angles (see the curves above the pentagons in Fig. 380). Thus in the irregular model there arises, through the joint activity of the rhabdomeres, the same effect as with an insect eye in a sensory cell with its single visual rod. In the spider's eye, as with insects, the analysis of the plane of vibration would have to be made by the cooperative activity of several, differently oriented sensory cells.

Thus the fine structure in the principal eye of the spider is not in contradiction with our theory. It is merely that Nature has apparently found a different way for these different eyes. From the statements and figures of Baccetti and Bedini it may be judged that in the accessory eyes the requirement for the analysis of polarized light is not met, because the fine structure of the rhabdomeres has in general the same direction over the entire eye (Fig. 379), while in the principal eyes there very likely exists a possibility of analysis, by means of the unequal and differently oriented pentagonal rhabdomeres (Fig. 378). This finding is in harmony with the results of the behavioral experiments.

The electrophysiological results of Magni, Papi, Savely, and Tongiorgi (1962, 1964) are not comprehensible. By rotating a polarizing sheet placed in front of the eyes of *Arctosa variana* Koch they obtained maxima and minima of the excitatory potentials in the electroretinogram of the principal eye and of the posterior median accessory eye, but not from the rest of the accessory eyes. It is also not clear why only dark-adapted eyes responded to the rotation of the interposed polarizing sheet, while after being light-adapted they no longer reacted to a change in the plane of vibration. Since these spiders are diurnal animals and orient in bright daylight in accordance with the polarization of the sky, one has to enquire whether the electrophysiological phenomena described warrant a definite decision regarding the analysis of polarized light by the sensory cells. Recording from individual sensory cells, as is done with the insect eye, could provide information. In any case, this

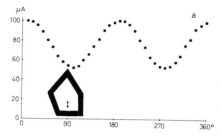
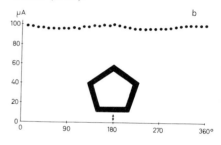

FIGURE 380. (*a*) Model of the spider's principal eye: strips cut from a polarizing sheet have been joined together to form an irregular pentagon; when light was projected through the model onto a photocell, rotation of a polarizing sheet in front of the model produced the variation in current depicted by the curve. (*b*) Results of the same experiment with a regular pentagon. Detailed explanation in text. After Görner (*in lit.*).

[44] Theoretically it should remain constant. Görner suspects that his pentagon was not precisely equiangular.

432 / The Orientation of Bees

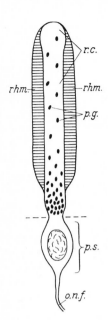

FIGURE 381. A visual cell from the retina of *Octopus*, schematic: the receptor cell *r.c.* has two rhabdomeres *rhm*; *p.g.*, pigment granules; in the lower, pigment-free portion of the cell *p.s.* is the nucleus; *o.n.f.*, optic nerve fiber. After Moody and Parriss 1961.

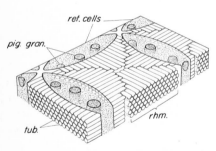

FIGURE 382. Structure of the rhabdom in *Octopus*, schematic: *ret. cells*, retinula cells; *rhm*, rhabdomere; *tub*, microtubuli; *pig. gran*, pigment granules. In the center is a complete rhabdom in cross section; it is formed of four rhabdomeres, the tubelets of which are perpendicular to each other. The second rhabdomere of the same cells participates in the formation of a neighboring rhabdom. After Moody and Parriss 1961.

interesting problem demands additional electrophysiological and morphological investigation.

In Crustacea the perception of polarized light is widespread; unfortunately, electron microscopic and electrophysiological studies of their visual sense cells do not seem to have been made.

But far afield from the arthropods, in the molluscan phylum, with octopuses and squids there were found perception of polarized light (see pp. 440ff) and a fine structure of the visual rods that is astoundingly similar, in the presence and arrangement of very fine tubelets, to conditions in the eye of the bee (Wolken 1958; Moody and Robertson 1960; Moody and Parriss 1961; Jander, Daumer, and Waterman 1963).

Octopuses and squids have eyes with lenses; in their construction and functioning they resemble the lensed eyes of vertebrates, but their development is different. The sensory cells of their retina have the free end turned toward the incident light. Figure 381 shows schematically such a visual cell in *Octopus*. It has two rhabdomeres (*rhm*) situated opposite one another, and microvilli that, in their strictly parallel arrangement perpendicular to the incoming light and in their size (diameter of a tubelet about 60 mμ), correspond to the relations in insects. The direction of the tubelets differs by 90° in adjacent visual cells, so that in each rhabdom, surrounded by four visual cells, are represented two directions perpendicular to one another (Fig. 382; see in this connection also Fig. 372, on the bee). Thus each visual cell in *Octopus* develops two visual rods (in insects there is only one) and takes part in the production of two rhabdoms. With squids the arrangement is somewhat different, but there is agreement in the points that are important here (Zonana 1961). The different construction of the sensory cells is understandable, inasmuch as the phylogenetic origin of octopuses and squids is wholly different from that of insects. It is all the more striking that they have developed the same kind of structure for perception of polarized light.

The connection between the fine structure described and the capacity for analyzing the plane of vibration becomes even more convincing because—so far as present knowledge extends—the absence of that structure goes hand in hand with lack of the ability to analyze polarized light. Above all this applies to the eyes of vertebrates. Their visual elements, the outer ends of the rods and cones, appear in the electron micrograph as columns of very fine platelets or flattened, double-walled sacs that—like the microvilli of arthropods and mollusks—are parallel to one another and are arranged perpendicular to the direction of illumination (Sjöstrand 1959; Moody and Robertson 1960; De Robertis 1960; Brown, Gibbons, and Wald 1963). But one misses in them the characteristic that apparently is significant for the perception of the plane of vibration: the radial arrangement.

By means of the well-known phenomenon of Haidinger's brushes, under favorable circumstances the human eye too is capable of detecting the plane of vibration of polarized light, for instance in the blue sky. But this phenomenon is a by-product, due to the yellow pigment of the *macula lutea* and radially coursing fibers above the layer of visual cells. One can make out Haidinger's brushes readily by looking at a piece of brightly illuminated white paper through a polarizing sheet. At the point fixated there appears a dim, colored cross, consisting of a pair of yellow sheaves and two bluish spots lateral to their constriction. When the polarizing sheet is rotated the brush figure rotates correspondingly, in such a manner that a line through the blue spots perpendicular to the axis of the sheaves represents the plane of vibration of the polarized light. In the blue sky the brush figure is dim because the light of the sky is only partially polarized. Nevertheless it can be seen by particularly sensitive people. Such individuals are capable of determining the plane of vibration of the polarized light from the position of the colored cross—but only at the spot fixated at the time, since the phenomenon is confined to the *fovea centralis*.[45] Only a very few people know of it. Whether there may be in this respect better observers among aboriginal peoples, and whether they are able to use the possibility thus afforded for orienting in accordance with the polarization of the heavens, is not known to me.

The polarizing capacity of polarizing sheets depends on dichroism. Some birefringent crystals and tinted sheets rendered birefringent by means of tension are transparent to one plane of vibration of light, while the others are absorbed. Hence natural light incident on them is polarized linearly in a certain direction, if the individual crystals are arranged in parallel or if the tinted sheet has the same direction of birefringence over its whole extent. The attractive notion has been developed that the rigidly orderly arrangement of the microvilli in the visual rods constitutes the basis of an equally orderly arrangement of the molecules of the visual pigment, so that this molecule itself, by means of dichroic absorption, serves as an analyzer (De Vries, Spoor, and Jielof 1953; Stockhammer 1956, 1959; Moody and Parriss 1961).

Attempts to observe dichroic absorption in the rhabdomeres have been unsuccessful. But theoretically perception of the plane of vibration would require so slight an amount of dichroic absorption that direct demonstration of it should not be demanded (De Vries, Spoor, and Jielof 1953). If there are both dichroism and an ordered arrangement of the visual pigment,[46] the mechanism of the analysis of polarized light has been accounted for. The correctness of these assumptions is yet to be established.

According to the classical theory, in its important features still largely valid today, each ommatidium of the compound eye produces one point of the image, and thus functions in this respect as a unit. But the analysis of polarized light is provided by the differential excitation of the variously oriented rhabdomeres within the ommatidium, and thus requires a separate evaluation of the stimuli occurring within a single ommatidium. Such a thing is possible only with separate innervation of the

---

[45] "Peripheral polarization brushes" are visible only under special conditions (Boehm 1940).

[46] According to Autrum (1953), an ordered arrangement of the molecules of the visual pigment is to be assumed for the reason also that this is a prerequisite for the high resolving power in time of many insect eyes.

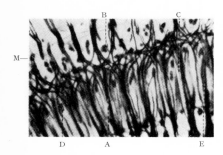

FIGURE 383. Section of the *lamina ganglionaris* of the housefly (*Musca domestica*), after Gros-Schultze impregnation: *M*, basement membrane; *A*, afferent fibers of the rhabdomeres; *B*, nucleus of a ganglion cell, with axon; *C*, ending of a short fiber from a receptor cell; *D*, nucleus of an interlaminar ganglion cell; *E*, two extended fibers in the region of the collaterals. After G. F. Meyer 1951.

individual sensory cells. The anatomical proof was first established long ago by Hesse (1901) and has since been confirmed repeatedly (for example, Phillips 1905; Cajal and Sanchez 1915; Eltringham 1919; Stockhammer 1956; Nolte 1961; Naka and Eguchi 1962).[47]

Since in neighboring ommatidia those rhabdomeres that have the same location also correspond in their planes of vibration, a grouping of the afferent fibers into bundles, each containing the axons of identically arranged receptor cells in a small group of ommatidia, would be meaningful for the analysis of polarization. Perhaps this is the significance of the crossovers of afferent axons that, coming from the receptor cells, are collected into tracts in the "neuro-ommatidia" (v. Frisch 1951a; see Fig. 383). The discussion of this striking finding, so far as I am aware, has not progressed beyond conjecture, which, indeed, has taken a somewhat different line (Cajal-Sanchez 1915; Hanström 1927; Kuiper 1962).

How the eye accomplishes its double task of perceiving not only images but polarization is an unsolved problem.

Summary: Orientation in Accordance with Polarized Light

1. In dances on a horizontal surface bees can point the direction to the goal even without a view of the sun if they can see the blue sky. A round blue patch the diameter of which corresponds to a visual angle of 10–15° is sufficient. When the sky is cloud covered and there is no free view of the sun's position, the dances are disoriented. If under such conditions one spot in the sky is conspicuously brighter than the surroundings, it can be accepted by the dancers as a replacement for the sun ("lamp effect"). The spot serves in lieu of the sun as a point of reference for indicating direction.

2. Orientation by the blue sky is possible because the bees are able to perceive the plane of vibration of the polarized light of the sky, which depends on the position of the sun (Figs. 329, 330).

3. That the bees perceive the plane of vibration of polarized light is to be deduced from the following facts:

(*a*) If bees dancing on the horizontal comb are shown in the north the polarization pattern of the southern sky, they indicate the mirror image of the true direction to the goal.

(*b*) If bees dancing on the vertical comb surface are shown blue sky, without being able to see the sun, the dances point along the bisector of the two directions the bees would have maintained (1) in transposing the solar angle relative to the direction of gravity or (2) in orienting directly to the sun's position (Fig. 195).

(*c*) Dancers on a horizontal comb when orienting in accordance with a patch of blue sky are not deranged by seeing, through an interposed polarizing sheet, light completely polarized in the same plane of vibration as exists (with only partial polarization) at that patch of sky. When

---

[47] In this connection it is noteworthy that in *Octopus* the two visual rods developed from each visual cell have the same orientation and are innervated in common by a single nerve fiber (see Fig. 381).

the polarization sheet is rotated, the direction of dancing is shifted in the direction of turning.

(d) Through a polarizing sheet so aligned that it does not change the plane of vibration of the light of the sky, the dancers are given a view of a laterally situated patch of blue sky. Then the sheet is rotated through a definite angle and the change in direction of dancing is measured. A star-shaped device, made from polarizing sheets as a model of the bees' eye (see item 15 of the Summary), is pointed toward the part of the sky that is visible to the dancers, and yields a corresponding polarization pattern. Then a polarization sheet, placed in front of the star-shaped model, is given the same orientation as such a sheet placed above the hive, and thus shows the altered polarization pattern to which the bees' eye is now exposed. By means of the star-shaped model we determine where in the sky this changed pattern is to be seen. The results of such experiments demonstrate the bees' orientation in accordance with the plane of vibration of the polarized light of the sky: if the interposed sheet produces a polarization pattern that is visible at that time at some other azimuth in the sky, the direction of dancing changes by the corresponding angle (Fig. 338). If at that time the artificially produced pattern is nowhere visible in the sky, the dances become disoriented. When the artificial pattern is to be seen at two places in the sky (as is the case when the sun is close to the horizon), then the dances are ambiguous. If the dancers see a blue patch of sky in the zenith above them, then, in accordance with the ambiguity of the pattern in the zenith, they indicate the direction partly correctly and partly incorrectly, shifting at random between the true line and its mirror image.

What determines the direction indicated is the plane of vibration of the polarized light. The degree of polarization is of importance only insofar as it must be adequate for perception of the plane of vibration (see item 8 of the Summary).

4. In an area strange to them bees were trained to a definite compass direction during the afternoon, in the shadow of a high mountain ridge. Once again displaced into an unfamiliar area, on the next morning they went searching for the feeding station in the direction to which they had been trained. Since during their training they had seen only blue sky and not the sun, they must be just as familiar with the diurnal course of the polarization pattern as with the diurnal path of the sun (Figs. 342–344).

5. The polarization pattern of the sky is linked for bees with the sun's position. Foragers that see only blue sky, in flying in a strange area to the feeding place, indicate in their dances on the vertical comb surface the correct solar angle—transposed to the direction relative to gravity. Thus they have deduced the sun's position from the polarization of the sky, which they could see when on their flight. The newcomers aroused by them have learned the solar angle they should maintain on their outward flight, and find the direction in accordance with the blue sky and without seeing the sun.

Bees that are frequenting a feeding station when the sky is completely covered with clouds orient their dances on a horizontal comb properly when a plane of vibration is produced artificially beneath a polarizing sheet. That is also the case with bees that first came to the feeding station at a time when the sky was completely cloud covered but could see the sun through the clouds, though not blue sky. This is further evidence that the polarization pattern of the heavens is linked in the bee's nervous centers with the position of the sun.

6. When for bees that are dancing on the horizontal comb the sun is made visible at the same time as an artificially changed plane of vibration, and the sun is thus put into competition with the polarization of the blue sky, the sun is the deciding factor. But if instead of the usual polarizing sheets, which absorb the ultraviolet almost completely, sheets that transmit ultraviolet are used, then the polarized light too has an evident influence on the direction of dancing. In a third experimental arrangement, the natural polarization of the entire expanse of the sky was made to conflict with the sun, by screening the latter from direct view and reflecting it at an angle of 90° onto the comb (Fig. 351). The bees then indicated a compromise direction, in which whether the polarized light or the sun had the stronger influence depended on the time of day. By way of explanation, attention is directed to the diurnal variations of the polarization pattern in the bees' line of flight and to the greater brightness of the sun in the ultraviolet at higher altitudes.

7. Tests with colored filters show that orientation in accordance with the plane of vibration may occur equally well in ultraviolet alone and in blue light alone. In the green and yellow, the plane of vibration of polarized light is not effective for orienting. This is biologically meaningful. In the ultraviolet region the plane of vibration of the light from the sky is least interfered with by atmospheric influences.

8. We measured the degree of polarization that is necessary for bees to orient in accordance with the blue sky. About 10 percent polarized light is sufficient. With less than 7 percent the dances were always disoriented; between 7 and 15 percent there is a transitional range in which orientation is in part deranged.

9. Perception of the plane of vibration is mediated by the compound eyes and not by the ocelli.

10. The orientation of the dances and the ability to find the way in accordance with polarized light depend on previous experience (menotactic orientation). There exists also an innate spontaneous orientation to the plane of vibration (fundamental orientation, *Grundorientierung*), in which there may be discerned certain preferentially adopted angular orientations. In many animals only a spontaneous orientation to polarization is known. In bees such a phenomenon is demonstrable in addition to menotactic orientation. Spontaneous orientation may be parallel or perpendicular to the plane of vibration, and may also be diagonal to it, at 45° to the right or to the left (Fig. 353).

11. Under certain laboratory conditions it is possible with polarized light to produce on the background a brightness pattern whose nature depends on the plane of vibration. This may indirectly cause an orientation of the animals according to this pattern of vibration. But corresponding conditions were not present in our experiments.

12. It can be shown that under natural circumstances the bees orient directly according to the polarized light of the sky.

(*a*) The orientation of the dances in accordance with the polarized light of the blue sky is deranged if parts of the upper portion of the eyes are capped with black; the dances are normal if the entire lower portion of the eyes is given a black cover. Hence the plane of vibration is read directly from the blue sky and not from the surface beneath.

(*b*) Bees were trained in an arena to walk in a given compass direction. In doing so they could see above them only the blue sky, and beneath them the substrate. For the latter, different papers were used; these differed greatly in the degree to which they favored the occurrence of a reflection pattern. The nature of the substrate did not affect the certainty with which the bees oriented. Besides, under the conditions of open-air experiments, brightness patterns that were related to the plane of vibration in the sky could not be demonstrated, even on papers that were especially favorable to their occurrence.

13. Up to now, wherever indirect orientation to polarized light by means of a brightness pattern in the environment has been found probable or has actually been demonstrated, this has been under laboratory conditions of a sort especially favoring the occurrence of such patterns. On the other hand, it has been shown that not only the bees but other insects, and moreover crustaceans, spiders, and mollusks, are able to orient directly in accordance with the plane of vibration of polarized light.

14. With bees and other Hymenoptera, as well as with flies, the analyzer for the plane of vibration is not in the dioptric system, but in the rhabdomeres of the sensory cells.

15. Analysis of the plane of vibration by the rhabdomeres is comprehensible if it is assumed that they themselves polarize in different directions the light transmitted, in consequence of their radial arrangement (Fig. 331*b*)—as the star-shaped polarizing model shows. Since the receptor cells of the ommatidium are innervated individually, there exists the basis for an analysis of the brightness patterns, which differ specifically from one another in accordance with the plane of vibration of the incident light (see Fig. 336).

16. This theory finds strong support in studies made by the methods of electrophysiology, polarization optics, and electron microscopy, which supplement one another harmoniously.

(*a*) In electrophysiological work it has been possible to insert microelectrodes into individual receptor cells in the eyes of flies and bees, and to record the excitatory potentials developed on illumination. When a polarizing sheet placed in front of the eye was rotated, maxima and minima occurred at 90° intervals. This demonstrates that the individual visual cell responds with a varying intensity of excitation to polarized light of differing planes of vibration. In different receptor cells the maximally effective plane of vibration differs, as theory would require.

(*b*) Investigations with polarization optics have shown that the visual rods are birefringent in the direction of their longitudinal axis. In each optical unit the planes of vibration of low refractive index are radial to the axis.

(c) With electron microscopy there was discovered in the rhabdomeres a fine structure that can be regarded as a basis for the analysis of polarized light. Perpendicular to the long axis of the rhabdomeres there run very fine, tightly packed tubelets (microvilli) that—like the planes of vibration of the polarized light produced by the rhabdomeres—are arranged radially (Figs. 364–370).

17. The arrangement of the fine structure corresponds to the eight-branched organization of the rhabdom only in some insects (Fig. 331b). With other species, each two adjacent rhabdomeres are united closely and their microvilli have the same direction. Thus with the bee there has been described a four-branched arrangement, in spite of her eight sensory cells. Since the microvilli in groups of sensory cells opposite one another run in the same direction, it results that here they follow just two lines perpendicular to one another (Fig. 372). At all events, it is still questionable whether this finding also applies fully in the dorsal portion of the eye, which is directed toward the sky (see Figs. 373 and 374). In case it does, the star-shaped polarizing model would require corresponding modification for the bee (Fig. 375, *lower*).

When the polarization of the sky is examined with this simplified model, the patterns seen are ambiguous in defining the plane of vibration. But the definition becomes unique as soon as the patterns produced by several neighboring optical units are taken into consideration. These patterns disclose how the plane of vibration is changing over the celestial field. Since the bees do not confuse the double significance of the pattern, they must be capable of evaluating appropriately the information from groups of optical units, assuming that the dorsal region of the eye has the four-branched arrangement.

18. The radial arrangement of the fine structures in the rhabdomeres has been confirmed over a very broad range of animals that are capable of analyzing polarized light: from the lowest to the most advanced insects, in spiders, in octopuses, and in squids. Contrariwise, vertebrates lack a radial arrangement of the fine structures in the visual cells, and are incapable of analyzing polarized light.

19. It is assumed that the rigidly orderly arrangement of the microvilli in the rhabdomeres provides the foundation for an equally orderly disposition of the molecules of the visual pigment, and that the latter by itself functions, through dichroic absorption, as the analyzer.

D. A GLANCE AT OTHER ANIMALS

Of recent years, the question how animals find their way about in space and time has attracted increasing interest. A number of symposiums and reviews have been devoted to this problem.[48] Hence it is un-

---

[48] J. Médioni, "L'orientation 'astronomique' des Arthropodes et des oiseaux," *Ann. Biol. 32*, 37–67 (1956); L. Pardi, "L'orientamento astronomico degli animali: risultati e problemi attuali," *Boll. Zool. 24*, 473–523 (1957); M. Renner, "Der Zeitsinn der Arthropoden," *Ergeb. Biol. 20*, 127–158 (1958); K. Stockhammer, "Die Orientierung nach der

necessary to summarize anew all the findings. But since as yet we have with no other creature so good an insight into its orientation to the sun and to polarized light as we have with the bee, we shall compare the main findings obtained in other animals.

Continuing from the preceding chapter, we first enquire how widespread in the animal kingdom is orientation to polarized light.

17. Orientation to the Plane of Vibration of Polarized Light

A large part of the species that have been examined at all were studied in the laboratory. Evidence for the perception of the plane of vibration was provided when, in the darkroom and illuminated from above with polarized light, they adopted preferentially certain definite angles to the plane of vibration. In such instances the angles are restricted to a few values established from the outset: 0°, 45°, 90°, ... around the circle. On p. 407 it was noted previously that this spontaneous orientation (*Grundorientierung*) is to be regarded as a relatively primitive attainment on the part of the nervous centers. As a more advanced achievement many arthropods have developed the ability to maintain, on the basis of experience, a chosen angle to the plane of vibration and thus, in accordance with the polarized light, to hold to a straight course (menotactic orientation). We found the greatest accomplishment in the ability to include in their calculation the shift in the plane of vibration that results from the sun's diurnal movement and to steer a true compass course at all times on the basis of polarized light alone.

With their goal-directed dances bees afford a unique opportunity of testing under a variety of conditions the orientation to polarized light. With other animals the experiments have to be accommodated to such possibilities as exist. A few examples will illustrate this.

Old observations directed initial attention to ants. Santschi had discovered that ants use the sun as a compass when proceeding through a territory poor in landmarks (see p. 134). But he noted also that they are able to maintain a direction when, in a shaded environment, they see nothing above them but the blue sky. Santschi (1923) supposed they could make out the stars in broad daylight and that they steered their course according to them. Our experiences with bees suggested a different interpretation. This suspicion was confirmed in the very first tests: ants running in the open beneath a polarizing sheet from the nest to the feeding place were diverted from the correct course by rotation of the

---

Schwingungsrichtung linearpolarisierten Lichtes und ihre sinnesphysiologischen Grundlagen," *Ergeb. Biol. 21,* 23–56 (1959); "Biological Clocks," *Cold Spring Harbor Symp. Quant. Biol. 25* (1960); A. Hasler, "Homing orientation in migrating fish," *Ergeb. Biol. 23,* 94–115 (1960); "Orientierung der Tiere (Animal orientation)," Symposium in Garmisch-Partenkirchen, *Ergeb. Biol. 26* (1963); E. Bünning, *Die physiologische Uhr* (2nd ed., Springer Verlag, Berlin, Heidelberg, New York, 1963); M. Lindauer, "Allgemeine Sinnesphysiologie, Orientierung im Raum," *Fortschr. Zool. 16,* 58–140 (1963); K. Schmidt-Koenig, "Über die Orientierung der Vögel; Experimente und Probleme," *Naturwissenschaften 51,* 423–431 (1964).

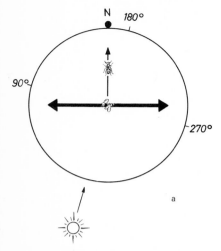

FIGURE 384. Demonstration in ants of orientation according to the plane of vibration of polarized light. (*a*) Ants beneath a circular polarizing sheet are carrying pupae from the center to their nest; they are running perpendicular to the plane of polarization of the light; the sun is visible to them simultaneously; (*b*) by rotating the sheet 45°, their course is diverted 30° to the right, on the average. Heavy arrow, plane of vibration of the polarized light; light arrow, ants' direction of running. Schematic, after Jander 1957.

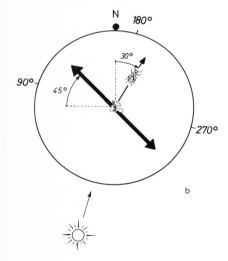

sheet, that is, by altering the plane of vibration (I. Schifferer; see v. Frisch 1950:220). Simultaneously and independently Vowles (1950, 1954) reached the same conclusion, which was also confirmed for other species of ants by Carthy (1951) and Jander (1957). In order to make ants in a confined space perform goal-directed runs under a polarizing sheet, it is best to give them, beneath the center of a circular sheet, a heap of their own pupae, which they transport to their nest at the circumference. Figure 384 shows as an example such an experiment by Jander with the red ant (*Formica rufa* L.). Although the sun was visible simultaneously to the ants, rotating the sheet 45° to the right while they were carrying the pupae into the nest clearly diverted them in this direction, on the average by 30°.

A biologically very fascinating method of orientation was discovered by Pardi and Papi (1952) in littoral animals. The little crustacean *Talitrus*, the "sand flea" (order Amphipoda), is to be found frequently in Italy on the damp sandy shore of the sea coast. If it is blown up onto dry sand during a storm or washed into the sea by waves, it always runs back toward its accustomed environment, the beach. In doing so it orients by the sun or, in shadow, by means of the blue sky. At all times of day it sets out on a path for the shore, and hence is familiar with the diurnal course of the polarization pattern. I am indebted to Prof. Pardi for the photograph in Fig. 385*b*, which he took at my request: the crustaceans were captured on the shore of San Rossore near Pisa (Fig. 385*a*) and were photographed from below in a dry glass dish on 28 April 1964 on the terrace of the Zoological Institute at Florence. The sun was screened off, so the animals could see only blue sky. From the dry place they strove toward the sea in a westerly direction (sector 13), which at home would have taken them to the water. To an artificial change in the plane of vibration they responded with a corresponding shift in the direction of movement (Pardi and Papi 1952).

With these crustaceans and other inhabitants of the shore the compass direction that leads to the goal is fixed by local circumstances. Matters are different with the funnel spider (*Agelena*), for example, with which Görner (1958), by taking advantage of its habits, succeeded in demonstrating an orientation to polarized light. This spider builds a tightly woven flat web and lurks at its margin in wait for prey. Since she will make her web inside an artificial wooden frame (Fig. 386), she can easily be subjected to various experimental conditions. She can be enticed to any chosen place on the web by throwing a fly into it there or by mimicking a fly's struggles by touching the web with a tuning fork. In the spider's straight return to the lookout post, polarized light, besides other signs, plays an important part in her orientation. If she is enticed out of the observation post beneath a polarizing sheet and this is then rotated so as to alter the plane of vibration, on her return she runs the wrong way. Experiments out of doors showed that the natural polarization of the light of the sky is taken account of similarly.

With *Octopus* use of the method of training provided information regarding the perception of polarized light. On being rewarded with the flesh of sardines or punished with an electric shock, the octopuses learned, in accordance with their training, either to emerge from their lair or not, in response to polarized light of different planes of vibration. The light source was a submerged lamp in a housing with a window equipped with a rotatable polaroid sheet. Well-thought-out variations of the experimental arrangement showed convincingly that the distinction was not attributable to brightness patterns in the surroundings (Moody and Parriss 1961; Rowell and Wells 1961; Moody 1962).

In these and similar ways many animals have been tested. In Table 41 are listed the species for which orientation to polarized light has been

FIGURE 386. (*a*) The horizontal web of the funnel spider *Agelena* in a wooden frame. (*b*) The spider lurking in front of her hiding place. After Görner 1958.

FIGURE 385. (*a*) Sketch map of experiment with *Talitrus saltator*: crustaceans were captured on the beach west of Pisa and were taken far inland to Florence. (*b*) Placed in a dry dish there, they jumped away to the west, a direction that at home leads from land to sea. In the experiment the animals could see only blue sky; the sun (in the ESE) was concealed by a screen, the reflection of which may be seen in the WNW on the side of the glass dish. Photograph by L. Pardi, 28 April 1964.

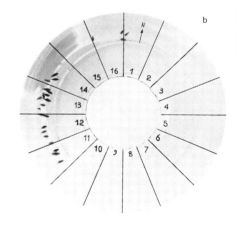

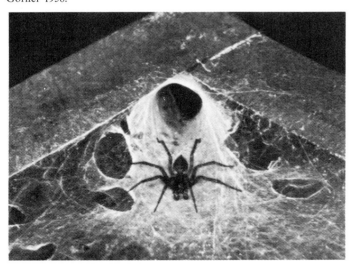

TABLE 41. List of animal species in which orientation to the plane of vibration of polarized light has been demonstrated.

Order	Species[a]	Common name	Reference
Insects			
Hymenoptera (bees, wasps, ants, etc.)	*Apis mellifera* L.	Honeybee	v. Frisch 1949, 1950
	Apis indica F.	Indian bee	Lindauer 1956
	Apis florea F.	Dwarf honeybee	Lindauer 1956
	Bombus terrestris L.	Bumblebees	
	Bombus agrorum F.		
	Bombus hypnorum L.		
	Trigona scaptotrigona Moure	Stingless bee	Jacobs-Jessen 1959
	Andrena sp.	Solitary bees	
	Halictus sp.		
	Vespa germanica F.	Wasp	
	Neodiprion larvae	Sawfly larvae	Wellington, Sullivan, and Green 1951
	Camponotus herculeanus L.		Jacobs-Jessen 1959
	Formica rufa L.		Jacobs-Jessen 1959
	Formica fusca Latr.		Jander 1957
	Lasius niger L.	Ants	Schifferer (v. Frisch 1950); Carthy 1951
	Myrmica ruginodis Nyl.		Vowles 1950, 1954
	Tetramorium caespitum L.		Jander 1957
	Tapinoma erraticum Latr.		Jander 1957
Diptera (flies, mosquitoes, etc.)	*Drosophila melanogaster* Meig.	Fruit fly	Stephens, Fingerman, and Brown 1952, 1953
	Sarcophaga aldrichi Parker	Flesh fly	Wellington 1953
	*mosquito larvae		Baylor and Smith 1953
Lepidoptera (butterflies, moths)	*Malacosoma* larvae		Wellington, Sullivan, and Green 1951
	Choristoneura larvae	Caterpillars	Wellington, Sullivan, and Green 1951
	Hyphantria textor Hass larvae		Wellington, Sullivan, and Henson 1954
Coleoptera (beetles)	*Geotrupes silvaticus* Panz.	Dung beetle	Birukow 1953a; Geisler 1961
	Dyschirius numidicus Putzeys	Ground beetle	Papi 1955
	Phaleria provincialis Fauv.	Darkling beetle (tenebrionid)	Pardi 1955/56
	Bidessus flavicollis Le Conte	Diving beetle (dytiscid)	Jander and Waterman 1960
Trichoptera (caddis flies)	*Neureclipsis bimaculata* L. larvae	Caddis worm	Baylor and Smith 1953; Stockhammer 1959:33
Heteroptera (true bugs)	*Velia currens* F.	Water strider	Birukow 1956

[a] In species marked with an asterisk, menotactic orientation at any chosen angle has been demonstrated; in other species only spontaneous orientation at certain given angles has been demonstrated.

TABLE 41 (*continued*).

Order	Species[a]	Common name	Reference
	Crustacea		
Anostraca	*Artemia salina* L.	Brine shrimp	Stockhammer 1959:32
Phyllopoda	*Sida*		
	Simocephalus		
	Ceriodaphnia		
	Moina	Cladocera	Baylor and Smith 1953
	Bosmina		
	Kurzia		
	Chydorus		
	Leptodora		
	Daphnia pulex de Geer	Water flea	Jander and Waterman 1960
Mysidacea	*Mysidium gracile* Dana		Bainbridge and Waterman 1957, 1958; Waterman 1960; Jander and Waterman 1960
Isopoda (wood lice)	*Tylos latreilii* Aud. and Sav.		Pardi 1953, 1954a, 1954
	Idotea baltica Basteri Audouin		Pardi 1962/63
	Oniscus		Birukow 1956, *fide* Bainbridge and Waterman 1957
	Porcellio		
Amphipoda	*Talitrus saltator* Montagu		Pardi and Papi 1952, 1953
	Hyalella azteca de Sauss.		Jander and Waterman 1960
Decapoda	*Uca tangeri* Eydoux		Altevogt 1963; Altevogt and v. Hagen 1964
	Goniopsis cruentata Latr.	Mangrove crab	Schöne 1963
	Ocypode ceratophthalma Pallas	Ghost crab	Daumer, Jander, and Waterman 1963
	Podophthalmus		Waterman 1961
	Arachnids		
Araneae (spiders)	*Arctosa* spp.	Shore spider	Papi 1955a, 1959; Papi and Syrjämäki 1963; Magni, Papi, Savely, and Tongiorgi 1964
	Agelena labyrinthica Cl.	Funnel spider	Görner 1958, 1962; Baylor and Smith 1953
Acari (mites)	*Hydracarina*	Water mites	Jander and Waterman 1960
	Arrenurus		
Cephalopoda	*Euprymna morsei* Verrill	Squids	Jander, Daumer, and Waterman 1963
	Sepioteuthis lessoniana Lesson		
	Octopus	Octopus	Moody and Parriss 1961; Rowell and Wells 1961; Moody 1962

[a] In species marked with an asterisk, menotactic orientation at any chosen angle has been demonstrated; in other species only spontaneous orientation at certain given angles has been demonstrated.

found up to the present. Some species (marked with an asterisk) are able, on the basis of experience, to orient menotactically at any chosen angle to the plane of vibration. Among these are those with the greatest attainment, finding the compass direction at any hour from the polarized light alone, which has been demonstrated for the honeybee (pp. 392ff), the shore spider *Arctosa* (Papi 1955a:232), the crustacean *Talitrus* (Pardi and Papi 1953), and the crabs *Goniopsis* (Schöne 1963) and *Uca tangeri* Eydoux (Altevogt and v. Hagen 1964), found probable for the ant *Formica rufa* L., the water strider *Velia*, and the wood louse *Tylos*, and is doubtless to be assumed in yet other instances. Evidence required is that the animals shall find a definite compass direction in accordance with the polarization of the light of the sky, even without having seen the sky and the sun simultaneously shortly beforehand.

The summary shows the broad distribution of orientation to polarized light among insects, crustaceans, spiders, and mollusks. Without doubt it will be found in other instances too if only attention is paid to it. Among the annelids (segmented worms) a search might well be rewarding. But this capacity is missing in vertebrates.

Its significance is that it contributes to orientation according to the sun's position, or replaces it if the sun becomes invisible. Since we know of an orientation to polarized light in a number of aquatic animals also, it may properly be assumed that this is an important factor in the broad realm of the ocean too. For even at a depth of 200 m in the sea, where the sun has long since been invisible, the presence of polarized light and the dependence of its plane of vibration on the sun's position were demonstrated clearly (Waterman 1954b, 1955, 1958; Ivanoff and Waterman 1958).

18. The Celestial Compass

The sun and the polarized light of the sky constitute the compass that guides the bees. For their own orientation and for exchange of information they use it in such a multitude of ways as no other animal does. But in one respect they are surpassed: as pronounced diurnal animals they pay attention only to the course of the sun, and leave it to others to steer according to the moon and stars at night.

Initially we shall confine our comparative survey to the arthropods, and afterwards shall glance at the vertebrates. What role the celestial compass may play in the lives of mollusks is yet to be determined.

(a) Arthropods

We are indebted to the Italian zoologists Pardi and Papi for fundamental studies concerning orientation to compass directions. Their demonstration that the little crustacean *Talitrus* is able to orient in accordance with the plane of vibration of polarized light in order to find its way back to the moist strand after it has been displaced has been mentioned (p. 440). Pardi and Papi investigated this rewarding creature from many points of view, and sought systematically on the sandy shore and in the littoral zone for other inhabitants among which similar responses might be expected. They were astonishingly successful. The most important results may be summarized approximately as follows.

The sand flea *Talitrus saltator* Montagu and other amphipods too (*Talorchestia deshayesei* Aud. and *Orchestia mediterranea* A. Costa), the terrestrial isopod *Tylos latreillii* Aud., a number of beetles that inhabit the shore (*Phaleria provincialis* Fauv., a tenebrionid, and the carabids *Omophron limbatum* Fabr., *Scarites terricola* Bon., and *Dyschirius numidicus* Putz., as well as the staphylinid *Paederus rubrothoracicus* Goeze), the shore spiders *Arctosa perita* Latr., *A. variana* C. L. Koch, and *A. cinerea* Fabr. (Fig. 387)—all have in common that from a dry place they set out promptly for the water and, if put into the water, in exactly the opposite direction, toward land. The isopod *Idotea baltica* C. L. Koch seeks the open sea if it gets into water that is too shallow. In doing this these animals, taking into account the time of day and the course of the sun, orient according to the sun's position and the polarized light of the sky.[49] Their orienting by the sun was demonstrated in part by Santschi's mirror experiment and partly (in the laboratory) by means of orientation relative to a lamp, which determines the direction of flight as though the sun were shining from its direction. Biologically, properly directed escape is of great significance to the inhabitants of the shore if they are carried out of their constricted environment by wind or waves.

FIGURE 387. A shore spider (*Arctosa*) that lives on the north bank of a river running east and west will run to the north away from open water and thus reach the shore. Here the spider hastens approximately northward from the middle of a dish of water. From within the dish she can see only the sky and the sun. Beneath the dish is a circular scale for reading her direction. After Papi 1955b.

There have been numerous elegant demonstrations that these animals find the way to the shore through their internal clock and the celestial compass. Such a one is vividly before my eyes now. During a trip to Italy more than 10 years ago Pardi and Papi showed me their discovery. We collected *Talitrus* on the beach and took them with us to Pisa. Placed in a dry dish under the clear sky, they all jumped toward the west. There lay the sea—10 km distant (see also Fig. 385).

Other crustaceans were displaced from the Italian west coast to the shore of the east coast and there too all jumped westward—this time inland, away from the water, in the direction that at home would have guided them out of the dryness to the moist beach. If one tests animals from other areas, where the coast runs in a different direction, there is a corresponding difference in their direction of flight, which is always such that from a dry place they head for the water and from the water for the land.

Long-distance displacement experiments, such as those Renner carried out with bees (see pp. 355ff), bear witness here too to the capacity of orienting according to an "internal clock," independently of local external factors. Both sand fleas (*Talitrus*) and isopods (*Tylos*) were carried from Italy to South America. There they adopted an angle of orientation to the sun that at that hour (according to local time at home) would have taken them to the water on the Italian coast.

In yet another way one may convince himself of the existence of an inner clock; it can be set wrongly, with all the consequences of steering

---

[49] So far as I know, it is only for *Omophron*, *Scarites*, and *Paederus* that this last has not yet been demonstrated.

in an erroneous direction in accordance with a clock that is going incorrectly. As Hoffmann had done with birds (see p. 453), *Talitrus* was exposed in the darkroom to an artificial day-night cycle that was 6 hr later than local time. The animals gradually adjusted to it, and if they were then put in a dry place under the open sky, say at 3 P.M., they maintained the direction relative to the sun that would have been proper at 9 A.M. and went jumping off at an angle that was wrong by 90°, parallel to the shore. After they had been accustomed to a light-dark rhythm that was 12 hr late, they oriented toward the rising sun as though it were setting.

Since the direction of flight differs in different populations in accordance with the coastline at home, one suspects that the proper response is learned in the course of individual life. But experiments have shown that in the amphipods *Talitrus, Talorchestia,* and *Orchestia* it is innate. Even laboratory animals that had never seen sky and sun and that were exposed throughout the day only to a fixed light source would set off in the direction appropriate to their home; the scatter was indeed increased, whence it may be deduced that individual experience too has a little to contribute.

That the direction of escape is inherited is not a common fact. Thus, after overwintering, the beetle *Paederus* (Staphylinidae) has at first to learn to find the shore in accordance with the celestial compass—a behavior that we shall encounter again in ants (p. 448). The spider *Arctosa,* too, requires individual experience.

The isopod *Tylos* and the amphipod *Talitrus* are active predominantly at dusk and at night. Hence their nocturnal orientation was also tested. That led to the discovery that they can find the proper direction for fleeing at night in accordance with the moon, as they do by the sun in the daytime. Since the diurnal course of the moon is by no means the same as that of the sun, in these animals one has to assume two internal clocks operating at different rates, so to speak a "sun clock" and a "moon clock," whose differing phsyiological rhythms are available for orientation by day and by night, respectively.

The total of littoral species is not large. But they afford fine possibilities for the experimenter, and have provided information that is general in its significance. For this reason I have considered them at the beginning here. For details and additional findings reference is made to: Pardi and Papi (1952); Papi and Pardi (1953, 1954, 1959); Pardi (1953/54, 1953/54a, 1954, 1955/56, 1960, 1962/63); Papi (1955, 1955a, 1959, 1960); Marchionni (1962); Papi and Tongiorgi (1963); Ercolini and Badino (1961); Ercolini (1962-63, 1963, 1964); Altevogt and v. Hagen (1964); Pardi and Ercolini (1965).

According to Birukow (1956), the water strider (*Velia*), too, runs off in a definite compass direction when it is put on dry land. But whereas the other inhabitants of the littoral zone regularly take the direction that leads to their home area, the water strider behaves senselessly: at all times of day it heads southward and in doing so it adheres to the sun and the polarized light of the sky as strongly as if its salvation depended on it. Specimens from North and South Germany, as well as from Spain, were alike in this "flight to the south." On the other hand, recent repetitions of the experiments often have turned out otherwise, even at the original site of investigation (Emeis 1959; Birukow 1960). Heran (1962), with veliids from the environs of Graz, was altogether unable to find any compass-true southerly course when he paid critical attention to possible sources of error. In the open the animals oriented preferentially in relation to the wind; in a closed space they were guided predominantly phototactically in a given situation. Further studies are needed to clarify the matter. Perhaps the southerly course seen by Birukow (occasionally also fleeing toward the north, Emeis 1959) is related to the finding of Papi and Tongiorgi (1963) that young shore spiders (*Arctosa*) reared in the laboratory without a view of the sky, when in the open maintain—after initial irregular responses—a northerly direction (see also "nonsense orientation" in birds, p. 453). The capacity to compensate for the movement of the sun is innate with them, but they have to learn the direction of fleeing appropriate to the place where they live. The spiders are able to accomplish this very rapidly, the water striders seem never to get it done. What provokes astonishment in both instances is the at times obstinate orientation to the southerly or northerly direction, for which no basis is evident. It reminds one of very recent reports regarding orientation in some animals to a magnetic field (see pp. 459f).

For the inhabitants of the beach, their very existence depends on their not losing the narrowly bounded moist environment to which they are adapted. The two established compass directions for fleeing from dryness or from the water meet this requirement. A freer use of their celestial compass is made by those animals that possess lairs; from their safe hollows they circulate at will in all directions, seeking prey; for them it is vital to be able to find the way home or, when danger threatens, to disappear into their hole by the shortest possible route. The celestial compass has potential use for them in this connection. Sample illustrations of this are available from the most varied areas of systematics. Pardi (1955–56) found solar orientation in the mole crickets (*Gryllotalpa*), which live in holes in the ground. The funnel spider (*Agelena*), which, guided by the blue sky, finds her way back to her hiding place from anywhere on her web, has already been discussed (pp. 440f). Daumer, Jander, and Waterman (1963) found that the ghost crab *Ocypode* runs from a bait near its hole directly back to its hiding place in accordance with the celestial compass (in the experiments, in accordance with the polarization pattern of the blue sky). Ants and bumblebees belong here, along with honeybees, whose hive entrance in a hollow tree or in the beehive is indeed no more than the opening of the cavity they occupy.

Regarding the honeybee, we know through Lindauer that the young specimens have to learn individually to incorporate time of day and the sun's position in their calculations (see pp. 364f). In this respect there are differences among the several species of animals. Whereas with

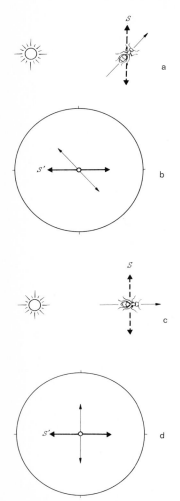

FIGURE 388. Demonstration in ants of the linkage between orientation to sun and to polarized light; in the laboratory experiment a lamp replaces the sun. (*a*) The ant is trained to maintain the direction of the slender arrow relative to the light of the "sun"; the dashed double-headed arrow indicates the plane of vibration *S* of the polarized light that the ant beneath the open sky would see in the zenith with the sun in that position. (*b*) Without light from the lamp, and illuminated from the zenith with polarized light, the ant adopts the angle to the plane of vibration *S'* that she would have maintained to the plane of vibration of the polarized light during training, if this had been visible then. In agreement with the ambiguity of polarization in the zenith, running may occur in either of two directions (see p. 391). (*c, d*) A similar experiment, with a different direction of running. Schematic, after Jander 1957.

Talitrus this capacity is inherited and the spider *Arctosa* is indeed able from birth to compensate for the sun's movement but has to learn the compass direction to the shore, the beetle *Paederus* after overwintering must initially obtain individual experience for both. Through Jander (1957) we know of comparable behavior in ants.

Brun struck upon the good notion of confining ants (*Lasius niger* L.) in the dark for some time on their way to the nest. After they were freed, they ran off in a direction that was incorrect by the angle by which the sun's position had changed meanwhile (p. 348 and Fig. 303). Initially Jander could not confirm this. After their imprisonment in the dark his *Lasius niger* continued their course in the previous compass direction; the ants had taken into account the movement of the sun during the interval of waiting. When in later years Jander repeated the same experiments with *Formica rufa* L., in summer tests he obtained the same result. But in March and April the outcome was quite different; as Brun had found, the movement of the sun was not taken account of and with angularly correct orientation an incorrect compass direction was adopted. Not until a somewhat later season of the year did the ants' celestial compass work properly. Evidently, as with the beetle, compass orientation has to be relearned after the winter (Jander 1957:230–234).

Bees are able, after they have flown a detour, to inform their comrades of the airline to the goal (pp. 173ff). Such accomplishments are foreign to the ant. But confronted with a detour she too—if given the opportunity—promptly manages to choose the airline direction. The funnel spider, having once traversed an angular path, embarks on the airline when she returns; in doing so she is able to orient according to the optical impressions received in the course of the detour (Görner 1958). In elaborate experiments Jander (1957) studied in the laboratory the process in ants and succeeded in showing that the optical impressions obtained during the detour are integrated in the central nervous system to enable finding the direct route. With this a basis is created that brings nearer an understanding of the bees' performance in the detour experiment. Naturally, relations are much more complex in the open countryside.

As with bees, so too with ants, orientation to the sun and to the polarized light of the sky are linked together, and one can be represented by the other. In the darkroom, ants can be trained to maintain in walking a certain angle to an artificial sun (Fig. 388). Under the open sky this position relative to the sun would correspond to a definite plane of vibration in the zenith (dashed arrow, Fig. 388*a*). In the experiment, the lamp is now extinguished and is replaced by diffuse light. Simultaneously the ant is illuminated from above with polarized light. Now she maintains relative to the plane of vibration *S'* the angle that would have been correct relative to the plane of vibration in the zenith for the position of the sun during her training (Fig. 388*b*). During training she saw only the "sun," during the experiment only polarized light. In spite of this she made the proper adjustment. Figures 388*c, d* provide an additional example of this (Jander 1957). In these experiments the ants ran in two different directions, mirror images of one another, in correspondence with the ambiguity of polarization in the zenith—just as we have seen already with bees (p. 391).

The flying bees and running ants agree also in that both make use simultaneously of the celestial compass and of landmarks; for ants, too, the relative importance of the landmarks depends on their conspicuousness. I call to mind our experiments in which the celestial and terrestrial signposts for orientation were placed in competition for bees; after the bee colony had been trained to a definite direction, the bees were transported into a landscape where the guidelines had a different course (pp. 339ff). Jander's method consisted of training ants to a given direction, for example east, in an arena with a view of landscape and sky. Then he put the arena on a flat roof where the ants could see only the sky and trained them to the north. If thereafter they were brought back to the first place, the landmarks there were still in their memory; by them they were directed eastward, but by the celestial compass, toward the north. The conflict led to a compromise solution: the ants ran toward the northeast. In our experiments the bees selected one of the two alternatives. But they had no other choice, because the bait stands were set out at intervals of 90° (see for instance Fig. 289). If we had placed these at separations of 30° or 15°, perhaps an influence of the competitive direction would have become evident. Experiments to complete this work have been planned for a long time, but have not yet been carried out.

During nocturnal journeys ants orient by the moon also. In the mirror experiment they can be led in the wrong direction in moonlight (Jander 1957). Whether the moon serves them as a compass only temporarily, or whether they too, like the crustaceans, are familiar with the diurnal course of both sun and moon, remains to be investigated.

Bees do not go flying out at night. But at times there is dancing at night in the hive by "marathon dancers" (*Dauertänzerinnen*), and then one sees that they indicate the direction of the goal to which they flew during the day—as in a dream, one might say—in accordance with the nocturnal azimuth of the sun, concerning which they have no knowledge from experience (pp. 351f). They "calculate" the course of the sun's movement, from the west across the north to the east. It has been possible to test other insects and even vertebrates in this regard by observing their nocturnal orientation to an artificial sun. A remarkable finding emerged: some animals act as bees do and let the sun continue its path at night from the west across the north to the east (lizards, fish, some birds, mammals). But others orient as though the sun were to travel in a reversed course from the west across the south to the east. It is thus with *Talitrus* and *Velia*, with the beetle *Phaleria,* and with the wolf spider *Arctosa* and some birds. We have no explanation for this contrast.

The selected examples have been studied well and are evidence, in distantly related members of the arthropods, for an orientation according to the celestial compass. It is scarcely to be doubted that numerous species, concerning which such things are unknown, merely represent gaps in our knowledge. Even forms with less sharply localized nesting or dwelling areas steer, somewhat more vaguely, in accordance with the solar compass, in order after excursions in the surroundings to return to the region of their homes. Observations by Marianne Geisler (1961) with dung beetles, which, after their morning wanderings, maintained during the afternoon the opposite course in a more or less constant compass direction, can be so interpreted. Here also belong the June bugs (*Melolontha*), whose behavior was tested by F. Schneider and by Couturier and Robert during several years' experimentation. As is well known, the larvae (grubs) live below ground and feed on the roots of plants. When after their hatching in the open field the beetles come out of the ground, they fly at dusk to trees whose foliage they consume. On this first flight they are guided by landmarks. They go toward any elevated dark area on the horizon and thus reach the nearest woods. But on this trip they pay heed to the celestial compass. After about 2 weeks, the mature females fly back for oviposition to the open field, to their birthplace, from which they had come. In doing so they orient according to the celestial compass, by transposing 180° the direction of their outward flight—which they have retained in their memory. Thus they get back into the area of their origin (F. Schneider 1956). They also set out again on this compass direction if they are captured en route and taken into another region that they are unacquainted with. They do this also when they are kept confined initially and set free at a different time of day. When the sky was cloud covered, they were disoriented. When the sun was visible, they could be led astray in the mirror experiment. With a clear sky they found the proper direction even without the sun. Nevertheless, a direct demonstration of orientation in accordance with polarized light has not been made, likely as this is on the basis of the experiments performed. The concept of a purely "astronomical" orientation adopted initially (Couturier and Robert 1955, 1955a, 1956) was later narrowed down, because in some experiments the beetles found the correct direction even under a completely cloudy sky (Couturier and Robert 1957, 1958). Clarification of this contradiction by means of further experiments would be desirable. Participation of other possible means of orientation has been discussed (see p. 459).

Some orders of insects, containing a wealth of species, have scarcely been examined as to their mode of orientation, although they directly invite such study. People have long been astounded by the enigma of the impressive migrations of butterflies, and basically are still so today (Williams 1958, 1961; Nielsen 1964). Thistle butterflies (*Vanessa cardui* L.) annually fly from North Africa to Europe, at times as far as Finland; even in crossing the Mediterranean each one of them maintains individually a northward course (observations from shipboard). In the spring, mass flights of the cabbage butterfly (*Pieris brassicae* L.) are seen directed to the north, in summer to the south, at times as dense "as a snowstorm." The monarch (*Danaus plexippus* L.) migrates in autumn from Canada to Florida or southern California (in round numbers 3000 km) to its winter quarters, and in spring flies back over the same distance. Lindauer (1963a:82) saw in Florida such a spring flight in which it was perfectly clear that there was active compensation for the lateral drift from the northerly direction of progress, caused by a strong east wind.[50] Migratory swarms of *Ascia monuste* L. (fam. Pieridae) often fly in opposite directions, close to one another. Frequently they hold to the same course for many hours, and may resume flight in the same direction the next morning. This was observed in America over uniform prairie also, far from the coast and other optical guidelines (Nielsen 1961 and *in lit.*). It can happen that the insects are borne along by the wind at times, but without compass orientation according to the sun directed active migration is hardly comprehensible. Demonstrations of this are yet to be furnished, just as other insect migrations offer a wealth of material for the research-minded biologist.

With vertebrates, in contrast, the more recent investigations of their capacity for orientation have found in their migrations a fruitful starting point.

(b) Vertebrates

It was the problem of bird migration to which Gustav Kramer gave new impetus in a series of fine studies. Fundamentally, the same steps, taken independently, led with birds, as with bees, to the solar compass. After Kramer had seen that migratory birds, when their time for departure has come, indicate clearly even when caged the direction they intend to take, he put starlings into a round cage under the open sky. Any sight of landmarks was screened off all around. Nevertheless the birds still strove toward the direction of migration—as long as they had a view of the sun (Fig. 389). When the sky was covered with clouds they were disoriented. If the sun was screened and reflected to them from another angle, they pressed toward a correspondingly altered direction. Magnetic disturbances were without effect. It proved possible also, by feeding the starlings, to train them to a given compass direction and thus to free the experiment from the seasonal limitations on bird migration (Kramer 1950). In a tent the sun could be replaced by a fixed lamp (Kramer and Ursula v. Saint Paul 1950). If the bird had been trained to the west under the open sky, in the tent it sought its food opposite the artificial sun in the morning, 90° to the right of it at noon, and in the direction of the lamp in the evening. Thus it no longer aimed for a given compass direction, but instead oriented with respect to the artificial light in the way it would have had to orient to the sun at the same hour if it wanted to go to the west (Fig. 390). Hence it evidently was able to include in its calculations the diurnal course of the sun and the time of day.[51]

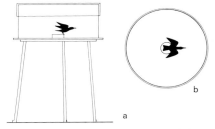

FIGURE 389. A starling on a perch in a round cage shows clearly the direction in which it will try to fly: (*a*) side view; (*b*) from above. The bird can see no landmarks, only the sky and sun above it. Schematic, after Kramer 1951.

The birds—like other vertebrates too—showed themselves inferior to bees in that they were not able to use the polarized light of the blue sky to learn the position of the sun. Also they become unable to detect the sun as soon as clouds obscure it from our eyes. So far as is yet known, bees alone are exceptional in this respect.

---

[50] According to Nielsen (1964), there is as yet no exact proof that the same individuals migrate back from the southern winter quarters to the northern habitat.

[51] Demonstration of the time sense was possible in training experiments with birds also. As with bees, training was successful only in a 24-hr rhythm (Hedwig Stein 1951).

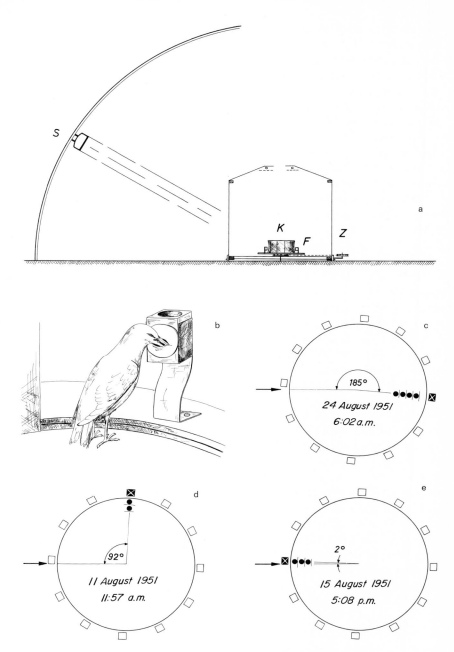

FIGURE 390. (*a*) Arrangement for training starlings to a chosen direction, under an artificially motionless sun, *S*; *Z*, experimental tent; *K*, cage in the middle of which the starling sits; from here it seeks the proper direction relative to the "sun" and gets food from one of the 12 food containers *F* around the circumference. (*b*) A starling getting food. (*c, d, e*) Three examples from a series of experiments: after being trained under the open sky to the western direction, in the morning the starling chose the container opposite the artificial sun, at noon the one 90° to the right of it, and in the evening the one toward the sun. Arrow, direction of light from the artificial sun; dots: starling's choices. The boxes around the circumference are the food containers (black with white cross, the expected direction). After Kramer 1953.

However, in another way birds give evidence of having greater mental flexibility than bees. With starlings, Kramer succeeded in an experiment with which we had nothing but failure with bees (see p. 338). The birds learned to overlook the diurnal movement of the sun and, for instance, to get their food at all times in the sun's direction. To be sure, they learned this reluctantly, and thus gave evidence that the procedure was unnatural for them (Kramer 1952, 1953).

The significance of the "internal clock" in finding direction was demonstrated with starlings by K. Hoffmann (1953, 1954), by means of an experiment that has since proved useful with various other animals (for instance *Talitrus*; see p. 446). Birds retain for months, without renewed practice, the direction to which they have been trained. This fact provided the following possibility. A starling with a view of the sun was trained to the south, for example, and was then put for some 2 weeks into a closed room, where it was exposed to artificial illumination from exactly 6 hr after sunrise to exactly 6 hr after sunset. When it was once again tested under the open sky, it searched for the feeding station 90° too far to the right. It chose the angle relative to the sun as though the latter were where it had been 6 hr previously. The internal clock demonstrated its existence by letting itself be set back by the altered lighting rhythm.

Even when the birds had been allowed for 2 hr before the experiment to observe the sun's movement in the open, which now conflicted with their own "clock," there was no change in the direction of their choice. From this it is evident that neither the elevation of the sun, nor its apparent motion, but only its azimuth determines the finding of the direction (Hoffmann 1960). In this the bird behaves just like bees.

That in their migrations out of doors in nature birds are able to use the sun as a compass, in order to maintain a hereditarily fixed line of flight, is borne witness to by many observations. One of the first indications (1933) was provided by the elegant experiment of displacing to West Germany young storks taken from the nest in East Prussia; they were held back until after the old storks had departed. Dependent entirely upon themselves, with no guidance from older storks that know the route to fly, they set out for their winter quarters on a course parallel to the route taken by eastern stork populations to their winter quarters, and thus flew over the Alps to Italy, instead of finding the way across Palestine to Africa (Schüz 1949).

Matthews (1961, 1962) described a senseless-sounding kind of orientation ("nonsense orientation") in mallards (*Anas platyrhynchos* L.). When they were caught, and set free in another vicinity, they flew off toward the northwest, irrespective of the direction and distance they had been transported; they steered according to the solar compass. In southern England and Sweden populations of the same species were found that took a course to the southeast. This performance is somewhat reminiscent of the southerly and occasionally northerly course of water striders (*Velia*) and of the young shore spiders (*Arctosa*) that press stubbornly toward the north (see p. 447).

Birds traveling at night also are able to orient by the sky, being guided by the images of the stars instead of by the sun, as has been shown by the experiments of Sauer (1957, 1960) with warblers in a planetarium. This again is an area in which birds show themselves superior to arthropods (even to those of nocturnal habits)—in all probability not because the arthropods are unable to "comprehend" the movement of the stars but simply because they lack the sharp eyes of birds to see them with.

The orientation of birds has become more comprehensible since we have learned of their celestial compass. But there is still a great deal that is unexplained. The behavior described above for young storks has indeed been confirmed in other young birds, for example starlings, but mature starlings showed that they could set out directly toward the goal even after they had been displaced laterally from the line of their migration (Perdeck 1958). Likewise birds that had been carried in closed containers to an unfamiliar region several hundred or even much more than 1000 km from their nesting sites often found their way back in an astonishingly short time. The solar compass permits them to maintain a given compass direction. But in a completely strange landscape, how do they know what is the direction to the goal? That under such circumstances they can find the way is understandable only if they are able to determine the geographical locations of the place where they are and of their objective. Hypotheses have been developed that are intended to assist comprehension of such accomplishments. So far they have found no general acceptance.

Another difficulty arises from the length of migration routes and the high speed of travel of many birds. Very swiftly they may reach areas where their internal clock is at considerable variance with local time, where the sun takes a course different from the one to which they are accustomed and at noon may stand in the zenith or in the north instead of in the south. Under such changing circumstances, to be able to travel in a true direction in accordance with the celestial compass presupposes a rapidity of adaptation that is hard to imagine and that is unknown among arthropods. The latter of course are not subjected to any such great demands.

There are not merely theoretical considerations for assuming additional means of orientation beside the celestial compass. Recent experiments strengthen this too (for example, Gerdes 1962 with blackheaded gulls). According to Fromme (1961), European robins (*Erithacus rubecula* L.) and whitethroats (*Sylvia communis* Lath.), during their nocturnal efforts to migrate, were able to find the natural direction of migration with statistically significant frequency, even in a cage that afforded them no optical points of reference and no view of the sky. Perdeck (1963) rechecked this observation and was unable to find that European robins preferred any direction. He criticized the statistics used by Fromme. Merkel, Fromme, and Wiltschko (1964) attribute the negative results obtained by Perdeck to differences in his technique and, supported by new statistical evaluation, restate Fromme's result. Merkel and Wiltschko (not yet published experiments at the Zoologisches

Institut in Frankfurt a. M.) confirmed the tendency of European robins to a certain direction in rooms without any sky light and were able to influence it with artificial magnetic fields (see also pp. 459ff). Yet one cannot tell whether these preliminary observations will clarify these complex matters.

The vertebrates offered the same surprise as the arthropods did: what initially was regarded as an isolated peak performance proved to be the widespread rule.

Some fishes, like eels or salmon, undertake just as extensive journeys to distant goals as the migratory birds. As yet people have wondered in vain how these fish find their way in the ocean. Now it is of course more difficult to experiment with fish swimming in the sea than with birds flying over land. One must be more modest initially, thought A. Hasler. His zoological laboratory at the University of Wisconsin is on Lake Mendota. In this inland lake the white bass (*Roccus chrysops* Raf.) assemble annually at spawning time at certain places on the north shore and afterward distribute themselves over the lake. At spawning time Hasler and his associates transported southward into the middle of the lake tagged fish caught on the north shore, and after a short time were able to retake them at the former place, 2–3 km distant. On a small scale, this is the same problem as in the ocean: how do they find their way?

When plastic floats were attached to the displaced fish by nylon threads (Fig. 391), it became evident that on sunny days the fish immediately proceeded northward from the middle of the lake, whereas when the sky was covered with clouds they dispersed in all directions. Laboratory experiments provided complete proof of orientation by the sun. The technique had to be accommodated to the habits of the fish. Since they like to conceal themselves, a large basin of water was equipped with 16 hiding places that could be reached from as many different directions. But the fish could get into only one of these recesses; the others were closed. In this way a fish, placed in the middle of the basin, learned to seek out at once that hiding place that was accessible from a certain compass direction. Once the fish had grasped that much, it could do the task successfully at any time of day, provided it could see the sun. When the sun was hidden behind clouds, the fish was helpless. It was unable to orient in accordance with the polarized light of the sky. Here too the sun could be replaced successfully by a laboratory lamp. At all hours fish would orient to it just as they would have had to orient to the sun at the same time in order to find the direction to which they had been trained (Hasler 1956, 1960; Hasler, Horrall, Wisby, and Braemer 1958).

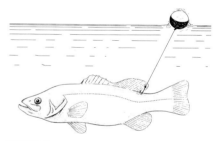

FIGURE 391. A white bass pulls a plastic float along behind it with a nylon thread and thus shows the observer its direction of travel in the water. After Hasler 1960.

Investigations by Hasler and his co-workers also afforded clarification of a point regarding which little is as yet known with other animals. They compared the innate capacity of orienting to the sun's course in fish from the north temperate zone with that of fish that are at home under the tropical sun (Braemer and Schwassman 1963). In the tropics the direction must of course be compensated for differently with the seasons, for at one time the sun moves clockwise and stands in the south at noon, at another time it moves counterclockwise and is in the north at noon. Under the simpler conditions in northern latitudes the ability to calculate the sun's course proved to be innate with the sun perch (*Lepomis cyanellus* Raf., fam. Centrarchidae). A tropical species (*Aequidens portalegrensis* H., fam. Cichlidae) is indeed able from the beginning to compensate for the movement of the sun, but has to learn by experience the direction in which it moves. So both possibilities are provided for. When twice a year the course of the sun is shifted from one side of the sky to the other, the change seems to be by no means easy for the fish. For it takes several weeks before they master the difference. We have learned similar things about the behavior of tropical bees (p. 363).

In contrast with all the findings in bees and in other animals too, with fish the elevation of the sun has an effect on their finding a direction. When for instance in mirror experiments the morning sun was displayed from the opposite direction, they swam wrong by 180°; if the sun's position was raised artificially, by appropriate placement of the mirror, they deviated toward the direction they would have had to maintain at a later hour in accordance with a higher sun.

When the zenith distance of the sun was less than 5°, orientation to a direction was no longer possible. In this respect bees are a little more capable (p. 160).

Today it is definite that numerous fresh-water fishes are accustomed to use the sun as a compass. This is so too for Salmonidae, to which the salmon belongs (Hasler 1960:109). Whether in its extensive journeys in the sea this species actually steers by the celestial compass it has not yet been possible to observe. But it is very likely, for salmon (*Oncorhynchus nerka* Walbaum) that have spent the first year of their life in an extensive, complexly divided system of Canadian lakes (Babine Lake) and at the appointed time set out on their journey to the sea maintain on all sections of their route the compass direction that will eventually bring them to the outlet of the lake. They do this too if they are moved into a large tank and can perceive no landmarks but only the sky above them. If the light of the sky is excluded, they become disoriented. Thus it is proved for these salmon that they use the sun as a compass and compensate for its diurnal course.[52] What the young fish are able to do may be imputed also to the mature salmon that set out from distant parts of the sea on their return to the coast (Brett and Groot 1963; Johnson and Groot 1963).

As for Amphibia, it is known that frogs, toads, and salamanders, which following oviposition have dispersed far over the countryside, at the next spawning time return loyally to their accustomed spawning ground in a given pool. For the North American grasshopper frog (*Acris gryllus* Lec.) it was shown experimentally that it uses the sun as a compass. The same is probable for toads (Ferguson 1963). Earlier

---

[52] Whether in addition they make use of other aids to orientation, and what the nature of these might be, is still unclarified. According to available observations, such aids would be of only subordinate significance.

suspicions that the animals were guided by humidity, olfactory stimuli, or the inclination of the terrain were overthrown by an astonishing observation in Switzerland: pools that had provided spawning places were filled in during the construction of a highway. In the following spring the toads came looking for their hereditary spawning ground, which should have been there, and sought it so obstinately that at some places there was an almost continuous layer of squashed toads that had been run over. They hunted for the place on the dry roadbed, which had been raised 2.5 m, in an area that certainly smelled entirely different from before (Heusser 1960).

The salamander *Taricha* (*Triturus*) *rivularis* Twitty enters the water at spawning time in the spring, but spends the rest of the year on land and returns to its former spawning place only at the next reproductive period. In the California mountains salamanders were captured in a brook at spawning time and were liberated in another brook as much as 4 km away. During their succeeding stay on land about half of them journeyed back to their native brook and most of these returned approximately to the places of their origin, although they had to surmount a ridge 300 m high. Even blinded animals accomplished this (Twitty 1959; Twitty, Grant, and Anderson 1964). Such observations indicate that with Amphibia too, in addition to the solar compass, there are at work means of orientation as yet unknown that guide them to the accustomed geographic location.

Among reptiles, green turtles (*Chelonia mydas* L.) undertake regular journeys between their feeding areas and breeding places. In doing so they swim for 1800 km and more over the open sea. With young specimens an orientation by solar compass was demonstrated (Klaus Fischer 1964). But among the vertebrates by no means only the pronounced migratory species make use of the solar compass. Lizards also (*Lacerta viridis* Laur., *L. sicula* and *L. muralis* Laur.) are quite adept at this method of finding direction. Birukow discovered that they are suitable as experimental animals. Once again technique had to accommodate itself to the material. Lizards are warmth-loving creatures. In laboratory experiments they were trained to seek out a warm spot (hot plate) in a certain compass direction, in accordance with an artificial sun. By learning to do this they demonstrated that they use the sun for orienting in space and that they take account of its course as well as of the time of day. They appeared to be honestly "convinced" of the correctness of their procedure. For in a test they would lie flat on their stomachs at the proper place for warming themselves, even if the hot plate were not there (Fischer and Birukow 1960).

Lizards can be reared readily from the egg and thus excluded completely from any experience of the sky and sun. If such animals are trained under an artificial sun to a definite direction at a given time of day, then at other hours they will seek the goal at the accustomed angle to the lamp. Their response is to a constant angle. Not until they have been trained to three discrete points of the sun's course at corresponding times do they master the solar compass and respond at all hours to a fixed direction. An inverse solar course, running from the west across the south to the east, was much harder for them to learn (Birukow, Fischer, and Böttcher 1963). Thus here too we see the interplay of hereditary disposition and individual experience, such as we have been confronted with on several previous occasions.

No doubt everyone has either heard of or experienced marvelous accomplishments of the gift of orientation among mammals; in particular, it is often related of dogs that they are able to find their way home from a distant strange territory. There has been also no dearth of experiments. But unfortunately there is, more than with other vertebrates, a dearth of definite results. In one instance—with a mouse—the successful demonstration of orientation to the sun is reported (Lüters and Birukow 1963); the experiment was accomplished by training the mouse with food to a given compass direction under an artificial sun. The sun's course and the time of day were taken into account by the animal.

Neither sun nor moon and stars could have been helping bats, whose eyes not only are poor but were covered with opaque blinders, when they came flying home at night without delay from a distance of 8 km (Mueller and Emlen 1957). In other displacement experiments into unfamiliar territory, bats have returned from as far away as 300 km. In many instances they got home so fast that they could not have made any great detours. There are species that fly regularly in autumn to the narrowly circumscribed location of a cave for hibernating, and return in spring to their summer quarters, separated by some 200–300 km (Griffin 1958). It is hard to imagine that in such extensive journeys they are guided by their marvelous echo location, for this functions only over distances of a few meters. If for birds, after one has taken away the known means of orienting, there is left as an unexplained residuum their directed return home from a strange territory, with bats the same enigma is present and awaiting solution.

Man too belongs among these insufficiently investigated creatures. One has only to go to primitive peoples to find great accomplishments in the area of orientation, for civilization has rendered us incompetent in such matters. I quote from A. Gabriel (1958:202) a passage from the description by the desert explorer N. St. J. B. Philby of his crossing of the great sand desert of Rub al-Khali in South Arabia (1932), the participants in which were close to dying of thirst:

"The sun stabbed down at us, and my companions slumbered as they rode, an easement that was denied to inexperienced me. Salim dozed like the rest, pulled himself together now and then and growled the direction. I wondered how under such conditions, half asleep and with nothing to guide him, he was able to hold to a direction at all, but frequently I checked his course with the compass and was astonished again and again by the precision with which he steered. I asked him how he accomplished this, but he simply did not know. Perhaps there is among men and animals of the desert an inborn feeling for direction, that is supervised by unconscious perception of the movement of the sun and stars. Salim also lacked the slightest inkling of time as such—an hour counted for exactly nothing—but he seemed never to lose the way, at least not when the sky was clear."

19. ORIENTATION TO A MAGNETIC FIELD

It was reported above (p. 450) that female June bugs, returning from their feeding places to oviposit in the open field, steer by the celestial compass. Their orientation to the sun has been demonstrated in the mirror experiment. But in various instances a correct orientation was seen also when all sight of the sky was excluded. F. Schneider (1957) took from the soil June bugs that were ready to fly, fastened them to aluminum wires, and at dusk suspended them under the open sky with the head in a definite direction, to which they became "imprinted." If a few days later he released them to fly, they preferred the "imprinted" compass direction or its exact opposite, even in a completely closed room or in a windowless basement. Additional experiments (Schneider 1961, 1963, 1963a) were suggestive of an orientation to electric and magnetic fields, though the results were very variable and not regularly reproducible. Unknown factors ("ultraoptical points of reference") seemed also to be involved.

Couturier and Robert too determined in several experiments that June bugs found the direction even under a completely cloud-covered sky (see p. 450). Efforts to demonstrate an orientation to the earth's magnetism turned out negative. They too found no explanation for the beetles' performance without assuming additional, still unknown reference signs (Couturier and Robert 1962; Robert 1963).

On the basis of physical measurements and calculations Kimm (1960) regards it as impossible that June bugs should be able to orient to the earth's magnetic field. But according to numerous very recent reports it seems nevertheless as though something of this sort might exist.

G. Becker (1963) noted with termites (sexual forms of species of *Macrotermes* and *Odontotermes*) in his laboratory in Berlin that they regularly came to rest in an east-west direction. "When the earth's magnetic field was excluded, in a suitable iron box, the orientation of their bodies to the compass direction ceased." If a permanent magnet was placed above or below the specimens they assumed "in the course of fifteen minutes up to several hours without exception a definite position relative to the field." Other species too behaved in this way, insofar as they were not too restless. They may also align themselves in a north-south direction. Termite queens in their natural habitat are said to be found preferentially with such orientation. Hitherto these reports were doubted but now no longer seem so incredible.

According to Becker (1963a), flies (*Calliphora erythrocephala* Meig., *Musca domestica* L., species of *Sarcophaga* and *Lucilia,* and *Tubifera pendula* L.) almost always land on horizontal surfaces in an east-west or a north-south direction when it is calm. These positions may, however, be interfered with by the sun or other stimuli. Positions parallel or perpendicular to the lines of force were seen in magnetic fields that were, say, 100 times as strong as the natural field of the earth. When the magnetic field is compensated for, this ceases. The preference for the north-south and east-west directions on landing and when sitting was confirmed in windowless rooms under artificial light. Dead flies suspended by long threads show the same phenomena (Becker and Speck 1964).

According to correspondence from Becker, comparable orientations to a magnetic field have meanwhile been noted also with beetles, cockroaches, crickets, grasshoppers, wasps, and honeybees. Doubtless additional critical experiments are needed to establish these findings.

Many years ago I tried without success to influence magnetically the direction of dancing of bees (see p. 134). Any orientation to the earth's magnetism should become evident in dances on the horizontal comb if the bees can see no optical guides. But under such conditions they are completely disoriented. In his long-distance displacement experiment, Renner came to the conclusion that the earth's magnetic field has no effect on the direction of flight of bees that are en route to their goal (see pp. 360f).

In view of the positive findings with other insects and of Becker's indication in reference to bees also, a re-examination of the latter was called for. Consequently, in cooperation with K. Daumer, in the summer of 1964 I did some experiments that were designed to answer three questions for us: (1) whether the direction of dancing of bees on the vertical honeycomb surface is influenced by changes in the earth's magnetic field; (2) whether when they land at the feeding station they prefer certain directions with respect to either the earth's magnetic field or an artificially induced magnetic field; and (3) whether their position when sitting at the food source shows any definite relation to the magnetic field.

At our experimental area in Brunnwinkl the total strength of the earth's magnetic field H_t is 0.47 oersted, and θ is 64° (Fig. 392). In order to exclude the earth's magnetic field we used a pair of Helmholtz coils 1 m in diameter, which were adjusted in such a way that the magnetic field H_s induced artificially by the coils just canceled (within about 3 percent) the earth's magnetic field between the coils. In the center of the pair of coils a small observation hive was set up; for our purposes it was constructed without any iron parts (Fig. 393). It contained only a single (unwired) comb, and was occupied by a small Carniolan colony. The flight funnel pointed to the east. A group of about 20 marked bees were coming and going at the feeding station 200 m to the west. The hive arrangement was housed in a fiberboard hut, so that the angle of dancing could be measured in diffuse light. Without the observer's knowledge, the magnetic field of the coils was switched on (earth's field compensated) or off (earth's field normal). The shift was made either after each five dances measured, or from dance to dance, or during a single dance.

Now, if one compares the deviations from the expected direction of dancing (that is to say, the "residual misdirection") in the unchanged and in the compensated terrestrial magnetic field, then any influence of the earth's magnetic field on the indication of direction in the dance would have to become apparent. But, as appears from Table 42 (I) the difference between the two mean values in the unchanged and compensated fields (3.9° ± 4.2° and 3.6° ± 4.2°, with $P = 0.7$) is not in

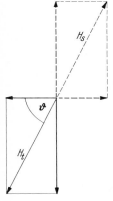

FIGURE 392. Compensation of the earth's magnetic field ($H_t = 0.47$ oersted, angle of dip $\theta = 64°$) by an opposed coil field H_s of equal intensity.

FIGURE 393. Helmholtz coils positioned to compensate for the terrestrial magnetic field; in the center is mounted the observation hive with a single comb.

TABLE 42. Deviation from the expected direction of dancing in various magnetic fields.

Date, 1964	Time	Mean deviation ± scatter (deg)	Number of dances	Significance, P
31 July	8:46–11:07	3.9 ± 4.2[a]	65	0.7
	8:50–11:07	3.6 ± 4.2[b]	67	
3 August	11:34–12:07	1.0 ± 3.1[a]	31	0.9
	11:31–12:03	0.5 ± 3.6[c]	35	
3 August	9:07–10:17	4.1 ± 6.0[a]	43	0.6
	9:09–10:21	4.9 ± 5.7[d]	40	
1 August	8:41–10:31	−1.0 ± 3.0[a]	69	1
	8:40–10:31	−1.0 ± 3.4[e]	72	
1 August	10:34–11:07	2.0 ± 2.9[a]	14	0.8
	10:34–11:07	1.9 ± 2.6[f]	18	

[a] Earth's field normal.
[b] Earth's field compensated.
[c] Resultant field equal but opposite to that of the earth.
[d] Resultant field equal to that of the earth, but at 90° to it.
[e] Resultant field 1.5 times that of the earth, and at 45° to it; inclination 71°.
[f] Resultant field 3 times that of the earth, and at 64° to it; inclination 70°.

the least significant. Thus, the direction of orientation in the dance is not influenced by excluding the magnetic field.

In additional experiments we examined the question whether perhaps magnetic fields of differing direction and intensity in comparison with that of the earth might exert any influence on the dances. Here we adjusted the field of the coils in such ways that the field resulting from interaction of the earth's field and the coils' field satisfied the following conditions (Table 42):

II. The resultant field is antiparallel to that of the earth; that is, it is directed upward and to the south and is equal to the earth's field in intensity.

III. The resultant field is at 90° to that of the earth and has the same inclination and intensity. The lines of force are parallel to the comb surface.

IV. The resultant field is at 45° to that of the earth, has an inclination of 71°, and is 1.5 times as intense as the earth's field.

V. The resultant field is at 64° to that of the earth, has an inclination of 70°, and is 3 times as intense as the earth's field.

As may be seen from the table, under none of the conditions specified did there occur a statistically significant difference between the mean values of the deviation from the expected direction of dancing in the natural and altered magnetic fields.

In still further experiments we tested the directions at which bees alighted at a feeding station set up 20 m from the hive; the feeding station was now placed at the center of the pair of coils. Becker had seen that in an optically neutral environment flies land or sit preferentially in angular areas ±20° to either side of the N-S and E-W axes,

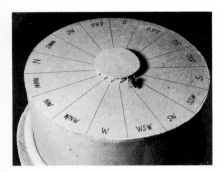

FIGURE 394. Arrangement for observing the directions of landing of bees at the feeding station.

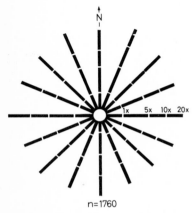

FIGURE 395. Frequency distribution of the directions of landing, in groups of 22.5° each: (*above*) in the natural terrestrial magnetic field; (*below*) in the resultant field rotated 90° to the west. The segments of the lines bounded by the white cross-strokes indicate the results obtained with increases in the horizontal field intensity of 1, 5, 10, and 20-fold; the length of each segment is proportional to the number of landings. The upper figure shows the corresponding control experiments. 12–19 August 1964.

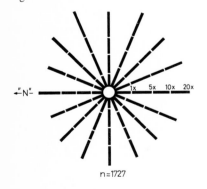

and that this preference for certain directions disappears when the earth's magnetic field is compensated for. As already mentioned, he thought he had observed comparable performances with bees also. But we were unable to confirm such behavior.

Initially we did experiments out of doors in "calm" weather. But since even a gentle breath of air proved to be a disturbing factor, the experiments were continued by Daumer in Munich in a room protected from the wind.

The apparatus for observing the direction of landing at the feeding station consisted of a cardboard cylinder 20 cm in diameter and 10 cm high, on the cover of which 16 sectors with the compass directions were marked out (Fig. 394). In the center is an entrance hole through which the bees can get to the sugar-water dish situated beneath it. The opening is roofed with a circular sheet of cardboard, 5 cm in diameter, that is supported on the boundaries of the sectors by 16 thin wooden strips. The apparatus stood in a small, half darkened, artificially lighted room, into which the bees were able to fly via a roundabout path. Note was made of all approaches that led straight from the margin of the cylinder to a landing at one of the 16 openings. The performance of the bees was tested alternately in the unchanged terrestrial magnetic field and in a field at 90° to that of the earth. The horizontal intensity of the resultant field was varied so that it amounted to 1, 5, 10, and 20 times the horizontal intensity of the terrestrial field (0.21 oe). After each 5 landings, the experiment leader switched the field of the coils on or off without the observer's knowledge.

Figure 395 shows the distribution recorded for 3487 landings on the 16 compass sectors during the time from 12 to 19 August 1964; on the left are those in the terrestrial magnetic field, on the right those in the resultant field which was rotated 90° to the west ("N"). The segments of the lines bounded by white cross-strokes depict separately the results at different field intensities. The length of each segment is proportional to the number of landings. Even without statistics it is evident that both in the natural terrestrial magnetic field and in the rotated field, up to as much as a 20-fold increase in intensity, no single compass direction is significantly preferred in landing.

By way of contrast, in an experiment on 13 August 1964 there were clearly more landings from directions between N and W (Fig. 396). This experiment was conducted in the open. Out of the southeast there fluctuated a scarcely perceptible breath of air with an average velocity of 0.15 m/sec. It was evident that the bees landed preferentially against the weak motion of the air, independently in fact of whether the terrestrial magnetic field remained unaltered (left-hand figure) or was rotated 90° with a tenfold increase in intensity (right-hand figure). A comparable influence of wind on the direction of landing was also confirmed clearly in other experiments out of doors.

Observation of the positions taken in sitting at the feeding place were made with individual bees, again in the inside room protected from the wind, but this time with an altered experimental arrangement (Fig. 397). At the center of the sectored circle there was placed a little aluminum-foil plate, 4 cm in diameter, into which there opened a copper capillary, 1 mm in inside diameter. Sugar water rose in the capillary from a glass container set beneath it. The arrangement was surrounded by a cardboard cylinder 50 cm in diameter and 50 cm high (not visible in the picture). By means of it the trained bee was compelled to approach the capillary in spiral flight from above. After landing she sometimes briefly changes her direction of sitting slightly; this position is then maintained throughout the process of drinking and can be observed very precisely.

In an experiment on 12 September 1964 from 8:00 A.M. to 4:30 P.M. there were measured in turn 220 directions of sitting in the unchanged terrestrial magnetic field and 211 in a field of equal intensity rotated 90° to the west. The outcome (Fig. 398) gives not the slightest indication of a preferred direction of sitting relative to the magnetic lines of force.

Thus, in agreement with our previous experience, no orientation to a magnetic field could be demonstrated with this experimental arrangement.

It is not intended thus to dispute its occurrence with other insects. For mollusks and flatworms, too, there are positive reports. According to Brown and his co-workers (for instance Brown, Webb, Bennett, and Barnwell 1959; Brown, Bennett, and Webb 1960; Brown, Webb, and Barnwell 1964), aquatic snails can be influenced in their direction of creeping by magnetic fields. Comparable phenomena, though of fundamental significance in the life of the animals, have become known with regard to some kinds of fish. Electric fields, which they produce themselves, serve in their orientation, in their defense, and in capturing prey (see Lindauer 1963a:122, 123 and the literature cited there).

Summary: A Glance at Other Animals

1. The ability to orient in accordance with the plane of vibration of polarized light is widespread (Table 41, pp. 442f). Vertebrates lack it.
2. A distinction is to be made between the spontaneous assumption of certain preferred positions, mostly parallel to and perpendicular to the plane of vibration, and the ability to maintain, on the basis of previous experience, any chosen angle ("menotaxis"). In doing the latter the diurnal shift in the plane of vibration may be taken into account and then—as the highest level of accomplishment—a certain compass course may be followed solely by reference to the polarized light of the sky. Methods of demonstrating the perception of polarized light have to be accommodated to the different experimental animals.
3. As a rule, the sun and the polarized light of the sky together constitute the celestial compass that guides so many animals on their paths. Inhabitants of the sandy shore and other littoral zones may be carried inland by wind or waves, or may be washed into the water. Various species of beetles, many little crustaceans, and shore spiders then find their way back to their home area by means of the celestial compass alone. They also set out on the proper compass course that

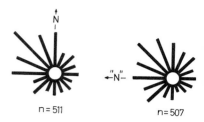

FIGURE 396. Frequency distribution of the directions of landing in an open-air experiment: (*left*) in the natural terrestrial magnetic field; (*right*) with 10-fold horizontal intensity, rotated 90°. The influence of a very gentle breeze from the SE is clear. The length of each bar is proportional to the number of landings. 13 August 1964.

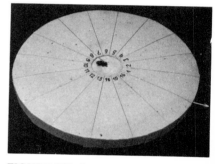

FIGURE 397. Arrangement for observing the directions in which individual bees sit at the feeding place. The bee drinks sugar water from a capillary placed in the center of the sectored plate.

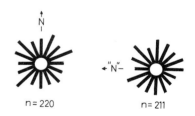

FIGURE 398. Frequency distribution of the directions of sitting: (*left*) in the natural terrestrial magnetic field; (*right*) in a resultant field of equal intensity, rotated 90° to the west. The length of each bar is proportional to the number of sitting positions.

would have led them to their beach at home when they are transported into a strange territory (Fig. 385), even when they are carried to another part of the globe. Knowledge of the direction that, according to the celestial compass and their internal clock, they must adopt in order to get to the shore from a dry place or from the water may sometimes be innate. But in some species each animal has to learn this during its individual life, or must improve rough inherent knowledge through experience. There are little crustaceans that in the daytime steer by the sun and at night by the moon, and that direct their course in accordance with their internal sun clock or moon clock as the need may arise. Steering toward the littoral zone affords many striking examples of such methods of orienting. But in other regions, too, many arthropods employ the celestial compass, for example in order to find their way back to their habitat or into their hiding place after excursions through the surroundings.

4. As far as can be seen at this time, the same rules regarding the celestial compass seem to hold for other animals as with bees. Thus, it was possible with ants to show in a competitive experiment the importance of the compass relative to landmarks, and to demonstrate the linkage between the position of the sun and the polarization pattern. Like the bee, the spider *Agelena* is able to integrate the air line over an angular path from starting point to goal. But there are differences also: bees dancing at night—although they have no experience of the nocturnal position of the sun—relate their indication of direction to a sun that continues its course from the west across the north to the east; some other animals behave in the same way, but yet others act as though the sun were to return from the west across the south to the east.

5. Among vertebrates, orientation on the sun, in which the time of day and the solar path are taken into account, was discovered in birds, simultaneously with and independently of the findings with bees. With migratory birds the celestial compass is an important factor. Polarized light is not involved. But birds en route at night are able to orient to the stars—and with this observation Santschi's erroneous notion in regard to desert ants finds confirmation in the birds half a century later.

Not only birds are capable of finding their way in accordance with the sun and their internal clock. The same has been demonstrated with fish, reptiles, and amphibia, and seems to be valid for some mammals also. Still, vertebrates must use other, as yet unknown, means of orientation. Not all their accomplishments are to be accounted for by the celestial compass.

6. Among insects there are observations regarding the orientation of June bugs that point to the participation of unknown factors. The involvement of an orientation to the earth's magnetic field has been both claimed and disputed. Such an effect could not be confirmed with bees.

XIII / Orientation When Near the Goal

Close to the goal stimuli emanating directly from it become effective. With bees that are setting forth, as a rule the goal is food-producing flowers; for those returning home it is their own hive.

Young bees are occupied with housekeeping duties. When after about 10 days they fly out for the first time, their only fixed objective is the return to the home hive. To impress its location and its identifying characteristics upon themselves is the significance of the orientation flights.

A. THE ORIENTATION FLIGHTS

In his studies with bees the physiologist Bethe reached the conviction that an unknown force is concerned in their finding the home hive (1902, 1950). This may hold true for birds, but for bees there is no reason to assume enigmatic means of orientation. All observations indicate that young bees learn during their orientation flights to find the way home according to landmarks and the signs of the heavens.

Bees that have never left their hive do not find the way home if they are liberated more than about 50 m away. Even from a shorter distance only a portion returns, and often not until hours later; during this time their behavior indicates clearly that in flying about searching they are attracted eventually by the odor of the colony, when by chance they happen into its vicinity. But when after their first orientation flight they are transported out into the surroundings of the hive, there are already many that find the way home from distances of several hundred meters, and after repeated orientation flights even several thousand meters. Among older foragers the ability to return home is better from good sources of food, and extends to a greater distance, than when they are liberated from a direction from which there is nothing for foragers to fetch. From places that are unknown to them only a few find the way back, and this after long delay; hence one may assume that by means of searching flights they finally happen upon a familiar region in which they are able to orient. From a distance of about 8 km or more, beyond the limits of their foraging area, none returns (Wolf 1926, 1927; Uchida and Kuwabara 1951; Lore Becker 1958).

All this shows that during their first outward flights bees gradually acquire knowledge of the appearance and location of their home hive, in accordance with optical and likely also olfactory cues.[1] Their characteristic behavior during the orientation flights also shows this; it is described as follows, for example by v. Buttel-Reepen (1900:215). As soon as they have taken flight the bees turn with the head toward the hive and in this position hover up and down in front of it; later they make orienting circles, first smaller and then larger ones, and then vanish into the surroundings. This "playing about" of bees hovering up and down in front of their home is a conspicuous phenomenon during warm noontime hours, and is familiar to every beekeeper. Yet the process goes somewhat differently than described hitherto.

In order to be certain of the first flight and of being able to observe just how it takes place, Lore Becker (1958) numbered the bees as soon as they emerged. When several days later they flew off for the first time, they did indeed hover back and forth in an undulant flight in front of the hive, with their heads turned toward it, but not for longer than a few seconds; thereupon they shot away and disappeared from view. After only 3–7 min they returned, and not until then did they play about in front of the hive for a few minutes before they went in again.

This is just the way reorientation goes when a colony is transported into a strange neighborhood.

The unanticipated performance became particularly obvious when the entrance of a displaced colony was opened and, after a batch of 20–30 foragers came out, was closed again with a screen. All of them disappeared at once and the silence of the grave reigned around the hive. About 4 min later the air began to fill with those returning; these hovered before the hive for another 1–3 min. The deceptive impression that bees in a strange area will fly about their hive for 0.5–1 hr in order to become acquainted with its location arises from the fact that after the entrance is opened new batches continually emerge, while the earlier ones are coming back, so that the composition of the cloud of bees hovering in front of the hive is changing constantly, though this is unnoticed.

The surprising thing about the finding is that the bees learn the orientation cues so rapidly. Now we would like to know what it is that they assimilate. That is shown when hive markers are changed or merely when attention is paid to the conditions under which often the hive is not regained.

[1] E. Wolf (1926) thinks it possible that besides this an additional factor plays a part in orientation: a noting by the antennae of turns and twistings. Foragers returning home were captured at the hive entrance, shut up in a cage, and released at a point several hundred meters away. When they and their cage were rotated continuously while they were being transported, they needed on the average somewhat longer for the flight home than when they were transported normally. From this it is concluded that they had taken note of the passive rotation, and in fact with the antennae, since antennaless bees were not similarly deranged. As yet it has not been possible to confirm this finding. Kuwabara (1952) repeated the experiments and found the same duration of the return flight for normal and rotated bees. Baerends (1941), in his investigations of the orientation capacities of the digger wasp *Ammophila* (see p. 470), was also unable to determine any influence of passive rotation during transportation on the subsequent return flight to the nest.

This last occurs not infrequently when beehives—as is customary in Germany and Austria, for instance—stand in rows close together. The inhabitants of hives in the middle of a row go astray more often than those at the ends. A terminal colony is characterized more effectively by its location. Where the placement of similarly arranged beehives is repeated, it happens often that bees confuse their hive with a strange one several places distant, if this occupies the same position in another group (for example Free 1958a; Mathis 1960). It follows that the bees have taken note of the hive's surroundings. In doing this they pay close attention also to the height of the hive entrance above the ground. With hives that stand alone they remember which side the hive entrance is on (Ribbands and Speirs 1953; Free and Spencer-Booth 1961).

From this last-mentioned fact it may be deduced that the bees returning home steer toward their hive in the proper compass direction, the celestial compass apparently being decisive in this. Whether this notion is truly correct with bees returning home has not been tested hitherto, so far as I know. But it has been demonstrated for the approach to feeding stations, and hence it should apply also in the approach to the hive, that the celestial compass plays a leading role even directly before the goal is reached.

Lindauer (1960:372) trained bees to a feeding dish F at the southern corner of a black square (Fig. 399). The feeding stand was east of the beehive, so that the approach was from the west (arrow in Fig. 399, *above*). Feeding was during the afternoon only. For the experiment, the colony, after several days' training, was transported in the morning to a region unfamiliar to it and the feeding stand was set up south of the hive; now the approach was from the north. At all four corners of the square there were empty dishes. Figure 399, *below* shows the result. Visits were predominantly to the southern station. Landmarks could not play any part, since they were completely different. Relative to the direction of approach, the bees should have flown to the right-hand corner of the square, now situated on the west. But they oriented to the compass direction toward the feeding station, and behaved in this way in the morning, that is when the sun was in a different position from the one they had become familiar with during their afternoon training.

A variation on this experiment led to the same result (Lindauer 1960:375). Wholly comparable observations had already been made by Elsbeth Wiechert (1938), only under less stringent conditions, and of course she could know nothing of a celestial compass at that date (see v. Frisch 1953a:70, 71).

Bees give much attention also to the color of the home hive. Beekeepers often make use of this, by painting their hives different colors. This is particularly useful when large numbers of hives are standing close together. In order to be certain of success, one must bear in mind that bees see colors differently than we do (see pp. 171ff and 475n).

If one pushes a beehive only 1–2 m to one side, the bees returning home customarily hover obstinately about the spot where the hive entrance was before. It was mainly this phenomenon that misled Bethe into assuming an "unknown force" in his investigations of the capacity

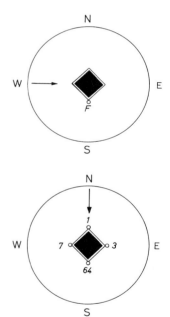

FIGURE 399. Arrangement for demonstrating orientation in the compass direction at the feeding station: (*above*) training to a feeding dish F at the south corner of a black square; approach flight from the hive to the west (arrow); (*below*) experiment after displacing the bees into a different territory, carried out at a different time of day; approach flight from another direction of the hive (from the N, arrow); empty dishes at all four corners; the numbers of approach flights during the experiment (the majority to the southern station) are given. After Lindauer 1960.

of bees for returning home. It seems odd to us that they do not immediately recognize as their home the conspicuous and only slightly displaced hive. But they are no doubt guided more than we are by other signs when they bear down on the goal. In favor of this view is the circumstance that moving the hive forward or back along the line of flight is less disturbing than displacing it laterally by the same amount. Also, a displacement is not deranging initially when the hive has been moved into a strange neighborhood. On the first day the bees fly unhesitatingly into the hive, even after it has been shifted; it is familiar to them, whereas its relation to the new surroundings first has to be learned. But after some 48 hr of flying, the bees then returning jam up at the old place if the hive is now moved (Heran 1958:216).

The significance of learned signs in the recognition of the home hive may be demonstrated most convincingly in experiments of the following kind.

In a large beeyard, the hives of which all look alike, there are a few adjacent hives empty. One of these (No. 4 in Fig. 400, *above*) is marked on the front and on the flightboard with blue tin plates and is supplied with a colony; the adjacent hive on the right is given yellow plates, and the left one remains white. We wait a few days, until through their orientation flights the bees have familiarized themselves with their new dwelling. Then we change the location of the colors. In doing this it is important to take into account that a "bee smell" remains clinging to the plates of the occupied hive, and that this would become a source of error if they were transferred to another hive. Therefore we have painted the blue plates yellow on the rear surface, and the yellow ones blue on the back. Now we need only reverse the plates in order to turn the blue hive into a yellow one. The yellow plates are likewise reversed and transferred from position 5 to position 3, thus from one empty hive to another. The result is startling: the throng of returning bees that has piled up in front of the hive during the short time while the plates were being rehung enters the wrong hive quantitatively and without hesitation (Fig. 400, *below*). Thus matters remain during the succeeding few minutes. All bees flying off emerge from the yellow box, all those returning home fly into the blue one. Since this is empty, they soon come out again and then gradually, guided by the olfactory sense, find their way into their true home.

If the experiment is performed in such a way that after the yellow plates have been reversed (and thus changed to blue) they are put back in the same place (thus are left hanging on hive No. 5), the result is not so uniform. Many bees hover about the empty blue hive and also go crawling into it, but others turn away to the inhabited box in spite of its yellow color. The bees returning home are disturbed and confused. They accumulate out in front of the blue hive, and thus sooner than in the other case the olfactory sense is called on for a decision. From this behavior it is clear that the surroundings of the home hive too play a part in orientation. In the first experiment the relative location of the adjacent colors was preserved, since even after the plates were reversed a white hive stood to the left of the blue one and a

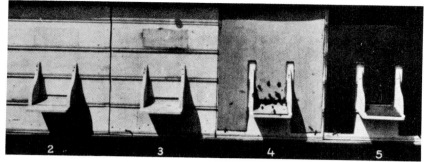

FIGURE 400. The significance for orienting of the color of the home hive: (*above*) hive No. 4 of a large shelf of beehives was occupied by a colony, while Nos. 2, 3, and 5 were left empty; No. 4 is marked with blue, No. 5 yellow, Nos. 2 and 3 lack colored tin plates (white); the tin plates (for the front of the hive and the flight-board) are painted the complementary color on the back; (*below*) hive No. 4 is made yellow by reversing the plate, and the reversed plate (blue) from No. 5 placed on hive No. 3; all bees returning home go into the uninhabited hive No. 3, which is now blue.

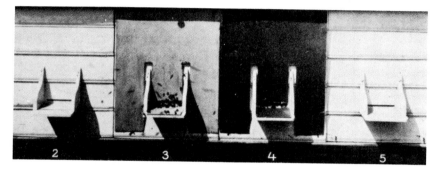

yellow hive to the right of it (Fig. 400), but in the second experiment the order was shifted (v. Frisch 1914; see also Rauschmayer 1928; Engländer 1941; Beer 1957).

From experiments concerned with the learning process during training to food we know that during their first approaches the bees quickly become acquainted with the color and shape of the food source, as also with its characteristic scent. When they first fly away from the feeding place they hover about it in circles and loops for a considerable time, and it looks as though they were closely observing the characteristic features of the food source. But, contrary to expectation, it has appeared that in this phase of orienting the characterization of the food source and its immediate environs are quite inconsequential for the later behavior, and that only the broader surroundings, the general location of the training place, are impressed upon the memory at that time. This becomes evident if, for example, one makes a bee fly to a certain color and take off from another one. If on her return she is presented with both side by side, she always selects the one previously flown to (Opfinger 1931, 1949; see also Lindauer 1963:164).

Apparently the bees behave no differently when during their first orientation flights they learn the characteristic features of the home hive. As soon as they leave the hive entrance they turn toward the hive and examine it, just for seconds, from the viewpoint of their return flight. Then they go darting off like an arrow and impress upon themselves the direction from which they will return in a few minutes, now to acquire from longer and more exact study, as they fly toward their hive, its features and those of its surroundings.

For the experimenter, a beehive and its environment are no very wieldy object. Solitary digger wasps, which build a nest in the soil, hunt for prey in the surrounding territory, bring in the paralyzed insects from an appreciable distance, and find their inconspicuous nest again with great certitude, are in many respects more favorable for the study of orientation. The problem that confronts the animal is at bottom the same as that of the bee, and it seems that in principle it is solved similarly. For informative experiments on this matter we are indebted to Tinbergen and Baerends. The former studied the digger wasp *Philanthus,* the bee wolf, the latter the digger wasp *Ammophila* (Tinbergen 1932; Tinbergen and Kruyt 1938; Tinbergen and van der Linde 1938; Baerends 1941).

For finding the entrance to the nest again, optical cues in its immediate vicinity, such as are prevalent in the countryside, are of decisive importance: stones, small plants, pine cones, and the like. If the nest entrance is surrounded by a circle of pine cones and these are moved to one side while the wasp is absent in her hunting grounds, then on her return she seeks the nest with great obstinacy at the wrong place, in the middle of the displaced circle of cones. The precise number and shape of the cones do not matter, the stimulus-complex "circle of cones" is the effective cue. By such experiments it was possible also to test the relative importance of different signs. Objects with a conspicuous shape rising above the ground were far more effective than flattish structures were, in agreement with the observations of Opfinger and others who investigated bees.

When a wasp leaves her nest in order to go hunting in the vicinity, she may fly about searching over sinuous paths. If she has caught prey, she returns to the nest almost linearly, along the air line. If *Ammophila* has captured a small caterpillar, she can carry this home in flight; large, heavy caterpillars have to be dragged home on foot. If the view is impeded by weeds, she leaves the prey lying here or there on the ground and climbs up on a plant until she has a free outlook; afterward she picks up the caterpillar again and continues on her way. At the time of the experiments, nothing was known about the insects' celestial compass. Today it seems likely that the climbing of vantage points is done in order to orient according to the sun and polarized light.

But aside from this, landmarks along the way are of great importance. That was demonstrated with pine trees that were stuck into the ground. By shifting them the wasps could be misled into an incorrect route. As the nest is approached more closely, smaller and smaller objects gain in significance, for instance small plants instead of trees, until in the immediate vicinity of the entrance pebbles, depressions in the ground, patches of sand, and the like serve as guideposts.

Wasps that were captured and transported to another place in the neighborhood often took a long time to find their way home, and the less an acquaintance with the area of release through earlier excursions could be assumed, the more time they took. Through these and many additional observations, the investigators came to the conviction that the capacity of the wasps for returning home is founded on their acquired knowledge of optical signposts.

B. OPTICAL ORIENTATION NEARBY

1. THE BEES' COLOR SENSE

That bees returning home are guided by the color of their hive (p. 468 and Fig. 400) is understandable only if they are able to distinguish colors. About this question there raged an old conflict, the historical development of which is described by A. Forel (1910). He seemed already to have decided in favor of color vision in insects when C. v. Hess (1912, 1913, 1916, 1918), on the basis of impressive but not conclusive experiments, proposed the thesis that bees, like all other invertebrates and fish, are totally color-blind. This induced me to undertake an experiment that Forel had already proposed in principle but had not carried out: bees were fed on blue paper interspersed in a checkerboard arrangement with shades of gray of the most varied brightness. After adequate training, they will fly to a clean blue paper, free of food, even when it is in a different position and all the fields are covered with a glass plate, so that any olfactory recognition is excluded (Fig. 401).

FIGURE 401. Demonstration of color vision: a blue paper among gray papers of various brightnesses, beneath a glass plate. All fields are provided with clean, empty glass dishes. The bees, trained to blue, distinguish with certainty the blue field from all shades of gray, and settle on it.

Since training to gray plates of different brightness on the gray scale is unsuccessful, they must perceive the blue field as a color.

After being trained to a red paper, in the test they confuse red fields with black ones and dark gray ones; they are red-blind. In other ways, too, their sense of color seemed inferior to that of man, since they confused orange-red, yellow, and yellow-green papers on the one hand, and blue, violet, and purple-red ones on the other, while training to blue-green papers (very unsaturated, to be sure) was altogether unsuccessful (v. Frisch 1914, 1914–15). Knoll (1926) confirmed this finding and extended it to other insects.

As happens so often, an improvement in technique led appreciably further. Kühn and Pohl (1921) and Kühn (1927) trained bees in a semidarkened room, into which they flew through a little window, to the lines of a mercury spectrum or to sharply defined regions of a continuous spectrum. With these saturated colors, well defined in respect to their wavelength, training to blue-green too was successful. The bees' red-blindness was confirmed. For bees the spectrum is shortened on the long-wavelength end; the limit of perception lies at about 650 mμ (with a certain breadth of variation depending on the intensity of the light; see Bertholf 1931); but on the other hand it extends far into the ultraviolet. The region from about 300 to 400 mμ is a special color for them. In the experiments with spectral colors too, red, yellow, and green on the one hand, and blue and violet on the other, were not distinguished.

That the color sense of bees is capable of appreciably more did not appear until our old plan of testing their behavior toward mixed[2] colors was carried out by Daumer (1956). The prerequisite was an apparatus suitable for training bees to a mixture of spectral colors; it is shown in Fig. 402.

A high-pressure xenon lamp served as a light source. It delivers white light (including ultraviolet) with the energy distribution of sunlight, and is fixed in the center of a drum, the cover of which contains four star-shaped windows; they are covered with ultraviolet-transparent WG$_7$ glass and support quartz feeding dishes (Figs. 403–405). The windows are illuminated from below with white light or with selected pure colors or color mixtures; the light from the lamp passes through the lateral openings, into which interference filters are set, into reflection-mixing chambers (see Fig. 402) and from there up to the training surface. The mixing chambers are lined with crumpled aluminum foil. Two of them receive white light or light of different colors from three windows that face the lamp; the intensity of the light is regulated by diaphragms and can be measured. A meter permits comparison of the energy of the white and colored light. In order to exclude training to a certain position, the apparatus rotates slowly about its axis. It is set up in a room moderately illuminated by daylight, into and out of which the bees fly through a window. For details of the technique reference is made to the original publication.

[2] In what follows, the meaning is always the *additive* mixture of colors, not their *subtractive* mixture as practiced by the artist on his palette.

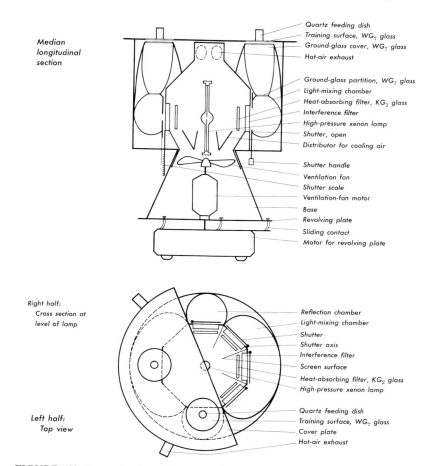

FIGURE 402. Apparatus for mixing spectral colors, for investigating the color vision of bees by means of the training method. Construction diagram, at a scale of 1:7.3. After Daumer 1956.

With this arrangement the bees were trained to the desired color or mixture of colors (Fig. 405), and were then tested experimentally as to how far they were able to distinguish these from colors or color mixtures produced in the same way and presented to them nearby. The experiments gave the following results.

Of all spectral regions tested, the ultraviolet (360 mμ) is the most stimulating to the eye of the bee. This is determined in comparison with other colors by training the bees to a spectral color of the same energy and then, in the experiment, reducing the intensity step by step until the color is no longer distinguished from an unlit window. In other spectral regions the stimulating effectiveness is appreciably less. When compared at equal intensities of radiation their effectiveness diminishes as follows: 5.6 (ultraviolet 360 mμ):1.5 (blue-violet 440 mμ):1.0 (green 530 mμ):0.8 (yellow 588 mμ):0.5 (blue-green 490 mμ):0.3 (orange 616 mμ). Participating in the stimulatory effectiveness are the subjective (to bees) brightness and saturation of the colors. The degree of satura-

FIGURE 403. The color-mixing apparatus in the opened experimental container. After Daumer 1956.

FIGURE 404. The experimental container closed, showing the ground-glass training surfaces with the quartz feeding vessels, illuminated from within: (*above left*) light meter for measuring the intensity of illumination in the room; (*above right*) end of the scale of the mirror galvanometer for comparative energy measurements. After Daumer 1956.

FIGURE 405. A bee is drinking from the quartz feeding vessel on the illuminated ground-glass training plate. After Daumer 1956.

tion can be determined by measurement of the amount of colored light that must be mixed with white light in order that the color quality may be detected by bees.

In earlier experiments, when colored pigments or spectral colors were tested, orange, yellow, and green were confused ("yellow region"). With Daumer's method these colors proved to be similar for bees, but distinguishable. In Fig. 406, the heights of the bars in the first column of diagrams show (from left to right) the number of flights to orange, yellow, green, and blue-green; in each of them the arrow points to the training color for the pertinent experimental series. One sees that only blue-green, which is included for comparison, is distinguished sharply. If the brightness of the training color is decreased (indicated by closer hatching in the series from left to right in Fig. 406), the discrimination becomes still worse. Correspondingly, in the blue region blue and violet prove to be similar but distinguishable.

Earlier failures of these experiments are to be attributed to the use of mass training. Kühn and Fraenkel (1927), after training the bees to segments of the spectrum, had already succeeded in demonstrating a weak capacity for discriminating rather narrow steps of color within the yellow and blue regions, if they prevented mass approaches and evaluated only individual flights. Experiments by Bertholf (1931) pointed in the same direction. Daumer worked with only a few, well-trained bees. With this method, training to differences in the yellow and blue regions is successful with colored papers also (Ostwald colored papers).[3]

In comparing the color sense of bees and man the scheme given earlier must therefore be modified, in that "yellow" and "blue" are not individual qualities for the bee, but rather regions of similarity (Fig. 407). The narrow blue-green region is discriminated sharply from neighboring areas. Within the ultraviolet, training to the difference between 375 mμ and 360 mμ was successful. Probably bees can see still more gradations of color here, but there was no technical possibility of testing this.

The color-mixing apparatus allowed us to extend still further the comparison with human color vision. For ourselves, purple colors, which are not contained in the spectrum, are formed by mixing red and violet. In a wholly comparable way, by mixing the terminal portions of the bees' spectrum, namely yellow and ultraviolet, there is produced a new fundamental quality of color, which they do not confuse with any shade of gray nor with any other shade of color. This "bee purple" converts the spectrum into a complete color circle for bees too (Fig. 408). Yet another quality that is distinguished well from other colors is obtained by mixing blue-violet with ultraviolet ("bee violet"). It is identical with the spectral region around 400 mμ. Within their "purple" and "violet," bees can discriminate at least three color tones in each.

We assume that a mixture, in the proportions of sunlight, of all wavelengths the bees can perceive is colorless white for them. What

---

[3] In my old experiments only the less highly saturated Hering colored papers were available. Discrimination of orange, yellow, and green has been confirmed by Mazochin-Proschnjakow (1959a) also.

they experience we of course do not know; but that mixture is different for them from every spectral color, and it is harder to train them to "white" than to any other color.

If the ultraviolet is filtered out of such a "white," there is no noticeable change for our ultraviolet-insensitive eyes, but doing this imparts color to the mixture for the bees; in fact it becomes similar to blue-green, to the point that the two are confused. Thus there are two kinds of white: one that is neutral for bees too ("bee white"), and another that is white only for us but that is blue-green to bees, because the mixture lacks ultraviolet. In analogy with the well-known laws of color vision, it may thence be deduced that blue-green and ultraviolet are complementary colors for the bee. For man, such colors mixed in a definite proportion yield the sensation of white. Bees confuse a mixture of 85 percent blue-green and 15 percent ultraviolet with "bee white." Thus the latter can be produced for them too, just as with man, not only by mixing all the spectral colors but likewise by mixing complementary colors. Other complementary colors for bees are: yellow and "bee violet" (65 percent + 35 percent = "bee white"), and blue and "bee purple" (30 percent + 70 percent = "bee white"). In order to produce "bee white" from all three primary colors, these must be mixed in the proportions 55 percent yellow + 30 percent blue-violet + 15 percent ultraviolet. Thus the bees' color system proves to be trichromatic and closely similar to that of man. The important difference is in the shift of sensitivity toward the short-wavelength end of the spectrum.[4]

Mazochin-Proschnjakov (1962) obtained the same results in color measurements with bumblebees, by means of electrophysiological methods. With them, sensitivity to long-wavelength radiation extends somewhat further than with the honeybee.

[4] By painting the hives with color, the finding of their home can be facilitated greatly for the returning bees (see pp. 468f). But attention must be paid to the fact that they see the colors differently than we do. The following paints are readily distinguished by them: pure red (not yellow red or blue red), or black (but not red and black side by side), yellow (*Echtgelb 51 BN*), true light blue (no ultraviolet reflection, and hence blue for bees too), cobalt blue 660 (ultraviolet-reflecting, and hence "bee violet"), zinc white (no ultraviolet reflection, and hence "blue-green" for bees), "Satolith White" (strong ultraviolet reflection, and hence "bee white"). Such paints can be obtained from the Siegle Company, Stuttgart-Feuerbach.

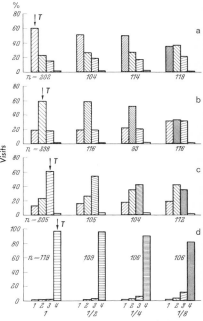

FIGURE 406. Differentiation of colors in the yellow range: distribution of visits (percent) over the four colored lights (1–4, numbers below the axis of abscissas), after training to (*a*) orange (1), (*b*) yellow (2), (*c*) green (3), and (*d*) blue-green (4); *n*, number of visits. The arrow above indicates in each case the color of training, *T*. From left to right: experimental results with decreasing intensity of the training light, in steps from $\frac{1}{2}$ to $\frac{1}{8}$ of the initial energy (abscissa). The increasing weight of hatching indicates the reduction during the experiment of the brightness of the training light. The results show weak differentiation of orange, yellow, and green, and sharp differentiation in the blue-green. After Daumer 1956.

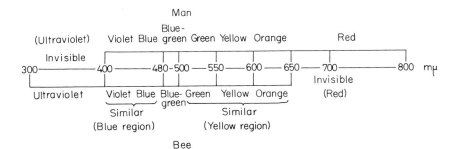

FIGURE 407. The colors of the spectrum: (*above*) for the human eye; (*below*) for the eye of the bee.

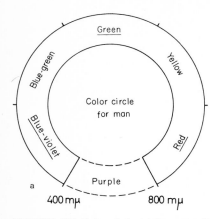

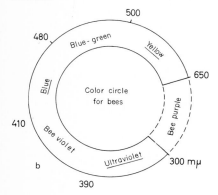

FIGURE 408. Color circle (*a*) for man; (*b*) for the bee. The three primary colors are underlined; by mixing them the intermediate colors can be produced. Complementary colors are opposite each other in the circles. Somewhat modified, after Daumer.

The effectiveness of ultraviolet light on insects has repeatedly impressed other observers. Lubbock in 1883 had already noticed the sensitivity of ants to ultraviolet and posed the question, "Whether for these insects white light differs from our white light?" (p. 185). C. v. Hess (1919, 1920, 1920a) found a great sensitivity to ultraviolet in caterpillars and other insects, and sought to explain this as due to fluorescence of the constituents of the eyes. Bertholf (1931a) made measurements on dark-adapted, positively phototactic bees and found their sensitivity in the ultraviolet very high. Lutz (1933), in experiments with the stingless bee *Trigona cressoni parastigma* Cockerell, noticed that it could distinguish white with and white without ultraviolet. Mathilde Hertz (1938, 1939) obtained the same result with honeybees, and realized that ultraviolet and blue-green are complementary colors for them—but was mistaken in her assumption that (as Kühn (1927) also had suspected) this was so for yellow and blue too. Engländer (1941) confirmed that white with and without ultraviolet could be distinguished in paint on beehives.

Investigation from quite another direction led also to the assumption of trichromatic color vision in bees. Autrum and v. Zwehl (1964), with quantally equal monochromatic stimuli, succeeded in recording the receptor potentials of individual receptor cells from the ommatidia of worker honeybees. From the relative effectiveness it was possible to calculate the sensitivity curves. Sensory cells were found that were most sensitive to green (at the boundary with yellow-green, around 530 mμ), and others with a sensitivity maximum in the blue (around 430 mμ)[5] or in the ultraviolet (around 340 mμ). This furnishes proof that in the bee's eye there are present three different kinds of visual cells, which respond maximally to the three primary colors, as the Young-Helmholtz color theory assumes.[6] The course of the sensitivity curves (Fig. 409) is in very close agreement with Daumer's findings. It makes understandable the poor quality of discriminating ability in the "yellow region," where in part only a single receptor is involved, in part a second one to a small extent. The outstanding power of discrimination in the blue-green, violet, and ultraviolet evidently is to be attributed to the fact that here, even with only a slight shift in wavelength, different receptors respond with a strong change, opposite in sign.

In the drone's eye the blue and ultraviolet receptors predominate: green receptors were found only in the most ventral portion of the eye. Dorsally the ultraviolet receptors are dominant (Autrum and v. Zwehl 1962, 1963). One is tempted to interpret these relations as having a biological significance. Drones are not interested in flowers. Their sole vital task is fulfilled in the nuptial flights, and these take place only in fine weather and require good orientation to the blue sky.

[5] In two instances the value was around 460 mμ; since as yet this value and 430 mμ have never been found in the same eye, individual variation may have been concerned.

[6] Approximately simultaneously, by a different method (absorption measurements in individual retinal cells), proof was provided of the existence in the human eye of three types of cones, adapted to different wavelengths (Brown and Wald 1964; Marks, Dobelle, and MacNichol 1964). What had been contested theory for a century and a half has today become visible, measurable actuality.

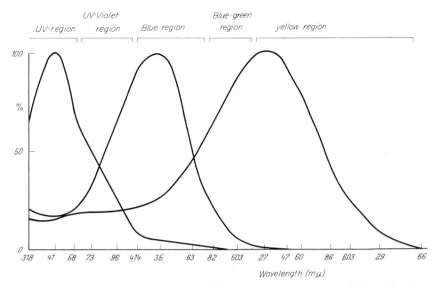

FIGURE 409. Sensitivity curves for three different receptors in the eye of the worker bee, maxima at 340, 430, and 530 mμ; ordinate, percent of maximal sensitivity. After Autrum and v. Zwehl 1964.

Similar differences within the eye, presumably of similar importance, have also become known in other insects. Thus Mazochin-Porschnjakov (1959b) found the dorsal region of the eye color-blind in dragonflies (*Libellula quadrimaculata* L.). But with the back swimmer (*Notonecta glauca* L.) the ventral portion of the eye, here directed skyward, is color-blind, while the dorsal part can discriminate colors (Rokohl 1942).

The eye of *Ascalaphus macaronius* Scop. (order Neuroptera) is divided by a groove into an upward and a lateral portion. In electrophysiological experiments the portion facing the sky shows a maximum of sensitivity in the ultraviolet; it does not respond to wavelengths above 500 mμ. The lateral portion of the eye has a secondary maximum at approximately 546 mμ; its sensitivity extends to around 650 mμ. The animals fly in bright sunshine (Gogala and Michieli 1965).

With the cockroach (*Periplaneta americana* L.), Walther (1958, 1958a) concluded, from the course of the sensitivity curves during selective adaptation, that there were two types of receptors in a dorsally situated area of the eye, for which therefore a certain amount of color perception can be assumed, while in a lateroventral part of the eye only one type of receptor could be demonstrated. "But among the results with both groups there are not infrequently deviations toward the findings from the other part of the eye." Beyond this, it was not possible to analyze systematically over the entire extent of the eye the distribution of the two functional modes (1958:95). Hence for the present it is impossible to say whether there is here a fundamental difference between the upper and lower halves of the eye. It must also be taken into consideration that *Periplaneta* has a nocturnal way of life and on this account is not to be placed in the same category with the insects mentioned above.

2. Form Vision

In the compound eye of insects a picture of the surroundings results from the fact that each individual ommatidium projects a point of the image in its line of sight. Since the axes of the ommatidia diverge slightly, and very regularly (Figs. 410, 411), these image points combine like stones in a mosaic to form a complete picture of the surroundings. As with a mosaic, this picture, under otherwise similar conditions, will allow more perception of detail, thus will be sharper, the larger the number of individual ommatidia that participate in the perception of a given visual field. Numerous ommatidia with small angles of divergence are able to resolve the picture more finely than a few greatly divergent ommatidia. These relations can be measured. They provide a certain basis for our conceptions of the visual acuity of insects. The bee's eye has an irregular elliptical circumference (Fig. 412); the divergence of the ommatidia is less in a dorsoventral section (Fig. 410) than in a horizontal cross section (Fig. 411); according to del Portillo (1936), in the zones of smallest divergence it amounts to 1° 20′ in the one case and to double this much, about 2° 40′, in the other. From comparison with the grid of retinal cells in the human, lens-bearing eye, one concludes that the visual acuity of the bee should be approximately a hundred times worse. Experimental findings are in good agreement with this conclusion (Zerrahn 1934; Hecht and Wolf 1929; Wolf 1931).

Recently some doubts have been expressed regarding the classical conception of the origin of a picture in the compound eye. A point of light that lies in the axial line of one ommatidium effects a brightening not only in this eye but also in its neighbors. Thus the visual fields of the ommatidia overlap, and to different extents in different insects. In the bee, around the image that corresponds to the line of sight there lie six secondary images, which however are less intense (Autrum and Wiedemann 1962; Ingrid Wiedemann 1965); it is possible that the secondary images are suppressed still further by nervous inhibition. In the eye of the fly *Calliphora,* with a deviation of the direction of illumination equivalent to the anatomical angle of divergence of the ommatidia, the brightness diminishes to $\frac{1}{5}$; that means that an individual ommatidium sees a point of light five times as bright as a neighboring ommatidium (Washizu, Burkhardt, and Streck 1964). The theory of the mode of picture formation is not rendered untenable by superpositions of this sort. As long ago as 1891 Exner indicated that the sensory cells of the individual ommatidia are reached by light not solely from the geometrically corresponding portion of the visual field.

Repeatedly there has been discussion of the question whether images can be formed within the individual ommatidia. In the eyes of flies, with their loosely built retinula, a light beam of slightly oblique incidence stimulates only a part of the rhabdomeres. If in view of the small number of sensory cells in an ommatidium one can hardly speak of an "image" even here (see Hocking 1964), then in the compact rhabdom of a hymenopteran eye the formation of an image cannot be considered at all.

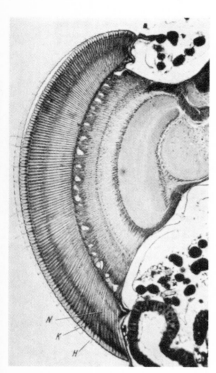

FIGURE 410. Longitudinal section through the eye of the bee (worker), in a dorsoventral direction (see Fig. 411): *H*, cornea; *K*, crystalline cone; *N*, retina (rhabdoms). In the upper portion of the eye, a small segment of the cornea has become separated from the crystalline cone layer during fixation. Photograph by Agnes Langwald.

Burtt and Catton (1962) studied the eyes of a grasshopper (*Locusta*) and of two species of flies (*Calliphora* and *Phormia*) and found three successive images at different depths. Their significance is problematic.

Regarding the degree of actual accomplishment in the recognition and discrimination of flowerlike forms, I (1914–15) attempted to get information by means of training experiments. With the shapes arranged horizontally, training succeeded for a yellow paper star contrasted with a yellow oval (Fig. 413). The reciprocal training was possible too, but was mastered with greater difficulty by the bees. When the patterns were exposed in vertical position, the training was done in two wooden boxes, whose flight entrances were encircled by blue, gentian-shaped models, flanked by empty wooden boxes bearing figures of blue rays (Fig. 414). If now new empty boxes were offered in the experiment, the bees flew uniformly to the shape to which they had been trained and crept into the box there (Fig. 415). In contrast, even with long-continued training they did not learn the difference between geometric figures of equal area that had square, circular, elliptical, or triangular outlines.

In extending similar experiments, Mathilde Hertz (1929, 1930, 1931, 1935) made the astonishing discovery that in distinguishing shapes bees are guided by other characteristics than we. The outline of a figure is unimportant to them, but the degree to which the outline is dissected is of fundamental significance. Bees will not learn to distinguish the figures in the upper row of Fig. 416. With those in the lower series too, training to a difference does not succeed. On the other hand, they grasp swiftly and surely the distinction between each figure in the upper row and the ones below.

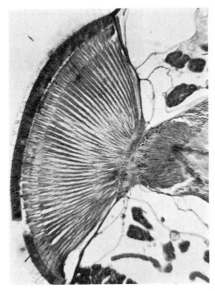

FIGURE 411. Cross section of the bee's eye, approximately in a horizontal direction. The cornea is damaged in one place. Silver impregnation according to Davenport; greater magnification than in Fig. 410. Preparation by V. Neese.

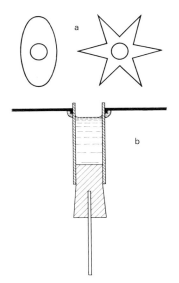

FIGURE 413. (*a*) Oval and star shapes (⅓ natural size), used horizontally for training in discrimination of form. (*b*) Longitudinal section, with the feeding vial in the center of the model.

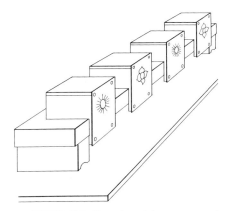

FIGURE 414. Experimental arrangement for form training with the shapes vertical; size of the boxes, 10 × 10 × 5 cm.

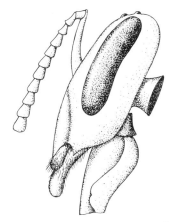

FIGURE 412. Side view of the bee's head.

FIGURE 415. After training to a gentian shape rather than a star shape, the trained bees are presented with new, empty boxes without food and with clean models in a different order. They come flying up to the shape to which they had been trained, and creep into the boxes.

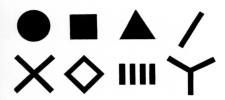

FIGURE 416. The bees will not learn to distinguish from one another the figures in the upper row, nor those in the lower row; on the other hand, for the bee's eye every figure in the upper row is clearly distinct from every figure in the lower row, and training to such differences is successful.

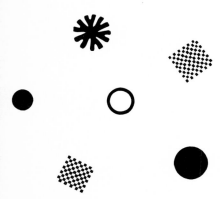

FIGURE 417. Among these figures, the bees showed a spontaneous preference for the checkerboard pattern; $\frac{1}{5}$ natural size. After Mathilde Hertz 1930.

Without preceding training, after bees have been fed on a homogeneous surface, when on the training table they are presented with different figures they assemble spontaneously on those that are most strongly dissected, for instance on the two checkerboard patterns of Fig. 417. According to Hertz, training to a less dissected shape in comparison with a more strongly dissected one is hard to achieve, and according to Zerrahn (1934) is altogether impossible. She says the bees are compelled to seek the more complex structure. But that certainly is incorrect. Figures 413 and 415 alone show two instances in which the bees learned to choose the less highly structured figure. Besides, when they are in the mood for their return home, they prefer spontaneously just the opposite, namely the least structured figure with the simplest contour (Jacobs-Jessen 1959:623–626). Such figures have the most similarity to the dark hive entrance, for which they now are aiming.

It is impossible for us to form any conception of so different a way of appreciating shapes. But it was possible to determine what is decisive in the bees' preference for the highly dissected pattern when they are in the mood for seeking food. According to Zerrahn (1934), the length of the contours surrounding a given area is what counts. With a highly dissected figure this length is greater and for the approaching bees entails a more frequent alternation between light and dark stimuli. Wolf (1934) obtained similar effects by presenting flickering fields. In their spontaneous choices the bees preferred the higher flicker frequency.

But this alone will not suffice to explain all reactions. The bee still sees in the patterns "more than the sum of different frequency components." This conclusion was reached by Kunze (1961) when analyzing in greater detail the perception of motion by optomotoric reactions. He used bees which were attached to a small piece of wood that kept them during flight in the center of a rotating drum.

These phenomena evidently are connected with the fact that it is during flight, that is, when in motion, that bees see the flowers and other shapes that are significant to them.[7] Their mere spatial proximity is

---

[7] In his fundamental book on the physiology of the compound eye Sigmund Exner (1891:183) expressed the opinion that the vertebrate eye is superior in the recognition of shapes, the compound eye better in the recognition of changes in the object viewed.

not of such great consequence as for us, but rather the continually changing pattern of the alternation of light and darkness. The eye of rapidly flying insects is well equipped for such vision during movement, since it is able to respond to about ten times the frequency of stimulation that we are. A motion picture for bees would have to have a tenfold greater frequency of succession of pictures to prevent the image from flickering. The deficient power of spatial resolution of their eyes is compensated for by a high resolving power in time (Autrum and Stoecker 1950, 1952; Autrum 1952). That the visual angles of the bee's eye are about twice as great in the horizontal as in the vertical direction is connected with this. The individual stimulus must be guaranteed a certain duration of action. In rapid flight things slip past so fast in a horizontal direction that if the visual angles were too small the otherwise individual stimuli would fuse (Autrum 1949). The coarser visual grid in the horizontal direction makes possible the resolution of finer patterns during flight.

3. Vision in Bees and the Appearance of Flowers

In the year 1793 there appeared Christian Konrad Sprengel's book, *Das entdeckte Geheimnis der Natur im Bau und in der Befruchtung der Blumen* ("Nature's Secret Discovered in the Structure and Fertilization of Flowers"). The "secret discovered" was that bees and other guests of the flowers, attracted by the color and fragrance of the blossoms, fetch nectar and pollen from them as food, and in doing so pollinate the flowers, the structure of which is designed wholly to this end and is comprehensible only in the light of this reciprocal relation. Sprengel recognized the mutually significant features clearly: wind-pollinated flowers have inconspicuous blooms, while plants with insect-pollinated flowers mostly have strikingly colored, fragrant blossoms, which consequently can be found easily. They alone provide their guests with sugary sap as food, and by this means serve their own interests, because the visitors bring about their pollination. Plants blooming at night are distinguished by having pale flowers, and in this way make themselves particularly noticeable. In their finer floral structure, in the form of and the design on the petals, or in the presence and arrangement of hairs that protect the nectar from dilution by rain, Sprengel, full of a discoverer's pure enthusiasm, finds one ingenious apparatus after another. How he would have exulted to learn that the flowers attract the insects in a perfect manner yet in a way invisible to us, such as can be meant only for the eyes of their pollinators.

Flowers bring their colors effectively into play. Basal to the colored cells, a light-reflecting layer, formed mostly of air-filled clefts and often of grains of starch, gives them special luminosity and plays the part of a foil beneath mounted gems. Among all the colors that we come to see in our daily lives, the floral colors belong to those with the greatest saturation (F. and S. Exner 1910).

The close relation between the colors and the guests of the flowers is demonstrated convincingly in the first instance by the kinds of color.

Pure red as a flower color is a rarity among insect-attracting flowers, but occurs very often in bird-attracting flowers, which are visited and pollinated by hummingbirds and honeybirds (Porsch 1931). The eyes of birds are sensitive to this color, but the bees' eyes are red-blind.

Red poppies are visited zealously by bees; but they are only apparently an exception. Their petals reflect strongly in the red and in the ultraviolet (300–400 mμ). The flowers, which look red to us, are ultraviolet to the bees, as can be shown by training experiments (Lotmar 1933).

That an ultraviolet reflection that we cannot see occurs in flowers has been noticed repeatedly, and in part has been documented photographically (Lutz 1924, 1933; Lotmar 1933; Seybold and Weissweiler 1944; Ziegenspeck 1955; Kullenberg 1956; Mazochin-Proschnjakov 1956, 1959). Several sources have also pointed out that this may influence the appearance and patterns of flowers for the ultraviolet-sensitive eyes of insects. But Daumer, in his analysis of the color sense of bees, was the first to establish a basis for judging more precisely the biological significance of floral colors. To these questions he has devoted an additional investigation (Daumer 1958).

With 204 different species of plants he determined the reflectance of the flowers, by taking photographs of them (with a quartz lens) through three different interference filters prepared for the purpose by Schott and Co. (Mainz). The transmission bands of the filters corresponded to the three ranges for primary colors in the bee's eye. For a quantitative evaluation of the reflectance, simultaneously with each flower a surface smoked with magnesium oxide was photographed; this picture coincided with a 10-step series of grays built into the camera (Fig. 422). By comparing photometrically with the gray series the darkening of the film by the flower, the reflectance of any part of the flower could be determined for each of the three primary colors of the bee, as a percentage of the magnesium oxide reflectance ($=$ 100 percent). For technical details reference is made to Daumer's paper.

First we shall consider flower colors in general, and then flower patterns. The high effectiveness of ultraviolet in stimulation proves to be significant with both. For instance, this was expressed in Daumer's experiments on color mixtures, where even 1 percent of ultraviolet mixed with 99 percent of yellow (percentages in terms of energy) suffices to change the color tone of the mixed light perceptibly for bees, in comparison with pure yellow.

By adding ultraviolet to yellow the "bee purples" are produced. A great many plants have yellow flowers that reflect ultraviolet as well but no blue. With a weak ultraviolet reflection (up to about 5 percent ultraviolet:95 percent yellow) there occurs a purple tone that for bees is distinguishable both from yellow and from a purple with more ultraviolet; but if the ultraviolet content exceeds 50 percent, then the bees confuse the color with pure ultraviolet—the yellow is no longer perceptible to them. As one example among many, Fig. 418 shows three flowers that are yellow for us but that are clearly different for bees. Blue is not reflected by these flowers, so that here only the photographs

FIGURE 418. Photographs of the flowers of (*left*) *Erysimum helveticum* (Jacq.) DC. (wormseed mustard), (*middle*) *Brassica napus* L. (rape), and (*right*) *Sinapis arvensis* L. (field mustard), which to us are yellow: (*top*) through the ultraviolet filter; (*bottom*) through the yellow filter; natural size. Because of differences in ultraviolet reflectance the flowers appear to the bees (*a*) pure yellow and (*b, c*) of two readily distinguished hues of "purple." Through the blue filter all three flowers appear equally dark; these last photographs are not reproduced here. After Daumer 1958.

taken through the yellow and ultraviolet filters are reproduced. In the left-hand set (yellow filter) all three flowers look bright; they reflect the yellow light strongly and uniformly. In the right-hand set (ultraviolet filter), differences in the ultraviolet reflection are revealed: the flowers of the wormseed mustard (*Erysimum helveticum* (Jacq.) DC.) reflect no ultraviolet and hence are yellow for bees; rape (*Brassica napus* L.) and field mustard (*Sinapis arvensis* L.) reflect ultraviolet unequally and, as demonstrated in training experiments, appear to the bees as two clearly distinguishable "purple tones." Thus, kinds of flowers that are so similar in color and form that we confuse them and that grow in the same locality are clearly separable for the bees' eye. With yellow flowers similar relations occur frequently. Daumer (1958:Tables 2 and 3, pp. 62, 63) gives quantitative data for the reflectance in the three regions of primary colors, for 53 species of plants with flowers in the two purple tones.

With other flowers the yellow color shifts for the human eye, in the absence of ultraviolet reflection and with increasing reflectance in the blue regions, through yellow-green to white; for the eye of the bee these are blue-green flowers. With great uniformity Daumer found, for 48 white-flowered species of plants investigated, a strong reflection in the yellow and blue ranges, combined with minimal reflection in the ultraviolet. The steep reduction is at about 410 mμ (see his Table 4, pp. 64–65). As an example, Fig. 419 shows a narcissus flower, the deli-

FIGURE 419. Narcissus flowers (*Narcissus poeticus* L.), photographed through yellow, blue, and ultraviolet filters; natural size. In consequence of minimal ultraviolet reflection the white flowers appear "blue-green" to the bees. The red margin of the short accessory corolla reflects ultraviolet and—like the color of poppy flowers—is "ultraviolet" for the bees. After Daumer 1958.

cate design of which we pass over here (see p. 488). Neutral white (with ultraviolet) is only slightly attractive to bees. Training them to it is much harder than with colors. It is therefore significant that in the bees' variegated garden restaurant all napery is colored, and that "white" tablecloths are lacking.

With nocturnal flowers, luminosity is more important than color. And it is precisely these flowers that seem not to fit the rule that Daumer determined for the white flowers that bloom by day. Kugler (1963:308) mentions that the white flowers of *Melandrium album* (Mill.) Garcke, of *Silene nutans* L. and *S. inflata* Sm., which are night-blooming and are pollinated by moths (*S. inflata* also by bumblebees in the daytime), reflect ultraviolet. Thus they will seem to their visitors more deeply veiled in white, perhaps even neutral white. Unfortunately, precise measurements are not available. Helga Schardt of the *Institut für spezielle Botanik der Universität Tübingen* (unpublished) found that the moth-attracting flowers she investigated invariably reflected some ultraviolet. They lack special nectar markers and designs, and they have a uniformly light color. The nocturnal visitors themselves document the importance to them of light-colored flowers. The noctuid moth *Plusia gamma* L. flies on starry, moonlit nights, but also in bright sunshine during the day. Whereas it visits colored flowers (for example *Dianthus cartusianorum* L.) in the daytime, at night it flies to predominantly white flowers, like *Melandrium* and *Silene* (Schremmer 1941).

Blue, violet, and purple-red flowers may appear to the bees in four readily differentiated color tones, two of which belong to their blue range and two—as the result of different admixtures of ultraviolet—to the "bee violet." Data are to be found in Daumer's report (Tables 5–8, pp. 67–69). An example is shown in Fig. 420. The little bouquet consists half of (*a*) garden forget-me-not (*Myosotis silvatica* (Erh.) Hoffm.), half of (*b*) Caucasus forget-me-not (*Brunnera macrophylla* M. B. Johnst.). For us, both have the same tone of blue. As the photograph through the ultraviolet filter shows (*right-hand picture*), *M. silvatica* reflects no ultraviolet, and hence is blue for the bees too, while *Brunnera,* because of

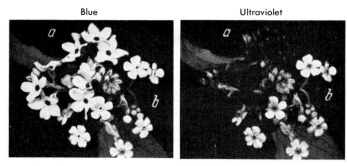

FIGURE 420. The left upper half (*a*) of the bouquet consists of garden forget-me-not (*Myosotis silvatica* (Erh.) Hoffm.), the right lower half (*b*) of Caucasus forget-me-not (*Brunnera macrophylla* M. B. Johnst.): (*left*) photographed through the blue filter; (*right*) through the ultraviolet filter; natural size. For us the flowers have an identical blue color, but for the bee's eye they are completely different because of the unequal reflection of ultraviolet. After Daumer 1958.

strong ultraviolet reflection, is "violet" for them. Training experiments have confirmed their ready differentiation.

Poppies have already been mentioned as "ultraviolet" flowers attractive to bees. Figure 421 shows the strong reflectance in the ultraviolet, while that in the yellow and blue is almost zero.

Flowers appearing black to bees have been found in a single wild plant (*Nonnea pulla* L.), and in a few cultivated ones.

The biological function of the floral colors, to serve as a visually conspicuous signal for sources of food, is reinforced considerably by the fact that green leaves have a weak and fairly uniform reflectance in all three primary ranges of the bees' color system, so that bees see leaves as almost colorless gray, in a highly unsaturated yellowish shade. Hence the flowers must stand out all the more vividly against such a background. In Fig. 422, besides the flowers of the creeping cinquefoil (*Potentilla reptans* L.) the leaves are photographed also, in all three primary colors. Their almost uniform reflectance is evident. Daumer found these relations confirmed in the foliage of 60 species of plants he investigated.

The flower of this *Potentilla* bears in the center a striking special nectar marker (Fig. 422, *lower*), such as Sprengel had already recognized

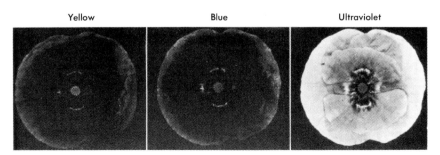

FIGURE 421. Poppy blossom (*Papaver rhoeas* L.), a flower that is "ultraviolet" to bees, photographed through the yellow, blue, and ultraviolet filters; $2/3$ natural size. After Daumer 1958.

FIGURE 422. *Potentilla reptans* L. (cinquefoil), photographed through yellow, blue, and ultraviolet filters. Below is the gray scale used for photometric comparison. The leaves, photographed at the same time, give only weak and almost uniform reflections in the bee's three primary color ranges, so that they are almost colorless for bees. Through the ultraviolet filter there is revealed a special nectar marker, invisible to us. After Daumer 1958.

and named as a signpost pointing to the source of food. He did not indeed see any on cinquefoil, for the marker exists here because the outer parts of the petals reflect ultraviolet strongly, the inner parts not at all. Consequently the flower, uniformly yellow for us, for the bees is "purple-red" with a yellow nectar marker.

Nectar markers invisible to us, originating in different ultraviolet reflections, are even more widespread in the world of flowers than are visible nectar markers.[8] The long list of instances ferreted out by Daumer was enriched further by Kugler (1963) and by Helga Schardt (unpublished). A few examples extracted from Daumer's report will serve to show how they are manifested.

[8]Concerning the origin of such patterns via deposits, invisible to us, of tanning substances in the petals, see Vogel (1950).

FIGURE 423. Blossom of *Althaea officinalis* L., photographed through yellow and ultraviolet filters; natural size. This is a "bee violet" flower with a blue nectar marker. After Daumer 1958.

FIGURE 424. *Buphthalmum salicifolium* L., photographed through yellow and ultraviolet filters; natural size. The inner fourth of the "bee purple" ligular flower and the disk-shaped flower reflect no ultraviolet and appear as "bee yellow." After Daumer 1958.

Althaea officinalis L. has violet flowers with a nectar marker invisible to us (Fig. 423). The outer parts of the petals reflect ultraviolet; hence the flowers are "bee violet" with a "bee blue" (ultraviolet-free) nectar marker.

In the composite *Buphthalmum salicifolium* L. (Fig. 424), which is pure yellow for us, the flowers of the disks and the inner parts of the flowers of the ligulae are yellow for bees also, while the outer parts of the ligular flowers are prominently "purple colored" for them because of reflection of yellow + ultraviolet. With the yellow flowers of *Chelidonium majus* L. (Fig. 425), the pistils and stamens strike the eye because of the absence of ultraviolet reflection (*right-hand picture*). Here it is noteworthy that the yellowish sepals, which enclose the buds until they open, in contrast with the petals reflect no ultraviolet. Consequently the open flowers, which have something to offer, stand out as being a different color from the buds.

Where nectar-marking spots arise because of unequal ultraviolet reflection, customarily the marker contrasts with its more strongly reflecting surroundings because its ultraviolet reflection is weaker or lacking;

FIGURE 425. *Chelidonium majus* L., photographed through yellow and ultraviolet filters; natural size. Stamens and pistils as well as sepals reflect no ultraviolet. They appear "bee yellow" against the "bee purple" petals. After Daumer 1958.

only exceptionally is the ultraviolet reflected more strongly by the marker itself, as in *Saxifraga rotundifolia* L. according to Kugler (1963). In several species there is strong ultraviolet reflection by stamens, pistils, and stigmas (Daumer 1958; Helga Schardt unpublished).

The nectar marker is by no means restricted to the place of nectar production and its immediate surroundings. Where the sweet nectar lies hidden deep in the base of the flower, the marker can obviously be at the entrance. Papilionaceous flowers (Fig. 426) provide an excellent example of this. In *Narcissus poeticus* (Fig. 419), also, the red border of the short crown of the corolla, which strongly reflects ultraviolet, lies like a luminous ring around the approach to the food.

Different species of the genus *Potentilla* have unequal distributions of the nectar markers, which to us are invisible (Fig. 427). Thus arises a species-specific pattern. Since training to differentiate between these flowers of the same form and color but of different patterns have had only slight success, such graduated characteristics seem to be of only limited significance for the differentiation of flowers.

This is of course not so in the many cases in which flower patterns differ conspicuously in quality of colors and in their arrangement. As early as 1914 I showed by training experiments how effective such patterns are as signs for the bees.

Very many flowers also reflect red light, which is invisible to bees. It is significant that nectar markers and other conspicuous pattern formations are lacking in this spectral region. This emphasizes the importance of nectar markers visible to insects as a meaningful adaptation (Helga Schardt, not yet published).

Whether the nectar markers were noticed by visitors to the flowers as guides to the food was argued frequently, with varied answers, since Sprengel's time, until experimental work furnished a decision. The first demonstration was provided by Knoll (1926) with the hawk moth *Macroglossum stellatarum* L. This moth, which, hovering over the flowers in bright sunshine, draws nectar from their deep corolla tubes with its long proboscis, brings the tip of the proboscis deliberately against the nectar marker, even when during its individual existence it could have had no experience whatever with flowers.

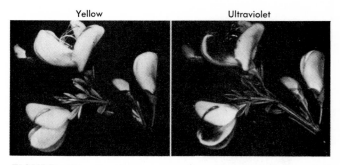

FIGURE 426. *Sorothamnus scoparius* Wimmer, photographed through yellow and ultraviolet filters; natural size. An upper area on the wings is free of ultraviolet. This "bee yellow" patch flanks the access to the nectar in the "bee purple" flower. After Daumer 1958.

FIGURE 427. Ultraviolet photographs of the yellow flowers of different species of *Potentilla*: (*a*) *P. erecta* (L.) Hampe; (*b*) *P. reptans* L.; (*c*) *P. verna* L.; (*d*) *P. aurea* L.; (*e*) *P. fruticosa* L.; natural size. After Daumer 1958.

With more extended observations the importance of the nectar markers in the nearby orientation of bumblebees also emerged clearly (Kugler 1943, 1954, 1963; Manning 1956). In general, this consists in the fact that the way to the food is pointed out to the visitors by the conspicuous contrast of color or luminosity or—with linear or dotted designs—by a richly dissected pattern. When in an experiment we wish to make the bees drink sugar water at a sharply circumscribed spot, for instance from a capillary, we have long employed the device of placing a differently colored "nectar marker" at just this place, because in this way we save much time, not only for the bees but for ourselves. With some blossoms, for instance those of the horse chestnut (*Aesculus hippocastanum* L.), a change in color of the nectar marker occurs as soon as the secretion of nectar ceases. Honeybees and bumblebees quickly learn, in accordance with this guidance, to head for the flowers that have nectar (Kugler 1936); Vogel (1950) made similar observations with other plants.

Experiments by Daumer (1958) with bees provided yet better insight into the mechanism of action of the nectar markers. Yellow ligular flowers (such as *Helianthus rigidus* (Cass.) Defr.), which reflect no ultraviolet at the base and thus bear a nectar marker that we cannot see, were cut up into lanceolate pieces and these were joined together to make rosettes, in one of which the nectar markers were inside, as in the natural arrangement, and outside in the other (Fig. 428). Bees drank sugar water from a centrally inserted capillary in one of these two arrangements (Fig. 429), and were thus trained to it. In the experiment, the bees were able to distinguish the two arrangements well. For this, even very small markers are sufficient. A flower pattern 1–2 mm² in area and invisible to man is clearly effective with bees.

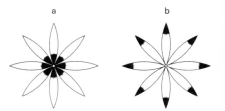

FIGURE 428. The individual petals or ligular flowers of a blossom with an ultraviolet-free nectar marker are arranged in a cluster: (*a*) normally (ultraviolet-free basal areas central) or (*b*) reversed (ultraviolet-free areas peripheral). After Daumer 1958.

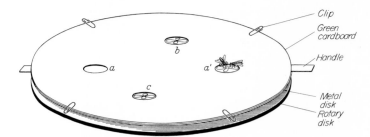

FIGURE 429. Arrangement for training: bees drink sugar water from the capillary in the center of the training flower a'. The flowers b and c that they are trained against can be seen. During training, flower a for testing the result is covered by a movable lid. By moving the two handles, the training arrangement is converted into the testing arrangement. After Daumer 1958.

If they were trained to having the nectar markers located centrally and had settled onto the correct arrangement, they mostly ran from the margin of a ligular flower toward its center; at the edge of the nectar marker they suddenly stopped running, bobbed the head forward jerkily, and often extended the proboscis simultaneously. But if they settled onto the wrong arrangement (Fig. 428), they either turned at once to an ultraviolet-free patch at the outer tip of a ligular flower and—although no capillary was there—went through the head and proboscis reactions, or they ran to the center, frequently passing over it without any response, then to extend the proboscis at the nectar marker of the ligular flower that was situated opposite.

As Knoll had already noticed with his hawk moths, with bees too no individual experience is needed for an appropriate response to the nectar marker. There was available a colony that for several years had been kept in a bee room in the Munich Zoological Laboratory under continuous artificial light and that had been fed at an artificial feeding station with sugar water, never on flowers. On their first visit to floral patterns, when the bees stepped across the circumference of the nectar marker they showed the jerky head movement and frequently too the extension of the proboscis. Thus their behavior is firmly established in their inheritance. This is true not merely of the nectar marker itself but to a certain extent in relation to its location in the flower. In the experiments that concerned the distinguishing of the patterns reproduced in Fig. 428, training to the natural arrangement (a) was very easy, but to the opposite arrangement (b) appreciably more difficult.

Demonstrating the efficiency of nectar markers that we could not see was a particularly alluring problem. The experiments just mentioned, by Daumer, are pertinent. But by comparative tests he showed also that such markers do not have any exceptional status. Nectar markers that are conspicuous to our eyes, for instance the dark red patches (black for bees) at the base of the yellow ligular flowers of *Coreopsis bicolor* Bosse, release the head-proboscis response just like those nectar markers that depend on the absence of ultraviolet reflection.

Since ancient times man has delighted in the colors of flowers, and this pleasure is rendered deeper for the biologist by a glimpse of hitherto hidden relations. Scouting bees must discover food-yielding blossoms (p. 247). This is easy for them because the saturated floral colors stand out effectively against the background of foliage, which is almost colorless for them. Foraging bees must be able swiftly and surely to distinguish from other flowers the species for which they have specialized. The great development of their color sense makes them better able to do this than we thought not long ago. Visible nectar markers, and frequently invisible ones, point the way to the food.

Clear as these adaptations are, we know very little about the role of floral shape. The fundamental attractiveness **of** complexity of shape and the ease with which bees are trained to strongly dissected patterns raise the suspicion that the form of many blossoms, such as those among the Compositae, or in other instances the bundles of delicate pistils, contribute their share to attracting the scouts and to holding the foragers. Hitherto analysis has been limited almost entirely to artificial models. The study of these relations in natural flower forms is a wish for the future.

Not all flowers that insects like have large, brightly colored petals. Many are conspicuous only for a strong fragrance. With hound's-tongue (*Cynoglossum officinale* L.), which has little, unimpressive blooms, Manning (1956a) observed that visiting bumblebees, and honeybees too, used as a guide the strongly dissected appearance of the whole plant (Fig. 430), and from a distance of several meters flew to hound's-tongue, which was new to them in the neighborhood, giving the plants a thorough inspection even when they bore no flowers. Quite different was their behavior toward foxglove (*Digitalis*), where they flew only toward conspicuous flowers and never learned the shape of the plant. Thus hound's-tongue is an example showing that a highly complex configuration of the plant as a whole is able to substitute for what is lacking in this instance in its flowers.

FIGURE 430. Hound's-tongue (*Cynoglossum officinale* L.) substitutes a high degree of dissection of the entire plant for the showiness lacking in its little flowers. After Manning 1956a.

C. ORIENTATION NEARBY BY MEANS OF THE SENSE OF SMELL AND TASTE

In man, as with all vertebrates, two separate organs serve for smell and taste. Both respond to chemical stimuli, but differ histologically in that they are equipped with primary and secondary sense cells, respectively, and anatomically in their different location, nerve supply, and central connections. They respond to different groups of chemical agents, "olfactory substances" and "taste substances." Different, too, are their biological functions. With its extraordinary sensitivity the sense of smell is able to give warning of a distant enemy or to provide guidance to a goal, whereas fundamentally the coarser sense of taste[9] controls the food during feeding. Invertebrate animals are built according to another scheme. Comparative anatomy, a trustworthy guide in the vertebrate stem, leaves us in the lurch here; and histology the same, since insects possess only primary sense cells. Where the other criteria fail as well,

---

[9] For further details of the comparison of threshold values see v. Frisch 1934:4–6.

the boundaries become confused and one speaks of a chemical sense rather than senses of smell and taste. With many of the lower animals this terminology is appropriate, without our being certain today whether it corresponds to reality or only to our deficient knowledge.

With highly organized insects, and particularly with bees, there is no cause for such resignation. In complete analogy with vertebrates, they have two kinds of chemical sense organs, whose preferred location is in different places on the body, with differing sensitivity and differing biological functions. We need have no hesitancy in speaking of smelling and tasting with reference to them[10] (v. Frisch 1926; Dethier and Chadwick 1948).

4. Olfactory Discrimination in Bees

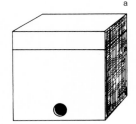

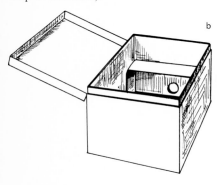

FIGURE 431. (*a*) Cardboard box for training bees to odor, 10 × 10 × 10 cm. (*b*) Same, with cover raised. Drops of odorant are placed on the inserted "odor shelf" (a strip of cardboard).

Fifty years ago it was a tempting but undemonstrated assumption that bees in their foraging flights distinguish the kind of flower they are seeking from other flowers by means of its species-specific fragrance. At that time we regarded their ability for discriminating colors as deficient. All the more important, therefore, was a knowledge of their sense of smell. That impelled me to undertake training experiments, for which the techniques of training the bees to color were altered slightly in view of the differing objective (v. Frisch 1919).

For training I used cardboard boxes with a hinged top and a flight entrance in the front (Fig. 431). One box was supplied with the training odor and a feeding dish; beside it three others remained empty. Frequent change in the position of the training box prevented the bees from becoming acquainted with a definite location. For the test, four clean (usually new) boxes were set out in a different order, one of them with the fragrance but none with food, and the number of bees entering was counted. The intrinsic odor of the cardboard boxes was not a disturbing factor, as control experiments with earthenware boxes showed. As odorants I used partly the fragrances of natural flowers in mineral oil (obtained by means of enfleurage), and in part essential oils (p. 20), and as a control fresh-cut flowers also.

In a test the trained bees would fly even to a clean perfumed box without food. That shows that they perceive the odor and use it as a cue.

In order to examine the extent to which the bees could distinguish between different odors, during the experiment boxes with other scents were exposed beside the box with the training odor. The bees flew only to the odor to which they had been trained. Even when this was left out of the assembly they did not visit the boxes with the other odors;

---

[10] For the organs of taste the designation "contact chemoreceptors" has come into quite prevalent use. In using this expression, one has to remain aware that during distance perception organs of smell too respond only to contact with (evaporated) particles of the effective substance, and that the concept becomes completely hazy with organisms that live in water. Thus for instance with dytiscids (water beetles) no closer contact is required for taste substances than for odors; with the one as with the other the molecules may be carried by the water to the sense organs from some distance. And yet with these beetles too there is a duplicity of the chemical sense, with spatially separate receptors for odors and taste substances, as Schaller (1926) demonstrated.

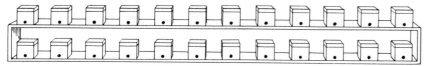

FIGURE 432. Experiment for testing the capacity for distinguishing odors. One box contains the training odor; the rest are supplied with 23 different fragrances.

rather they would occasionally inspect an odorless box. Finally, after training them to oil of bitter orange peel (Messina), in two successive experiments I set out in each instance 23 (thus a total of 46) other essential oils for them to choose from (for the arrangement see Fig. 432). In both tests, in addition to the fragrance used in training, which was visited most, only a total of three other boxes received many visits; these boxes had been supplied with Spanish oil of bitter orange peel, oil of citron, and oil of bergamot.[11] Precisely these odors, and only they, were similar too for the human nostril to the training odor, which has its objective foundation in the fact of their similar composition and origin (expressed from the fruit of various *Citrus* species).

At that time I was able quickly and with good success to train the bees to a total of 28 different floral odors, essential oils, or isolated odorants. These were spicy, flowery, or fruity smells that we sense as being close to floral fragrances. In contrast, training was difficult and incompletely successful with the fetid odor of skatole and carbon disulfide, and to oil of patchouli with its camphorlike smell. In some experiments the skatole box was not distinguished at all from odorless boxes. The most likely suspicion, that bees perceive these odorants scarcely or not at all, proved to be wrong. For after they have been trained to a flowery smell they shun it if skatole is added. Also they showed that they perceived the putrid stench by ventilating the training box actively, by fanning at the opening and thus attempting to reduce the concentration of odor within. One gets the impression that they do not "care for" this category of smells. But that cannot account for the failure of training. Lysol (a mixture of cresols) is so much disliked by bees that beekeepers use it to drive them out of a place. In training, too, they show unequivocally by their behavior that its scent is repugnant to them. They interrupt their drinking frequently to go and sit on the outside of a box for a while or to hover in front of it. And yet training is successful. They do not display so overpowering a preference for the lysol box as is the case with floral scents, but nevertheless they visit it in much greater strength than the odorless boxes. Hence, paradoxical as it sounds, they are attracted by an odor that repels them. Schwarz (1955) had the same experience with butyric acid.

That the bees learn floral scents quickly, but other kinds of odors with greater difficulty or not at all, can thus be attributed neither to a defective ability to perceive the latter nor to a repellent effect. Probably a failure of the ability to comprehend is concerned. In other sensory areas this ability is also manifested better in the face of biologically

[11] Concerning the origin and composition of the odorants see v. Frisch 1919:37–42.

familiar tasks than with those that for countless generations have been useless in the life of bees (v. Frisch 1956).

Relative to the ability of bees to recognize a certain fragrance in a mixture of odors, there have been few experiments.

I trained bees to oil of tuberose flowers (the fragrance of the natural flowers in paraffin oil) and found in testing that a mixture of oils of tuberose and jasmine flowers was visited up to a ratio of tuberose to jasmine of 1:5. If the bees were trained to a mixture containing equal parts of jasmine and tuberose oils, and thus accustomed to both scents from the very beginning, not until a mixture contained a tuberose:jasmine ratio of 1:24 was the tuberose odor no longer perceived and the mixture not distinguished from pure jasmine fragrance. Bees trained to methylheptenone, on addition of ω-bromstyrol recognized the training odor up to a methylheptenone:bromstyrol ratio of 1:10; bees trained to bromstyrol still detected the odor in a mixture of bromstyrol:methylheptenone of 1:24. Ribbands (1955) trained bees to a mixture of linalool and benzyl acetate in equal parts (1-percent solutions) and tested how great a shift in proportions was just detected. After three days of training the mixture of linalool:benzyl acetate 10:10 was still preferred to one 9:11, but no longer to a mixture of 9.5:10.5. In another series of experiments the bees learned to distinguish a mixture of five odorants (benzyl acetate, oil of citronella, linalool, methyl benzoate, and phenylethyl acetate) from the same mixture to which 0.5 percent of geraniol had been added. The contamination with geraniol was still detected when only 0.2 percent was added, but not with only 0.1 percent.

The result of such tests will depend strongly on the odorants used and on the experimental technique. Yet from these experiences and comparable experiments with people, it seems that the bees are superior to us in the analysis of mixed smells.

That there is actually an analysis of the components, rather than a new, different odor quality for the bees resulting from a mixture of two odorants, emerges from experiments by Gubin (1957). He trained bees

FIGURE 433. An antennaless bee is seeking the training odor. In the left foreground is the plate with the training odor; the other plates are supplied with the fragrance trained against; all the plates bear clean, empty watch glasses. The bee is trying to smell the plates from as close by as possible. She is unable to distinguish the training odor from the other smell.

successively to oils of lavender and of peppermint. When thereafter he provided the bees with both training substances and with a mixture of them besides, they visited the latter even more strongly than the pure odorants. In the mixture they must have recognized both of the scents. The reciprocal experiment was also successful: after the bees had been trained to a mixture of two fragrances, both were visited actively when presented separately.

In general, substances that are odorless for us are not smelled by bees. However, they can detect carbon dioxide (see p. 499). They also have sense organs for perceiving moisture. Some time ago, Hertz (1934) showed that they can be trained to water. Ribbands (1955), as well as Kuwabara and Takeda (1956), observed the same, and Kiechle (1961) was able to confirm this with a refined technique. Even in free flight the bees can still detect relative-humidity differences of 5–10 percent. According to Kiechle's experiments, the humidity receptors would have to be located on the antennae, where indeed they were found (Lacher 1964, see p. 499). Moisture in the air and carbon dioxide are perceived by receptors other than those that respond to odors. Whether one chooses to classify such performance under the sense of smell is a matter of definition.

5. The Location of the Sense of Smell

The olfactory organs of insects are located where they will serve orientation best: on the protruding antennae, and with some species on the palpi. In the honeybee they are confined to the antennal flagellum. That was long a subject of controversy, but was settled by amputation of the antennae in combination with training to odors. As an example, we cite a few bees trained to oil of verbena, drops of which on filter paper are placed about the feeding dish. Three odor plates with oil of marjoram are alternated with three empty dishes. After the antennae have been cut off the bee is no longer able to find the training odor. It is impressive to see how she again and again inspects the plates and hovers close above them to "smell" them (Fig. 433). Although in the course of doing so she is surrounded by clouds of fragrance, her behavior toward the training odor and the other odor plates is precisely the same, and when finally she settles and hunts out a dish she chooses according to the laws of chance. But if she has been trained for color, after amputation of the antennae she decides unhesitatingly in accordance with her training; therefore the absence of the response to odor cannot be attributed to an effect of shock or to general damage from the operation, but only to elimination of the olfactory organs (v. Frisch 1921). Hence it is demonstrated that the latter are located on the antennae.

The antennae are thickly sown with sense organs (sensilla), which is in harmony with our results (Fig. 434). But the amazed microscopist is presented with numerous different structures, and it becomes a question which of these are the olfactory receptors.

In order to make some headway here, I undertook partial amputation

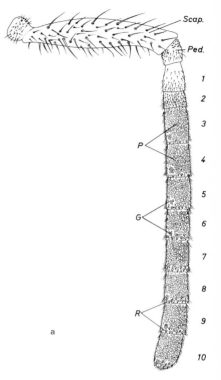

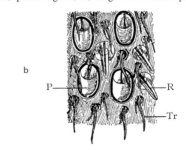

FIGURE 434. (*a*) Worker-bee's antenna: *Scap.*, scape (basal stalk); *Ped.*, pedicel (second member); 1–10, first to tenth segments of the flagellum; *P*, plate organs; *G*, pit organs (*sensilla coeloconica* and *s. ampullacea*); *R*, large thin-walled pegs (*sens. basiconica*). Schematic; actually the plate organs are more numerous and relatively smaller; here the *sensilla trichodea* are shown only at the margin. (*b*) Small part of the antennal surface, near the distal margin of a flagellar segment, more highly enlarged; schematic: *Tr*, *sensilla trichodea*; *P*, plate organs; *R*, large thin-walled pegs.

of the antennae with bees trained to an odor. The clear result was that they are still capable of detecting the training odor when one antenna is cut off completely and only the basal three flagellar segments of the other are preserved. Such bees can also be trained successfully to a new scent. But if one removes yet another segment, so that only the first two flagellar segments are left (see Fig. 432a), the capacity for smelling is destroyed (v. Frisch 1921).

Since then this finding has been confirmed repeatedly (Marshall 1935; Frings 1944; Fischer 1957; Dostal 1958). Just like the olfactory organs, the humidity receptors (according to Kiechle 1961) are confined to the eight distal segments of the antennal flagellum.

If now one compares with the eight excised flagellar segments the microscopic picture of the antennal portion that was preserved in the experiments just mentioned, one is impressed by a dense population in the former of sense organs that are lacking on the more proximal segments. Integumental pores, easily visible as light dots, indicate their location and distribution. The pores permit the distal processes of the sensory cells to make contact with the environment. They are covered over by attenuated cuticular structures, which may protrude above the antennal surface as hairs or cones. Morphological analysis, extended recently by means of electron-microscopic studies, shows seven types of sensilla. In Fig. 435 they have been characterized merely through illustration of their cuticular structure and not of the softer parts.[12]

In types 1–3 are compared mutually similar hair-shaped structures. Types 4–7 clearly are different from them, as they are from one another also.

1. *Sensillum trichodeum* (*small thick-walled hair*,[13] Fig. 435:1). Diameter about 3 μ. The tip bent toward the antennal surface. Apparently innervated by a single receptor cell. Evenly distributed in great numbers over the eight distal flagellar segments (according to Lacher 1964; Vogel 1923:315, 316 says they are sparse, being numerous only at the tips of the antennae).

2. *Sensillum trichodeum* (*thick-walled peg*, Fig. 435:2). Length more than 30 μ, diameter about 5 μ. Long and slender, with a rounded tip; probably innervated by a single receptor cell. Not very numerous; on all flagellar segments; on segments 3 to 10 predominantly at the proximal and distal ends.

3. *Sensillum trichodeum olfactorium* (*slender thin-walled peg*, Figs. 435:3, 436). Length about 15 μ, diameter up to 5 μ, slightly S-shaped, conical, rounded at the end. With the electron microscope numerous pores (diameter about 0.1 μ) can be discerned in its wall (Slifer and Sekhon 1961). Whether they are closed off from the exterior by a thin

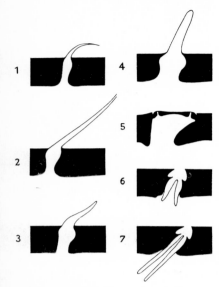

FIGURE 435. Types of sense organs (sensilla) from the antenna of a worker bee; only cuticular structures are shown; each sensillum corresponds to a pore in the integument (black), which is closed off externally by a thin cuticular structure of specific form: 1, *sensillum trichodeum* (small thick-walled hair); 2, *s. trichodeum* (thick-walled peg); 3, *s. trichodeum olfactorium* (slender thin-walled peg); 4, *s. basiconicum* (olfactory cone, large thin-walled peg); 5, *s. placodeum* (pore plate, plate organ); 6, *s. coeloconicum* (pit organ, Forel's champagne-cork organ, coeloconic sense organ); 7, *s. ampullaceum* (pit organ, Forel's bottle-shaped organ, ampullaceous sense organ). From Lacher 1964.

FIGURE 436. Longitudinal section through a *sensillum trichodeum olfactorium* (slender thin-walled peg) from the antenna of a worker bee. Schematic sketch, from an electron micrograph by Slifer and Sekhon 1961.

---

[12] In what follows I adhere generally to the most recent analysis by Lacher (1964), but omit his type *a*, since with it neither innervation nor a definite integumental pore are to be seen; it may not be a sense organ and certainly is not an olfactory receptor. Also I shall not consider his type *i*, with an arrow-shaped sensory peg, since it was found only very seldom and, because of its structure, likewise does not come into question for olfaction.

[13] The English designations in italics have been taken over from Slifer and Sekhon (1961).

membrane is uncertain. In any event the pores facilitate the penetration of stimulants. Innervated by 5–10 nerve cells. Numerous, and distributed evenly over the eight distal flagellar segments. Number of sensilla trichodea on an antenna, about 8000–9000.

4. *Sensillum basiconicum* (olfactory cone, *large thin-walled peg*, Fig. 435:4). Length as much as 25 μ, diameter about 5 μ. Tubular, with a rounded tip, thin-walled, with pores. Only at the distal end of the eight distal flagellar segments. Innervated by 16–20 sensory cells (Krause 1960). Only 100–150 on each antenna.

5. *Sensillum placodeum* (pore plate, *plate organ*, Figs. 435:5, 437). Long diameter 12 μ. An oval plate with an attenuated ring-shaped margin. Innervated by 16–20 receptor cells, which end as tubular processes beneath the plate (*tubuli T* in Fig. 437); these lie immediately below the plate and seem also to protrude into the surrounding ring-shaped groove; the electron microscope reveals a structure in this groove that is produced by extremely thin, radially arranged cuticular strips (C_s in Fig. 437). In each of these thin strips there is a row of about 20 very fine pores (diameter about 20 mμ). Each pore plate has about 2400–3000 of them (Richards 1952; Slifer 1961). The pore plates are distributed evenly among the eight distal flagellar segments, but are absent from the posterior side of the antenna (with respect to antennal posture during flight). Number of pore plates, about 3000 on an antenna.

6. *Sensillum coeloconicum* (pit cone, Forel's (1874) champagne-cork organ, *coeloconic sense organ*, Fig. 435:6). Length of the cone 10–20 μ, breadth up to 10 μ. The conical sensory peg is depressed in a pit. Innervation by a single receptor cell. Situated in several groups at the distal end of each of the eight distal flagellar segments, principally on the upper and lower antennal surface (with respect to the antennal position during flight).

7. *Sensillum ampullaceum* (pit cone, Forel's bottle-shaped organ, *ampullaceous sense organ*, Fig. 435:7). Length of the cone 15–30 μ, width up to 10 μ. A sensory hair sunk in a deep pit. Innervated by a single sensory cell. Located on the eight distal flagellar segments, more numerous than the sensilla coeloconica and mostly together with them. According to Lacher (1964), the total numer of pit cones on an antenna is 236. Their number per flagellar segment increases distally.

Among the organs named, the sensilla trichodea "olfactoria" and the "olfactory cones" (Nos. 3 and 4) are already dubbed organs of smell by the use of these terms. But these designations bring to mind an opinion that is still without definite proof. The pore plates also can be counted as olfactory organs; but several people claim that they perceive pressure, sound, or heat.

That the pore plates are organs of smell is quite likely, if only that they are heavily predominant on the drones' antennae. For the drones a well-developed olfactory sense is of fundamental importance when they are seeking a queen during the nuptial flight. On one antenna the worker bee has 3,000 pore plates, the drone 15,000. In the drone, pit organs are not increased in number, and hairlike sense organs are scanty.

In 1921 I thought I had demonstrated the olfactory function of the

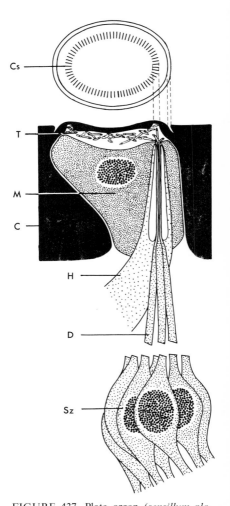

FIGURE 437. Plate organ (*sensillum placodeum*): (*above*) surface view; (*below*) section; *C*, sclerotized integument; *M*, membrane cell; *H*, sheath cell (recently Lacher has corrected this figure somewhat; the sheath cell does not reach the plate, but ends approximately where the dendrites narrow down); *Sz*, receptor cells; *D*, dendrites; *T*, tubuli; *Cs*, cuticular structure of the plate organ. Only a few of the receptor cells and dendrites are shown; the receptor cells are located somewhat deeper beneath the integument. Schematic, from Lacher 1964, somewhat altered.

pore plates, when with three bees I succeeded in excising the last eight flagellar segments on one side and on the other side the last seven and the distal one-third of the preceding segment. These specimens could still discriminate scents, although all they had left were "tactile hairs" and pore plates. But the conclusion was not binding, because R. Vogel (1923), in a fine morphological study, showed that the sensilla trichodea do not have identical structures. He explained only some of them as tactile hairs; for the rest he coined the concept of sensilla trichodea olfactoria, because he regarded them as organs of smell. Further aspirants for the function are the "olfactory cones." In the three bees mentioned, which had lost eight flagellar segments on one side and seven and one-third on the other, all the olfactory cones had been removed, since the latter are located only at the ends of the eight distal flagellar segments. Since a remnant of the olfactory ability was preserved, it is possible to smell even without "olfactory cones." But it was undecided whether under normal conditions these might not also be concerned in olfaction.

Given the dense proximity of the different sensilla, any attempt at a further experimental clarification seemed hopeless. In this state of uncertainty, the advance of electrophysiology provided decisive assistance.

Lacher and Schneider (1963) and Lacher (1964) were able with drones and worker bees to record the excitatory potentials from individual receptor cells (receptor potentials). The finding was that the pore plates —and only the pore plates—respond to odorants, mostly with an increase in the background firing, more rarely with inhibition. They reacted to all 33 odorants tested; these were predominantly flowery-aromatic substances. According to verbal communication since the publications appeared, the odors from the worker's scent gland and from the mandibular gland of the queen have been tested on the pore plates with positive results.

An individual receptor cell of a pore plate responds to various odorants, but not to all. Another sensory neuron will respond to the same materials, in part, and in part to others. Hitherto, no two sensory cells have been found that were adapted to precisely the same selection of odorants. That is to say, each one had a different "reaction spectrum." This utterly inexhaustible combination of reaction spectra is no doubt the physiological foundation for discrimination of such a multiplicity of odors.

Pit cones, sensilla trichodea (including Vogel's olfactory hairs, sensilla trichodea olfactoria), and sensilla basiconica (olfactory cones) gave no response to olfactants. Hence, according to Lacher, the pore plates are the sole olfactory organs of the bee. That is surprising. For demonstrably sensilla basiconica serve as olfactory organs in other insects, for instance in grasshoppers (Slifer, Prestage, and Beams 1959), beetles (Boeckh 1962), and moths (Schneider, Lacher, and Kaissling 1964), and both they and the sensilla trichodea "olfactoria" seem suited by the pores in their wall for the reception of chemical stimulants. Since hitherto no other sensory function has been demonstrated for them,[14] and since

---

[14] Odorants, damp air, carbon dioxide, mixtures of air with oxygen or nitrogen, temperature, air currents at 1–11 m/sec, light, sound, and other mechanical stimuli were tested.

nevertheless they doubtless have one, they might be serving as organs of taste or for smelling objects in the immediate vicinity, as I assumed in 1921 for the olfactory cones that protrude beyond the pore plates and little hairs. One may look forward with keen anticipation to the results of further investigations.

When tested electrophysiologically, the pit cones proved to be receptors for moisture and carbon dioxide, as well as for temperature stimuli. Whether morphologically distinguishable organs are concerned with these three kinds of stimuli, and particularly how the *sensilla coeloconia* and *ampullacea* differ functionally, is an open question. For these structures stand too close together for even the electrophysiologist, and on external examination are too much like one another to have been analyzed separately hitherto.

6. THE CAPACITY TO LOCALIZE BY SMELLING

We have seen that bees are surprisingly like us in their olfactory accomplishments. But in one important point they are decidedly superior: when odorants penetrate to the sensory epithelium within our nostrils, they have been whirled around so turbulently that their relation to the shape of the odorous source and their distribution there have been lost. A bee bears its organs of smell outside on the antennae, which can be waved about like arms and can simultaneously smell and palpate an odorous body. August Forel (1910) long ago assumed that on this account it is possible for bees and other insects to establish a very close relation between the scents and form of an odorous object, to smell "plastically," as we see plastically. For this he coined the designation "topochemical olfactory sense."

His concept was reasonable. But that a bee can actually localize, separately and correctly, odorous impressions received by the two antennae has just recently been demonstrated (Lindauer and Martin 1963; H. Martin 1964).

Martin had his bees run in a Y-tube, one arm of which led into a perfumed box, the other into an odorless box (Fig. 438), and trained them to choose the fragrant side at the fork. As soon as they had mastered this exercise, they were tested without food in a clean apparatus. Here too they ran to the scent. They did so even when both antennae were glued fast at the root and hence were immovable. But when Martin fastened the antennae in a crossed position (Fig. 439), the bees chose the wrong direction at the fork. Hence they are able to orient (tropotactically) in accordance with the simultaneous but differing olfactory impressions that are received by the two antennae. In doing so they are led astray when the position of the antennae is exchanged—as does not occur under normal conditions and is not realized by the bees. The central connections indeed remain unaltered, and when they now turn to the wrong side they are responding correctly in the physiological sense.

The bee can localize the scent in yet another way. If one of her antennae is cut off, despite this she usually finds her way into the

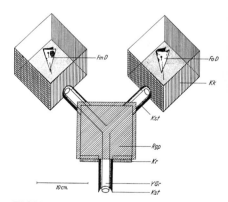

FIGURE 438. Experimental arrangement for stimulation by odor from one side; the bee runs through the Y-shaped glass tube (*YGr*), and at the fork has to decide in favor of one of the sides: *FmD*, scented filter paper; *FoD*, odorless filter paper; *Kk*, cardboard box with lid removed; *Kr*, cardboard holder for the ruby-glass plate, *Rgp*; *Kst*, cardboard sheath for darkening the tube. After H. Martin 1964.

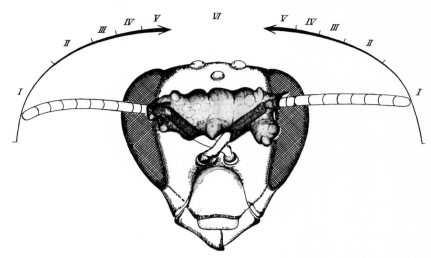

FIGURE 439. A bee with antennae crossed and fastened in this position with glue at the base. The distance between the tips of the antennae is varied in different animals (positions I–VI); see text. After H. Martin 1964.

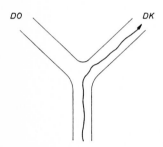

FIGURE 440. Course in the Y-tube of a bee trained to an odor, with the left antenna movable and the right antenna amputated: *Dk*, scented box; *Do*, odorless box. The bee chooses correctly, without hesitation. After H. Martin 1964.

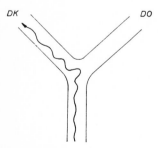

FIGURE 441. Course in the Y-tube of a bee trained to an odor, with the right antenna amputated and the left antenna glued in its normal position, and hence immovable: *Dk*, perfumed box; *Do*, odorless box. The bee makes a correct choice, after oscillating progress. After H. Martin 1964.

proper arm, by waving her remaining feeler actively from right to left and thus comparing the odors of the two sides in rapid succession (klinotactic orientation, Fig. 440). She also finds the correct way unhesitatingly when the one remaining antenna is glued fast at the base and can no longer be swung to and fro. Then the bee substitutes oscillation of the entire body for what she had accomplished previously by waving the antenna, and with oscillating progress chooses the proper side (Fig. 441). It is only this last way of localizing odors, by means of successive comparisons, that we possess to a modest extent, when we smell of an object at various places.

Using simultaneous spatial comparison, it was of interest to learn what difference of concentration between right and left is necessary if it is to be noticed by the bee. When both antennae are fastened in the crossed position shown in Fig. 439, the tips of the antennae are 9 mm apart. Now with different specimens, the antennae were stuck fast at the base so that the interval between the tips was reduced stepwise. With positions I–V (arrows in Fig. 439) the choices became only insignificantly worse. Not until the distance between the antenna tips amounted to 2 mm was there a sudden reversal: the spatial separation of the two odor concentrations was no longer recognized, the bee went over to oscillatory progression, and now on the basis of testing the smell successively decided for the objectively correct side, whereas previously with simultaneous odor sampling—in consequence of the crossed antennae—she had made a subjectively correct but objectively false choice. By this means, the threshold value for recognition of a difference in odor can be determined very precisely.

But in this way one learns only the required spatial separation of the antennae for a given experimental arrangement, and nothing about the actual difference in concentration. A different method provides information about the latter. Once again a bee is trained in a feeding box to an odor, then during carbon dioxide narcosis a slender glass tube is fitted over each antenna and sealed airtight at the base. Into the distal open end of the capillary is placed a little drop consisting of a precisely measured mixture of an odorous oil with odorless paraffin oil (Fig. 442). The bee is fastened to a slip of cardboard by head and thorax, and when she begins to walk she rotates beneath her a cork hemisphere. Pointers are attached to the latter so that its rotation can be measured (Fig. 443). If now the two capillaries are supplied with differing concentrations of odor, the trained bee attempts to go to the side of higher concentration so long as she can perceive a difference. With rather low initial concentrations, a concentration difference of 1:2.5 sufficed for the proper choice (Martin 1964). Such differences can be assumed to occur in the immediate vicinity of the antennae both in the comb area of the hive and during visits to flowers out of doors (see pp. 508f).[15]

In pursuing this problem further, Martin (1965) found that not only quantitative but also qualitative olfactory differences between the right and left sides were heeded. Normal bees that are walking to the source

---

[15] Stimulated by Martin's report, Neuhaus (1965) undertook to calculate relations involved in the diffusion of odorants and obtained the result that near a source of odor, at a distance of 0.5–2 cm, the prerequisites exist for a tropotactic orientation by bees.

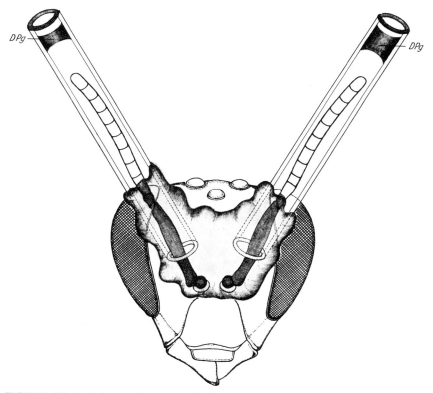

FIGURE 442. Bee's head, with glass capillaries fitted over the antennae and glued fast at the base: *DPg*, mixture of fragrant oil and paraffin oil. After H. Martin 1964.

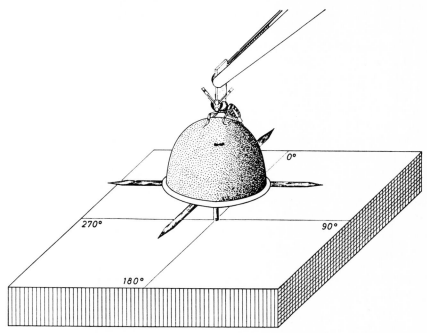

FIGURE 443. Experimental arrangement for determining the difference threshold in an odor field. The cork hemisphere is easily rotated; the pointers facilitate measurement of the rotation. After H. Martin 1964.

of food through a tube that is scented with methylheptenone and geraniol notice when it smells more strongly of methylheptenone on the left and of geraniol on the right. In an experiment, they avoid tubes with the opposite distribution of smells. On the way to the food source they also pay attention to spatially successive qualitatively different olfactory cues. When tested, they expect to meet the succession of odors they have learned, and turn around if it deviates. For this olfactory analysis of the immediate environment, the extreme tips of the antennae, with which the ground is examined directly or from very close by while the bees are en route, prove to be indispensable. If the final segment of both antennae is even partially excised, the topographic olfactory analysis previously learned is completely abolished. Apparently the sense organs for topochemical orientation are developed to perfection on the antennal tips. One may look forward expectantly to a precise morphological and physiological investigation of this favored region of the antennae.

That the prerequisites for the use of such capacities are provided in many flowers will be discussed on pp. 508f.

7. The Bees' Olfactory Acuity

At the time of my old experiments there existed only guesses as to the olfactory acuity of bees. Their performance during training did not support the widespread opinion that they could perceive odorants at appreciably greater dilutions than we. In order to make a more precise test of this, I trained bees to a floral odor (oil of tuberose flowers) in paraffin oil, and in the experiments offered it at progressively greater dilutions until the perfumed box could no longer be distinguished from an odorless box. For dilution I used pure paraffin oil, which is demonstrably odorless for bees too. The limit was reached with a dilution between 1:100 and 1:200, and it was found to be approximately the same in simple experiments by a person of average olfactory capacity, who had the task of sniffing at the cover, opened a crack, of the perfumed box. The possibility was to be reckoned with that the bees had indeed perceived the weak scent, but that they had no longer flown to it because they had been accustomed to a higher concentration of odor. However, the result was unchanged when in training instead of the concentrated odorant there was chosen a degree of dilution that was only slightly above the threshold value that had been found.

Experiments with oil of jasmine flowers and with the chemically pure odorants ω-bromstyrol and methylheptenone confirmed the fact that the threshold value for these fragrances is approximately the same for the bee and for man. There is also agreement with man in that there are appreciable individual differences in olfactory acuity.

Ribbands (1955), on the basis of a few observations, thought that the olfactory acuity of bees surpassed that of man 40-fold with methylheptenone, and 100-fold with geraniol. On a broader basis, three other investigators have confirmed the fact that the threshold values agree approximately in the human and the bee.

Schwarz (1955) improved the method significantly, so that in comparative threshold determinations he was able to estimate the required number of molecules of odorant per cubic centimeter of air per unit time. He applied in a somewhat simplified form the dynamic olfactometer that Neuhaus had worked out with dogs. By this means it was possible to regulate the strength of the stimuli precisely and the desired concentration of the scent could be maintained for a long time.

The principle of the apparatus may be seen from Fig. 444. An electrically driven turbine Tu produces a constant air stream that flows through a rotameter Ro (airflow meter) into a distribution vessel Vf; in order to avoid confusing the diagram, only one of the three further attachments is shown. From the evaporator Vd (a glass capillary containing the scent) the airstream is supplied with a measured amount of odorant. In order that the flow of perfume may reach the bees uniformly at the feeding place, they drink sugar water on the training stand from a narrow opening in a glass tube, sitting meanwhile directly above the outlet for the scent on a screen cover (Fig. 445).

In the experiment three training stands were set up side by side; they could be supplied as desired with either pure or scented air—and correspondingly the drinking vessel with either water or sugar water. Reference is made to the original report for technical details and control experiments.

The experiments were carried out with synthetic substances of known chemical constitution. The results may be seen in Table 43. The last

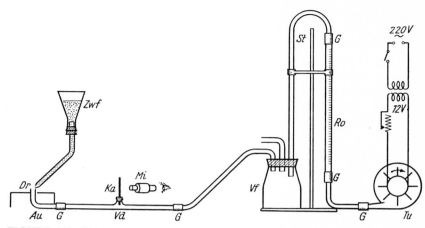

FIGURE 444. Olfactometer for bees: *Dr*, training place; *Zwf*, sugar-water flask; *Au*, outlet for scent; *G*, rubber connections; *Vd*, evaporator; *Ka*, capillary (scent container); *Mi*, microscope for reading the height of the meniscus in the capillary; *Vf*, distribution vessel; *Ro*, rotameter; *St*, ring stand; *Tu*, turbine. After Schwarz 1955.

FIGURE 445. A bee drinking at the glass tube. She is sitting on the scent outlet, which extends beyond the glass base. After Schwarz 1955.

column shows the relative accomplishments of the bees' olfaction: n times better (+) or worse (−) than man's. One sees that in general the olfactory acuity of bees corresponds approximately to that of man. They detect flowery scents somewhat better, fatty acids and their ethyl esters somewhat worse than we.

The noteworthy finding that the bees' olfactory organ is more sensitive toward biologically significant odorants than toward those that

TABLE 43. Some olfactory thresholds and their differences in man and in the bee; the factor in the last column shows how much worse (−) or better (+) the bees smelled the substance than man did. (After Schwarz 1955.)

Odorant	Threshold concentration (molecules/cm³ air)				Relative acuity factor
	Man	Author	Bee	Difference	
Propionic acid	$4.2 \cdot 10^{11}$	v. Skramlik	$4.3 \cdot 10^{11}$	$1.0 \cdot 10^{10}$	1.02 −
Butyric acid	$7.0 \cdot 10^{9}$	v. Skramlik	$1.1 \cdot 10^{11}$	$10.3 \cdot 10^{10}$	15.7 −
Valeric acid	$6.0 \cdot 10^{10}$	v. Skramlik			
iso-Valeric acid	$4.5 \cdot 10^{10}$	Schwarz	$1.6 \cdot 10^{11}$	$11.5 \cdot 10^{10}$	3.5 −
Caproic acid	$2.0 \cdot 10^{11}$	v. Skramlik	$2.2 \cdot 10^{11}$	$2.0 \cdot 10^{10}$	1.1 −
Ethyl caproate	$1.3 \cdot 10^{11}$	Schwarz	$3.8 \cdot 10^{11}$	$2.5 \cdot 10^{11}$	2.9 −
Ethyl caprylate	$3.7 \cdot 10^{10}$	Schwarz	$5.4 \cdot 10^{10}$	$1.7 \cdot 10^{10}$	1.5 −
Ethyl pelargonate (ethyl nonanoate)	$3.1 \cdot 10^{10}$	Schwarz	$3.7 \cdot 10^{10}$	$6.0 \cdot 10^{9}$	1.2 −
Ethyl caprate (ethyl decanoate)	$4.2 \cdot 10^{9}$	Schwarz	$5.6 \cdot 10^{9}$	$1.4 \cdot 10^{9}$	1.2 −
Ethyl hendecanoate	$1.4 \cdot 10^{10}$	Schwarz	$1.8 \cdot 10^{10}$	$4.0 \cdot 10^{9}$	1.3 −
Methyl anthranilate	$2.6 \cdot 10^{10}$	v. Skramlik	$1.9 \cdot 10^{9}$	$24.1 \cdot 10^{9}$	13.6 +
Phenyl propyl alcohol	$6.5 \cdot 10^{9}$	Schwarz	$2.2 \cdot 10^{9}$	$4.3 \cdot 10^{9}$	3.0 +
Nerol	$5.7 \cdot 10^{9}$	Schwarz	$3.2 \cdot 10^{9}$	$2.5 \cdot 10^{9}$	1.8 +
α-Ionone	$3.1 \cdot 10^{8}$	Zwaardemaker	$1.5 \cdot 10^{10}$	$146.9 \cdot 10^{8}$	48.5 −
Eugenol	$8.5 \cdot 10^{11}$	Ohma	$2.0 \cdot 10^{10}$	$8.3 \cdot 10^{11}$	42.5 +
Citral	$4.0 \cdot 10^{11}$	v. Skramlik	$6.0 \cdot 10^{10}$	$3.4 \cdot 10^{11}$	6.6 +

are of no consequence in their daily life was confirmed and extended by Fischer (1957). In choosing odors, he mostly gave up chemically pure agents and returned to floral perfumes and other natural odorants, so that there was no possibility of applying molecules of scent quantitatively. Fischer was forced to limit himself to an approximate comparison with human subjects. But his setup also provided a very uniform presentation of the odorant. It was placed in a petri dish beneath the feeding dish. Beside it there were, in a shifting arrangement, three odorless dishes (Fig. 446). In these experiments too the bees' olfactory acuity corresponded approximately to that of normal human olfaction, but for the most part was somewhat superior with natural floral odors and essential oils, and surpassed man's capacity manyfold for the smell of wax and for the biologically significant aroma of the scent gland (Table 44).

That the fragrance of the scent organ is much more intense for bees than for us I too had found, in a comparison with methylheptenone, on an earlier occasion (1923:171, 172).

With a new method, H. Martin (1964) succeeded in retesting the olfactory acuity once again. His technique has already been described

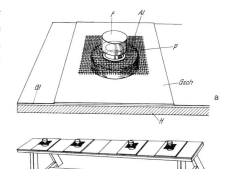

FIGURE 446. (*a*) Experimental arrangement for determining the relative olfactory acuity. (*b*) Training setup: *F*, feeding vessel; *P*, petri dish with perfume; *Al*, aluminum-wire screen; *H*, wooden baseboard; *Gl*, glass base; *Gsch*, glass plate. After Fischer 1957.

TABLE 44. Olfactory thresholds for man and bee; the factor in the last column shows how much worse (−) or better (+) the bees smelled the substance than man did. (After Fischer 1957.)

Type of odorant	Odorant	Threshold value found		Relative acuity factor
		Man[a]	Bee	
Floral oil	Jasmine	1:10000	1:20000	2 +
Essential oils	Geranium	1:10000	1:20000	2 +
	Lavender	1:200000	1:500000	2.5 +
	Rosemary	1:500000	1:100000	5 −
	Peppermint	1:200000	1:200000	1 ±
Pure chemicals	Bromstyrol	1:200000	1:100000	2 −
	Methylheptenone	1:200000	1:200000	1 ±
	Nerol R	1:200000	1:500000	2.5 +
	Nitrobenzol	1:200000	1:100000	2 −
Natural floral scent	*Solidago canadensis* L. (goldenrod)	2 flowers	1 flower	2 +
	Polygonum baldschuanicum Regel	1 flower	0.5 flower	2 +
	Sambucus nigra L. (elder)	1 flower	1 flower	1 ±
Other biologically important odorants	Wax	500 mg	100 mg	5 +
	Scent gland	Piece of filter paper exposed about 10 min	Piece of filter paper exposed 15 sec	Manyfold +

[a] Recognition threshold.

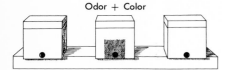
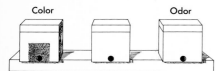

FIGURE 447. Experimental arrangement for comparison of the effect of color and odor: (*above*) training; (*below*) test. The darkened area signifies blue color.

on pp. 500f, when we discussed what differences in concentration the two antennae could just perceive on simultaneous exposure. If now a known concentration of odorant is presented in one glass capillary (see Fig. 442) and odorless paraffin oil in the other, the absolute threshold value for response to an odorant can be determined. In the series of experiments the concentration of the scent is reduced progressively until the bees no longer show a statistically significant bias toward the scented side. For methylheptenone and bromstyrol the threshold value was $1:200,000$. Fischer had obtained the same value for methylheptenone, and for bromstyrol $1:100,000$. With the same compounds I had found the threshold at dilutions between $1:2,000$ and $1:20,000$. The difference is due to the difference in method. In my experiments the bees had to make their choice out in front of the perfumed boxes, where the odorant was already diluted by the ambient air. With both experimental arrangements the performance of man was tested under corresponding conditions and was found to be approximately the same as that of the bees.

The question has often been asked whether their fragrance or their color is the more effective attractant and cue for flowers. Hermann Müller (1873:429) considered the floral odor the more potent attractant, because among adjacent plants those that are inconspicuous but have a strong fragrance are much more heavily frequented than those with striking but odorless blossoms. This way of regarding matters can be deceptive, because the amount of visitation by insects depends primarily on the food available. Flowers may have ever so alluring a fragrance and color, but if they have nothing to offer they will not be much visited, since their visitors neither tarry nor fetch assistance. In order to be able to judge the relative significance of fragrance and color, we must make them compete with one another experimentally.

If, for instance, one trains bees to a box bearing a blue plate and the odor of jasmine, and in the experiment offers odor and color separately (Fig. 447), the bees coming back from their home hive fly with certainty from a distance of several meters to the blue box. In front of the entrance they falter, only a few crawl in hesitantly, most of them begin to fly about questingly, and when they have approached within a few centimeters of the entrance to the scented box they customarily crawl into it quickly despite the absent color. In 1919 I carried out such experiments with many variations and with various scents. The proportion of visits to fragrance and color varies according to the quality and intensity of the odor employed. But nothing about the fundamental result is changed: color exerts its effect at a greater distance; from nearby the fragrance carries the greater "conviction" and even in the absence of the training color impels the bees to settle much more readily than the color is able to do in the absence of the odor.

With field flowers too one can see that the color is effective at a greater distance. Not infrequently one will note a foraging bee that is looking for a certain sort of flower flying up to other flowers also, if these resemble for the bee's eye the kind sought; not until she is in their immediate vicinity does she become aware of her mistake, because of the foreign fragrance, and without settling she veers away to where the next patch of color catches her attention. Jörgensen (1939) reports similar observations. Nothing else is to be expected if the bees' olfactory acuity is of the same order of magnitude as our own. But with her as with us, exceptions may occur, as when the fragrant intensity of a flowering linden hedge makes itself felt at a greater distance than does the appearance of its unimpressive flowers.

On the whole, a more extensive agreement is seen in the olfactory performance of the bee and man than would have been expected in creatures with such different organizations. All the odorants to which bees could be trained also have odors for us. Flowers that are odorless to us are so for bees too. Only a few scents that are important to bees have an appreciably more intense odor for them than for us (wax, scent gland) and only the secretion of the mandibular gland seems to have an odor that, in contrast with the bee, man cannot perceive (Nedel 1960:169). With most materials olfactory threshold is approximately the same in man and the bee. Scents that are similar for us are confused by bees also. That is true even for many pairs of odorants that smell the same to us in spite of their wholly differing chemical composition (for instance, the methyl ester of anthranilic acid and nerolin), whereas such closely related substances as the methyl ethers of *para-* and *meta-*cresol smell different for both man and bee (v. Frisch 1919:168-192). Threshold determinations by Schwarz (1954) in the homologous series of the ethyl esters of fatty acids and in the fatty acids yielded similar regular relations with bees and with man. All this points to a conformity in the physiological basis of the olfactory sense, such as we have already met in color vision.

8. The Biological Significance of Floral Odor

Inasmuch as the assumption of a fantastic olfactory acuity in bees has been relegated to the realm of fancy, the principal importance of their sense of smell is in the area of nearby orientation. Possibly an exception is furnished by the drones, whose antennae bear five times as many plate organs as the workers'. Nothing is known of the olfactory acuity of drones. But since a mature queen that is allowed to fly about in the air fastened with a nylon thread frequently is swarmed about after a few minutes by hundreds of drones (Gary 1963, see p. 289), the latter must be extremely sensitive to the sexual attractant of the queen.

The olfactory accomplishments of the workers are of course important for their performance in the dark beehive. Here especially, in addition to their capacity for qualitative discrimination, their ability for localizing by smelling will be of use. Only we know practically nothing about this.

And so our interest turns now to that mutual relation that has fascinated equally the scientist and the layman and that has already occupied us in connection with color vision: the interplay between bees and flowers.

It cannot be doubted that in a general way the evolution of floral odor has been an adaptation to the visitation of flowers by insects. The broad distribution of floral odors among plants with flowers attractive to insects and its absence in most wind-pollinated flowers, as well as in flowers attractive to birds, are otherwise incomprehensible. Frequently the exposure of fragrance is limited to that time of the day or night when the pollinators are on the wing. The quality of the scent, too, is appropriate to the guests of the flowers, as is shown for instance by the carrionlike smell of flowers that in their structure also are adapted to visitation by carrion-loving flies and beetles.

But our consideration is limited to the flower-frequenting bees. For them the floral odor has a fourfold significance.

Upon the scout bees, few in number but nevertheless so important in the economy of the colony, flowery scents are effective primarily as attractants. On a fine autumn day I set out in a field three odor plates bearing drops of oil of peppermint, oil of geranium, and oil of patchouli. As expected, the unflowerlike scent of the oil of patchouli remained unheeded, as did another control plate that was supplied with water, whereas the peppermint and geranium plates were visited in the course of the hour's observation by six and two bees, respectively, and in part were examined at length (v. Frisch 1919:164). These odorants could not have been known to the bees through experience. It was possible to follow the performance of scout bees more exhaustively with colonies that had been set up in a closed courtyard, and that hence could find only what was presented to them intentionally. Natural floral fragrances, including some from species of plants that are not pollinated by bees, proved to be the most effective attractants. What was decisive was not the specific odor, but the odor category.

A scent clinging to the body of a dancer that has been visiting flowers serves as a means of communication; by means of it the hive-mates learn what is the goal they must seek (see pp. 46f).

Only exceptionally do forager bees on a collecting flight visit several kinds of flowers. As a rule they remain loyal to a single species of flower throughout the day. For them the specific floral odor is a sign by which they are able surely to distinguish from other blossoms the one sought. That in orienting from nearby the significance of the scent predominates over that of color has been discussed above (p. 506), and has also been emphasized by Manning (1957).

A fourth function of the floral odor has just recently been recognized. By way of topographic olfactory fields it supplies the guests of the flowers with scented guideposts to the source of food. In her outstretched motile antennae the bee possesses the equipment required to make use of these. Three studies from our laboratory have been devoted to these relations.

The starting point was the question whether perhaps a visual nectar marker might also display an olfactory contrast with the surrounding floral parts. Mathilde Huber (1943) found this notion actually confirmed in 68 species of flowers. In order to detect this, it was necessary, of course, to present the floral parts separately to the human nostril, which does not have at its disposal a topochemical olfactory sense.

The nectar markers are cut out of the petals and kept for a few minutes in a covered glass box; their odor is then compared with that of similarly treated areas of equal size from neighboring flower parts. In this way local differences in scent are easily determined. The visual nectar marker proves to be simultaneously an olfactory marker.

With 48 species of plants Therese Lex (1954) made a more precise analysis of this difference in scent. She found confirmation that the visual marker always is also an odorous marker, but noted that the boundaries of the two do not invariably coincide. The scent of the odorous marker may differ qualitatively from that of the surroundings and may even have a stronger smell than the latter (for instance in the spotted dead nettle, *Lamium maculatum* L., Fig. 448) or may smell less strong (as in the bee nettle, *Galeopsis speciosa* Miller, Fig. 449). For the same surface area, the nectar marker may also smell qualitatively like but more intense than the adjacent parts of the petals (for example in the horse chestnut, *Aesculus hippocastanum* L., Fig. 450). Further, it developed that odor markers occur also in blossoms without a visual marker, and are hence more widespread than the latter. Thus in the meadow Canterbury bell (*Campanula patula* L., Fig. 451) the proximal part of the corolla smells stronger than the distal portion. In other cases the quality of the scent also differs. Finally, odorous areas in the flower, just like colors, may form characteristic patterns outside the nectar marker. With the nasturtium (*Tropaeolum majus* L., Fig. 452) the nectar markers smell different and more intense than the other parts of the corolla petals, the sepals have a different odor, and the calcar (spur) yet another. By means of such olfactory multiplicity many flowers are still further adapted to a localizing sense of smell on the part of their visitors.[16]

An additional step was to test on bees also the olfactory differences for the human nostril determined within the blossoms. In sample tests with eight species of plants these differences were confirmed successfully by means of training experiments.

Alexandra v. Aufsess (1960) repeated many of the earlier experiments with similar results and extended them to nectar markers, invisible to us, that resulted from unequal ultraviolet reflection. In all 14 cases studied, these were simultaneously odor markers. The great variety of the olfactory patterns is increased yet further by the fact that in entomophilous flowers the pollen for the most part smells stronger and qualitatively different than the petals (see in this connection p. 35). By means of training experiments, it was possible to confirm this also for the bees' olfactory sense.

Thus flowers provide a rich field of action for the bees' capacity for localizing smells. That there is logic in these arrangements and that they are taken advantage of by bees seems to me beyond doubt.[17] The biological correlation becomes even clearer when one includes in the

FIGURE 448. Spotted dead nettle (*Lamium maculatum* L.): *U*, lower lip; *O*, upper lip; *K*, corolla tube. The nectar marker of the lower lip smells stronger and qualitatively different than the corolla tube. After Therese Lex 1954.

FIGURE 449. Bee nettle (*Galeopsis speciosa* Miller). The violet nectar marker *S* on the lower lip smells weaker, but qualitatively different, than the rest of the floral parts. After Therese Lex 1954.

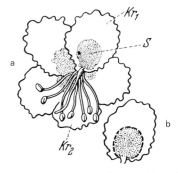

FIGURE 450. (*a*) Horse chestnut (*Aesculus hippocastanum* L.); the smell of the nectar marker is more intense than but qualitatively just like that of the rest of the corolla petal. (*b*) Upper corolla petal with nectar marker; broken line, plane of section. After Therese Lex 1954.

---

[16] As a rule the scent is produced diffusely in the epidermis, even where local differences exist. But scent glands of restricted extent may also occur beneath the epidermis, as in some flowers that smell of carrion and that are attractive to flies, and in orchids (St. Vogel 1962).

[17] See also p. 247, n. 16.

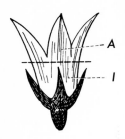

FIGURE 451. Meadow Canterbury bell (*Campanula patula* L.), with no visual nectar marker. The inner part *I* of the flower smells stronger than the outer part *A*. After Therese Lex 1954.

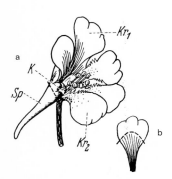

FIGURE 452. Nasturtium (*Tropaeolum majus* L.): Kr_1 and Kr_2, corolla petal with and without a nectar marker; *K*, sepal; *Sp*, spur (calcar). All the parts mentioned have different scents, and in addition the nectar marker has the strongest odor. After Therese Lex 1954.

comparison flowers attractive to birds. Birds are adjusted predominantly to vision and do not have a good sense of smell. When 27 kinds of flowers, with and without a visual marker but attractive to birds, were tested with the human nostril, it was found that there was only an unusually weak production of fragrance. Here, too, in most instances, there were differences in scent between the inner and outer floral parts, but, in contrast to flowers attractive to insects, the outer parts were characterized by a relatively stronger odor just as often as the inner ones. The biologically meaningful connection with the source of food was lacking. The pollen too is only weakly fragrant in flowers attractive to birds (and in those that are wind-pollinated).

Thus the local differences in fragrance are but slight and of random arrangement in flowers attractive to birds. Yet it is interesting that there are such differences, for similar relations must be assumed for the ancestors of the entomophilous flowers, if selection is to have had a starting point for adaptation. With ornithophilous flowers a driving force in favor of olfactory differentiation is lacking from the bird's side. That such differences are absent here is apparent not only to the human nose. In attempts at training bees also, with ten randomly chosen species of ornithophilous flowers, v. Aufsess was able to demonstrate that here the difference in fragrance between the nectar marker and its surroundings or between the inner and outer parts of the flower were too slight for successful training to odor.

In the fragrance of flowers we have seen a factor of such great biological significance that it is only with astonishment that we recognize the existence also of flowers that are both odorless and inconspicuous yet are visited zealously by bees. Admittedly, they are rare. In spite of an extensive search (1919:103ff) I know of only three examples in our flora: the flowers of the wild grape (*Parthenocissus quinquefolia* (L.) Planchon), the huckleberry (*Vaccinium myrtillus* L.), and the red currant (*Ribes rubrum* L.). Kerner v. Marilaun (1898, vol. 2:186) could think of no other way of interpreting the heavy visitation of the grape flowers than that they must produce an odor that we could not perceive but that was widely detected by bees. This seemed a likely assumption, one that could be supported more strongly today than then, now that we know that in order to attract the males many lepidopterous females produce in special scent glands an odorant that we cannot perceive. With the silk moth (*Bombyx mori* L.) the structure has been worked out and the compound prepared synthetically (Butenandt *et al.* 1961). Yet those flowers really are odorless for bees. Training experiments produced completely negative results (v. Frisch 1919:88–102).[18] We have already spoken of the fact that communication relative to the sort of flower visited is astonishingly effective, by way of the fragrance that clings to the dancer's body, even where the production of scent is rather weak, but that it breaks down with those flowers in which we, too, discern no smell.

[18]After an hour's training to flowers of *Ribes rubrum* L., Ribbands (1955) in a short series of experiments with six bees obtained a positive result. Unfortunately, he does not say whether he had determined that his *Ribes* flowers were odorless for man also. It is possible that there are differences in this respect, according to site and race of the plants.

If the blossoms mentioned are visited so assiduously by bees despite the lack of readily perceived sensory cues,[19] this is surely to be ascribed in good part to the circumstance that these flowers occur in great numbers in dense stands. The wild grape covers appreciable areas with its luxuriant growth, huckleberries spread over extensive patches, and with currants too numerous blossoms occur together in a restricted space. Discovery of a single flower rich in nectar may at one stroke tap the abundant honey supply of thousands for the finder's colony. The pioneers send other bees to the spot and by using their scent glands make it easier for the newcomers to locate the unimpressive source of food.

A person who wishes to deny the existence of reciprocal biological adaptations of bees and flowers may say: the process works without floral fragrance also. These three examples demonstrate that. But one must not draw from them the conclusion that all other plants too could without disadvantage dispense with their floral scent. No one can survey the multiplicity of factors whose totality determines the capacity of a species for self-preservation. What nature has treated like a stepmother in one respect, she does better for in some other way, so that it may win the bitter battle for existence.

9. THE SENSE OF TASTE

A liking for sweetness is extraordinarily widespread in the animal kingdom. The flowers take advantage of this by offering nectar as bait to their pollinators, whether birds, mammals, or insects. It is secreted by nectaries in the base of the flowers. In order to reach them the visitors to the flowers have to take a course by which pollination is brought about.

The secretion of sugar by nectaries also occurs elsewhere than in flowers, for instance in many leaves, generally when, after growth is concluded, the stream of nourishment accumulates because it no longer is used up. Such processes in the regulation of sap pressure probably were the starting point for the phylogenetic evolution of the nectar glands in the flowers of entomophilous plants. These are true glands with their own active metabolism (Frey-Wyssling, Zimmermann, and Maurizio 1954).

As principal components the nectar contains the disaccharide sucrose and its two structural constituents glucose and fructose, in varying proportions (Beutler 1930; Beutler and Schöntag 1940; Maurizio 1960). Recently with the method of paper chromatography it has been possible to demonstrate traces of yet other sugars, among them the disaccharides maltose and melibiose as well as the trisaccharides raffinose and melezitose. The small quantities of these other sugars result mainly secondarily from enzyme action on sucrose in the secreted nectar. The enzymes are also produced in the nectaries (Wykes 1952, 1953; Maurizio 1960; see these papers for additional references).

---

[19] But in this connection the hound's-tongue (Fig. 430) should be recalled. With the wild grape and the currant the stands of flowers are highly dissected, and on this account no doubt make some degree of visual impression on the bees, because of their form.

In many areas the sweet excretions of plant lice constitute an important part of the bees' food. In this honeydew the trisaccharides melezitose and trehalose may also occur copiously.

Most flowers produce a nectar of high sugar concentration. According to Beutler (1930), in 18 flowers attractive to bees that she examined, it lay between 8 and 70 percent. From additional studies, the data of which I have taken from the report by Beutler and Schöntag (1940) and from the summaries[20] in v. Frisch (1934, Table 28, pp. 96, 97) and in Maurizio (1960), with a total of 65 species of plants average values for the concentration of floral nectar were found as a rule to be from 20 to 70 percent; exceptionally they were greater or less. The quantity of sugar secreted per flower in 24 hr (the "sugar value") is mostly about 1 mg, but may amount to more than 4 mg, as with some lime trees (*Tilia* spp.), the broad bean (*Vicia faba* L.) and the gooseberry (*Ribes grossularia* L.). In individual instances the quantity of sugar and its concentration depend strongly on various external conditions.

Bees obtain information about the presence and concentration of sugar in the food sap through their sense of taste.

Whereas their sense of smell resembles ours in its main features, the capacities of their sense of taste are surprisingly different in both quantitative and qualitative respects. They differ quantitatively in that the taste organs at the mouth parts respond to the vital sugar only at quite high concentrations, and qualitatively in that most of the substances that we find sweet have no taste to the bees. Both matters gain in interest when we look more closely at the relations (v. Frisch 1934).

If bees are supplied with sucrose in a series of stepwise decreasing concentrations, a limiting value is reached below which the solution is no longer accepted (acceptance threshold). Depending on circumstances, this value varies between about 1 M/lit (34.3 percent) and $\frac{1}{8}$ M/lit (4.25 percent).[21] Apart from significant individual differences, the prevailing state of the outside harvest is influential. In times of famine the bees make fewer demands with respect to the concentration of the sugar solutions (see p. 243). This factor can be excluded by letting the experimental bees starve for a few hours before the taste test. Then the acceptance threshold falls to a lower and very constant value of $\frac{1}{8}$–$\frac{1}{16}$ M/lit (about 4–2 percent). It does not change further even when the bees are close to starvation. Since it is very unlikely that in this condition they would refuse a solution that still had a detectable sweet taste, this solution, which is barely accepted, probably represents the dilution at which a sweet taste is still just perceived (specific threshold).

[20] Both of which contain additional references.
[21] Kunze (1928) indicates that in experiments in the open his bees would still accept a 2-percent sucrose solution, though with hesitation. The difference probably depends on the difference in his technique. I do not speak of acceptance of a solution until bees, observed individually, drink it uninterruptedly for more than 20 sec (1934:9). Kunze measured in mass experiments the length of time required to empty the feeding dish, which was surrounded by numerous bees. It would necessarily be emptied in time even if the solution were merely sipped at. Besides, there are some indications that water foragers were included among his bees (see v. Frisch 1934:36, 37).

For man this value is about $1/80$ M/lit (0.4 percent). Thus for the sweet taste of sucrose we are some tenfold more sensitive than the bee. The same is approximately true for other kinds of sugar that occur in quantities worth mentioning in nectar and honeydew—glucose, fructose, and melezitose.

Scout bees and foragers have to find and fetch home sugar solutions; thus fondness for sweets is a part of their calling. That nevertheless they are less sensitive than we to "sweetness" is comprehensible biologically. In nectar and honeydew they have available sugary solutions that mostly are of rather high concentration. If they should happen upon dilute nectar, it would not be advantageous to gather it. Honey as a winter store needs a high concentration (70–80 percent) in order to keep. If the bees were to fetch in dilute nectar with a sugar content of, say, 10 percent, then they would later have to remove from the hive as water vapor 85 percent of what had been laboriously brought in. Only 10 percent of what had been collected could be stored as honey (Beutler 1935). A bar to such squandering of energy is set by the high threshold of the sweet taste. Bumblebees, which do not lay in any winter stocks, and store up honey only for short rainy periods, will accept somewhat more dilute sugar solutions ($1/16$–$1/32$ M/lit according to Sakagami 1950), while ants, flies, and Lepidoptera are much more sensitive than bees to a sweet taste and some of them surpass man in this respect by many times.[22] They do not produce any honey and are able to make immediate use of even the most dilute sugar solutions.

Qualitative differences in the sense of taste become evident as soon as bees are presented with a rather broad choice of compounds that taste sweet to us. With 19 substances, not belonging to the sugars, that are sweet for us, in no instance was a sweet taste demonstrable with bees. Among these were both saccharin and dulcin, which we employ as artificial sweeteners. In higher concentrations they are more or less repellent to bees, as becomes clear when they are offered in mixture with sucrose. When more dilute their effect is no different from that of pure water; thus they then are evidently tasteless. In general it may be said that these substances in concentrations that have a definite bitter aftertaste for us are only repellent to bees, while dilutions that are purely sweet for us have no effect on their sense of taste.

With the kind support of chemists it was possible for me to offer the bees 34 different, partly rare, sugars and substances closely related to sugars. Of these, 30 taste sweet to man, but for bees only 7 were definitely sweet, with 2 the sweet taste for them was at the limits of perception, and with the 25 others such a taste was either far below threshold or altogether undemonstrable. Thus most of the sugars that are sweet for us are tasteless for bees.

In reference to the experimental technique it may be said that when a solution is declined by the foraging bees it may be tasteless, it may be not sweet enough, or it may have a bad taste. The last two possibilities can be tested by adding the substance to a sucrose solution of

---

[22] Further details in v. Frisch 1934:81, Table 25.

such low concentration that it is barely accepted. A slight sweet taste, subthreshold in itself, makes itself apparent by improving the acceptance of a sugar solution that is near threshold. A bitter taste causes a reduced acceptance of the sucrose solution that is near threshold. If neither is the case, thus if the effect of the solution is not different from that of pure water, then we regard the material as being tasteless for bees. For details, reference is made to my paper of 1934:39–45.

The results of those tests are brought together in Table 45. Substances, l-molar solutions of which are above the threshold or at the

TABLE 45. The degree of sweetness of sugars and sugarlike compounds for bees and for man. (From v. Frisch 1934.)

	Sweet for bees			Not sweet for bees		
		Molar degree of sweetness			Molar degree of sweetness	
Group	Substance	Bee	Man	Substance	Bee	Man
Sugar alcohols						
4 OH-groups				Erythritol	0	1/4–1/8
5 OH-groups (cyclic)				Quercitol	0	1/4
6 OH-groups				Mannitol	0	1/4
				Sorbitol	0	1/4
				Dulcitol	0	1/4
(cyclic)	Inositol	1/8–1/16	1/4			
Methyl glucosides	α-Methyl glucoside	1/4	1/4	β-Methyl glucoside	0[a]	1/8[b]
				β-Methyl fructoside	0[c]	1/8–1/16[d]
				α-Methyl galactoside	1/32	1/4–1/8[d]
				α-Methyl mannoside	0[c]	1/4–1/8[b]
Trimethyl glucose				Trimethyl glucose	0[c]	1/8–1/16[e]
Monosaccharides						
Pentoses				*l*-Arabinose	ca. 1/32	ca. 1/4
				d-Arabinose	ca. 1/32	ca. 1/4
				Xylose	0	1/4
Methyl pentoses	Fucose	ca. 1/8	1/4–1/8[e]	Rhamnose	0	1/2–1/4
Hexoses	Glucose	1/2–1/4[f]	ca. 1/4	Galactose	1/16–1/32	ca. 1/4
				Mannose	ca. 1/64	ca. 1/4[e]
	Fructose	1/2–1/4[f]	ca. 1/2	Sorbose	0	1/2–1/4
Disaccharides	Sucrose	1	1	Lactose	1/32–1/64	ca. 1/4
	Maltose	1–1/2	ca. 1/2	Melibiose	1/24	1/2–1/4
	Trehalose	1/2–1/4	1/2–1/4[g]	Cellobiose	0[a]	1/2–1/4
				Gentiobiose	0[a]	0[a]
Trisaccharides	Melezitose	ca. 1/2	ca. 1/2	Raffinose	0	ca. 1/2
Tetrasaccharides				Tetraglycosan	0[c]	0
				Tetralaevan	0[c]	0
Polysaccharides				Glycogen	0	0

[a] Repellent. [b] Repellent as well. [c] Slightly repellent. [d] With a slight bitter aftertaste. [e] With a bitter aftertaste. [f] Closer to 1/2; according to Wykes (1952a), fructose is somewnat less acceptable than glucose. [g] Closer to 1/2.

limit of the bees' perception are shown separately as "sweet substances" in the left part of the table. In the column headed "molar degree of sweetness," the numbers denote the molar concentration of a sucrose solution that tastes equally sweet with a 1-molar solution of the substance in question. Thus the values indicate the relative sweetness of the material with reference to sucrose, or, in other words: "degree of sweetness = $\frac{1}{4}$" means that the substance is four times less sweet than sucrose when the same number of molecules are contained in an equal volume of solution. A degree of sweetness of 0 means that a sweet taste could not be demonstrated.

As an important result of numerous experiments it is to be emphasized that of seven sugars that clearly are sweet to bees, five occur in appreciable quantities in nectar or honeydew: sucrose, glucose, and fructose in nectar, and in honeydew, melezitose and trehalose as well. But there are also sugars that are sweet for bees but—so far as is known—play no part in their natural food: maltose and α-methyl glucoside. Of the sugars and sugarlike substances that are scarcely or not at all sweet for bees, not one is known to occur in quantities worth mentioning as a constituent of nectar or honeydew. Some of them are formed in trace amounts secondary to the action of enzymes. Thus the bees' sense of taste is adapted to the food provided by nature, in that all sugars that are of significance in their natural food taste definitely sweet to them and that the modality "sweet" is in general limited for them to these few compounds.

The relation goes deeper yet. Bees are able to use in their metabolism all sugars that taste sweet for them. Imprisoned bees can be kept alive about as long when fed exclusively with fructose, glucose, or α-methyl glucoside, with maltose, trehalose, or melezitose, as with sucrose. In general, sugars that are not sweet for bees have little or no nutritive value for them. There are, to be sure, a few exceptions to this rule. Bees can even be kept alive longer with the sugar alcohol sorbitol than with sucrose (Vogel 1931).[23]

A few sugars that have little or no sweetness for bees are toxic to them. After they have been fed mannose their movements become sluggish and, with increasing signs of paralysis, they presently succumb (v. Frisch 1934:55–57). This seems to depend on enzymatic processes. The breakdown of glucose is blocked by mannose because the latter preempts and holds fast to the phosphorylating enzyme that initiates glucose metabolism (Staudenmayer 1939; Sols, Cadenas, and Alvarado 1960). Galactose and, to a slight extent, rhamnose curtail the life of bees. It is suspected that the occasional occurrence of such sugars in nectar or pollen causes the repeatedly observed poisoning of bees that have frequented certain food-bearing plants (Geissler and Steche 1962). Here the sense of taste fails to caution against dangerous food.

Neither the findings in regard to the distribution of the sweet taste nor those concerning the utilization of sugars in metabolism and their toxic effects can be generalized for all insects. Scattered assays have

[23] Regarding the breakdown of sugar by enzymes of the pharyngeal gland and the midgut, see Maurizio (1957, 1957a, 1962) and the additional literature cited there.

revealed many differences that we do not understand (v. Frisch 1934; Haslinger 1935; Schmidt 1938; Hassett, Dethier, and Gans 1950; Dethier 1955; Dethier, Evans, and Rhoades 1956).

The gustatory effect of a sweet sugar may be enhanced by adding a different sugar. In this way, with suitable concentrations two subthreshold sugar solutions may interact to raise the over-all effect above threshold. "The simplest interpretation, though not necessarily binding, is that for bees, as for us, the different sugars have the same taste modality" (v. Frisch 1934:147).[24] Wykes (1952a) found that mixtures of sucrose, glucose, and fructose in a proportion of 1:1:1 are accepted better by bees than the individual sugars or than other mixtures of them at an equal total concentration. Whether that depends on the bees' ability to distinguish the tastes of these sugars and preferring the taste of a certain mixture, or whether it is to be attributed to other conditions unknown to us, has not been clarified. The likely assumption, that the bees prefer the mixture that is found in the natural nectar of flowers, cannot be correct, for in them the relative proportions of the three sugars vary greatly (see for instance Maurizio 1960:71, Table 1). Jamieson and Austin (1958) did not find any preference for certain mixtures of sugar. But, from their Table V, p. 1061, one sees that they did not test the mixture of sucrose : glucose : fructose = 1:1:1 that was preferred according to Wykes. Additional experiments are needed to clarify this question.

Jamieson and Austin also investigated what differences in concentration bees just noticed in sucrose solutions. In the range of 30–55 percent they found that sugar solutions were preferred if they contained 5 percent more sugar, whereas differences of 2.5 percent were not distinguished. But their method was not suited for precise determinations. One who wants such determinations will have to observe the behavior of individually marked bees toward different solutions. In this way, for instance, I found concentration differences of about 1 percent still effective, though, to be sure, with near-threshold sucrose solutions (1934:41f). At that time I did not pursue the matter further. A more thorough test of it would be interesting.

From experiments with the taste of other substances it appears that bees are much less sensitive to bitter substances than we.[25] For acids and the salty taste the threshold values agree approximately with ours. By methods that need not be detailed here (see v. Frisch 1934) it can be shown that sugar, hydrocholoric acid, salt, and quinine taste qualitatively different for bees. Whether they possess more than these four taste modalities cannot be decided certainly.

[24] Wykes (1952a:516) says I ascribed to the bees only a single "sweet" taste modality. As the quotation shows, I expressed myself more cautiously.

[25] Hence it is possible to contaminate sugar with a bitter substance that does not interfere with its being taken up by bees but that renders it unacceptable to man. Since in some countries sugar is given out at a reduced price for feeding bees, the authorities would like to impede its misuse by making it bitter. Acetyl sucrose proved to be suitable for this purpose. Compounding the sugar molecule with acetic acid renders the sugar tasteless for bees and bitter for people. The product came into commerce under the name "Oktosan." It is quite harmless and in honey breaks down again into its constituents, sugar and unnoticeable traces of acetic acid (see Wahl 1937).

It is a good thing that the world is so rich in different kinds of animals. For where one species treacherously conceals a secret from the eager investigator, at times another shows itself as so much the more readily accessible—if he has the skill required to find the right experimental subject. Thus Minnich (1926, 1931) discovered that blowflies (*Calliphora vomitoria* L., *Phormia regina* Meigen) have at the edge of the proboscis on the labella relatively large hairs that can be touched individually with a fine paintbrush. If one of them is dampened with sugar water, the fly immediately protrudes the proboscis. These hairs are taste organs. Since then, they have been studied more closely and have been found likewise in some other species of flies. They are also an unusually favorable object for the electrophysiologist. A hair is innervated by three receptor cells (Fig. 453), one of which ends at the hair base. This cell responds to tactile stimuli. The other two send their processes all the way to the tip of the hair and are receptive to chemical stimuli (Dethier 1955; Wolbarsht and Dethier 1958; Hodgson and Roeder 1956; Hodgson 1957). Recently different types of these sensory hairs have been reported; the number of receptor cells may vary from three up to six (Peters 1961; Larsen 1962; Brunhild Stürckow 1962).

Such a union of two chemoreceptors with a mechanoreceptor in one sensillum seems to be a peculiarity of flies. Meanwhile, taste hairs have been found and studied individually in butterflies (*Vanessa*, Morita, Doira, Takeda, and Kuwabara 1957) and in the potato beetle (*Leptinotarsa decemlineata* Say, Stürckow and Quadbeck 1958; Stürckow 1959). In the potato beetle every hair is supplied by five sensory neurons that respond only to chemical stimuli.

As far as the discrimination of taste substances is concerned, bees are no doubt more capable than those flies, but they are less cooperative and do not present the investigator with any protruding taste bristles. With them, as with many other insects, on taste-sensitive parts of the body there are to be found thin-walled cones or pegs or hairs or bristles that mostly are so densely intermingled with other sense organs that it has not yet been possible to investigate them in isolation. Apparently, thornlike, extremely thin-walled cones (*sensilla basiconica*), about 15 μ tall, on the bee's mandibles serve as taste organs (Fig. 454). Each sensillum is supplied by two or three receptor cells, whose processes extend to the tip. A field of some 50 such sensilla is located almost immediately below the orifice of the mandibular glands and possibly checks the secretion discharged. Other similarly constructed sensilla are on the cutting surface of the mandibles. During drinking they are spread somewhat apart, in such a position that they could be wetted by the solution (Nedel 1960). But no sure conclusions can be drawn from morphological considerations alone. At present we can only give a few crude anatomical facts as to the location of the sense of taste in the bee.

Taste organs certainly are located about the mouth, probably on the proboscis as well as on the mandibles and inside the mouth. For recognizing the taste of a solution, submerging the proboscis in it is sufficient. Understandably, taste organs in the mouth and in its immediate vicinity are distributed very widely—not alone in insects.

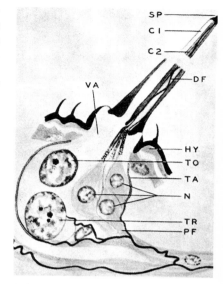

FIGURE 453. Typical sensory hair of the blowfly *Phormia regina* Meigen, with three sensory cells: a mechanoreceptive cell terminates at the hair base, the other two extend to the tip of the hair (shortened in the illustration) and subserve the sense of taste: *N*, nerve cells; *Pf*, their proximal processes; *DF*, distal processes that penetrate the hair. After Dethier 1955.

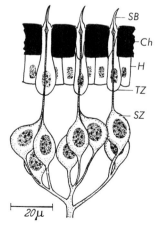

FIGURE 454. Sense organs (*sensilla basiconica*), probably organs of taste, from the mandible of the honeybee, at the orifice of the mandibular gland. Schematic. After Nedel 1960.

But many insects are able to taste with other parts of the body also. Minnich (1921, 1922) discovered this with Lepidoptera, which perceive the sweet taste of a sugar solution with the tips of their legs. In addition to sucrose, they are able to taste a number of other kinds of sugar with their legs (Ilse Weis 1930). Minnich (1929) succeeded in demonstrating the same capacity in the blowfly, and Haslinger (1935) confirmed it for a large number of other sugars. With flies and Lepidoptera it is quite sensible that in running about over the ground they should respond with a directed proboscis reaction as soon as they step into something edible. With the honeybee, too, Minnich (1932) was able to demonstrate taste organs on the forelegs. This finding was confirmed in several quarters (for instance by H. and M. Frings 1949). The hind legs respond more rarely to taste stimuli. That taste organs are concerned in all these instances is not to be doubted. The experimental specimens were fastened by the back to a little block of wax or paraffin and were under precise observation. Thirsty specimens also extend the proboscis when the legs touch water. Hence one has to find out, before each experiment, whether water is effective. Even with bees (or Lepidoptera or flies) that are not thirsty, touching a foreleg with sucrose solution causes an immediate extension of the proboscis, whereas touching it with lactose of the same concentration is without effect. Since these two sugar solutions have equal viscosity and osmotic pressure, the difference in behavior can be ascribed only to their chemical dissimilarity.

In his report of 1932 Minnich describes how with the bee a proboscis reaction can be released also by touching the antennae with sugar solution. Contact with the tip of the antenna suffices. That, too, is certainly a good place for bees to have instruments of taste, both when visiting flowers and during their activities in the hive. Just like the organs on the mouth and on the legs, these distinguish with complete certainty between sucrose and lactose.

Thus in the distribution of their equipment of taste organs the bees prove to be superior by far to man.

The astonishing discovery that the bees' antennae, already gifted in so many ways, serve further as the seat of organs of taste has since then been confirmed several times and supplemented in important respects. Whereas Minnich had regularly stimulated the antenna with a single concentration of sucrose (0.64M), Marshall (1935) looked for the threshold value and found that for sucrose on the antenna it averaged $\frac{1}{12}$M, being frequently even less, down to $\frac{1}{24}$M. For the tarsi of the foreleg it was about 1M, or often about 0.5M.

Kunze (1933) tested antennae with different sugars and found that the proboscis reaction could be released most frequently by contact with dextrose, sucrose was not quite so effective, and fructose clearly gave fewer positive results. Mannitol in 10-percent solution caused a proboscis reaction. Since this sugar alcohol is ineffective on the bee's oral taste organs, a retesting of this brief statement of Kunze's would be desirable. In agreement with Marshall's later report, according to Kunze also the antennae are more sensitive than the mouthparts, in fact to a considerably higher degree: for sucrose he indicates the threshold value of 2.5–0.3 percent, thus as a lower limit 0.01 molar. For dextrose the antennae are said to be similarly sensitive, and for fructose less so.

Even these values were surpassed in our laboratory in an investigation by Kantner; unfortunately, this work was not completed and hence was not published. In a table (1934:81) I have made use only of his principal result. He obtained lower threshold values by training his bees by touching the antennae with sugar solution; they were fastened with paraffin.[26] After each antennal contact with sucrose solution they were fed with sugar water. In this way they learned to react to dilutions to which they would not have responded spontaneously. The threshold value was about 0.0001 molar (0.0034 percent). At the time that seemed improbable to me. I kept urging new control experiments on Kantner and to this day do not know whether my skepticism was the actual reason for his giving up the work. His finding no longer seems incredible when one recalls that Minnich (1922a) with the starved admiral butterfly *Pyrameis atalanta* L. found the threshold value for the tarsal taste organs at 1/12,000 molar and with starving flies (*Calliphora*) at a lower limit of 1/25,000 molar (Minnich 1929). These values determined for the tips of the legs are of the same order of magnitude as Kantner reports for the bee's antennae. According to this, these organs are about 300 times more sensitive to the sweet taste of sucrose than the human tongue.

We attributed a biological significance to the greatly inferior sensitivity of the oral taste of bees: they are not tempted to ingest and fetch home a nectar that, because of its excessively low concentration, would dilute the honey or that would cause unnecessary labor because of the requirement of condensing it later. This consideration does not apply to the antennal organs, for even less concentrated traces may be useful in discovering a source of food. The final decision as to profitability does after all reside in the oral taste sense.

It would be of interest to know whether these highly sensitive antennal taste organs also differ morphologically from the oral receptors. Yet in order to learn that we must first recognize them. And so we end with an unanswered question—of which there are many in this book.

That does not matter. It is the unanswered questions that give the young an enthusiasm for science. It would be a pity if one day they had to say: now we know everything and there is nothing left for us to do. Have no fear! That day will never come. For the mind of man is limited, but the miracles of living nature are without end.

Summary: Orientation Flights; Nearby Orientation by the Senses of Sight, Odor, and Taste

1. During their orientation flights bees make themselves familiar with the appearance and location of the home hive. That is quickly accomplished. Even after their initial outward flight, they find the way home from a distance of several hundred meters. After repeated orien-

[26] With training experiments, Anneliese Schmidt (1938) was able to demonstrate taste organs on the antennae of ants also.

tation flights, this distance may be extended to a few thousand meters. Immediately after she has first taken wing the bee heads toward the hive, but a few seconds later she flies off like an arrow. About 5 min afterward she returns and hovers in front of the hive ("playing about"), oscillating to and fro for a few minutes before she settles; in the course of doing this she gains a firmer impression of the hive's characteristics. It is to be assumed that with the help of the celestial compass and of the more distant landmarks she grasps the hive's location in the landscape during her very first flight. For nearby orientation, the coloration of the hive and its position relative to its immediate neighborhood are important (Fig. 400).

In experiments with digger wasps, whose orientation is based on the same fundamentals, it was possible to determine the effective landmarks especially clearly. With these wasps, as with bees, there is no reason for assuming in the capacity for returning home any contribution from other unknown factors besides the learning of visual signposts. That the familiar odor of the home hive plays a part also is beyond doubting.

2. The bees' color sense is well developed. To a great extent it corresponds to the trichromatic color vision of man, but differs in a displacement of the sensitivity toward the short-wavelength end of the spectrum. Of all colors the bees are able to perceive, ultraviolet has the greatest stimulating effectiveness. In terms of equal intensities of radiation, their sensitivity decreases as follows: 5.6 (ultraviolet 360 mμ) : 1.5 (blue-violet 440 mμ) : 1.0 (green 530 mμ) : 0.8 (yellow 588 mμ) : 0.5 (blue-green 490 mμ) : 0.3 (orange 616 mμ). They are insensitive to red of longer wavelengths than 650 mμ. Orange, yellow, and green are similar for them, and so are blue and violet (Fig. 407). By mixing the end colors of their visible spectrum there is produced for them—in analogy with human color vision—a new color quality ("bee purple") that is not contained in the spectrum, and that closes the latter to form a color circle (Fig. 408). By mixing ultraviolet with blue-violet, "bee violet" is produced. In the latter, as in the ultraviolet and in the region of purple colors, the bee's eye is able clearly to distinguish several different hues. For bees—as for us—the mixture of all perceptible colors that occur in composite sunlight yields neutral "white," which looks different from any spectral color and is less easily learned than colors in a training experiment. "White" is produced also when certain pairs of colors (complementary colors) are mixed in the proper proportions. Complementary colors for bees are: blue-green and ultraviolet, yellow and bee violet, blue and bee purple.

With electrophysiological methods it is possible to determine the sensitivity curves for single receptor cells in the bee's eye (Fig. 409). The results are in excellent agreement with the outcome of behavioral experiments, and constitute a direct demonstration of the trichromatic nature of the sense of vision.

3. If the capacity for resolution of the bee's eye is gauged by the angles of divergence of the ommatidia, their form vision must be adjudged about 100 times inferior to our own. But in training experiments it becomes evident that in the perception of shapes the bees give heed

to other characteristics than we. Not the course of the contours but the degree of dissection of a figure is its most important feature for bees. Greatly divided patterns have a fundamental attraction for them and are more easily remembered than closed geometric figures, such for instance as a circle, triangle, or square, which are not distinguished from one another (Fig. 416). In contrast with the bees' small capacity for spatial resolution, they have a high resolving power in time. Bees are able to register ten times as many stimulus changes per unit time as we are. This favors the perception of dissected patterns during flight (Fig. 417). But it is not the case that bees are guided under all circumstances by the most strongly divided figures. They can be trained successfully to weakly dissected patterns versus more strongly dissected ones, and when they are in the mood to return home they show a basic preference for patterns that in their simple contours bear a resemblance to their goal, the dark flight entrance of the home hive.

4. The colors of flowers show a definite relation to the color vision of their visitors. Flowers visited by birds are frequently red, whereas with flowers attractive to bees pure red is a great rarity. Yellow flowers that look alike to us may be distinguished readily by bees, owing to differences in the admixture of ultraviolet light; the same is true of some blue flowers (Figs. 418 and 420). Flowers that are white for us are colored blue-green for bees because they reflect no ultraviolet (Fig. 419). One might expect that the red poppy would be colorless black for bees, but because of strong ultraviolet reflection it appears to them as an ultraviolet flower (Fig. 421). Foliage gives a rather uniform reflection in all three of the bee's primary color ranges and hence is nearly colorless gray for them, so that the floral colors are so much the more conspicuous against such a background (Fig. 422).

5. Nectar markers in flowers as guideposts to the source of food are of much greater frequency for bees than for us, because these cues often depend on inequalities in the ultraviolet reflection and then are invisible to us (Figs. 422–426). Experiments with honeybees and bumblebees have permitted demonstration of the significance of the nectar markers. The reaction of bees to a nectar marker is innate.

6. The effect of floral colors on scout bees is primarily one of attraction. For the foragers the colors are a guidepost that facilitates their finding the species of flower sought and discriminating it from other flowers. Designs in the pattern may also contribute to the visibility, and at times the form of the plant as a whole is put to use in this way (Fig. 430).

7. The bees' sense of smell enables them to discriminate among such a multiplicity of fragrances as is in accord with our own experience. The bees are trained readily to flowery scents. They have a remarkable capacity for recognizing the training odor even among a mixture of scents. They are indeed able to perceive nonfloral odors, but they have difficulty in learning to use them as cues.

8. The bees' sense of smell is located on the eight distal segments of the antennal flagellum (Fig. 434). Here are found seven clearly distinguishable sense organs (sensilla, see Fig. 435). The plate organs

function as olfactory receptors. According to electrophysiological investigations, a single sensory neuron of a plate organ responds to several different odorants, but as yet no two receptor cells with a precisely identical reaction spectrum have been found. This multiplicity of response renders comprehensible their almost unlimited capacity for qualitative discrimination. Hitherto it has not been possible to demonstrate an olfactory function with any of the other sensilla. The coeloconic sense organs serve for perception of relative humidity, and also of carbon dioxide and temperature stimuli.

9. Through the location of the olfactory organs on the protruding motile antennae the bees are capable of localizing by smelling (topochemical olfactory sense). It was possible to show experimentally that they can refer odors to the correct side. With the antennae crossed (Fig. 439), bees turn toward the wrong side in a choice experiment. It could be determined what difference in concentration is required for making a proper lateral choice. The result justifies the assumption of a tropotactic orientation with reference to local differences in odor, such as occur in respect to nearby orientation with flowers and in the hive. In addition to tropotactic orientation by means of simultaneous comparison of the olfactory impressions from the two sides, there is also a klinotactic orientation by sampling in rapid temporal succession both sides with the same sense organs. After losing one antenna, a bee can still make the proper selection in a choice experiment, by moving her remaining antenna rapidly to left and right.

10. The olfactory acuity of bees is approximately the same as that of man. In general it is somewhat better than with us for flowery smells, somewhat worse for odors that lack biological significance (Tables 41 and 42). In perceiving the smell of wax and that of their scent organ bees are severalfold superior.

If color and smell are put into competition, bees orient to the color when they are at a considerable distance, while from near at hand the fragrance carries more "power of conviction" for them.

11. The physiological basis of olfaction in bees apparently resembles ours to a great extent. All odorants to which they can be trained also have a smell for us. Flowers that are without fragrance for us are odorless for bees also. Odors that are similar for us are confused by bees too. In general there is no important difference in olfactory acuity between bees and man.

12. The evolution of floral fragrance is clearly allied with the visitors to the flowers. As an attractant the fragrance is adapted to the kind of visitor (for example, the carrionlike odor of flowers attractive to flies). The flowery scent of blossoms attractive to bees allures the scout bees; for the foragers it is the most certain feature by which the various kinds of flowers may be distinguished and, clinging to the body, is a means of communication in the dance. Differentiations in the scent within a flower serve as olfactory guideposts. Odor markers in flowers are of even more frequent occurrence than visual nectar markers. They designate olfactorily the way of access to the source of food. In flowers attractive to birds they are lacking. Birds have a weakly developed olfactory sense

and they also do not possess the gift of localizing odors, as bees do with their antennae.

13. The sense of taste of bees differs from ours in several respects. Its foremost importance for bees is in connection with the sweet taste by which they are oriented with respect to the presence and concentration of nectar. Otherwise than might have been expected, the bees are not very sensitive to the sweet taste of sugar. The threshold value for sucrose is about $\frac{1}{8}$–$\frac{1}{16}$ molar (4–2 percent). Thus they are ten times less sensitive than we for this sugar. That is sensible, because the nectar of flowers mostly has a high concentration (as a rule between 20 and 70 percent) and because ingestion of sap with a low sugar content would be unfavorable for preparing honey that would keep well as a store for the winter.

14. Of the many compounds that taste sweet to us most are tasteless to bees. That is true not only of sweeteners that are entirely different in chemical structure, such as saccharin, but also of sugars. Of 34 different sugars and closely related substances that were tested, only 7 proved to be definitely sweet for the bees. Five of these are known to be important constituents of nectar or honeydew (sucrose, glucose, fructose, melezitose, and trehalose). Of the sugars that are scarcely or not at all sweet for bees, not one has been found in quantities worth remarking in the bees' natural sources of food. Bees can use in their metabolism all sugars that taste sweet for them. The majority of sugars that are not sweet for bees have little or no nutritive value for them. But these findings cannot be generalized to all insects.

15. As a rule, bees are only slightly sensitive to bitter substances. Bitter, sour, salty, and sweet constitute different taste modalities for them. Whether they are able to distinguish yet other modalities is not known.

16. Certain definite sensory hairs have been found to be organs of taste in flies, beetles, and Lepidoptera. In the bee, thin-walled thornlike *sensilla basiconica* on the mandibles probably are to be regarded as taste cones. Neither these nor other sensilla have been demonstrated with certainty to be taste organs in bees.

17. The sites of the organs of taste are the mouth and its vicinity, the forelegs, and the antennae. Even spontaneously, and to a much greater extent after appropriate training, the antennae respond to lower concentrations of sugar than the oral receptors. Following training, a threshold value of 0.0001 molar (0.0034 percent) for sucrose was determined. The lower threshold value may be useful in the detection of sources of sugar. Whether the sugar is ingested is determined by the oral receptors, whose high threshold bars the bees from bringing home solutions of insufficient concentration.

Retrospect

The reader will find summaries at the ends of the individual sections. Reference to them may be made here. They may be found easily from the table of contents and should assist one in preserving a general view in the face of a multitude of details. How much of all that may be most worth noting now?

Bees are insects; with respect to anatomical development their nervous centers are paltry in comparison with the human brain. In spite of this, these creatures can inform their comrades of a goal that is of importance for their colony and can describe its location so exactly that the hivemates find it independently in flight, without being led there, by the most direct route—even at a distance of kilometers. The information is transmitted by rhythmic movements that we have dubbed a dance. The tempo of the dance announces the distance. In particular, the duration of the straight run of the tail-wagging dance seems to be the decisive signal. It is emphasized sharply by the tail-wagging movements and by a buzzing noise—the greater the distance the longer the duration of the tail wagging and the accompanying sound during each run. The direction of the tail-wagging segment designates relative to the position of the sun the line to be followed to the goal. In horizontal dances under the open sky this indication is given directly, as if by pointing a finger. On the vertical honeycomb in the dark hive, what cannot be said directly here is told in a simile: the bee transposes from the visual sense to the gravitational sense and announces the solar angle in a paraphrase, as the angle with the vertical. The specific floral scent that still clings to the dancer's body from her visit to the flowers informs the comrades in the hive of the sort of blossoms they must seek. The more valuable the discovery, the more lively and persistent and hence too the more effective are the dances of encouragement.

When there is simultaneous flowering of different species of plants, a reasonable distribution of the labor forces is attained most simply in this way. The better the nectar that one kind of blossom has to offer, the livelier are the dances and the more are new working groups guided to this very kind of flower—until the time when, with exhaustion of the excess, the supply grows sparse and the intensity of invitation diminishes, like the influx of additional workers. In this process the quality of the source of nourishment is by no means judged merely according to the sweetness and quantity of the food. An astounding multiplicity of factors is concerned in the profitability of a source of nourishment, is given heed, and regulates the intensity with which it is exploited.

When at the time of swarming suitable nesting places are reported by scout bees, and frequently with a rather wide choice, the swarm, unlike the foragers for nectar, cannot distribute itself over various objectives. The decision has to fall on a single nesting site, and the best one is chosen. Each scout advertises with graded intensity, in accordance with an innate scale of evaluation, depending on the advantages and disadvantages of the dwelling. In addition to tempestuous dancers one sees some that support their finding less energetically; faced with the former, the latter scouts gradually diminish the urgency of their invitation or let themselves be converted in favor of the better home—until agreement is reached and the swarm flies off to its new quarters.

Equivalent accomplishments are not known among other animals, not even among other social insects. How they may have evolved in the course of phylogenesis can only be guessed at. However, interesting indications are given by comparative studies with closer or more distant relatives of the honeybee.

To return to a distant goal, still less to describe one in such a way that others may find it, is unthinkable in the absence of prominent gifts for orientation. Such gifts exist here. Landmarks play an important part in them, but, in addition, orientation to the sun and to the blue sky are of outstanding importance. The sun serves the bees as a compass. They know its diurnal course and, thanks to their "internal clock," are capable, at any hour of the day, of setting out at that angle with the sun that will permit them to follow a given compass direction. They can still perceive the sun through a cloud cover that makes it invisible to us. This capacity they owe to their great sensitivity for ultraviolet. But they also orient with certainty to the blue sky, in accordance with the plane of polarization of the light, which depends on the position of the sun. The two methods of orienting, to the sun and to the polarized light of the sky, are closely related and can substitute for one another.

During their first outward flights young bees have to acquire individual experience with the sun's course in order that their compass orientation shall function properly. But in this process, becoming acquainted with the sun's course over a few hours suffices for them to find the proper direction later, at another time of day. The bees fill out the fragment to make an entire diurnal course; in fact, they even complete the semicircle between sunset and sunrise to form a full circle they have never seen. Some dancers will reiterate their invitation for hours, under some circumstances even at night. Without any renewed orientation to the sky the dancer's direction in the tail-wagging run compensates for the advancing position of the sun and demonstrates before our eyes her time sense—as it were an hour hand of her "internal clock."

In accord with well-founded theory, perception of the plane of vibration of polarized light is mediated by the radial arrangement of the receptor cells, in which their rhabdomeres act like a radial Nicol (polarizing) prism. The basis for this analysis of the plane of vibration is considered to be the radially arranged fine structure of the rhabdomeres, which the electron microscope has revealed to us.

Orientation according to the sun's position and according to polarized light has proved to be a capability that is widely distributed among animals. But exploiting it for purposes of communication and building it into their "language" as the principal supporting element remains the unique domain of bees.

That strikes me as the quintessence of the knowledge to which a long route has led. I cherish the secret hope that people will read not this section alone. The charm of a hike lies in the totality of the impressions gained. When once the goal has been reached, one would not wish to overlook one's memories of flowers and of the shapes of trees that no longer are in the field of vision, nor of the many views that opened temporarily before one. In this the trails resemble each other, whether they lead across mountains or into scientifically new terrain.

References

ALLEN, M. DELIA, The occurrence and possible significance of the "shaking" of honeybee queens by the workers, *Animal behaviour 7,* 66–69 (1959).

——— The "shaking" of worker honeybee by other workers, *Animal behaviour 7,* 233–240 (1959a).

ALTEVOGT, R., Wirksamkeit polarisierten Lichtes bei *Uca Tangeri, Naturwissenschaften 50,* 697–698 (1963).

——— and H. V. HAGEN, Über die Orientierung von *Uca Tangeri* Eydoux im Freiland, *Z. Morph. Ökol. Tiere 53,* 636–656 (1964).

ARMBRUSTER, L., Über Bienentöne, Bienensprache und Bienengehör, *Archiv f. Bienenkunde 4,* 221–259 (1922).

ASCHOFF, J., Zeitgeber der tierischen Tagesperiodik, *Naturwissenschaften 41,* 49–56 (1954).

——— Tagesperiodik bei Mäusestämmen unter konstanten Umgebungsbedingungen, *Pflügers Arch. ges. Physiol. 262,* 51–59 (1955).

AUFSESS, ALEXANDRA V., Geruchliche Nahorientierung der Bienen bei entomophilen und ornithophilen Blüten, *Z. vergl. Physiol. 43,* 469–498 (1960).

AUTRUM, H., Neue Versuche zum optischen Auflösungsvermögen fliegender Insekten, *Experientia (Basel) 5,* 271–277 (1949).

——— Über zeitliches Auflösungsvermögen und Primärvorgänge im Insektenauge, *Naturwissenschaften 39,* 290–297 (1952).

——— Elektrobiologie des Auges, *Verhandl. Gesellsch. Deutscher Naturf. u. Ärzte zu Essen 1952,* 104–108 (Berlin, Göttingen, Heidelberg: Springer, 1953).

——— and N. METSCHL, Beziehungen zwischen Lichtreiz und Erregung im Ocellusnerven von *Calliphora erythrocephala. Z. Naturforsch. 16b,* 384–388 (1961).

——— ——— Die Arbeitweise der Ocellen der Insekten, *Z. vergl. Physiol. 47,* 256–273 (1963).

——— and WILFRIEDE SCHNEIDER, Vergleichende Untersuchungen über den Erschütterungssinn der Insekten, *Z. vergl. Physiol. 31,* 77–88 (1948).

——— and MARIELUISE STOECKER, Die Verschmelzungsfrequenzen des Bienenauges, *Z. Naturforsch. 5b,* 38–43 (1950).

——— ——— Über optische Verschmelzungsfrequenzen und stroboskopisches Sehen bei Insekten, *Biol. Zbl. 71,* 129–152 (1952).

——— and HILDEGARD STUMPF. Das Bienenauge als Analysator für polarisiertes Licht, *Z. Naturforsch. 5b,* 116–122 (1950).

——— and INGRID WIEDEMANN, Versuche über den Strahlengang im Insektenauge (Appositionsauge), *Z. Naturforsch. 17b,* 480–482 (1962).

——— and VERA V. ZWEHL, Zur spektralen Empfindlichkeit einzelner Sehzellen der Drohne (*Apis mellifica* ♂), *Z. vergl. Physiol. 46,* 8–12 (1962).

——— ——— Die Sehzellen der Insekten als Analysatoren für polarisiertes Licht, *Z. vergl. Physiol. 46,* 1–7 (1962a).

────── ────── Ein Grünrezeptor im Drohnenauge (*Apis mellifica* ♂), *Naturwissenschaften 50,* 698 (1963).

────── ────── Die spektrale Empfindlichkeit einzelner Sehzellen des Bienenauges, *Z. vergl. Physiol. 48,* 357–384 (1964).

BACCETTI, B., and C. BEDINI, Research on the structure and physiology of the eyes of a Lycosid spider. I. Microscopic and ultramicroscopic structure, *Arch. ital. Biol. 102,* 97–122 (1964).

BAERENDS, G. P., Fortpflanzungsverhalten und Orientierung der Grabwespe *Ammophila campestris* Jur., *Tijdschrift voor Entomol. Deel 84,* 68–275 (1941).

BAINBRIDGE, R., and T. H. WATERMAN, Polarized light and the orientation of two marine crustacea, *J. exp. Biol. 34,* 342–364 (1957).

────── ────── Turbidity and the polarized light orientation of the crustacean *Mysidium, J. exp. Biol. 35,* 487–493 (1958).

BALTZER, F., Einige Beobachtungen über Sicheltänze bei Bienenvölkern verschiedener Herkunft, *Arch. Jul. Klaus-Stiftung 27,* 197–206 (1952).

BAUMGÄRTNER, H., Der Formensinn und die Sehschärfe der Bienen, *Z. vergl. Physiol. 7,* 56–143 (1928).

BAYLOR, E. R., The responses of snails to polarized light, *J. exp. Biol. 36,* 369–376 (1959).

────── and D. KENNEDY, Evidence against a polarizing analyzer in the bee eye, *Anat. Rec. 132,* 411 (1958).

────── and F. E. SMITH, The orientation of Cladocera to polarized light, *Amer. Naturalist 87,* 97–101 (1953).

BAYRAMOGLU-ERGENE, SAADET, Untersuchungen über den Einfluß der Ocellen auf die Fluggeschwindigkeit der Wanderheuschrecke *Schistocerca gregaria, Z. vergl. Physiol. 48,* 467–480 (1964).

────── Die Funktion der Ocellen bei *Anacridium aegyptium, Z. vergl. Physiol. 49,* 465–474 (1965).

BECKER, G., Ruheeinstellung nach der Himmelsrichtung, eine Magnetfeldorientierung bei Termiten, *Naturwissenschaften 50,* 455 (1963).

────── Magnetfeld-Orientierung von Dipteren, *Naturwissenschaften 50,* 664 (1963a).

────── and U. SPECK, Untersuchungen über die Magnetfeld-Orientierung von Dipteren, *Z. vergl. Physiol, 49,* 301–340 (1964).

BECKER, LORE, Untersuchungen über das Heimfindevermögen der Bienen, *Z. vergl. Physiol. 41,* 1–25 (1958).

BEER, INGEBORG, Neue Anstrichfarben für Bienenstöcke zur besseren Orientierung am Stand (Lehramtsarbeit Naturw. Fak. Univ. München, 1957).

BELING, INGEBORG, Über das Zeitgedächtnis der Bienen: *Z. vergl. Physiol. 9,* 259–338 (1929).

BENNETT, MIRIAM F., and M. RENNER, The collecting performance of honey bees under laboratory conditions, *Biol. Bull. 125,* 416–430 (1963).

BERGER, PAULETTE, and M. J. SEGAL, La discrimination du plan de polarisation de la lumière par l'oeil de l'abeille, *C. R. Acad. (Paris) 234,* 1308–1310 (1952).

BERLEPSCH, A. v., *Die Biene und die Bienenzucht in honigarmen Gegenden* (Mühlhausen, 1860).

BERTHOLF, L. M., Reactions of the honeybee to light, *J. agric. Res. 42,* 379–419 (1931).

────── The distribution of stimulative efficiency in the ultraviolet spectrum for the honeybee, *J. agric. Res. 43,* 703–713 (1931a).

BETHE, A., Die Heimkehrfähigkeit der Ameisen und Bienen, *Biol. Zbl. 22,* 193–215, 234–238 (1902).

────── Das Finden des Weges—eine kritische Betrachtung, *Studiun generale 3,* 75–87 (1950).

BEUTLER, RUTH, Biologisch-chemische Untersuchungen am Nektar von Immenblumen, *Z. vergl. Physiol. 12,* 72–176 (1930).

——— Neue Untersuchungen über den Zuckergehalt des Blütennektars, *Leipziger Bienenzeitung 50, Jg.,* 271–273 (1935).

——— Zeit und Raum im Leben der Sammelbiene, *Naturwissenschaften 37,* 102–105 (1950).

——— Über die Flugweite der Bienen, *Z. vergl. Physiol. 36,* 266–298 (1954).

——— and ADELE SCHÖNTAG, Über die Nektarabscheidung einiger Nutzpflanzen, *Z. vergl. Physiol. 28,* 254–285 (1940).

——— and O. WAHL, Über das Honigen der Linde in Deutschland, *Z. vergl. Physiol. 23,* 301–331 (1936).

BIRUKOW, G., Photo-Geomenotaxis bei *Geotrupes silvaticus* Panz., *Naturwissenschaften 40,* 61–62 (1953).

——— Menotaxis im polarisierten Licht bei *Geotrupes silvaticus* Panz., *Naturwissenschaften 40,* 611–612 (1953a).

——— Photo-Geomenotaxis bei *Geotrupes silvaticus* Panz. und ihre zentralnervöse Koordination, *Z. vergl. Physiol. 36,* 176–211 (1954).

——— Lichtkompaßorientierung beim Wasserläufer *Velia currens* F. (Heteroptera) am Tage und zur Nachtzeit. I. Herbst- und Winterversuche, *Z. Tierpsychol. 13,* 463–484 (1956).

——— Angeborene und erworbene Anteile relativ einfacher Verhaltenseinheiten, *Verh. d. Deutschen Zool. Ges. in Erlangen 1955,* 32–48 (Leipzig, 1956a).

——— Innate types of chronometry in insect orientation, *Cold Spring Harbor Symposia on quantitative Biology 25,* 403–412 (1960).

——— K. FISCHER, and H. BÖTTCHER, Die Sonnenkompaßorientierung der Eidechsen, *Ergebn. Biol. 26,* 216–234 (1963).

——— and HANNE OBERDORFER, Schwerkraftorientierung beim Wasserläufer *Velia currens* F. (Heteroptera) am Tage und zur Nachtzeit, *Z. Tierpsychol. 16,* 693–705 (1959).

——— and R. L. DE VALOIS, Über den Einfluß der Höhe einer Lichtquelle auf die Lichtkompaßorientierung des Mistkäfers *Geotrupes silvaticus* Panz., *Naturwissenschaften 42,* 349–350 (1955).

BISETZKY, A. RUTH, Die Tänze der Bienen nach einem Fußweg zum Futterplatz, *Z. vergl. Physiol. 40,* 264–288 (1957).

BLEST, A. D., The evolution, ontogeny and quantitative control of the settling movements of some new world Saturnid moths, with some comments on distance communication by honey-bees, *Behaviour 16,* 188–253 (1960).

BLINOW, N. M., Dressur der Bienen auf Rotklee, *Ptschelowodstwo,* Nr. 2, 50–54 (1959) und Nr. 7, 35–36 (1960).

BOCH, R., Die Tänze der Bienen bei nahen und fernen Trachtquellen, *Z. vergl. Physiol. 38,* 136–167 (1956).

——— Rassenmäßige Unterschiede bei den Tänzen der Honigbiene (*Apis mellifica* L.), *Z. vergl. Physiol. 40,* 289–320 (1957).

——— Food handling of honeybees within the hive, *Symposium food gathering behavior of Hymenoptera* (Cornell University, 1959).

——— and D. A. SHEARER, Identification of Geraniol as the active component in the Nassanoff Pheromone of the honeybee, *Nature (London) 194,* 704–706 (1962).

——— ——— Production of geraniol by honey bees of various ages, *J. Ins. Physiol 9,* 431–434 (1963).

——— ——— Identification of nerolic and geranic acids in the Nassanoff Pheromone of the honey bee, *Nature (London) 202,* 320–321 (1964).

——— ——— and B. C. STONE, Identification of Iso-Amyl-Acetate as an active component in the sting pheromone of the honey bee, *Nature (London) 195,* 1018–1020 (1962).

BOECKH, J., Elektrophysiologische Untersuchungen an einzelnen Geruchsrezeptoren auf den Antennen des Totengräbers (*Necrophorus,* Coleoptera), *Z. vergl. Physiol. 46,* 212–248 (1962).

BOEHM, G., Über maculare (Haidingersche) Polarisationsbüschel and über einen polarisationsoptischen Fehler des Auges, *Acta ophthal. (Kbh.) 18,* Fasc. 2, 109–142 (1940).

——— Über ein neues entoptisches Phänomen im polarisierten Licht. "Periphere" Polarisationsbüschel, *Acta ophthal. (Kbh.) 18,* Fasc. 2, 143–169 (1940a).

BOLWIG, N., The role of scent as a nectar guide for honeybees on flowers and an observation on the effect of colour on recruits, *Brit. J. Anim. Behav. 2,* 81–83 (1954).

BRAEMER, W., A critical review of the sun azimuth Hypothesis, *Cold Spring Harbor Symposia quantitative Biology 25,* 413–427 (1960).

——— The effect of experimentally changed photoperiod on the sun-orientation rhythm of fish, *Physiol. Zoology 34,* 273–286 (1961).

——— and H. O. SCHWASSMANN, Vom Rhythmus der Sonnenorientierung am Äquator (bei Fischen), *Ergebn. der Biol. 26,* 182–201 (1963).

BRÄUNINGER, H. D., Über den Einfluß meteorologischer Faktoren auf die Entfernungsweisung im Tanz der Bienen, *Z. vergl. Physiol. 48,* 1–130 (1964).

BRANDENBURG, J., Die Feinstruktur des Seitenauges von *Lepisma, Zoologische Beiträge N. F. 5,* 291–300 (1960).

BRETSCHKO, J., Der Einfluß der Morphologie des Bienenauges auf die Wahrnehmung des Sonnenstandes im Flug und seine Wiedergabe bei der Richtungsweisung, *Inaugural-Dissertation Philosoph. Fak. Universität Graz* (1952).

BRETT, J. R., and C. GROOT, Some aspects of olfactory and visual responses in Pacific salmon, *J. Fish. Res. Bd. Canada 20,* 287–303 (1963).

BRIAN, ANNE D., The foraging of bumble bees, *Bee World 35,* 61–67, 81–91 (1954).

BROWN, F. A., M. F. BENNETT, and H. M. WEBB, A magnetic compass response of an organism, *Biol. Bull. 119,* 65–74 (1960).

——— H. M. WEBB, and F. H. BARNWELL, A compass directional phenomenon in mud-snails and its relation to magnetism, *Biol. Bull. 127,* 206–220 (1964).

——— H. M. WEBB, F. M. BENNETT, and F. H. BARNWELL, A diurnal rhythm in response of the snail *Ilyoanassa* to imposed magnetic fields, *Biol. Bull. 117,* 405–406 (1959).

BROWN, P. K., I. R. GIBBONS, and G. WALD, The visual cells and visual pigment of the Mud-puppy, *Necturus, J. cell. Biol. 19,* 79–106 (1963).

——— and G. WALD, Visual pigments in single rods and cones of the human retina, *Science 144,* 45–52 (1964).

BRUN, R., *Die Raumorientierung der Ameisen* (Jena, 1914).

BUDDENBROCK, W. v., Die Lichtkompaßbewegungen bei den Insekten, insbesondere den Schmetterlingsraupen, *Sitzungsber. Heidelberger Akadem. d. Wissensch. Math.-naturw. Kl. VIII B,* 3–26 (1917).

——— *Vergleichende Physiologie,* I, *Sinnesphysiologie* (Basel, 1952).

BÜCKMANN, D., Die Leistungen der Schwereorientierung bei dem im Meeressande grabenden Käfer *Bledius bicornis* Grm. (Staphylinidae), *Z. vergl. Physiol. 36,* 488–507 (1954).

——— Zur Leistung des Schweresinnes bei Insekten, *Naturwissenschaften 42,* 78–79 (1955).

——— Das Problem des Schweresinnes bei den Insekten, *Naturwissenschaften 49,* 28–33 (1962).

BÜDEL, A., Das Mikroklima in einer Blüte, *Z. Bienenforsch. 3,* 185–190 (1956).

——— Réaumur und die Bienenphysik, *Z. Bienenforsch. 4,* 23–37 (1957).

——— Das Mikroklima der Blüten in Bodennähe, *Z. Bienenforsch. 4,* 131–140 (1959).

BÜNNING, E., *Die physiologische Uhr* (2nd ed., Berlin, Göttingen, Heidelberg: Springer, 1963).

BULLMANN, O., Über die Größe des Sammelbereichs bei Apiden, *Inaug. Dissertation Philos. Fak. Universität Graz* (1952).

BURDON-JONES, G., and G. H. CHARLES, Light reactions of littoral gastropods, *Nature* (London) *181*, 129–131 (1958).

——— ——— Light responses of littoral gastropods, *XV Internat. Congress of Zoology* (London, 1958), 889–891 (1959).

BURKHARDT, D., and LOTTE WENDLER, Ein direkter Beweis für die Fähigkeit einzelner Sehzellen des Insektenauges, die Schwingungsrichtung polarisierten Lichtes zu analysieren, *Z. vergl. Physiol. 43*, 687–692 (1960).

BURTT, E. T., and W. T. CATTON, A diffraction theory of insect vision. I. An experimental investigation of visual acuity and image formation in the compound eyes of three species of insects, *Proc. Roy. Soc. B 157*, 53–82 (1962).

BUTENANDT, A., R. BECKMANN, and E. HECKER, Über den Sexuallockstoff des Seidenspinners. I. Der biologische Test und die Isolierung des reinen Sexualstoffes Bombykol, *Hoppe-Seylers Z. physiol. Chem. 324*, 71–83 (1961).

——— ——— and D. STAMM, Über den Sexuallockstoff des Seidenspinners. II. Konstitution und Konfiguration des Bombykols, *Hoppe-Seylers Z. physiol. Chem. 324*, 84–87 (1961).

——— and E. HECKER, Synthese des Bombykols, des Sexuallockstoffs des Seidenspinners, und seiner geometrischen Isomeren, *Angewandte Chemie 73*, 349–353 (1961).

——— B. LINZEN and M. LINDAUER, Über einen Duftstoff aus der Mandibeldrüse der Blattschneiderameise *Atta sexdens rubropilosa*, *Arch. Anat. micr. Morph. exp. 48*, 13–19 (1959).

BUTLER, C. G., The choice of drinking water by the honeybees, *J. exp. Biol. 17*, 253–261 (1940).

——— The influence of various physical and biological factors of the environment on honeybee activity and nectar concentration and abundance, *J. exp. Biol. 21*, 5–12 (1945).

——— The behaviour of bees when foraging, *J. Roy. Soc. Arts 93*, 501–511 (1945a).

——— The importance of perfume in the discovery of food by the worker honeybee, *Proc. Roy. Soc. B 138*, 403–413 (1951).

——— *The world of the honeybee* (London, 1954).

——— *Die Honigbiene* (Düsseldorf-Köln. 1957).

——— The method and importance of the recognition by a colony of honeybees (*A. mellifera*) of the presence of the queen, *Trans. Roy. Entomol. Soc. London 105*, 11–29 (1954a).

——— Some further observations on the nature of "queen substance" and of its role in the organisation of a honeybee community, *Proc. Roy. Entomol. Soc. London (A) 31*, 12–16 (1956).

——— The control of ovary development in worker honeybees (*Apis mellifera*), *Experientia 13*, 256–257 (1957).

——— Bee department, *Rep. Rothamsted exp. Station*, 146–153 (1958).

——— The scent of queen honeybees (*A. mellifera* L.) that causes partial inhibition of queen rearing, *J. Ins. Physiol. 7*, 258–264 (1961).

——— R. K. CALLOW, and NORAH C. JOHNSTON, The isolation and synthesis of queen substance, 9-oxodec-trans-2-enoic acid, a honeybee pheromone, *Proc. Roy. Soc. B 155*, 417–432 (1961).

——— and ELAINE M. FAIREY, Pheromones of the honeybee: biological studies of the mandibular gland secretion of the queen, *J. apic. Res. 3*, 65–76 (1964).

——— J. B. FREE, and J. SIMPSON, Some problems of red clover pollination, *Ann. applied Biol. 44*, 664–669 (1956).

——— E. P. JEFFREE, and H. KALMUS, The behavior of a population of honeybees on an artificial and on a natural crop, *J. exp. Biol. 20*, 65–73 (1943).

——— and J. SIMPSON, Bees as pollinators of fruit and seed crops, *Rep. Rothamst. exp. Station*, 167–175 (1953).

────── ────── The source of the queen substance of the honey-bee (*Apis mellifera* L.), *Proc. Roy. Entomol. Soc. London (A) 33,* 120–122 (1958).

────── ────── Pheromones of the honeybee (*Apis mellifera* L.), An olfactory pheromone from the Koschewnikow gland of the queen, *Scientific Studies,* vol. 4, University of Libčice, Czechoslovakia, 33–36 (1965).

BUTTEL-REEPEN, H. v., *Sind die Bienen Reflexmaschinen?* (Leipzig, 1900); also in *Biol. Zbl. 20,* 97–109, 130–144, 177–193, 209–224, 289–304 (1900).

────── *Leben und Wesen der Bienen* (Brunswick, 1915).

BYERS, W., An unusual nest of the honey bee, *J. Kans. Entomol. Soc. 32,* 46 (1959).

CAJAL, S., and SANCHÉZ, Contribucion al conocimiento de los centros nervosos de los insectos, *Trabajos del laborat. de investig. biol. de la univers. de Madrid 14* (1915).

CARTHY, J. D., The orientation of two allied species of British ant, *Behaviour 3,* 275–318 (1951).

CHALIFMAN, T. A., New facts about foraging behaviour of bees, *Pschelovodstvo 8,* 415–418 (1950).

CHARLES, G. H., The orientation of *Littorina* species to polarized light, *J. exp. Biol. 38,* 189–202 (1961).

────── The mechanism of orientation of freely moving *Littorina littoralis* to polarized light, *J. exp. Biol. 38,* 203–212 (1961a).

────── Orientational movements of the foot of *Littorina* species in relation to the plane of vibration of polarized light, *J. exp. Biol. 38,* 213–224 (1961b).

CHAUVIN, R., Sur les possibilités d'adaptation chez les insectes sociaux et spécialement chez l'abeille, *Insectes sociaux 7,* 101–108 (1960).

COUTURIER, A., and P. ROBERT, Maintien de la direction de vol chez *Melolontha melolontha* L., *C. R. Acad. Sci. Paris 240,* 2561–2563 (1955).

────── ────── Recherches sur le comportement du Hanneton commun (*Melolontha melolontha*) au cours de sa vie aérienne, *Ann. des Epiphyties 1,* 19–60 (1955a).

────── ────── Principaux aspects de l'orientation astronomique chez le Hanneton commun, *Bull. Soc. d'Histoire Natur. Colmar 47, 4th ser., 4,* 27–42 (1956).

────── ────── Recherches sur la faculté d'orientation du Hanneton commun (*Melolontha melolontha* L.), *C. R. Acad. Sci. Paris 245,* 2399–2401 (1957).

────── ────── Recherches sur les migrations du Hanneton commun (*Melolontha melolontha* L.), *Ann. des Epiphyties,* 257–329 (1958).

────── ────── Observations sur le comportement du Hanneton commun, *Melolontha melolontha* L., *Revue de Zoologie agricole et appliquée,* Nr. 7–9 (1962).

DANNEEL, R., and B. ZEUTZSCHEL, Über den Feinbau der Retinula bei *Drosophila melanogaster, Z. Naturforsch.* 12b, 580–583 (1957).

DAUMER, K., Reizmetrische Untersuchung des Farbensehens der Biene, *Z. vergl. Physiol. 38,* 413–478 (1956).

────── Blumenfarben, wie sie die Bienen sehen, *Z. vergl. Physiol. 41,* 49–110 (1958).

────── Kontrastempfindlichkeit der Bienen für "Weiss" verschiedenen UV-Gehalts, *Z. vergl. Physiol. 46,* 336–350 (1963).

────── R. JANDER, and H. WATERMAN, Orientation of the ghost-crab *Ocypode* in polarized light, *Z. vergl. Physiol. 47,* 56–76 (1963).

DETHIER, V. G., The physiology and histology of the contact chemoreceptors of the blowfly, *Quart. Rev. Biol. 30,* 348–371 (1955).

────── Communication by insects, Physiology of dancing, *Science 125,* 331–336 (1957).

────── and L. E. CHADWICK, Chemoreception in Insects, *Physiol. Rev. 28,* 220–254 (1948).

—— D. R. Evans, and M. V. Rhoades, Some factors controlling the ingestion of carbohydrates by the blowfly, *Biol. Bull. 111*, 204–222 (1956).

Dietrich, W., Die Facettenaugen der Dipteren, *Z. wissensch. Zool. 92*, 465–539 (1909).

Dirschedl, Hannelore, Die Vermittlung des Blütenduftes bei der Verständigung im Bienenstock, *Dissert. Naturw. Fak. Univ. München* (1960).

Djalal, Ahmad Schah sei, Untersuchungen über die Ursache der Richtungsmißweisung bei den Tänzen der Honigbiene, *Dissert. Naturw. Fak. Univ. München* (1959).

Dostal, Brigitte, Riechfähigkeit und Zahl der Riechsinneselemente bei der Honigbiene, *Z. vergl. Physiol. 41*, 179–203 (1958).

Eckert, B., Orientující vliv polarisovaného světla na perloočky (Der orientierende Einfluß polarisierten Lichtes auf Daphnia), *Československ, Biol. 2*, 76–80 (1953).

Eckert, J. E., The flight range of the honey bee, *J. agric. Res. 47*, 257–285 (1933).

Eckstein, M., Über die Sichtbarkeit der Sonne für Bienen in ultraviolettem Licht bei bedecktem Himmel, *Diplom Arbeit Universitäts-Sternwarte München* (1959).

Eltringham, H., Butterfly vision, *Trans. Entomol. Soc. London 67*, 1, 1–49 (1919).

Emeis, D., Untersuchungen zur Lichtkompaßorientierung des Wasserläufers Velia currens, *Z. Tierpsychol. 16*, 129–154 (1959).

Engländer, H., Die Bedeutung der weißen Farbe für die Orientierung der Bienen am Stand, *Arch. f. Bienenkunde 22*, 516–549 (1941).

Ercolini, A., Ricerche sull'orientamento astronomico di *Paederus rubrothoracicus* Goeze (Coleoptera-Staphylinidae), *Monitore zoologico italiano 70/71*, 416–429 (1962/1963).

—— Ricerche sull'orientamento solare degli Anfipodi. La variazione dell'orientamento in cattività: *Archivo zoologico italiano 48*, 147–179, 1963.

—— Ricerche sull'orientamento astronomico in anfipodi litorali della zona equatoriale. I. L'orientamento solare in una popolazione somala di *Talorchestia martensii* Weber, *Z. vergl. Physiol. 49*, 138–171 (1964).

—— and G. Badino, L'orientamento astronomico di *Paederus rubrothoracicus* Goeze (Coleoptera Staphylinidae), *Bollett. di Zoologia 28*, fasc. II, 421–432 (1961).

Esch, H., Analyse der Schwänzelphase im Tanz der Bienen, *Naturwissenschaften 43*, 207 (1956).

—— Die Elemente der Entfernungsmitteilung im Tanz der Bienen, *Experientia (Basel) 12*, 439–441 (1956a).

—— Über die Körpertemperaturen und den Wärmehaushalt von *Apis mellifica*, *Z. vergl. Physiol. 43*, 305–335 (1960).

—— Ein neuer Bewegungstyp im Schwänzeltanz der Bienen, *Naturwissenschaften 48*, 140–141 (1961).

—— Über die Schallerzeugung beim Werbetanz der Honigbiene, *Z. vergl. Physiol. 45*, 1–11 (1961a).

—— Über die Auswirkung der Futterplatzqualität auf die Schallerzeugung im Werbetanz der Honigbiene, *Verhandl. d. Deutsch. Zoolog. Gesellsch. in Wien 1962* (Leipzig, 1963), 302–309.

—— Beiträge zum Problem der Entfernungsweisung in den Schwänzeltänzen der Honigbienen, *Z. vergl. Physiol. 48*, 534–546 (1964).

Exner, S., *Die Physiologie der facettierten Augen von Krebsen und Insekten* (Leipzig and Vienna, 1891).

Exner, F. and S., Die physikalischen Grundlagen der Blütenfärbungen, *Sitzungsber. Akad. d. Wissensch. Wien, math.-naturw. Kl., 119, Abt. I*, 1–55 (1910).

FERGUSON, D. E., Orientation in three species of anuran Amphibians, *Ergebn. Biol. 26,* 128–134 (1963).

FERNANDEZ-MORAN, H., Fine structure of the insect retinula as revealed by electron microscopy, *Nature (London) 177,* 742–743 (1956).

———— Fine structure of the light receptors in the compound eyes of insects, *Exp. Cell Res. 5,* 586–644 (1958).

FINKE, INGRID, Zeitgedächtnis und Sonnenorientierung der Bienen, *Lehramtsarbeit Naturw. Fak. Univ. München* (1958).

FIRSOW, I. G., On the training of bees to red clover, *Pschelovodstvo,* Nr. 4, 15 (1951), in Russian.

FISCHER, KLAUS, Spontanes Richtungsfinden nach dem Sonnenstand bei *Chelonia mydas* L. (Suppenschildkröte), *Naturwissenschaften 51,* 203 (1964).

FISCHER, K., and G. BIRUKOW, Dressur von Smaragdeidechsen auf Kompaßrichtungen, *Naturwissenschaften 47,* 93–94 (1960).

FISCHER, W., Untersuchungen über die Riechschärfe der Honigbiene, *Z. vergl. Physiol. 39,* 634–659 (1957).

FLOREY, E., Experimentelle Erzeugung einer "Neurose" bei der Honigbiene, *Naturwissenschaften 41,* 171 (1954).

FOREL, A., Les fourmis de la Suisse, *Neue Denkschr. allg. schweiz. Gesellsch. ges. Naturw.* 26 (Zürich, 1874).

———— *Das Sinnesleben der Insekten* (Munich, 1910).

FRANÇON, J., *L'esprit des abeilles* (Paris: Gallimard, 1938).

———— *Die Klugheit der Bienen* (Berlin, 1939).

FRANK, A., Eigenartige Flugbahnen bei Hummelmännchen, *Z. vergl. Physiol. 28,* 467–484 (1941).

FREE, J. B., Attempts to condition bees to visit selected crops, *Bee World 39,* 221–230 (1958).

———— The drifting of honey-bees, *J. agric. Sci. 51,* 294–306 (1958a).

———— The effect of moving colonies of honeybees to new sites on their subsequent foraging behaviour, *J. agric. Sci. 53,* No. 1, 1–9 (1959).

———— The transfer of food between the adult members of a honeybee community, *Bee World 40,* 193–201 (1959a).

———— The stimuli releasing the stinging response of honeybees, *Animal Behaviour 9,* 193–196 (1961).

———— The attractiveness of Geraniol to foraging honeybees, *J. agric. Res. 1,* 52–54 (1962).

———— N. W. FREE, and S. C. JAY, The effect of foraging behavior of moving honey bee colonies to crops before or after flowering has begun, *J. Econom. Entomol. 53,* 564–566 (1960).

———— and YVETTE SPENCER-BOOTH, Further experiments on the drifting of honey-bees, *J. agric. Sci. 57,* 153–158 (1961).

FREY-WYSSLING, A., M. ZIMMERMANN, and A. MAURIZIO, Über den enzymatischen Zuckerumbau in Nektarien, *Experientia (Basel) 10,* 490–491 (1954).

FRINGS, H., The loci of olfactory end-organs in the honey bee, *J. exp. Zool. 97,* 123–134 (1949).

———— and F. LITTLE, Reactions of honey bees in the hive to simple sounds, *Science 125,* 122 (1957).

———— and MABLE FRINGS, The loci of contact chemoreceptors in insects, *The American Midland Naturalist 41,* 602–658 (1949).

FRISCH, K. V., Demonstration von Versuchen zum Nachweis des Farbensinnes bei angeblich total farbenblinden Tieren, *Verhandl. d. Deutsch. Zool. Ges. in Freiburg* (Berlin, 1914).

———— Der Farbensinn und Formensinn der Bienen, *Zool. Jb. (Physiol.) 35,* 1–188 (1914/1915).

———— Über den Geruchssinn der Bienen und seine blütenbiologische Bedeutung, *Zool. Jb. (Physiol.) 37,* 1–238 (1919).

———— Zur Streitfrage nach dem Farbensinn der Bienen, *Biol. Zbl. 39,* 122–139 (1919a).

——— Über die "Sprache" der Bienen, *Münch. med. Wschr.* (1920), 566–569.
——— Über den Sitz des Geruchssinnes bei Insekten, *Zool. Jb. (Physiol.) 38*, 1–68 (1921).
——— Über die "Sprache" der Bienen. II., *Münch. med. Wschr.* (1921a), 509–511.
——— Über die "Sprache" der Bienen. III., *Münch. med. Wschr.* (1922), 781–782.
——— Über die "Sprache" der Bienen, eine tierpsychologische Untersuchung, *Zool. Jb. (Physiol.) 40*, 1–186 (1923).
——— Vergleichende Physiologie des Geruchs- und Geschmackssinnes, *Handb. norm. u. patholog. Physiol. 11*, 203–239 (Berlin, 1926).
——— *Ein Vorschlag für die Wanderimker: Bienenzucht und Bienenforschung in Bayern* (Neumünster, Holstein: Wachholts-Verlag, 1927).
——— Über den Geschmackssinn der Bienen, *Z. vergl. Physiol. 21*, 1–156 (1934).
——— Psychologie der Bienen, *Z. Tierpsychol. 1*, 9–21, 1937.
——— Die Tänze und das Zeitgedächtnis der Bienen im Widerspruch, *Naturwissenschaften 28*, 65–69 (1940).
——— Die Werbetänze der Bienen und ihre Auslösung, *Naturwissenschaften 30*, 269–277 (1942).
——— Christian Konrad Sprengels Blumentheorie vor 150 Jahren und heute, *Naturwissenschaften 31*, 223–229 (1943).
——— Versuche über die Lenkung des Bienenfluges durch Duftstoffe, *Naturwissenschaften 31*, 445–460 (1943a).
——— Die Tänze der Bienen, *Österr. Zoolog. Zeitschr. 1*, 1–48 (1946).
——— Die "Sprache" der Bienen und ihre Nutzanwendung in der Landwirtschaft, *Experientia (Basel) 2*, 397–404 (1946a).
——— *Duftgelenkte Bienen im Dienste der Landwirtschaft und Imkerei* (Vienna: Springer-Verlag, 1947).
——— Gelöste und ungelöste Rätsel der Bienensprache, *Naturwissenschaften 35*, 12–23, 38–43 (1948).
——— Die Polarisation des Himmelslichtes als orientierender Faktor bei den Tänzen der Bienen, *Experientia (Basel) 5*, 142–148 (1949).
——— Die Sonne als Kompaß im Leben der Bienen, *Experientia (Basel) 6*, 210–221 (1950).
——— Orientierungsvermögen und Sprache der Bienen, *Naturwissenschaften 38*, 105–112 (1951).
——— Spitzenleistungen tierischer Sinnesorgane und ihre biologische Bedeutung, *Vjschr. Naturf. Ges. Zürich 96*, 176–178 (1951a).
——— "Sprache" oder "Kommunikation" der Bienen? *Psychol. Rundschau 4/4* (1953).
——— Die Richtungsorientierung der Bienen, *Verhandl. d. Deutsch. Zool. Ges. in Freiburg 1952* (Leipzig, 1953a), 58–72.
——— Symbolik im Reich der Tiere, *Münchner Universitätsreden*, N. F., H. 7 (Munich, 1954).
——— Die Fähigkeit der Bienen, die Sonne durch die Wolken wahrzunehmen, *Sitzungsber. 1953 Bayer. Akad. Wiss. math. naturw. Kl.*, 197–199 (1954a).
——— *Sprechende Tänze im Bienenvolk* (Munich: Verlag d. Bayer. Akad. d. Wiss., 1955).
——— *Lernvermögen und erbgebundene Tradition im Leben der Bienen: L'instinct dans le comportement des animaux et de l'homme* (Colloqu. intern. 1954; Paris: Masson, 1956), 345–386.
——— The language and orientation of the bees, *Proc. Amer. Philos. Soc. 100*, 515–519 (1956a).
——— Über die durch Licht bedingte "Mißweisung" bei den Tänzen im Bienenstock, *Experientia (Basel) 18*, 49–53 (1962).
——— *Aus dem Leben der Bienen* (Verständliche Wissenschaft, vol. 1, 7th ed.; Berlin, Göttingen, Heidelberg: Springer, 1964).

―――― H. Heran, and M. Lindauer, Gibt es in der "Sprache" der Bienen eine Weisung nach oben oder unten? *Z. vergl. Physiol. 35,* 219–245 (1953).

―――― and R. Jander, Über den Schwänzeltanz der Bienen, *Z. vergl. Physiol. 40,* 239–263 (1957).

―――― and O. Kratky, Über die Beziehung zwischen Flugweite und Tanztempo bei der Entfernungsmeldung der Bienen, *Naturwissenschaften 49,* 409–417 (1962).

―――― and M. Lindauer, Himmel und Erde in Konkurrenz bei der Orientierung der Bienen, *Naturwissenschaften 41,* 245–253 (1954).

―――― ―――― Über die Fluggeschwindigkeit der Bienen und ihre Richtungsweisung bei Seitenwind, *Naturwissenschaften 42,* 377–385 (1955).

―――― ―――― Über die "Mißweisung" bei den richtungsweisenden Tänzen der Bienen, *Naturwissenschaften 48,* 585–594 (1961).

―――― ―――― and K. Daumer, Über die Wahrnehmung polarisierten Lichtes durch das Bienenauge, *Experientia (Basel) 16,* 289–301 (1960).

―――― ―――― and F. Schmeidler, Wie erkennt die Biene den Sonnenstand bei geschlossener Wolkendecke? *Naturwissensch. Rundschau,* 169–172 (1960).

―――― and G. A. Rösch, Neue Versuche über die Bedeutung von Duftorgan und Pollenduft für die Verständigung im Bienenvolk, *Z. vergl. Physiol. 4,* 1–21 (1926).

Fromme, H. G., Untersuchungen über das Orientierungsvermögen nächtlich ziehender Kleinvögel (*Erithacus rubecula, Sylvia communis*), *Z. Tierpsychol. 18,* 205–220 (1961).

Gabriel, A., *Das Bild der Wüste* (Vienna, 1958).

Gary, N. E., Queen honey bee attractiveness as related to mandibular gland secretion, *Science 133,* 1479–1480 (1961).

―――― Chemical mating attractants in the queen honeybee, *Science 136,* 773–774 (1962).

―――― Observations of mating behaviour in the honeybee, *J. apic. Res. 2,* 1–13 (1963).

Gebelein, H., and H. J. Heite, *Statistische Urteilsbildung* (Berlin, Göttingen, Heidelberg: Springer, 1951).

Geiger, R., *Das Klima der bodennahen Luftschicht* (4th ed.; Brunswick, 1961).

―――― *The Climate Near the Ground* (rev. ed.; Cambridge, Massachusetts: Harvard University Press, 1965).

Geisler, Marianne, Untersuchungen zur Tagesperiodik des Mistkäfers *Geotrupes silvaticus* Panzer, *Z. Tierpsychol. 18,* 389–420 (1961).

Geissler, G., and W. Steche, Natürliche Trachten als Ursache für Vergiftungserscheinungen bei Bienen und Hummeln, *Z. Bienenforsch. 6,* 77–92 (1962).

Gerdes, K., Richtungstendenzen vom Brutplatz verfrachteter Lachmöven (*Larus ridibundus* L.) unter Ausschluß visueller Gelände- und Himmelsmarken, *Z. wiss. Zool. 166,* 352–410 (1962).

Ghent, R. L., and N. E. Gary, A chemical alarm releaser in honey bee stings (*Apis mellifera* L.), *Psyche 69,* 1–6 (1962).

Gillard, A. De betekenis van de Honingbij voor de Landbouwproductie, *Mededelingen Landbouwhogeschool e. Opzoekingsstations Gant. XII,* 103–137 (1947).

Giltay, E., Über die Bedeutung der Krone bei den Blüten und über das Farbenunterscheidungsvermögen der Insekten. I, *Jb. wissensch. Botanik 40,* 368–402 (1904).

Glushkov, N. M., Problems of Beekeeping in the U.S.S.R. in relation to pollination, *Bee World 39,* 81–92 (1958).

Görner, P., Die optische und kinästhetische Orientierung der Trichterspinne *Agelena labyrinthica* Cl., *Z. vergl. Physiol. 41,* 111–153 (1958).

―――― Die Orientierung der Trichterspinne nach polarisiertem Licht, *Z. vergl. Physiol. 45,* 307–314 (1962).

Goetsch, W., *Vergleichende Biologie der Insektenstaaten* (Leipzig, 1953).

───── *Die Staaten der Ameisen* (Verständliche Wissenschaft, vol. 33, 2nd ed.; Berlin, Göttingen, Heidelberg: Springer, 1953a).

GOGALA, M., and S. MICHIELI, Das Komplexauge von *Ascalaphus,* ein spezialisiertes Sinnesorgan für kurzwelliges Licht, *Naturwissenschaften 52,* 217–218 (1965).

GOLDSMITH, T. H., The physiological basis of wave-length discrimination in the eye of the honeybee, in *Sensory Communication,* ed. W. A. ROSENBLITH (Massachusetts Institute of Technology, 1961), 357–375.

───── Fine structure of the retinulae in the compound eye of the honeybee, *J. Cell Biol. 14,* 489–494 (1962).

───── The visual system of insects, in *The Physiology of Insects,* ed. M. ROCKSTEIN, vol. 1 (New York, London, 1964), 397–462.

───── and D. E. PHILPOTT, The microstructure of the compound eyes of insects, *J. biophys. biochem. Cytol. 3,* 429–438 (1957).

GONTARSKI, H. Leistungsphysiologische Untersuchungen an Sammelbienen, *Arch. Bienenkunde 16,* 107–126 (1935).

───── Wandlungsfähige Instinkte der Honigbiene, *Umschau 49,* 310–312 (1949).

GRIFFIN, D. R., *Listening in the dark* (New Haven, Connecticut: Yale University Press, 1958).

GROUT, R. A., *The hive and the honey bee* (Hamilton, Illinois, 1949).

GUBIN, A. F., Bestäubung und Erhöhung der Samenernte bei Rotklee *Trifolium pratense* L. mit Hilfe der Bienen, *Arch. Bienenkunde 17,* 209–264 (1936).

───── *Pschelovodstvo 5,* 40–44, *7,* 15–17 (1938), in Russian.

───── Further considerations of clover pollination by bees in Stalin's third five-year plan, *Pschelovodstvo 6,* 31 (1939), in Russian.

GUBIN, W. A. Über die Geruchsempfindlichkeit bei Honigbienen, *Pschelovodstvo 7,* 17–19 (1957).

GUCKELSBERGER, H., Zur Orientierung der Bienen bei bedecktem Himmel, *Zulassungsarbeit Universitätssternwarte München* (1962).

HAAS, A., Neuere Beobachtungen zum Problem der Flugbahnen bei Hummelmännchen, *Z. Naturforsch. 1,* 596–599 (1946).

───── Arttypische Flugbahnen von Hummelmännchen, *Z. vergl. Physiol. 31,* 281–307 (1949).

───── Gesetzmäßiges Flugverhalten der Männchen von *Psithyrus silvestris* Lep. und einigen solitären Apiden, *Z. vergl. Physiol. 31,* 671–683 (1949a).

───── Die Mandibeldrüse als Duftorgan bei einigen Hymenopteren, *Naturwissenschaften 39,* 484 (1952).

───── Vergleichende Verhaltensstudien zum Paarungsschwarm solitärer Apiden, *Z. Tierpsychol. 17,* 402–416 (1960).

───── Das Rätsel des Hummeltrompeters, Lichtalarm. 1. Bericht über Verhaltensstudien an einem kleinen Nest von *Bombus hypnorum* mit Arbeiter-Königin, *Z. Tierpsychol. 18,* 129–138 (1961).

HALDANE, J. B. S., and H. SPURWAY, A statistical analysis of communication in "*Apis mellifera*" and a comparison with communication in other animals, *Insectes sociaux* (Paris) *1,* 247–283 (1954).

HAMMANN, ELEONORE, Wer hat die Initiative bei den Ausflügen der Jungkönigin, die Königin oder die Arbeiterinnen? *Insectes sociaux 4,* 91–106 (1957).

HAMMER, O., Investigations on the nectar-flow of red clover, *Oikos 1,* 34–47 (1949).

HANNES, F., *Vom Lernmechanismus der Insekten* (Freiburg i. Br.: L. Bielefelds Verlag, 1959).

───── *Vom Lernmechanismus der Insekten,* II (Freiburg i. Br.: L. Bielefelds Verlag, 1961).

HANSSON, ÅKE, Lauterzeugung und Lautauffassungsvermögen der Bienen *Opuscula entomologica,* Suppl. VI (Lund 1945).

HANSTRÖM, B., Über die Frage ob funktionell verschiedene Zapfen- und Stäbchen-artige Sehzellen im Komplexauge der Arthropoden vorkommen, *Z. vergl. Physiol. 6*, 566–597 (1927).
HASE, A., Über den "Verkehr" am Wespennest, nach Beobachtungen an einer tropischen Art, *Naturwissenschaften 23*, 780–783 (1935).
HASLER, ARTHUR, Influence of environmental reference points on learned orientation in fish (*Phoxinus*), *Z. vergl. Physiol. 38*, 303–310 (1956).
―――― Homing orientation in migrating fish, *Ergebn. Biol. 23*, 94–115 (1960).
―――― R. M. HORRALL, W. J. WISBY, and W. BRAEMER, Sun orientation and homing in fishes, *Limnology and Oceanography 3*, 353–361 (1958).
HASLER, A., and A. MAURIZIO, Über den Einfluß verschiedener Nährstoffe auf Blütenansatz, Nektarsekretion und Samenertrag von honigenden Pflanzen, speziell von Sommerraps (*Brassica napus* L.), *Schweiz. Landwirtsch. Monatshefte* (1950), 201–211.
HASLINGER, F., Über den Geschmackssinn von *Calliphora erythrocephala* Meigen und über die Verwertung von Zuckern und Zuckeralkoholen durch diese Fliege, *Z. vergl. Physiol. 22*, 614–640 (1935).
HASSENSTEIN, B., Die bisherige Rolle der Kybernetik in der biologischen Forschung, *Naturwissensch. Rundschau 13*, 349–355, 373–382, 419–424 (1960).
HASSETT, C. C., V. G. DETHIER, and J. GANS, A comparison of nutritive values and taste thresholds of carbohydrates for the blowfly, *Biol. Bull. 99*, 446–453 (1950).
HAYDAK, M., H., Some new observations of the bee life, *Cesky Vcelar 63*, 133–135 (1929).
―――― The language of the honeybee, *American Bee Journal 85*, 316–317 (1945).
HECHT, S., and E. WOLF, The visual acuity of the honeybee, *J. gen. Physiol. 12*, 727–760 (1929).
HEIN, G., Über richtungsweisende Bienentänze bei Futterplätzen in Stocknähe, *Experientia (Basel) 6*, 142 (1950).
―――― Der Rucktanz als wesentlicher Bestandteil der Bienentänze, *Experientia (Basel) 10*, 23–24 (1954).
HENKEL, CHR., Unterscheiden die Bienen Tänze? (Dissertation, University of Bonn, 1938).
HERAN, H., Untersuchungen über den Temperatursinn der Honigbiene (*Apis mellifica*) unter besonderer Berücksichtigung der Wahrnehmung strahlender Wärme, *Z. vergl. Physiol. 34*, 179–206 (1952).
―――― Versuche über die Windkompensation der Bienen, *Naturwissenschaften 42*, 132–133 (1955).
―――― Ein Beitrag zur Frage nach der Wahrnehmungsgrundlage der Entfernungsweisung der Bienen, *Z. vergl. Physiol. 38*, 168–218 (1956).
―――― Die Orientierung der Bienen im Flug, *Ergebn. Biol. 20*, 199–239 (1958).
―――― Wahrnehmung und Regelung der Flugeigengeschwindigkeit bei *Apis mellifica* L., *Z. vergl. Physiol. 42*, 103–163 (1959).
―――― Anemotaxis und Fluchtorientierung des Bachläufers *Velia Caprai* Tam. (= *V. currens* F.), *Z. vergl. Physiol. 46*, 129–149 (1962).
―――― Wie beeinflußt eine zusätzliche Last die Fluggeschwindigkeit der Honigbiene? *Verhandl. d. Deutsch. Zool. Gesellsch. in Wien 1962, Zool. Anz. 26, Suppl.*, 346–354 (Leipzig 1963).
―――― and M. LINDAUER, Windkompensation und Seitenwindkorrektur der Bienen beim Flug über Wasser, *Z. vergl. Physiol. 47*, 39–55 (1963).
―――― and L. WANKE, Beobachtungen über die Entfernungsmeldung der Sammelbienen, *Z. vergl. Physiol. 34*, 383–393 (1952).
HERTZ, MATHILDE, Die Organisation des optischen Feldes bei der Biene. I, *Z. vergl. Physiol. 8*, 693–748 (1929).
―――― Die Organisation des optischen Feldes bei der Biene, II, *Z. vergl. Physiol. 11*, 107–145 (1930).
―――― Die Organisation des optischen Feldes bei der Biene, III, *Z. vergl. Physiol. 14*, 629–674 (1931).

——— Eine Bienendressur auf Wasser, *Z. vergl. Physiol.* 21, 463–467 (1934).

——— Zur Physiologie des Formen- und Bewegungssehens. II. Auflösungsvermögen des Bienenauges und optomotorische Reaktion, *Z. vergl. Physiol.* 21, 579–603 (1935).

——— Zur Physiologie des Formen- und Bewegungssehens. III. Figurale Unterscheidung und reziproke Dressuren bei der Biene, *Z. vergl. Physiol.* 21, 604–615 (1935a).

——— Zur Technik und Methode der Bienenversuche mit Farbpapieren und Glasfiltern, *Z. vergl. Physiol.* 25, 239–250 (1938).

——— New experiments on colour vision in bees, *J. exp. Biol.* 16, 1–8 (1939).

HESS, C. V., Gesichtssinn, in WINTERSTEIN, *Handb. d. vergl. Physiol.* (Jena, 1912), vol. 4.

——— Experimentelle Untersuchungen über den angeblichen Farbensinn der Bienen, *Zool. Jb.* (*Physiol.*) 34, 81–106 (1913).

——— Messende Untersuchung des Lichtsinnes der Biene, *Arch. ges. Physiol.* 163, 289–320 (1916).

——— Beiträge zur Frage nach einem Farbensinn bei Bienen, *Arch. ges. Physiol.* 170, 337–366 (1918).

——— Über Lichtreaktionen bei Raupen und die Lehre von den tierischen Tropismen, *Arch. ges. Physiol.* 177, 57–109 (1919).

——— Die Bedeutung des Ultraviolett für die Lichtreaktionen bei Gliederfüßern, *Arch. ges. Physiol.* 185, 281–310 (1920).

——— Die Grenzen der Sichtbarkeit des Spektrums in der Tierreihe, *Naturwissenschaften* 8, 197–200 (1920a).

HESS, W. R., Die Temperaturregulierung im Bienenvolk, *Z. vergl. Physiol.* 4, 465–487 (1926).

HESSE, R., Untersuchungen über die Organe der Lichtempfindung bei niederen Tieren. VII. Von den Arthropoden-Augen, *Z. wissensch. Zool.* 70, 347–473 (1901).

——— and F. DOFLEIN, *Tierbau und Tierleben*, vol. 2 (2nd ed.; Jena, 1943).

HEUSSER, H., Über die Beziehungen der Erdkröte (*Bufo bufo* L.) zu ihrem Laichplatz II, *Behaviour* (*Leiden*) 16, 93–109 (1960).

HINGSTON, R. W. G., *Problems of instinct and intelligence* (London: Arnold, 1928).

HOCKETT, CH. F., Logical considerations in the study of animal communication, in *Animal sounds and communication*, Publ. No. 7, Amer. Inst. Biol. Sci., Washington, 6, 392–430 (1960).

HOCKING, B., Aspects of insect vision, *Canadian Entomologist* 96, 320–334 (1964).

HODGSON, E. S., Electrophysiological studies of arthropod chemoreception. II. Responses of labellar chemoreceptors of the blowfly to stimulation by carbohydrates, *J. insect. Physiol.* 1, 240–247 (1957).

——— and K. D. ROEDER, Electrophysiological studies of arthropod chemoreception. I. General properties of the labellar chemoreceptors of diptera, *J. Cell Physiol.* 48, 51–76 (1956).

HÖRMANN, MARIA, Über den Helligkeitssinn der Bienen, *Z. vergl. Physiol.* 21, 188–219 (1934).

HOFFER, E., *Die Schmarotzerhummeln Steiermarks* (Graz, 1889).

HOFFMANN, ELISABETH, F. KÖHLER, and W. WITTEKINDT, Photographische Methode zur Aufzeichnung der Tanzfiguren bei der Honigbiene, *Naturwissenschaften* 43, 405–406 (1956).

HOFFMANN, IRMGARD, Über die Arbeitsteilung in weiselrichtigen und weisellosen Kleinvölkern der Honigbiene (nach Beobachtungen von PETRONELLA GESCHKE †), *Z. Bienenforsch.* 5, 267–279 (1961).

HOFFMANN, KLAUS, Experimentelle Änderung des Richtungsfindens beim Star durch Beeinflussung der inneren Uhr, *Naturwissenschaften* 40, 608–609 (1953).

——— Versuche zu der im Richtungsfinden der Vögel enthaltenen Zeitschätzung, *Z. Tierpsychol.* 11, 453–475 (1954).

——— Experimental manipulation of the orientational clocks in birds, *Cold Spring Harbor Symposia on quantitative Biology 25*, 379–387 (1960).
HOLST, E. V. Die Tätigkeit des Statolithenapparats im Wirbeltierlabyrinth, *Naturwissenschaften 37*, 265–272 (1950).
HUBER, FR., *Nouvelles observations sur les abeilles* (2nd ed.; Paris and Geneva, 1814).
——— Neue Beobachtungen an den Bienen, deutsch von G. KLEINE (Einbeck, 1856–1859).
HUBER, H., Die Abhängigkeit der Nektarsekretion von Temperatur, Luft- und Bodenfeuchtigkeit, *Planta 48*, 47–98 (1956).
HUBER, MATHILDE, Über Saftmale bei Blumen, *Lehramtsarbeit Naturw. Fak., Universität München* (1943).

ISTOMINA-TSVETKOVA, K. P., Contribution to the study of trophic relations in adult worker bees, *XVII. internat. Beekeeping Congr. Bologna-Roma 1958*, 2, 361–368 (1960).
IVANOFF, A., and T. H. WATERMAN, Factors, mainly depth and wavelength, affecting the degree of underwater light polarization, *J. marine Res. 16*, 283–307 (1958).

JACOBS, W., Das Duftorgan von *Apis mellifica* und ähnliche Hautdrüsenorgane sozialer und solitärer Apiden, *Z. Morphol. u. Ökol. 3*, 1–80 (1925).
JACOBS-JESSEN, UNA F., Zur Orientierung der Hummeln und einiger anderer Hymenopteren, *Z. vergl. Physiol. 41*, 597–641 (1959).
JAMIESON, C. A., and G. H. AUSTIN, Preference of honeybees for sugar solutions, *X. Internat. Congress of Entomology, Montreal 1956*, 4, 1059–1062 (1958).
JANDER, R., Die optische Richtungsorientierung der roten Waldameise (*Formica rufa* L.), *Z. vergl. Physiol. 40*, 162–238 (1957).
——— Menotaxis und Winkeltransponieren bei Köcherfliegen (Trichoptera), *Z. vergl. Physiol. 43*, 680–686 (1960).
——— Grundleistungen der Licht- und Schwereorientierung von Insekten, *Z. vergl. Physiol. 47*, 381–430 (1963).
——— K. DAUMER, and H. WATERMAN, Polarized light orientation by two Hawaiian decapod Cephalopods, *Z. vergl. Physiol. 46*, 383–394 (1963).
——— and T. H. WATERMAN, Sensory discrimination between polarized light and light intensity patterns by Arthropods, *J. Cell. compar. Physiol. 56*, 137–159 (1960).
JÖRGENSEN, H., Das Anlocken der Honigbiene zu *Centaurea montana* und *Helenium* sp., *Dansk Botanisk Arkiv 9*, 1–18 (1939).
JOHNSON, W. E., and C. GROOT, Observations on the migration of young Sockeye Salmon (*Oncorhynchus nerka*) through a large complex lake system, *J. Fish. Res. Bd. Canada 20*, 919–938 (1963).
JORDAN, R., Versuche hinsichtlich des Reagierens der Bienen auf Königinnenduft-Extrakt, *Bienenvater*, 3–5, 45–48, 78–81, 110–112, 145–146, 177–178 (1961).

KAINZ, F., *Die "Sprache" der Tiere* (Stuttgart, 1961).
KALMUS, H., Vorversuche über die Orientierung der Biene im Stock, *Z. vergl. Physiol. 24*, 166–187 (1937).
——— Der Füllungszustand der Honigblase entscheidet die Flugrichtung der Honigbiene, *Z. vergl. Physiol. 26*, 79–84 (1939).
——— Finding and exploitation of dishes of syrup by bees and wasps, *Brit. J. Anim. Behav. 2*, 136–139 (1954).
——— Responses of insects to polarized light in the presence of dark reflecting surfaces, *Nature (London) 182*, 1526–1527 (1958).
——— Orientation of animals to polarized light, *Nature (London) 184*, 228–230 (1959).

——— and C. R. Ribbands, The origin of the odours by which honeybees distinguish their companions, *Proc. Roy. Soc. B 140,* 50–57 (1952).

Kaltofen, R. S., Das Problem des Volksduftes bei der Honigbiene, *Z. vergl. Physiol. 33,* 462–475 (1951).

Kappel, Irmgard, Die Form des Safthalters als Anreiz für die Sammeltätigkeit der Bienen, *Z. vergl. Physiol. 34,* 539–546 (1953).

Kapustin, *Pschelovodstvo,* No. 8/9, 37–38 (1938), in Russian.

Kaschef, A. H., Über die Einwirkung von Duftstoffen auf die Bienentänze, *Z. vergl. Physiol. 39,* 562–576 (1957).

Kennedy, D., and E. R. Baylor, Analysis of polarized light by the bee's eye, *Nature (London) 191,* 34–37 (1961).

Kerner von Marilaun, A., *Pflanzenleben* (2 vols., Leipzig and Vienna, 1896, 1898).

Kiechle, H., Die soziale Regulation der Wassersammeltätigkeit im Bienenstaat und deren physiologische Grundlage, *Z. vergl. Physiol. 45,* 154–192 (1961).

Kimm, I. H., Orientation of cockchafers, *Nature (London) 188,* 69–70 (1960).

Kleber, Elisabeth, Hat das Zeitgedächtnis der Bienen biologische Bedeutung? *Z. vergl. Physiol. 22,* 221–262 (1935).

Knaffl, Herta, Über die Flugweite und Entfernungsmeldung der Bienen, *Z. Bienenforsch. 2* (Zander-Festschr.), 131–140 (1953).

Knoll, F., Insekten und Blumen, *Abhandl. Zool.-Bot. Gesellsch. Wien 12* (1926).

Knuth, P., *Handbuch der Blütenbiologie* (3 vols., Leipzig, 1898–1905).

Kobel, F., Eine wertvolle Untersuchung über die Ortsstetigkeit der Honigbiene, *Schweiz. Bienenztg.* (1949), 67–69.

Köhler, F., Photographisch registrierte Autooszillogramme, *Glas-Instrumenten-Technik 3,* 151–157 (1959).

Körner, Ilse, Zeitgedächtnis und Alarmierung bei den Bienen, *Z. vergl. Physiol. 27,* 445–459 (1939).

Koller, S., *Graphische Tafeln zur Beurteilung statistischer Zahlen* (Dresden and Leipzig, 1940).

Komarow, How we obtained a heavy clover-seed harvest, *Pschelovodstvo 5* (1939), in Russian.

Komisarenko, E. M., Bienendressur verdoppelte den Samenertrag von Rotklee, *Pschelovodstvo 6,* 48 (1955).

Kramer, G., Orientierte Zugaktivität gekäfigter Singvögel, *Naturwissenschaften 37,* 188 (1950).

——— Eine neue Methode zur Erforschung der Zugorientierung und die bisher damit erzielten Ergebnisse, *Proc. X. internat. Ornithologen Congress (Uppsala 1950),* 269–280 (1951).

——— Experiments on bird orientation, *Ibis 94,* 265–285 (1952).

——— Die Dressur von Brieftauben auf Kompaßrichtung im Wahlkäfig, *Z. Tierpsychol. 9,* 245–251 (1952a).

——— Die Sonnenorientierung der Vögel, *Verhandl. Deutsch. Zool. Ges. in Freiburg 1952* (Leipzig, 1953), 72–84.

——— and Ursula v. Saint-Paul, Stare lassen sich auf Himmelsrichtungen dressieren, *Naturwissenschaften 37,* 526–537 (1950).

Krause, Barbara, Elektronenmikroskopische Untersuchungen an den Plattensensillen des Insektenfühlers, *Zool. Beiträge N. F. 6,* 161–205 (1960).

Kroeber, A. L., Sign and symbol in bee communications, *Proc. Nat. Acad. Sci. (Wash.) 38,* 753–757 (1952).

Kröning, F., Über die Dressur der Biene auf Töne, *Biol. Zbl. 45,* 496–507 (1925).

Krüger, E., Über die Bahnflüge der Männchen der Gattungen *Bombus* und *Psithyrus, Z. Tierpsychol. 8,* 61–75 (1951).

Kühn, A., *Die Orientierung der Tiere im Raum* (Jena, 1919).

——— Über den Farbensinn der Bienen, *Z. vergl. Physiol. 5,* 762–800 (1927).

——— and G. Fraenkel, Über das Unterscheidungsvermögen der Bienen

für Wellenlängen im Spektrum, *Nachr. Ges. d. Wissensch. Göttingen math.-physikal. Kl.* (1927).
——— and R. POHL, Dressurfähigkeit der Bienen auf Spektrallinien, *Naturwissenschaften 9*, 738–740 (1921).
KUGLER, H., Die Ausnutzung der Saftmalsumfärbung bei den Roßkastanienblüten durch Bienen und Hummeln, *Berichte d. Deutsch. Bot. Ges. 54*, 394–400 (1936).
——— Hummeln als Blütenbesucher, *Ergebn. Biol. 19*, 143–323 (1943).
——— Blütenfärbung und Insektenbestäubung, *Berichte physikal.-medizin. Ges. Würzburg N. F. 66*, 28–41 (1954).
——— UV-Musterungen auf Blüten und ihr Zustandekommen, *Planta 59*, 296–329 (1963).
KUIPER, J. W., The optics of the compound eye, *Symposia Soc. exp. Biol. 16, Biological receptor mechanisms* (Cambridge, 1962), 58–71.
KULLENBERG, B., Blommor och Insekter, *Svensk Naturvetenskap*, 81–136 (1956).
KUNZE, G., Einige Versuche über den Geschmackssinn der Honigbiene, *Zool. Jb. (Physiol.) 44*, 287–314 (1928).
——— Einige Versuche über den Antennengeschmackssinn der Honigbiene, *Zool. Jb. (Physiol.) 52*, 465–512 (1933).
KUNZE, P., Untersuchung des Bewegungssehens fixiert fliegender Bienen, *Z. vergl. Physiol. 44*, 656–684 (1961).
KUWABARA, M., Über die Funktion der Antenne der Honigbiene in bezug auf die Raumorientierung, *Mem. Faculty of Science Kyushu University Ser. E (Biology) 1*, 13–64 (1952).
——— and K. NAKA, Response of a single retinula cell to polarized light, *Nature (London) 184*, 455–456 (1959).
——— and K. TAKEDA, On the hygroreceptor of the honeybee *Apis mellifera*, *Physiology and Ecology 7*, 1–6 (1956).

LACHER, V., Elektrophysiologische Untersuchungen an einzelnen Rezeptoren für Geruch, Kohlendioxyd, Luftfeuchtigkeit und Temperatur auf den Antennen der Arbeitsbiene und der Drohne (*Apis mellifica* L.), *Z. vergl. Physiol. 48*, 587–623 (1964).
——— and D. SCHNEIDER, Elektrophysiologischer Nachweis der Riechfunktion von Porenplatten (Sensilla placodea) auf den Antennen der Drohne und der Arbeitsbiene (*Apis mellifica* L.), *Z. vergl. Physiol. 47*, 274–278 (1963).
LARSEN, J. R., The fine structure of the labellar chemosensory hairs of the blowfly *Phormia regina* Meig., *J. Insect Physiol. 8*, 683–691 (1962).
LATHAM, A., The mysteries of swarming, *Gleanings in Bee Culture 55*, 441–442 (1927).
LECOMTE, J., Le comportement agressif des ouvrières d'*Apis mellifica* L., *Annales de l'Abeille 4*, 165–270 (1961).
LEE, W. R., Food gathering behavior of honey bees on fruit crops, in *Symposium food gathering behavior of Hymenoptera*, (mimeo.; Ithaca, N. Y.: Cornell University, 1959).
LEUENBERGER, F., *Die Biene* (3rd ed.: Aarau and Frankfurt a. M., 1954).
LEX, THERESE, Duftmale an Blüten, *Z. vergl. Physiol. 36*, 212–234 (1954).
LINDAUER, M., Über die Einwirkung von Duft- und Geschmacksstoffen sowie anderer Faktoren auf die Tänze der Bienen, *Z. vergl. Physiol. 31*, 348–412 (1948).
——— Ein Beitrag zur Frage der Arbeitsteilung im Bienenstaat, *Z. vergl. Physiol. 34*, 299–345 (1952).
——— Temperaturregulierung und Wasserhaushalt im Bienenstaat, *Z. vergl. Physiol. 36*, 391–432 (1954).
——— Dauertänze im Bienenstock und ihre Beziehung zur Sonnenbahn, *Naturwissenschaften 41*, 506–507 (1954a).
——— Schwarmbienen auf Wohnungssuche, *Z. vergl. Physiol. 37*, 263–324 (1955).

—— Über die Verständigung bei indischen Bienen, *Z. vergl. Physiol. 38,* 521–557 (1956).

—— Zur Biologie der stachellosen Bienen, ihre Abwehrmethoden, *Bericht 8. Wanderversammlung Deutscher Entomologen* (Berlin, 1957), 71–78.

—— Sonnenorientierung der Bienen unter der Aequatorsonne und zur Nachtzeit, *Naturwissenschaften 44,* 1–6 (1957a).

—— Angeborene und erlernte Komponenten in der Sonnenorientierung der Bienen, *Z. vergl. Physiol. 42,* 43–62 (1959).

—— Time-compensated sun orientation in bees, *Cold Spring Harbor Symposia on quantitative Biology 25,* 371–377 (1960).

—— *Communication among social bees* (Cambridge, Mass.: Harvard University Press, 1961).

—— Kompaßorientierung, *Ergebn. Biol. 26,* 158–181 (1963).

—— Allgemeine Sinnesphysiologie, Orientierung im Raum, *Fortschr. Zool. 16,* 58–140 (1963a).

—— and W. KERR, Die gegenseitige Verständigung bei den stachellosen Bienen, *Z. vergl. Physiol. 41,* 405–434 (1958).

—— and H. MARTIN, Über die Orientierung der Biene im Duftfeld, *Naturwissenschaften 50,* 509–514 (1963).

—— and J. O. NEDEL, Ein Schweresinnesorgan der Honigbiene, *Z. vergl. Physiol. 42,* 334–364 (1959).

—— and B. SCHRICKER, Über die Funktion der Ocellen bei den Dämmerungsflügen der Honigbiene, *Biol. Zbl. 82,* 721–725 (1963).

LINEBURG, B., Communication by scent in the honeybee—a theory, *The American Naturalist 58,* 530–534 (1924).

LITTLE, H. F., Reactions of honey bees to oscillations of known frequency, *Anat. Rec. 134,* 601 (1959).

—— Reactions of the honey bee *Apis mellifera* L. to artificial sounds and vibrations of known frequencies, *Ann. entom. Soc. Amer. 55,* 82–89 (1962).

LOCKE, M., Pore canals and related structures in insect cuticle, *J. biophys. biochem. Cytol. 10,* 589–618 (1961).

LOPATINA, N. G., M. A. KUANJEZOWA, and S. W. PANKOWA, On the physiologic nature of the dances of bees, *Zh. Obshch. Biol.* (Akad. Nauk USSR) *19,* No. 1 (1958), in Russian.

LOTMAR, RUTH, Neue Untersuchungen über den Farbensinn der Bienen mit besonderer Berücksichtigung des Ultravioletts, *Z. vergl. Physiol. 19,* 673–723 (1933).

LOTZ, J., Speech and language, *J. acoust. Soc. Amer. 22,* 712–717 (1950).

LUBBOCK, J., *Ants, Bees and Wasps* (London, 1882).

—— *Ameisen, Bienen und Wespen* (Leipzig, 1883).

LUDWIG, LUISE, Ist den Suchbienen ein Schema ihrer Trachtquellen angeboren? *Lehramtsarbeit Naturw. Fak. Univers. München* (*1956*).

LÜDTKE, H., Beziehungen des Feinbaues im Rückenschwimmerauge zu seiner Fähigkeit, polarisiertes Licht zu analysieren, *Z. vergl. Physiol. 40,* 329–344 (1957).

LÜSCHER, M., and B. MÜLLER, Ein spurbildendes Sekret bei Termiten, *Naturwissenschaften 47,* 503 (1960).

LÜTERS, W., and G. BIRUKOW, Sonnenkompaßorientierung der Brandmaus (*Apodemus agrarius* Pall), *Naturwissenschaften 50,* 737–738 (1963).

LUTZ, F. E., Apparently non-selective characters and combinations of characters including a study of ultraviolet in relation to the flower-visiting habits of insects, *Ann. N. Y. Acad. Sci. 29,* 181–283 (1924).

—— "Invisible" colors of flowers and butterflies, *Natural History 33,* 565–576 (1933a).

—— Experiments with "stingless bees" (*Trigona cressoni parastigma*) concerning their ability to distinguish ultraviolet patterns, *American Museum Novitates,* No. 641 (New York, 1933b).

MAGNI, F., F. PAPI, H. E. SAVELY, and P. TONGIORGI, Electroretinographic responses to polarized light in the Wolf spider *Arctosa variana* C. L. Koch, *Experientia* (*Basel*) *18,* 511 (1962).
—— —— —— —— Research on the structure and physiology of the eyes of a Lycosid spider. II. The role of different pairs of eyes in astronomical orientation, *Arch. ital. Biol. 102,* 123–136 (1964).
—— —— —— —— Research on the structure and physiology of the eyes of a Lycosid spider. III. Electroretinographic responses to polarized light, *Arch. ital. Biol. 103,* 146–158 (1965).
MANGER, Über das Tanzen der Honigbiene, *Bayer. Bienenztg. 42,* 167–168 (1920).
MANNING, A., The effect of honey-guides, *Behaviour 9,* 114–139 (1956).
—— Some aspects of the foraging behaviour of bumble-bees, *Behaviour 9,* 164–201 (1956a).
—— Some evolutionary aspects of the flower constancy of bees, *Proc. Roy. physical. Soc. 25,* 67–71 (1957).
MARCHIONNI, VALERIA, Modificazione sperimentale della direzione innata di fuga in *Talorchestia Deshayesei* And. (Crustacea Amphipoda), *Bollett. dell'Istit. e Museo di Zoologia dell'Univers. di Torino 6,* No. 3 (1962).
MARKL, H., Borstenfelder an den Gelenken als Schweresinnesorgane bei Ameisen und anderen Hymenopteren, *Z. vergl. Physiol. 45,* 475–569 (1962).
—— Die Schweresinnesorgane der Insekten, *Naturwissenschaften 50,* 559–565 (1963).
—— Geomenotaktische Fehlorientierung bei *Formica polyctena* Förster, *Z. vergl. Physiol. 48,* 552–586 (1964).
MARKS, W. B., W. H. DOBELLE, and E. F. MACNICHOL, Visual pigments of single primate cones, *Science 143,* 1181–1183 (1964).
MARSHALL, J., On the sensitivity of the chemoreceptors on the antenna and fore-tarsus of the honey-bee, *Apis mellifica* L., *J. exp. Biol. 12,* 17–26 (1935).
MARTIN, H., Zur Nahorientierung der Biene im Duftfeld, zugleich ein Nachweis für die Osmotropotaxis bei Insekten, *Z. vergl. Physiol. 48,* 481–533 (1964).
—— Leistungen des topochemischen Sinnes bei der Honigbiene, *Z. vergl. Physiol. 50,* 254–292 (1965).
MARTIN, P., Die Steuerung der Volksteilung beim Schwärmen der Bienen. Zugleich ein Beitrag zum Problem der Wanderschwärme, *Insectes sociaux* (*Paris*) *10,* 13–42 (1963).
MASCHWITZ, U., Gefahrenalarmstoffe und Gefahrenalarmierung bei sozialen Hymenopteren, *Z. vergl. Physiol. 47,* 596–655 (1964).
MATHIS, M., Erreurs d'orientation de la reine abeille en retour de son vol nuptial, *Insectes sociaux* (*Paris*) *7,* 213–219 (1960).
MATTHEWS, G. V. T., "Nonsense" orientation in mallard *Anas platyrhynchos* and its relation to experiments on bird navigation, *Ibis 103a,* 211–230 (1961).
—— The astronomic bases of "nonsense" orientation, *Proc. XIII. intern. ornithol. Congr.* (1963), 415–429.
MAURIZIO, ANNA, Weitere Untersuchungen an Pollenhöschen, *Beihefte z. Schweiz. Bienenztg. 2,* 20 (1953).
—— Breakdown of sugars by inverting enzymes in the pharyngeal glands and midgut of the honeybee. 1. Summer bees, preliminary report, *Bee World 38,* 14–17 (1957).
—— Zuckerabbau unter der Einwirkung der invertierenden Fermente in Pharynxdrüsen und Mitteldarm der Honigbiene (*Apis mellifica* L.), 1. Sommerbienen der Krainer- und Nigra-Rasse, *Insectes sociaux* (*Paris*) *4,* 225–243 (1957a).
—— Bienenbotanik, in A. BÜDEL and E. HEROLD, *Biene und Bienenzucht* (Munich, 1960), 68–104.
—— Zuckerabbau unter der Einwirkung der invertierenden Fermente in

Pharynxdrüsen und Mitteldarm der Honigbiene (*Apis mellifica* L.). 5. Einfluß von Alter und Ernährung der Bienen auf die Fermentaktivität der Pharynxdrüsen, *Annales des Abeilles* 5 (3), 215–232 (1962).

——— Das Pollenbild des Honigs einzelner Völker eines Standes, *Deutsche Bienenwirtschaft* (Schels-Festschrift), 235–239 (1962a).

MAZOCHIN-PORSCHNJAKOV, G. A., Über das Farbensehen der Insekten *Biophys.* (Akad. Nauk USSR) *1*, No. 1 (1956).

——— Die Ultraviolettreflexion der Blüten und das Sehen der Insekten *Entomol. Obozvenia* (Akad. Nauk USSR) *38*, 312–325 (1959).

——— Green, yellow and orange colour discrimination in bees *Biofizika* (Akad. Nauk USSR) *4*, 48–54 (1959a).

——— Colorimetric study of colour vision in the dragon-fly *Biofizika* (Akad. Nauk USSR) *4*, 427–436 (1959b).

——— Farbmetrischer Beweis der Trichromasie des Farbensehens der Bienen (am Beispiel der Hummeln) *Biofizika* (Akad. Nauk USSR) *7*, 211–217 (1962).

MEDER, E., Über die Einberechnung der Sonnenwanderung bei der Orientierung der Honigbiene, *Z. vergl. Physiol.* 40, 610–641 (1958).

MÉDIONY, J., L'orientation "astronomique" des Arthropodes et des oiseaux, *Ann. Biol.* 32, Fasc. 1–2, 37–67 (1956).

MENZER, G., and K. STOCKHAMMER, Zur Polarisationsoptik des Facettenauges von Insekten, *Naturwissenschaften* 38, 190–191 (1951).

MERKEL, F. W., H. G. FROMME, and W. WILTSCHKO, Nicht visuelles Orientierungsvermögen bei nächtlich zugunruhigen Rotkehlchen. *Die Vogelwarte* 22, 168–173 (1964).

METSCHL, N., Elektrophysiologische Untersuchungen an den Ocellen von *Calliphora*, *Z. vergl. Physiol.* 47, 230–255 (1963).

MEYER, G. F., Versuch einer Darstellung von Neurofibrillen im zentralen Nervensystem verschiedener Insekten, *Zool. Jb.* (*Anatom.*) 71, 413–426 (1951).

MEYER, WALDTRAUT, Die "Kittharzbienen" und ihre Tätigkeiten, *Z. Bienenforsch.* 5, 185–200 (1954).

——— Arbeitsteilung im Bienenschwarm, *Naturwissenschaften* 42, 350 (1955).

MILLER, W. H., Morphology of the ommatidia of the compound eye of Limulus, *J. biophys. biochem. Cytol.* 3, 421–428 (1957).

MILUM, V. G., Grooming dance and associated activities of the honeybee colony, *Illinois Academy of Sci. Transactions* 40, 194–196 (1947).

——— Honey bee communication, *Amer. Bee J.* 95, 97–104 (1955).

MINDERHOUD, A., Untersuchungen über das Betragen der Honigbiene als Blütenbestäuberin, *Gartenbauwissenschaft* 4, 342–362 (1931).

——— Over pogingen om de honigbij te dwingen de roode Klaver te bevliegen, *Mededeelingen Dir. van de Tuinbouw*, 70–76 (1946).

——— Over het leiden van Bijen naar bepaalde Drachtplanten, *Mededeelingen Dir. van de Tuinbouw*, 381–392 (1948).

MINNICH, D. E., An experimental study of the tarsal chemoreceptors of two nymphalid butterflies, *J. exp. Zool.* 33, 173–203 (1921).

——— The chemical sensitivity of the tarsi of the red admiral butterfly, *Pyrameis atalanta* L., *J. exp. Zool.* 35, 57–81 (1922).

——— A quantitative study of tarsal sensitivity to solutions of saccharose, in the red admiral butterfly, *Pyrameis atalanta* L., *J. exp. Zool.* 36, 445–457 (1922a).

——— The organs of taste on the proboscis of the blowfly *Phormia regina* Meigen, *Anat. Rec.* 34, 126 (1926).

——— The chemical sensitivity of the legs of the blow-fly, *Calliphora vomitoria* Linn. to various sugars, *Z. vergl. Physiol.* 11, 1–55 (1929).

——— The sensitivity of the oral lobes of the proboscis of the blow-fly, *Calliphora vomitoria* L., to various sugars, *J. exp. Zool.* 60, 121–139 (1931).

——— The contact chemoreceptors of the honey bee *Apis mellifera* Linn., *J. exp. Zool.* 61, 375–393 (1932).

Mirić, D., Propolis und seine Verwendung bei den Bienen, *Sbornik matice srpske sweska 8, Ser. tschlan. nautsch. saop.* (1955).

——— Untersuchungen über das Sammeln und den Verbrauch von Wasser bei den Bienen, *Sbornik matice srpske, Ser. prirodn. nauk-sweska 10* (1956).

Mittelstaedt, H., Probleme der Kursregelung bei frei beweglichen Tieren, *Aufnahme und Verarbeitung von Nachrichten durch Organismen* (Stuttgart, 1961), 138–148.

——— Control systems of orientation in insects, *Ann. Rev. Entomol. 7,* 177–198 (1962).

Moody, M. F., Evidence for the intraocular discrimination of vertically and horizontally polarized light by *Octopus, J. exp. Biol. 39,* 21–30 (1962).

——— and J. R. Parriss, The discrimination of polarized light by *Octopus,* a behavioural and morphological study, *Z. vergl. Physiol. 44,* 268–291 (1961).

——— and J. D. Robertson, The fine structure of some retinal photoreceptors, *J. biophys. biochem. Cytol. 7,* 87–92 (1960).

Morgenthaler, O., Ein "Putz-Tanz" der Bienen, *Schweiz. Bienenztg.* (1949), 198–199.

Morita, H., S. Doira, K. Takeda, and M. Kuwabara, Electrical response of contact chemoreceptor on tarsus of the butterfly *Vanessa indica, Mém. Fac. Sci. Kyushu Univ. Japan, Ser. E* (*Biol.*) *2,* 119–139 (1957).

Müller, H., *Die Befruchtung der Blumen durch Insekten und die gegenseitigen Anpassungen beider* (Leipzig, 1873).

Mueller, H. C., and J. T. Emlen, Homing in bats, *Science 126,* 307–308 (1957).

Naka, K., Recording of retinal action potentials from single cells in the insect compound eye, *J. Gen. Physiol. 44,* 571–584 (1961).

——— and E. Eguchi, Spike potentials recorded from the insect photoreceptor, *J. Gen. Physiol. 45,* 663–680 (1962).

Nedel, J. O., Morphologie und Physiologie der Mandibeldrüse einiger Bienen-Arten (Apidae), *Z. Morphol. u. Ökol. d. Tiere 49,* 139–183 (1960).

Neese, V., Zur Funktion der Augenborsten bei der Honigbiene. *Z. vergl. Physiol. 49,* 543–585 (1965).

Neuhaus, W., Zur Frage der Osmotropotaxis, besonders bei der Honigbiene, *Z. vergl. Physiol. 49,* 475–484 (1965).

New, D. A. T., Effects of small zenith distances of the sun on the communication of honey bees, *J. Insect Physiol. 6,* 196–208 (1961).

——— F. R. Burrowes, and A. J. Edgar, Honeybee communication when the sun is close to the zenith, *Nature* (*London*) *189,* 155–156 (1961).

——— and J. K. New, The dances of honey-bees at small zenith distances of the sun, *J. exp. Biol. 39,* 271–291 (1962).

Nielsen, E. T., On the habits of the migratory butterfly *Ascia monuste* L., *Biol. Meddelser kong. Danske Vidensk. Selskab 23,* Nr. 11, 1–81 (1961).

——— On the migration of insects, *Ergebn. Biol. 27,* 162–193 (1964).

Nixon, H. L., and C. R. Ribbands, Food transmission within the honeybee community, *Proc. Roy. Soc. B 140,* 43–50 (1952).

Nolte, D. J., Submicroscopic structure of the Drosophilid eye, *S. Afr. J. Sci. 57,* 121–125 (1961).

Oettingen-Spielberg, Therese zu, Über das Wesen der Suchbiene, *Z. vergl. Physiol. 31,* 454–489 (1949).

Opfinger, Elisabeth, Über die Orientierung der Biene an der Futterquelle, *Z. vergl. Physiol. 15,* 431–487 (1931).

——— Zur Psychologie der Duftdressuren bei Bienen, *Z. vergl. Physiol. 31,* 441–453 (1949).

Otto, F., Die Bedeutung des Rückfluges für die Richtungs- und Entfernungsangabe der Bienen, *Z. vergl. Physiol. 42,* 303–333 (1959).

PAIN, J., and M. F. RUTTNER, Les extraits des glandes mandibulaires des reines d'abeilles attirent les males, lors du vol nuptial, *C. R. Acad. Sci. (Paris) 256,* 512–515 (1963).

PALITSCHEK V. PALMFORST, E., Ein Beitrag zur optischen und statischen Orientierung der Biene, *Inaug. Dissertation Philos. Fak. Universität Graz* (1952).

PAPARIA, Training of bees for seeking and pollinizing crops, *Kolgosnje Bschiluizstwo 4,* 16 (1940), in Ukrainian.

PAPI, F., Orientamento astronomico in alcuni Carabidi, *Atti della societá toscana di Sci. nat. Pisa Mem. 62,* Ser. B, 83–97 (1955).

——— Astronomische Orientierung bei der Wolfspinne *Arctosa perita* Latr., *Z. vergl. Physiol. 37,* 230–233 (1955a).

——— Ricerche sull'orientamento astronomico di *Arctosa perita* Latr. (Araneae Lycosidae), *Pubbl. Staz. Zoolog. Napoli 27,* 76–103 (1955b).

——— Sull'orientamento astronomico in specie del gen. *Arctosa* (Araneae Lycosidae), *Z. vergl. Physiol. 41,* 481–489 (1959).

——— Orientation by night, the moon, *Cold Spring Harbor Symposia on quantit. Biology 25,* 475–480 (1960).

——— and L. PARDI, Nuovi reperti sull'orientamento lunare di *Talitrus saltator* Montagu (Crustacea Amphipoda), *Z. vergl. Physiol. 41,* 583–596 (1959).

——— ——— Ricerche sull'orientamento di *Talitrus saltator* (Montagu) (Crustacea-Amphipoda). II. Sui fattori che regolano la variazione dell'angolo di Orientamento nel corso del giorno; l'orientamento di notte. L'orientamento diurno di altre popolazioni, *Z. vergl. Physiol. 35,* 490–518 (1953).

——— ——— La luna come fattore di orientamento degli animali, *Boll. Istit. Mus. Zool. Universit. Torino 4,* 1–4 (1954).

——— L. SERRETTI, and S. PARRINI, Nuove ricerche sull'orientamento e il senso del tempo di *Arctosa perita* Latr., *Z. vergl. Physiol. 39,* 531–561 (1957).

——— and J. SYRJÄMÄKI, The sun-orientation rhythm of wolf spiders at different latitudes, *Arch. ital. Biol. 101,* 59–77 (1963).

——— and P. TONGIORGI, Innate and learned components in the astronomical orientation of wolf spiders, *Ergebn. Biol. 26,* 259–280 (1963).

PARDI, L., Esperienze sull'orientamento di *Talitrus saltator,* l'orientamento al sole degli individui a ritmo nicti-emerale invertito durante la "loro notte," *Boll. Istit. Mus. Zool. Universit. Torino 4,* No. 9 (1953/54).

——— L'orientamento diurno di *Tylos latreillii: Boll. Istit. Mus. Zool. Universit. Torino 4,* No. 11 (1953/54a).

——— Über die Orientierung von *Tylos latreillii* (Isopoda terrestria), *Z. Tierpsychol. 11,* 175–181 (1954).

——— Orientamento solare in un Tenebrionide alofilo, *Phaleria provincialis* Fauv. (Coleopt.). *Boll. Istit. Mus. Zool. Universit. Torino 5,* No. 1 (1955/56).

——— Modificazione sperimentale della direzione di fuga negli anfipodi ad orientamento solare, *Z. Tierpsychol. 14,* 261–275 (1957).

——— L'orientamento astronomico degli animali, resulati e problemi attuali, *Boll. Zool. 24,* fasc. II (1957a).

——— Esperienze sull'orientamento solare di *Phaleria provincialis* Faur. (Col.), il comportamento a luce artificiale durante l'intero ciclo di 24 ore, *Atti Accad. Sci. Torino 92* (1957/58).

——— Innate components in the solar orientation of littoral amphipods, *Cold Spring Harbor Symposia on quantit. Biolog. 25,* 395–401 (1960).

——— Orientamento astronomico vero in un isopodo marino, *Idotea baltica Basteri* (Audouin), *Monitore zoologico italiano 70/71,* 491–495 (1962/63).

——— and A. ERCOLINI, Ricerche sull'orientamento astronomico di Anfipodi litorali della zona equatoriale. II. L'orientamento lunare in una popolazione somala di *Talorchestia martensii* Weber, *Z. vergl. Physiol. 50,* 225–249 (1965).

——— and M. GRASSI, Experimental modification of direction-finding in

Talitrus saltator Montagu and *Talorchestia deshayesei* Aud. (Crustacea-Amphipoda), *Experientia* (*Basel*) *11*, 202 (1955).

—— and F. PAPI, Die Sonne als Kompaß bei *Talitrus saltator* (Montagu), Amphipoda, Talitridae, *Naturwissenschaften 39*, 262–263 (1952).

—— —— Ricerche sull'orientamento di *Talitrus saltator*. I. L'orientamento durante il giorno in una popolazione del litorale Tirrenico, *Z. vergl. Physiol. 35*, 459–489 (1953).

PARK, W., Flight studies of the honey bee, *Amer. Bee J. 63*, 71 (1923).

—— The "language" of bees, *Amer. Bee J. 63*, 227 (1923a).

—— Some "whys" of bee behavior, *Amer. Bee J. 63*, 399–400 (1923b).

—— Behavior of water-carriers, *Amer. Bee J. 63*, 553 (1923c).

PARKER, R. L., *The collection and utilization of pollen by the honeybee* (Ithaca, N. Y., 1925).

PASCHKE, I., Über das Zeitgedächtnis der Bienen, *Lehramtsarbeit Naturw. Fak. Univ. München* (1956).

PEER, D., The foraging range of the honey bee, Part 1 (Ph. D. thesis, University of Wisconsin, 1955).

PERDECK, A. C., Two types of orientation in migrating starlings, *Sturnus vulgaris*, and chaffinch, *Fringilla coelebs*, as revealed by displacement experiments, *Ardea* (*Leiden*) *46*, 1–37 (1958).

—— Does navigation without visual clues exist in robins? *Ardea* (*Leiden*) *51*, 91–104 (1963).

PETERS, W., Die Zahl der Sinneszellen von Marginalborsten und das Vorkommen multipolarer Nervenzellen in den Labellen von *Calliphora erythrocephala* Meig. (Diptera), *Naturwissenschaften 48*, 412–413 (1961).

PHILLIPS, E. F., Structure and development of the compound eye of the honey bee, *Proc. Acad. Nat. Sci. Philadelphia 57*, 123–157 (1905).

PISCITELLI, ANNEMARIE, Über die Bevorzugung mineralstoffhaltiger Lösungen gegenüber reinem Wasser durch die Honigbiene, *Z. vergl. Physiol. 42*, 501–524 (1959).

PORSCH, O., Grellrot als Vogelblumenfarbe, *Biologia generalis 7*, 647–674 (1931).

PORTILLO, J. DEL, Beziehungen zwischen den Öffnungswinkeln der Ommatidien, Krümmung und Gestalt der Insektenaugen und ihrer funktionellen Aufgabe, *Z. vergl. Physiol. 23*, 100–145 (1936).

PRITSCH, G., Versuche zur Duftlenkung der Bienen auf Rotklee zwecks Erhöhung der Samenerträge, *Arch. Geflügelzucht u. Kleintierkunde 8*, 214 (1959).

RADEMACHER, B., *Meine Erkenntnisse über die Zellstellungen der Bienenwaben* (Hannover, 1960).

RAU, P., *The jungle bees and wasps of Barro Colorado Island* (Kirkwood, Missouri; P. Rau, 1933).

RAUSCHMAYER, F., Das Verfliegen der Bienen und die optische Orientierung am Bienenstand, *Arch. Bienenkunde 9*, 249–322 (1928).

RENNER, M., Über die Haltung von Bienen in geschlossenen, künstlich beleuchteten Räumen, *Naturwissenschaften 42*, 539–540 (1955).

—— Neue Versuche über den Zeitsinn der Honigbiene, *Z. vergl. Physiol. 40*, 85–118 (1957).

—— Der Zeitsinn der Arthropoden, *Ergebn. Biol. 20*, 127–158 (1958).

—— Über ein weiteres Versetzungsexperiment zur Analyse des Zeitsinnes und der Sonnenorientierung der Honigbiene, *Z. vergl. Physiol. 42*, 449–483 (1959).

—— Das Duftorgan der Honigbiene und die physiologische Bedeutung ihres Lockstoffes, *Z. vergl. Physiol. 43*, 411–468 (1960).

—— and MARGOT BAUMANN, Über Komplexe von subepidermalen Drüsenzellen (Duftdrüsen?) der Bienenkönigin, *Naturwissenschaften 51*, 68–69 (1964).

RENSING, L., Beiträge zur vergleichenden Morphologie, Physiologie und Ethologie der Wasserläufer (Gerroidea), *Zoologische Beiträge N. F. 7*, 447–485 (1962).

RÉVÉSZ, G., Der Kampf um die sogenannte Tiersprache, *Psychologische Rundschau 4/2* (1953).

RHEIN, W. v., Über die Duftlenkung der Bienen beim Raps im Jahre 1952 und ihre Ergebnisse, *Die Hessische Biene 88*, H. 8/9 (1952/53).

——— Über die Bedeutung der Honigbiene für die Saatzuchtwirtschaft, *Saatgut-Wirtschaft*, 30–32 (1954).

——— Über die Duftlenkung der Bienen zur Steigerung der Samenerträge des Rotklees (*Trifolium pratense* L.), *Z. Acker- u. Pflanzenbau 103*, 273–314 (1957).

——— Referat d. Arbeit FREE (1958), *Arch. Bienenkunde 36*, 74 (1959).

RIBBANDS, C. R., The foraging method of individual honeybees, *J. Anim. Ecol. 18*, 47–66 (1949).

——— Division of labour in the honeybee community, *Proc. Roy. Soc. B 140*, 32–43 (1952).

——— The inability of honeybees to communicate colours, *Brit. J. Anim. Behav. 1*, 1–2 (1953).

——— *The behaviour and social life of honeybees* (London: Bee Research Association, Ltd., 1953a).

——— Communication between honeybees. I. The response of crop-attached bees to the scent of their crop, *Proc. Roy. Entom. Soc. London (A) 29*, 10–12 (1954).

——— Communication between honeybees. II. The recruitment of trained bees, and their response to improvement of the crop, *Proc. Roy. Entom. Soc. London (A) 30*, 1–3 (1955).

——— The scent perception of the honeybee, *Proc. Roy. Soc. B 143*, 367–379 (1955a).

——— and N. SPEIRS, The adaptability of the home coming honeybees, *Brit. J. Anim. Behav. 1*, 59–66 (1953).

RICHARDS, G., Studies on arthropod cuticle. VIII. The antennal cuticle of honeybees with particular reference to the sense plates, *Biol. Bull. 103*, 201–225 (1952).

——— Structure and development of the integument, in K. D. ROEDER, *Insect Physiology* (New York and London, 1953).

ROBERT, P., Les migrations orientées du Hanneton commun *Melolontha melolontha* L., *Ergebn. Biol. 26*, 135–146 (1963).

ROBERTIS, E. DE, Some observations on the ultrastructure and morphogenesis of photoreceptors, *J. gener. Physiol. 43*, 1–6 (1960).

ROBERTS, W. C., and O. MACKENSEN, Breeding improved honeybees, *Amer. Bee J. 91*, 292–294, 328–330, 382–384, 418–421, 473–475 (1951).

ROEPKE, W., Beobachtungen an indischen Honigbienen, insbesondere an *Apis dorsata, Mededeelingen van de Landbouwhoogeschool Deel 34* (Wageningen, 1930).

RÖSCH, G. A., Beobachtungen an Kittharz sammelnden Bienen, *Biol. Zbl. 47*, 113–121 (1927).

ROKOHL, RUTH, Über die regionale Verschiedenheit der Farbentüchtigkeit im zusammengesetzten Auge von *Notonecta glauca, Z. vergl. Physiol. 29*, 638–676 (1942).

ROTHSCHILD, F. S., Die symbolischen Tänze der Bienen als psychologisches und neurologisches Problem, *Schweiz. Z. Psychologie u. ihre Anwendungen 12*, 177–199 (1953).

ROWELL, C. H. F., and M. J. WELLS, Retinal orientation and the discrimination of polarized light by *Octopus, J. exp. Biol. 38*, 827–831 (1961).

RUTTNER, F., Die Sexualfunktionen der Honigbienen im Dienste ihrer sozialen Gemeinschaft, *Z. vergl. Physiol. 39*, 577–600 (1957).

——— and H. RUTTNER, Untersuchungen über die Flugaktivität und das

Paarungsverhalten der Drohnen, 2. Beobachtungen an Drohnensammelplätzen. *Z. f. Bienenforschung 8,* 1–9 (1965).

——— and O. MACKENSEN, The genetics of the honeybee, *Bee World 33,* 53–62 and 71–79 (1952).

SAKAGAMI, S. F., Einige Versuche über den Geschmackssinn der Hummeln, *Kontyu 18,* Pt. 5, 4–10 (1950).

SANTSCHI, F., Observations et remarques critiques sur le mecanisme de l'orientation chez les fourmis, *Rev. Suisse de Zoologie 19,* 303–338 (1911).

SAUER, F., Die Sternenorientierung nächtlich ziehender Grasmücken (*Sylvia atricapilla, borin* und *curruca*), *Z. Tierpsychol. 14,* 29–70 (1957).

——— and ELEONORE M. SAUER, Star navigation of nocturnal migrating birds, *Cold Spring Harbor Symposia on quantitative Biology 25,* 463–473 (1960).

SCHALLER, A., Sinnesphysiologische und psychologische Untersuchungen an Wasserkäfern und Fischen, *Z. vergl. Physiol. 4,* 370–464 (1926).

SCHICK, W., Über die Wirkung von Giftstoffen auf die Tänze der Bienen, *Z. vergl. Physiol. 35,* 105–128 (1953).

SCHIFFERER, GERTRAUD, Über die Entfernungsangabe bei den Tänzen der Bienen, *Lehramtsarbeit naturw. Fak. Univ. München* (1952).

SCHMID, J., Zur Frage der Störung des Bienengedächtnisses durch Narkosemittel, zugleich ein Beitrag zur Störung der sozialen Bindung durch Narkose, *Z. vergl. Physiol. 47,* 559–595 (1964).

SCHMIDT, ANNELIESE, Geschmacksphysiologische Untersuchungen an Ameisen, *Z. vergl. Physiol. 25,* 351–378 (1938).

SCHMIDT-KOENIG, K., Über die Orientierung der Vögel; Experimente und Probleme, *Naturwissenschaften 51,* 423–431 (1964).

SCHNEIDER, D., V. LACHER, and K. KAISSLING, Die Reaktionsweise und das Reaktionsspektrum von Riechzellen bei *Antheraea pernyi* (Leptidoptera, Saturniidae), *Z. vergl. Physiol. 48,* 632–662 (1964).

SCHNEIDER, F., Über die Vergiftung der Bienen mit Dinitrokresol, *Mitt. Schweiz. Entomol. Gesellsch. 22,* 293–308 (1949).

——— Zur Orientierung des Maikäfers beim Rückflug, *Mitt. Schweiz. Entomol. Gesellsch. 29,* 69–70 (1956).

——— Neue Beobachtungen über die Orientierung des Maikäfers, *Schweiz. Z. Obst- u. Weinbau 66,* 414–415 (1957).

——— Beeinflussung der Aktivität des Maikäfers durch Veränderung der gegenseitigen Lage magnetischer und elektrischer Felder, *Mitt. Schweiz. Entomol. Gesellsch. 33,* 223–237 (1961).

——— Orientierung und Aktivität des Maikäfers unter dem Einfluß richtungsvariabler künstlicher elektrischer Felder und weiterer ultraoptischer Bezugssysteme, *Mitt. Schweiz. Entomol. Gesellsch. 36,* 1–26 (1963).

——— Ultraoptische Orientierung des Maikäfers (*Melolontha vulgaris* F.) in künstlichen elektrischen und magnetischen Feldern, *Ergebn. Biol. 26,* 147–157 (1963a).

——— Systematische Variationen in der elektrischen, magnetischen und geographisch-ultraoptischen Orientierung des Maikäfers, *Vjschr. Naturf. Gesellsch. Zürich 108,* 373–416 (1963b).

SCHÖNE, H., Statozystenfunktion und statische Lageorientierung bei dekapoden Krebsen, *Z. vergl. Physiol 36,* 241–260 (1954).

——— Optisch gesteuerte Lageänderungen (Versuche an Dytiscidenlarven zur Vertikalorientierung), *Z. vergl. Physiol. 45,* 590–604 (1962).

——— Menotaktische Orientierung nach polarisiertem und unpolarisiertem Licht bei der Mangrovekrabbe *Goniopsis, Z. vergl. Physiol. 46,* 496–514 (1963).

SCHOLZE, E., H. PICHLER, and H. HERAN, Zur Entfernungsschätzung der Bienen nach dem Kraftaufwand, *Naturwissenschaften 51,* 69–70 (1964).

SCHREINDLER, ELFRIEDE, Der Einfluß des Flugalters auf die Entfernungsweisung der Honigbiene, *Lehramtsarbeit Univ. Graz* (1964).

SCHREMMER, F., Sinnesphysiologie und Blumenbesuch des Falters von *Plusia gamma* L., *Zool. Jb. (System. and Ökol.)* 74, 375–434 (1941).

SCHRICKER, B., Die Orientierung der Honigbiene in der Dämmerung, zugleich ein Beitrag zur Frage der Ocellenfunktion bei Bienen, *Z. vergl. Physiol.* 49, 420–458 (1965).

SCHUÀ, L., Untersuchungen über den Einfluß meteorologischer Elemente auf das Verhalten der Honigbiene, *Z. vergl. Physiol.* 34, 258–277 (1952).

SCHÜZ, E., Die Spät-Auflassung ostpreußischer Jungstörche in Westdeutschland 1933, *Vogelwarte 15*, 63–78 (1949).

SCHWARZ, R., Über die Riechschärfe der Honigbiene, *Z. vergl. Physiol.* 37, 180–210 (1955).

SCHWEIGER, ELISABETH M., Über individuelle Unterschiede in der Entfernungs- und Richtungsangabe bei den Tänzen der Bienen, *Z. vergl. Physiol.* 41, 272–299 (1958).

SEITZ, A., Die Paarbildung bei einigen Cichliden I., *Z. Tierpsychol.* 4, 40–84 (1940).

SEKERA, Z., Polarization of skylight, in T. F. MALONE, ed., *Compendium of meteorology* (Boston: American Meteorological Society, 1951), 79–90.

——— *Investigation of polarization of skylight* (Report AF 19(122); Los Angeles: Department of Meteorology, University of California, 1955).

——— Polarization of skylight, in S. FLÜGGE, ed., *Handbuch der Physik/Encyclopedia of Physics* (Berlin, New York: Springer-Verlag, 1957), vol. 48, pp. 288–328.

SELZER, R., Untersuchungen über den Feinbau und Bestimmung optischer Daten von Cornea und Kristallkegel des Rückenschwimmerauges, Staatsexamensarbeit Freiburg/Br. 1955.

SEYBOLD, A., and A. WEISSWEILER, Spektrophotometrische Messungen an Blumenblättern, *Botanisches Archiv 45*, 358–366 (1944).

SHAPOSHNIKOVA, N. G., On the factors determining the formation of the recruitment signal in honey bees, *Entomol. Obozvenia* (Akad. Nauk USSR) 37, 546–557 (1958).

SIMPSON, J., The mechanism of honey-bee queen piping, *Z. vergl. Physiol.* 48, 277–282 (1964).

SINGH, S., *Behavior studies of honeybees in gathering nectar and pollen* (Cornell Univ. Agric. Exp. Station, Ithaca; Mem. 288, 1950).

SJÖSTRAND, F. S., The ultrastructure of the retinal receptors of the vertebrate eye, *Ergebn. Biol.* 21, 128–160 (1959).

SLADEN, F. W., A scent-producing organ in the abdomen of the worker of *Apis mellifera, Entomol. Monthly Mag.* 38, 208–211 (1902).

SLIFER, ELEANOR H., J. J. PRESTAGE, and H. W. BEAMS, The chemoreceptors and other sense organs on the antennal flagellum of the grasshopper (Orthoptera: Acrididae), *J. Morphol.* 105, 145–192 (1959).

SLIFER, ELEANOR, and S. S. SEKHON, Fine structure of the sense organs on the antennal flagellum of the honey bee, *Apis mellifera* L., *J. Morphol.* 109, 351–362 (1961).

SMITH, F. G., Communication and foraging ranges of African bees compared with that of European and Asian bees, *Bee World 39*, 249–252 (1958).

SMITH, F. E., and E. R. BAYLOR, Bees, Daphnia and polarized light, *Ecology 41*, 260–363 (1960).

SOLS, A., E. CADENAS, and F. ALVARADO, Enzymatic basis of mannose toxicity in honeybees, *Science 131*, 297–298 (1960).

SOROKIN, Unsere Versuche über den Flug der Bienen auf Kleesaaten, *Pschelovodstvo*, H. 8, 9 (1938).

SPITZNER, M. J. E., *Ausführliche Beschreibung der Korbbienenzucht im sächsischen Churkreise, ihrer Dauer und ihres Nutzens, ohne künstliche Vermehrung nach den Gründen der Naturgeschichte und nach eigener langer Erfahrung* (Leipzig, 1788).

——— *Ausführliche theoretische und praktische Beschreibung der Korbbienenzucht* (Leipzig, 1810).

SPRENGEL, CHR. K., *Das entdeckte Geheimnis der Natur im Bau und in der Befruchtung der Blumen* (Berlin, 1793).
SSACHAROW, Dressur von Bienen auf Nektarsammeln, *Pschelovodstvo,* No. 7, 24 (1952).
STAPEL, CHR., Experiments on scent-feeding of honeybees with reference to red clover pollination, *Tidsskrift for Planteavl 65,* 477–518 (1961).
STAUDENMAYER, TH., Die Giftigkeit der Mannose für Bienen und andere Insekten, *Z. vergl. Physiol. 26,* 644–668 (1939).
STECHE, W., *Gibt es "Dialekte" der Bienensprache?* (Diss. Naturw. Fak. Univ. München 1954).
——— Beiträge zur Analyse der Bienentänze, *Insectes sociaux (Paris) 4,* 305–318 (1957).
——— Gelenkter Bienenflug durch "Attrappentänze," *Naturwissenschaften 44,* 598 (1957a).
STEIN, G., Beiträge zur Biologie der Hummel, *Zool. Jb. (System.) 84,* 439–462 (1956).
——— Untersuchungen über den Sexuallockstoff der Hummelmännchen, *Biol. Zbl. 82,* 343–349 (1963).
STEIN, HEDWIG, Untersuchungen über den Zeitsinn bei Vögeln, *Z. vergl. Physiol. 33,* 387–403 (1951).
STEINHOFF, HILDTRAUT, Untersuchungen über die Haftfähigkeit von Duftstoffen am Bienenkörper, *Z. vergl. Physiol. 31,* 38–57 (1948).
STEPHENS, G. C., M. FINGERMAN, and F. A. BROWN, A non birefringent mechanism for orientation to polarized light in Arthropods, *Anat. Rec. 113,* 559–560 (1952).
——— ——— ——— The orientation of *Drosophila* to plane polarized light, *Ann. Entomol. Soc. Amer. 46,* 75–83 (1953).
STOCKHAMMER, K., Die Wahrnehmung der Schwingungsrichtung linear polarisierten Lichtes bei Insekten, *Z. vergl. Physiol. 38,* 30–83 (1956).
——— Die Orientierung nach der Schwingungsrichtung linear polarisierten Lichtes und ihre sinnesphysiologischen Grundlagen, *Ergebn. Biol. 21,* 23–56 (1959).
STUART, A. M., Mechanism of trail-laying in two species of termites, *Nature (London) 189,* 419 (1961).
——— Origin of the trail in the Termites *Nasutitermes cornigera* (Motschulsky) and *Zootermopsis nevadensis* (Hagen), Isoptera, *Physiol. Zool. (Chicago) 36,* 69–84 (1963).
——— Studies on the communication of alarm in the termite *Zootermopsis nevadensis* (Hagen), Isoptera, *Physiol. Zool. (Chicago) 36,* 85–96 (1963a).
STÜRCKOW, BRUNHILD, Über den Geschmackssinn und den Tastsinn von *Leptinotarsa decemlineata* Say (Chrysomelidae), *Z. vergl. Physiol. 42,* 255–302 (1959).
——— Ein Beitrag zur Morphologie der labellaren Marginalborste der Fliegen *Calliphora* und *Phormia, Z. Zellforschung 57,* 627–647 (1962).
——— and G. QUADBECK, Elektrophysiologische Untersuchungen über den Geschmackssinn des Kartoffelkäfers *Leptinotarsa decemlineata* Say, *Z. Naturforsch. 13 b,* 93–95 (1958).
STUMPER, M. R., Sur l'évaluation ergométrique des distances chez les abeilles, *C. R. Acad. Sci. (Paris) 240,* 1936–1938 (1955).
SUDD, J. H., Interaction between ants on a scent trail, *Nature (London) 183,* 1588 (1959).
——— The foraging method of Pharaoh's ant, *Monomorium pharaonis* L., *Anim. Behav. 8,* 67–75 (1960).

TENCKHOFF-EIKMANNS, INGE, Licht- und Erdschwereorientierung beim Mehlkäfer *Tenebrio molitor* L. und einigen anderen Insekten, *Zool. Beiträge N. F. 4,* 307–341 (1959).
THAKAR, C. V., and K. V. TONAPI, Nesting behaviour of Indian honeybees. I.

Differentiation of worker, queen and drone cells on the combs of *Apis dorsata* Fab., *Bee World 42*, 61–62 (1961).
THORPE, W. H., Orientation and methods of communication of the honey bee and its sensitivity to the polarization of the light, *Nature (London) 164*, 11 (1949).
THURM, U., Die Beziehungen zwischen mechanischen Reizgrößen und stationären Erregungszuständen bei Borstenfeldsensillen von Bienen, *Z. vergl. Physiol. 46*, 351–382 (1963).
TINBERGEN, N., Über die Orientierung des Bienenwolfes (*Philanthus triangulum*), *Z. vergl. Physiol. 16*, 305–334 (1932).
―――― and W. KRUYT, Über die Orientierung des Bienenwolfes. III. Die Bevorzugung bestimmter Wegmarken, *Z. vergl. Physiol. 25*, 292–334 (1938).
―――― and R. J. VAN DER LINDE, Über die Orientierung des Bienenwolfes. IV. Heimflug aus unbekanntem Gebiet, *Biol. Zbl. 58*, 425–435 (1938).
TITOW, J., and A. KOWALJEW, Bienenbestäubung des Klees im Moskauer Gebiet, *Pschelovodstvo*, H. 6, 15 (1939).
TONGIORGI, P., Effects of the reversal of the rhythm of nycthemeral illumination on astronomical orientation and diurnal activity on *Arctosa variana* C. L. Koch, *Ann. ital. Biol. 97*, 251–265 (1959).
TSCHUMI, P., Über den Werbetanz der Bienen bei nahen Trachtquellen, *Schweiz. Bienenztg.*, 129–134 (1950).
―――― Über den Werbetanz der Bienen bei nahen Trachtquellen und seine richtungsweisende Bedeutung, *Rev. Suisse Zool. 57*, 584–590 (1950a).
TWITTY, V., Migration and speciation in newts. An Embryologist turns Naturalist and conducts field experiments on homing behavior and speciation in newts, *Science 130*, 1735–1743 (1959).
―――― D. GRANT, and O. ANDERSON, Long distance homing in the newt *Taricha rivularis*, *Proc. Nat. Acad. Sci. USA 51*, 51–58 (1964).

UCHIDA, T., and M. KUWABARA, The homing instinct of the honey bee, *Apis mellifica*, *J. Fac. Sci., Hokkaido Univ., Ser. VI, Zool. 10*, 87–96 (1951).
UNHOCH, N., *Anleitung zur wahren Kenntnis und zweckmäßigsten Behandlung der Bienen* (Munich, 1823).

VERHEIJEN-VOOGD, CHRISTINE, How worker bees perceive the presence of their queen, *Z. vergl. Physiol. 41*, 527–582 (1959).
VISWANATHAN, H., Note on *Apis dorsata* queen cells, *Indian Bee J. 55* (April 1950).
VOGEL, BERTA, Über die Beziehungen zwischen Süßgeschmack und Nährwert von Zuckern und Zuckeralkoholen bei der Honigbiene, *Z. vergl. Physiol. 14*, 273–347 (1931).
VOGEL, R., Zur Kenntnis des feineren Baues der Geruchsorgane der Wespen und Bienen, *Z. wissensch. Zool. 120*, 281–324 (1923).
VOGEL, ST., Farbwechsel und Zeichnungsmuster bei Blüten, *Österr. Bot. Z. 97*, 44–100 (1950).
―――― Duftdrüsen im Dienste der Bestäubung, *Abhandl. Akad. Wiss. Lit. Naturw. Kl. Mainz*, 601–763 (1962).
VOWLES, D. M., Sensitivity of ants to polarized light, *Nature (London) 165*, 282–283 (1950).
―――― The orientation of ants. I. The substitution of stimuli, *J. exp. Biol. 31*, 341–355 (1954).
―――― The orientation of ants. II. Orientation to light, gravity and polarized light, *J. exp. Biol. 31*, 356–375 (1954a).
VRIES, H. DE, and J. W. KUIPER, Optics of the insect eye, *Ann. N. Y. Acad. Sci. 74*, 196–303 (1958).
―――― A. SPOOR, and R. JIELOF, Properties of the eye with respect to polarized light, *Physica (Amsterdam) 19*, 419–432 (1953).

WAGNER, W., Psychobiologische Untersuchungen an Hummeln. I und II, *Zoologica 19,* H. 46, 1–78, 79–239 (Stuttgart, 1907).

WAHL, O., Neue Untersuchungen über das Zeitgedächtnis der Bienen, *Z. vergl. Physiol. 16,* 529–589 (1932).

——— Beitrag zur Frage der biologischen Bedeutung des Zeitgedächtnisses der Bienen, *Z. vergl. Physiol. 18,* 709–717 (1933).

——— Untersuchungen über ein geeignetes Vergällungsmittel für Bienenzucker, *Z. vergl. Physiol. 24,* 116–142 (1937).

WALTHER, J. B., Untersuchungen am Belichtungspotential des Komplexauges von *Periplaneta* mit farbigen Reizen und selektiver Adaptation, *Biol. Zbl. 77,* 63–104 (1958).

——— Changes induced in spectral sensitivity and form of retinal action potential of the cockroach eye by selective adaptation, *J. Insect Physiol. 2,* 142–151 (1958a).

WASHIZU, Y., D. BURKHARDT, and P. STRECK, Visual field of single retinula cells and interommatidial inclination in the compound eye of the blowfly *Calliphora erythrocephala, Z. vergl. Physiol. 48,* 413–428 (1964).

WATERMAN, T. H., Directional sensitivity of single ommatidia in the compound eye of *Limulus, Anat. Rec. 117,* 566 (1953).

——— Directional sensitivity of single ommatidia in the compound eye of *Limulus, Proc. Nat. Acad. Sci. 40,* 252–257 (1954).

——— Polarized light and angles of stimulus incidence in the compound eye of *Limulus, Proc. Nat. Acad. Sci. 40,* 258–262 (1954a).

——— Polarization patterns in submarine illumination, *Science 120,* 927–932 (1954b).

——— Polarization of scattered sunlight in deep water, *Deep-Sea Research Suppl. 3, Bigelow-Festschrift,* 426–434 (1955).

——— Polarized light and plankton navigation, in A. A. BUZZATI-TRAVERSO, ed., *Perspectives in marine biology* (Berkeley: University of California Press, 1958), 429–450.

——— Interaction of polarized light and turbidity in the orientation of *Daphnia* and *Mysidium, Z. vergl. Physiol. 43,* 149–172 (1960).

——— Polarized light orientation by aquatic arthropods, in B. C. CHRISTENSEN and B. BUCHMANN, eds., *Progress in Photobiology* (Proc. 3rd Intern. Congr. Photobiology; Amsterdam: Elsevier, 1961), 214–216.

WATSON, J. B., and K. S. LASHLEY, *Homing and related activities of birds* (Publ. 211; Washington: Carnegie Institution, 1915).

WEAVER, N., The foraging behavior of honeybees on hairy vetch. II. The foraging area and foraging speed, *Insectes sociaux 4,* 43–57 (1957).

WEIS, ILSE, Versuche über die Geschmacksrezeption durch die Tarsen des Admirals *Pyrameis atalanta* L., *Z. vergl. Physiol. 12,* 206–248 (1930).

WELLINGTON, W. G., Motor responses evoked by the dorsal ocelli of *Sarcophaga aldrichi* Parker and the orientation of the fly to plane polarized light, *Nature (London) 172,* 1177 (1953).

——— C. R. SULLIVAN, and G. W. GREEN, Polarized light and body temperature levels as orientation factors in the light reactions of some hymenopterous and lepidopterous larvae, *Canad. J. Zool. 29,* 330–351 (1951).

——— and W. R. HENSON, The light reactions of larvae of the spotless Fall Webworm *Hyphantria textor* Harr., *Canad. Entomol. 86,* 529–542 (1954).

WENNER, A. M., Sound production during the waggle dance of the honey bee, *Anim. Behav. 10,* 79–95 (1962).

——— Communication with queen honey bees by substrate sound, *Science 138,* 446–448 (1962a).

——— The flight speed of honeybees; a quantitative approach, *J. Apicult. Res. 2,* 25–32 (1963).

WERNER, GRETE, Tänze und Zeitempfinden der Honigbiene in Anhängigkeit vom Stoffwechsel, *Z. vergl. Physiol. 36,* 464–487 (1954).

WEYER, F., Cytologische Untersuchungen am Gehirn alternder Bienen und die Frage nach dem Alterstod, *Z. Zellforsch. mikrosk. Anat.* 14, 1–54 (1932).

WIECHERT, ELSBETH, Zur Frage der Koordinaten des subjektiven Sehraumes der Biene, *Z. vergl. Physiol.* 25, 455–493 (1938).

WIEDEMANN, INGRID, Versuche über den Strahlengang im Insektenauge (Appositionsauge), *Z. vergl. Physiol.* 49, 526–542 (1965).

WILLIAMS, C. B., *Insect migration* (London: Collins; New York: Macmillan, 1958).

——— *Die Wanderungen der Insekten,* trans. and ed. by H. ROER (Hamburg and Berlin, 1961).

WILSON, E. O., Communication by tandem running in the ant genus *Cardiocondyla, Psyche* 66, 29–34 (1959).

——— Chemical communication among workers of the fire ant *Solenopsis saevissima* (Fr. Smith) I, II, III, *Anim. Behav.* 10, 134–147, 148–158, 159–164 (1962).

——— The social biology of ants, *Ann. Rev. Entomol.* 8, 345–368 (1963).

WITTEKINDT, E., and W. WITTEKINDT, Entfernungsangaben beim Schwänzeltanz im Rahmen experimentell ausgelösten Tanzverhaltens der Honigbiene, *Naturwissenschaften* 47, 239 (1960).

WITTEKINDT, W., Experimentelle Auslösung von Tänzen bei der Honigbiene, *Naturwissenschaften* 42, 567–568 (1955).

——— Schwänzelbewegungen als Ausdruck gesteigerter Erregung innerhalb des Tanzverhaltens der Honigbiene, *Naturwissenschaften* 47, 335–336 (1960).

——— Tanzverhalten der Honigbiene bei Wechsel zwischen freiem Ausflug und Auslöseversuchen im geschlossenen Stock, *Naturwissenschaften* 48, 605–606 (1961).

WOJTUSIAK, R., Unveröffentlichtes Manuskript einer nicht abgeschlossenen Arbeit; Munich, 1934.

WOLBARSHT, M. L., and V. G. DETHIER, Electrical activity in the chemoreceptors of the blowfly. I. Responses to chemical and mechanical stimulation, *J. gen. Physiol.* 42, 393–412 (1958).

WOLF, E., Über das Heimkehrvermögen der Bienen, *Z. vergl. Physiol.* 3, 615–691 (1926).

——— Über das Heimkehrvermögen der Bienen. II., *Z. vergl. Physiol.* 6, 221–254 (1927).

——— Sehschärfeprüfung an Bienen im Freilandversuch, *Z. vergl. Physiol.* 14, 746–762 (1931).

——— Das Verhalten der Bienen gegenüber flimmernden Feldern und bewegten Objekten, *Z. vergl. Physiol.* 20, 151–161 (1934).

WOLKEN, J. J., Retinal structure. Mollusc Cephalopods, *Octopus, Sepia, J. biophys. biochem. Cytol.* 4, 835–838 (1958).

——— J. CAPENOS, and A. TURANO, Photoreceptor structures. III. *Drosophila melanogaster. J. biophys. biochem. Cytol.* 3, 441–448 (1957).

——— and P. D. GUPTA, Photoreceptor structures. The retinal cells of the cockroach eye, *J. biophys. biochem. Cytol.* 9, 720–724 (1961).

WYKES, G. R., An investigation of the sugars present in the nectar of flowers of various species, *New Phytologist* 51, 210–215 (1952).

——— The preferences of honeybees for solutions of various sugars which occur in nectar, *J. exp. Biol.* 29, 511–518 (1952a).

——— The sugar content of nectars, *Biochem. J.* 53, 294–296 (1953).

YASAMUZI, G., and N. DEGUCHI, Submicroscopic structure of the compound eye as revealed by electron microscopy, *J. Ultrastructure Research 1,* 259–270 (1958).

ZANDER, E., *Grundlagen und Fortschritte im Garten- und Weinbau,* part 20, *Bienenkunde im Obstbau* (3rd ed.; Stuttgart, 1946).

—— *Handbuch der Bienenkunde in Einzeldarstellungen,* vol. 4, *Das Leben der Biene* (5th ed.; Stuttgart, 1947).

ZERRAHN, GERTRUD, Formdressur und Formunterscheidung bei der Honigbiene, *Z. vergl. Physiol. 20,* 117–150 (1934).

ZIEGENSPECK, H., Die Farben- und UV-Photographie und ihre Bedeutung für die Blütenbiologie, *Mikroskopie 10,* 323–328 (1955).

ZMARLICKI, C., and R. A. MORSE, The effect of mandibular gland extirpation on the longevity and attractiveness to workers of queen honey bees, *Apis mellifera, Ann. entomol. Soc. Amer. 57,* 73–74 (1964).

ZONANA, H. V., Fine structure of the squid retina, *Bull. Johns Hopkins Hosp. 109,* 185–205 (1961).

Index

Acris gryllus, 456
Aequidens portalegrenis, 456
Aesculus hippocastanum, 489, 509
Agelena, 414, 429, 440, 447
Agriocnemis, 423
Allen, M. Delia, jerking dance, 281
Allium: cepa, 258; *porrum,* 263
Altevogt, R., orientation, 444
Althaea officinalis, 486, 487
Alvarado, F., toxic sugars, 515
Alyssum, 263
Ammophila, 333, 470
Anacridium aegystium, 406
Anax junius, 423
Anderson, O., orientation, 457
Andrena, 408
Antennae: drumming, 29, 30, 50, 243; Johnston's organs, 98, 187; gravity perception, 144; body control, 144; airstream perception, 187; use in dance, 281; humidity response, 494, 496; olfactory organs, 495–502, 508, 521–522; taste organs, 518
Ants: mandibular glands, 291; odor communication, 316–317, 320, 328; behavior communication, 323–324; orientation, 134, 148, 185, 219–220, 234, 326, 348, 407, 439, 440, 444, 448, 449; eye structure, 415; taste threshold, 513; taste organs, 519n
Apis: mellifera carnica (Carniolan), 27, 29, 54, 61, 100, 112n, 151, 179, 279, 282–283, 293–296, 318; *mellifera,* 51, 162, 325–326; *m. mellifera (nigra),* 54, 61, 125, 293, 296, 297, 302; *m. lingustica* (Italian), 18, 27, 54, 61, 66, 167–168, 170, 179–181, 291, 293–296, 303, 318, 359–361, 366–367; *indica* (Indian), 54, 125, 160, 172, 181, 301, 318; *florea* (dwarf honeybee), 54, 125, 126, 128, 301–305, 318–319, 324, 325, 328; *dorsata* (giant honeybee), 125, 301–306, 303, 319, 365; *trigona* (stingless), *see* Meliponini; *m. caucasica* (Caucasus), 294, 298; *m. fasciata* (Egyptian), 294–298; *m. intermissa* (African Punic), 294; *m. adansonii* (African honeybee), 294

Apple, 258
Arctosa, 429, 444, 445, 447, 448, 449
Aristotle, dance, 6
Armbruster, L., queen's sounds, 286
Artificial feeding experiments: preparation, 17–20, 22–23; effect on dance, 18–19, 31–32; bee scent, 22–23, 28–29
Ascalaphus macaronius, 477
Aschoff, J., diurnal rhythm, 359
Ascia monuste, 451
Asclepias curassavica, 32
Atta sexdeus rubropilosa, 291
Auditory perception: testing, 285–286; vibration response, 286, 287; dance sounds, 286, 291; queen's sounds, 286, 291–292
Aufsess, Alexandra von: pollen-odor tests, 35; sensory perception, 247; nectar markers, 509, 510
Austin, G. H., taste threshold, 516
Autrum, H.: vibratory sense, 98, 287; polarized light, 385, 386, 416–418, 420, 433n; ocelli, 406; visual acuity, 476–478, 481

Baccetti, B., polarized light, 429–431
Baerends, G. P., orientation, 331, 333, 470
Bainbridge, R., orientation, 414
Baltzer, F., sickle dances, 293–297
Bamberger, Guido, stimulatory feeding, 257, 262
Barnwell, F. H., orientation, 463
Bass, orientation, 455
Baumann, Margot, scent glands, 51, 289
Baumgärtner, H.: feeding vessel, 18, 20; eye structure, 162
Baylor, E. R., polarized light, 407, 409, 412, 415–416, 418
Bayramoglu-Ergene, Saadet, ocelli, 406
Beams, H. W., olfactory organs, 498
Bean, broad, 512
Becker, G., orientation, 459–461
Becker, Lore, orientation, 465, 466
Bedini, C., polarized light, 429–431
Bee glue. *See* Propolis
Beer, Ingeborg, hive color, 469

557

Beetles: gravity perception, 174–178, 217; solar orientation, 326, 445, 446, 448, 450, 463–464; polarized light, 407, 414; olfactory organs, 498; taste organs, 517
Beling, Ingeborg: temporal memory, 253, 254; time training, 353–356
Benndorf, Hans, solar orientation, 198
Bennett, Miriam F.: diurnal rhythm, 357; orientation, 463
Berlepsch, A. von, flight range, 66
Berthoff, L. M., color vision, 472, 474, 476
Bethe, A., orientation, 465, 467
Beutler, Ruth: honey stomach, 29n; flight range, 66; foraging factors, 251; diurnal productivity, 253; flight path, 240; nectar components, 511, 512; taste threshold, 513
Bidessus, 414
Bilberry, 249
Birds: orientation, 45, 331, 338, 347, 453, 454, 464; communication, 324–325; sensory perception, 510, 522–523
Birukow, G., orientation, 326, 407, 446, 447, 457, 458
Bisetzky, A. Ruth, pedestrian bees, 114, 116, 183–185
Blaberus giganteus, 422
Bledius bicornis, 147–148
Blest, A. D., dance, 322–323
Blinow, N. M., stimulatory feeding, 259, 262
Boch, R.: feeding vessel, 20; geraniol, 50–51; hive dance area, 138, 139; flight path, 240; antennae drumming, 243; noontime sluggishness, 244; effect of weather on dance, 245; dance message, 248; foraging factors, 250–251; colonies compared, 294–300
Boeckh, J., olfactory organs, 498
Bolwig, N., sensory perception, 247
Bombyx mori, 510
Böttcher, H., orientation, 458
Braemer, W., orientation, 455
Brassica napus, 258, 483
Bräuninger, H. D.: heated hives, 12, 13; dance tempo, 67–80, 118–120; flight path, 190, 251; noontime sluggishness, 244; solar orientation, 351
Brett, J. R., orientation, 456
Brian, Anne D., bumblebees, 314
Brown, P. K.: polarized light, 416; eye structure, 423; orientation, 463
Brun, R., orientation, 348, 448
Brunnera macrophylla, 484–485
Bückmann, D., gravity perception, 148
Buddenbrock, W. van: orientation, 134; ocelli, 406
Büdel, A., food temperature, 78
Bullmann, O., consistency, 249n
Bumblebees (Bombinae): consistency, 249n; pollination, 261, 484; nectar theft, 262; communication, 313–315, 319–320, 323; orientation, 327, 407, 408; color sensitivity, 475, 489; taste threshold, 513
Bünning, E., diurnal rhythm, 253, 359

Buphthalmum salicifolium, 487
Burdon-Jones, G., orientation, 414
Burkhardt, D.: polarized light, 417, 418, 420; visual acuity, 478
Burtt, E. T., eye structure, 479
Butenandt, A., odor, 291, 510
Butler, C. G.: eversion, 54; sensory perception, 247; consistency, 249; foraging factors, 251; water preference, 266; nuptial flight, 289; dance comparisons, 302–303
Buttel-Reepen, H. von: sound, 3, 285; flight range, 65–66; orientation, 466
Butterflies: tail-wagging, 322; eye structure, 422, 429; migrations, 450–451; taste organs, 517, 518; taste threshold, 519
Buzzing run: in swarming, 276–280, 283; mood indicator, 285; bumblebees, 314

Cadenas, E., toxic sugars, 515
Cajal, S., eye structure, 434
Calliphora, 417, 478, 479, 519
Callow, R. K., queen substance, 288
Campanula: medium, 36; *patula,* 509, 510
Camponotus: sericeus, 316, 317; *paria,* 317; *compressus,* 317
Canterbury bells, 35, 36, 509, 510
Capenos, J., eye structure, 421
Carathamus tinctorius, 258
Cardiocondyla: venustula, 316; *emeryi,* 316
Carthy, J. D., orientation, 440
Caterpillars, color vision, 476
Catton, W. T., eye structure, 479
Chadwick, L. E., chemical perception, 492
Chalifman, T. A.: night dance, 245; spontaneous dance, 350
Charles, G. H., orientation, 414
Chauvin, R., horizontal comb experiment, 131
Chelidonium majus, 487
Chelonia mydas, 457
Chestnut, 489, 509
Cinquefoil, 485–486, 488, 489
Cirsium oleraceum, 48, 258
Clover, 258, 261–263
Coccinella septempunctata, 217, 326
Cockroach, eye structure, 422, 429
Consistency, 6, 30, 32, 34, 37–39, 42, 43, 47–48, 153n, 249
Coreopsis bicolor, 490
Couturier, A., orientation, 450, 459
Cricket, orientation, 447
Crowfoot, 48
Currant, 249, 510, 511
Cyclamen, 47, 224
Cynoglossum officinale, 491

Djalal, Ahmad, "misdirection," 213n
Danaus plexippus, 450
Dandelion, 258
Danneel, R.: eye structure, 421; polarized light, 422
Daphnia, 414

Daumer, K.: sensory perception, 247, 374–376, 472–476, 482; eye structure, 432; orientation, 447, 460, 462; floral color, 483–485; nectar markers, 486–490
Deguchi, N., eye structure, 421
Dethier, V.: dance, 321–322, 324; chemical perception, 492; taste organs, 517
Detour experiment: "residual misdirection," 176; flight path, 178; pedestrian bees, 183–185; evaluation, 185, 223
Dianthus cartusianorum, 484
Dietrich, W., polarized light, 420
Digitalis, 491
Dirscheld, Hannelore, honey stomach, 224, 226, 227
Displacement experiment: flight path, 169–172; "misdirection," 169–170; effect on dance, 171–172; diurnal rhythm, 356–360
Dissosteria, 422
Diurnal rhythm: noontime sluggishness, 244, 247n, 255; temporal memory, 253, 256; productivity, 253–254; time index, 352–355, 379; displacement experiment, 356–360; importance of, 378
Doira, S., taste organs, 517
Dostal, Brigette, olfactory organs, 496
Dragonfly: eye structure, 132, 423, 429; color vision, 477
Drones: olfactory perception, 289, 292; assembly sites, 289; eye structure, 376, 476; antennae, 507
Drosophila, 416, 421, 422
Drugged feeding, 82–84, 127
D-VAV. *See* Jerking dance
Dyschirius: nitidus, 148; *numidicus*, 445

Eckert, B., crustaceans, 407
Eckert, John E.: flight range, 66; colony comparison, 296
Eckstein, M., ultraviolet, 372, 377
Eguchi, E., eye structure, 434
Elevation experiment: influence on direction, 165–167; Italian bees, 167, 168; round dance, 168; evaluated, 168
Ellmauer, M., communication, 250
Eltringham, H., eye structure, 434
Emeis, D., orientation, 447
Emlen, J. T., orientation, 458
Energy expenditure, distance index, 114, 127
Engländer, H., color vision, 469, 476
Erithacus rubecula, 454
Eruca sativa, 258
Erysimum helveticum, 258, 483
Esch, H.: recording dances, 25; dance sounds, 58–60, 237, 325; body temperatures, 76; temperature and dance, 79; vibratory sense, 98; sound frequency, 99; artificial bee, 104–105; mirror experiment, 105–106; disoriented dances, 135; dance message, 248; buzzing run, 280; stingless bees, 310, 313, 325; bumblebees, 314

Evans, D. R., toxic sugar, 516
Exner, F. and S.: visual acuity, 478, 480n; floral color, 481
Eye-bristle perception, 187

Fairey, E. M., nuptial flight, 289
Fan-shaped experiment: 156–161; evaluated, 164; "misdirection," 203–204
Ferguson, D. E., orientation, 456
Fernandez-Moran, H., eye structure, 421, 422
Fingerman, M., polarized light, 416
Finke, Ingrid, temporal memory, 254
Firsow, I. G., stimulatory feeding, 258, 262
Fish, orientation, 455–456, 463
Fischer, Klaus, orientation, 457, 458
Fischer, W.: feeding vessels, 22; olfactory acuity, 54, 506; olfactory organs, 496
Flea, orientation, 445, 446
Flies: searching movements, 321–322, 327; orientation, 406, 409, 416–417, 459–460; eye structure, 419, 421–423, 429, 437; visual acuity, 478; taste organs, 517, 518; taste threshold, 519
Flight path: distance curve, 65, 66, 121–126, 128; range, 65–67, 89, 171–172, 172n, 191; factors, 66, 82, 109, 188–190; effect on dance, 82, 173, 240–241; direction index, 165–167, 169, 182, 183; solar orientation, 170–171; detour experiment, 178
Florey, E., trembling dance, 283
Flowers: importance of odor, 49, 508, 509; importance of structure, 481, 482; importance of color, 481–482, 485, 506–507, 521; nectar markers, 486–491, 508–509, 521; nectar glands, 511; nectar components, 511–513
Foraging factors, 250–252
Forel, August: diurnal rhythm, 253; color vision, 471; olfactory organs, 497, 499
Forgetfulness concept, 122–128
Formica: rufa, 148, 440, 448; *polyctena*, 219–220
Forget-me-not, 484–485
Foxglove, 491
Fraenkel, G., color vision, 474
Fragrance communication: external, 43, 46, 50, 53, 224, 225, 242, 288; odorless flower experiment, 48–49; internal, *see* Honey stomach
Françon, Julian, bee sagacity, 22
Free, J. B.: scent gland, 51; stimulatory feeding, 262–263; hive food exchange, 266; olfactory perception, 290; hive placement, 467
Frey-Wyssling, A., nectar glands, 511
Frings, H.: vibratory sense, 98, 287; olfactory organs, 496; taste organs, 496
Fromme, H., migration, 454
Fundamental orientation, 407–409, 436, 439

Galeopsis speciosa, 509
Gary, N. E.: queen substance, 288; nuptial flight, 289; olfactory perception, 290, 507

Geiger, R., food temperature, 78
Geisler, Marianne, orientation, 450
Geissler, G., toxic sugars, 515
Geotrupes silvaticus, 326
Geraniol, 51
Gerdes, K., orientation, 454
Geschke, social organization, 7
Ghent, R. L., olfactory perception, 290
Gibbons, R., eye structure, 432
Gillard, A., stimulatory feeding, 263
Giltay, E., consistency, 249
Glushkov, N. M., stimulatory feeding, 259, 263
Goedart, J., buzzing, 314
Goetsch, W., odor trail, 316, 317, 324
Gogala, M., color vision, 477
Goldsmith, T. H., eye structure, 406, 421–424
Goniopsis, 414, 444
Gontarski, H.: social organization, 7; honey stomach, 119
Gooseberry, 512
Görner, P.: orientation, 414, 448; eye structure, 429; polarized light, 431, 440, 441
Götz, R., hive dance areas, 36–37, 138–139
Grant, D., orientation, 457
Grape, 48, 249, 510, 511
Grasshopper: eye structure, 406, 422, 429; olfactory organs, 498
Gravity perception: orientation, 137, 145–149; antennae, 144; sense organs, 144–145; disoriented dance, 154–155; direction index, 155; transition from solar orientation, 160, 202; vertical-comb dance, 204–205
Griffin, D. R., orientation, 458
Grooming dance, 280–281, 283
Groot, C., orientation, 456
Grout, R. A., orientation, 366
Gubin, A. F., stimulatory feeding, 257
Gubin, W. A., olfactory perception, 493
Guckelsberger, H., ultraviolet light, 376, 377
Gupta, P. D., eye structure, 422, 423

Hagen, H. von, orientation, 444
Haidinger's brushes, 432, 433
Haidl, "marathon" dance, 350, 351
Haldane, J. B. S.: historical references, 4n, 6; waggling, 104; distance curve, 121
Halictus, 408
Hammann, Eleonore, jerking dance, 281
Hammer, O., foraging factors, 252
Hannes, F., orientation, 185–186, 195
Hansson, A.: vibration perception, 3, 98; flight tone, 53; auditory testing, 285–287
Hanström, B., eye structure, 434
Hase, A., tail-wagging dance, 315
Hasler, A., orientation, 455, 456
Hass, A., communication, 314, 315
Hassenstein, B., neurophysiology, 186
Haydak, M., dance, 280–282
Hecht, S., visual acuity, 478
Hein, G.: "spasmodic dance," 279; "twitching segments," 279; colony comparison, 297

Helianthus: annus, 258; *rigidus,* 489
Henkel, Chr., dance, 4, 5
Heracleum spondylium, 258
Heran, H.: dance tempo, 68–69, 73, 76, 113–115; antennae, 98, 187; dance elements, 100–102; flight speed, 109; impeded bee, 110–111, 116; water flight, 112, 113; transport experiment, 117; distance curve, 121; displacements, 182–183; flight pattern, 186–187; foraging factors, 251; infrared radiation, 368; hive placement, 468
Hertz, Mathilde: visual acuity, 476, 479–480; humidity perception, 495
Hess, C. von: color blindness, 471; color vision, 476
Hess, W. R., hive temperature control, 265
Hesse, R., eye structure, 434
Heusser, H., orientation, 457
Hingston, R. W. G., odor trails, 316, 317
Hocking, B., visual acuity, 478
Hodgson, E. S., taste organs, 517
Hoehn, W., stimulatory feeding, 260, 263
Hoff, van't, rule of movement, 75, 77, 114, 127
Hoffmann, Klaus, migrations, 453
Hogweed, 258
Honey stomach: use of, 28, 29, 34, 59–60, 63, 134, 238, 243; identification and directional index, 224–227; effect of sugar solution, 236–237; effect on dance tempo, 243; foraging factors, 251; relation to water, 266, 276
Horrall, R. M., orientation, 455
Hound's-tongue, 491
Huber, H., foraging factors, 252
Huber, Mathilde, nectar markers, 508–509
Huckleberry, 510, 511

Identification index: scent glands, 224; body odor, 224, 225; honey stomach, 224–225
Insects, solar orientation, 134
Iris, 48
Istomina-Tsvetkova, K. P., honey stomach, 29n
Ivanoff, A., polarized light, 444

Jacobs, W., scent organ, 50, 51
Jacobs-Jessen, Una F.: "residual misdirection," 222; orientation, 314, 327, 407, 408, 413; visual acuity, 480
Jahn, W., solar orientation, 198
Jamieson, C. A., taste threshold, 516
Janders, Rudolf: observation hive, 24–25; consistency, 39–42; waggling run, 98–102, 104, 107; optic orientation, 185; "misdirection," 217–220; dance message, 248; solar orientation, 326; fundamental orientation, 407, 408, 414, 440, 447–449; eye structure, 432
Jay, S. C., stimulatory feeding, 261
Jeffree, E. P., consistency, 249
Jerking dance (D-VAV): pattern, 281; age of bees, 281; occurrence, 281–282; effect of sugar solution, 282

Jielof, R., eye structure, 433
Jordan, R., queen substance, 288
Jörgensen, H., color vs. odor, 507
Johnson, W. E., orientation, 456
Johnston, Norah C., queen substance, 288
Johnston's organs. *See* Antennae
Joos, G., 368n
Jostling run: characteristics, 278, 324; purpose, 278, 283
June bugs, orientation, 450, 459, 464

Kaissling, K., olfactory organs, 498
Kalmus, H.: odor, 54; weighted-bee experiment, 110; antennae, 144; consistency, 249; bee and wasp experiment, 315; solar orientation, 363–364; polarization, 409, 412
Kaltofen, R. S., odor, 54
Kantner, taste threshold, 519
Kappel, Irmgard: feeding vessel, 18–19; food-source shapes, 18–19, 242, 245
Kapustin, stimulatory feeding, 257
Kaschef, A. H., external fragrance, 20, 242
Kennedy, D., polarized light, 409, 418
Kerr, W.: vibratory sense, 98; stingless bees, 307–313
Kickhöffel, flight range, 66
Kiechle, H.: hive-temperature control, 265; water preference, 267–268; humidity perception, 495; olfactory organs, 496
Kiepenhauer, K., orientation, 400
Kimm, I. H., orientation, 459
Kleber, Elisabeth: diurnal productivity, 253–254; temporal memory, 254–255
Knaffl, Herta: use of controlled bee, 18; flight range, 66; dance tempo, 68–69; weather sensitivity, 245; colony comparison, 296
Kneissl, M., water-flight experiment, 111
Knoll, F.: foraging factors, 252; color vision, 472; nectar markers, 488, 490
Kobel, F., consistency, 249
Köhler, F., dance photography, 25
Komarow, stimulatory feeding, 257
Komisarenko, E. M., stimulatory feeding, 257, 262
Körner, Ilse, temporary memory, 255
Kowaljew, A., stimulatory feeding, 257
Kramer, Gustav, orientation, 338, 347, 451–453
Kratky, O., distance index, 99, 121, 124, 323
Krause, Barbara, olfactory organs, 497
Krüger, E., bumblebees, 315
Kruyt, W., orientation, 470
Kuanjezowa, M. A., dance tempo, 70
Kugler, H.: floral color, 484; nectar markers, 487, 488, 489
Kühn, A.: orientation, 407; color vision, 472, 474, 476
Kuiper, J. W., polarized light, 409, 418
Kullenberg, B., floral colors, 482
Kunze, G., taste threshold, 518–519
Kunze, P., visual acuity, 480
Kuwabara, M.: polarized light, 417; orientation, 465; humidity perception, 495; taste organs, 517

Lacerta, 457
Lacher, V.: humidity perception, 495; olfactory organs, 496n, 497, 498
Ladybird, orientation, 217, 326
Lamium maculatum, 509
Langwold, Agnes, eye structure, 478
Larsen, J. R., taste organs, 517
Lashley, K. S., orientation, 331
Lasius niger, 326, 448
Latham, A., swarm dancers, 269
Lecomte, J., olfactory perception, 290
Lee, W. R., flight range, 66
Lepisma, 422
Lepomis cyanellus, 456
Leptinotarsa decemlineata, 517
Leuenberger, F., resin, 268
Lex, Therese: sensory perception, 247; nectar markers, 509, 510
Libellula quadrimaculata, 477
Lice, excretions, 512
Lime tree, 512
Limnophilus, 217, 415
Lindauer, M.: observation hive, 12, 13; pollen, 49, 246; dance tempo, 65, 68–71, 74, 238, 241, 243, 245, 248–249, 281–282, 349–352; flight range, 66–67, 89–90, 125, 172, 173, 393; wind experiment, 80–82, 109, 186–187, 188; vibratory sense, 98; waggling, 107–108; water-flight experiments, 112, 113, 186–187, 267; horizontal-comb experiment, 131–133; hive dance areas, 138; gravity perception, 146; solar orientation, 160–163; "residual misdirection," 205–210; detour experiment, 181–183; taste experiments, 244; sensory perception, 247, 499; foraging factors, 251; temporal memory, 254; hive temperature, 265; water preference, 265; swarming, 269–276; buzzing run, 280; odor trail, 291; colony comparison, 296; *A. dorsata,* 302; *A. florea,* 304–305; stingless bees, 307–313; wasps, 315; orientation, 327, 338–347, 362–365, 373; light intensity, 375, 376; eye structure, 406; fish, 463; hive placement experiment, 467, 469
Linde, R. J. van der, orientation, 470
Linden, 32
Lineburg, B., direction index, 222, 223
Linnaeus, "floral clock," 253
Linzen, B., odor, 291
Little, H. F., vibration response, 97, 287
Littorina, 414
Lizards, orientation, 457
Locke, N., wax, 50
Locust, 53–54, 238
Locusta, 479
Lopatina, N. G., distance index, 70
Lotmar, Ruth, color sensitivity, 482
Louse, orientation, 444
Lubbock, J., color vision, 476
Lucilia caesar, 417
Lüdtke, H., polarized light, 428
Ludwig, L., sensory perception, 247
Lüscher, M., termites, 317
Lüters, W., orientation, 458

Lutz, F. E.: color vision, 476; floral colors, 482

Macroglossum stellatarum, 488
Magnetic-field experiments, 460–462
Magni, F., polarized light, 429, 431
Mammals, orientation, 458
Mandibular glands: queen's, 51, 288–289, 292; alarm substance, 290–291; odor, 311, 507; taste cones, 523
Manning, A.: nectar markers, 489; color vs. odor, 508
"Marathon" dance: swarming, 350, 378; solar orientation, 350, 351, 449
Mark, H.: "residual misdirection," 219–222; polarized light, 376
Marshal, J.: olfactory organs, 496; taste threshold, 518
Martin, H., olfactory acuity, 499, 500, 501–502, 505–506
Martin, P., buzzing run, 280
Maschwitz, U.: odor, 290, 314, 316; alarm substance, 291
Mathis, M., hive placement, 467
Matthews, V. G. T., orientation, 453
Maurizio, Anna: nectar collection, 3n, 252; pollen baskets, 240; nectar glands, 511; nectar components, 511–512
Mazochin-Proschnjakov, G. A.: color vision, 475, 477; floral colors, 482
Meder, E., solar orientation, 348–350
Melandrium album, 484
Meliponini (stingless bees), 98, 306, 307–308, 323, 324, 325, 328; *Trigona iridipennis,* 307, 309, 311; *Tr. jaty,* 308, 309; *Tr. silvestris,* 308; *Tr. droryana,* 308, 310; *Tr. ruficrus,* 311; *Nannotrigona,* 308; *Scaptotrigona,* 308–313; *Melipona quadrifasciata,* 309, 313; *M. scutellaris,* 309, 310, 313; *M. fasciata merillae,* 310
Melolontha, 450
Melosoma populi, 217
Merkl, F. W., migrations, 454–455
Messerschmidt, water-flight experiment, 111
Metschl, N., ocelli, 406
Meyer, Waltrand: propolis collection, 268–269; swarming, 269
Mice, diurnal rhythm, 359
Michieli, S., color vision, 477
Milkweed, 32
Miller, W. H., polarized light, 429
Milum, V.: propolis collection, 268–269; grooming dance, 280–281; jerking dance, 281, 282
Minderhoud, A.: consistency, 249; stimulatory feeding, 263
Minnich, D. E.: taste organs, 517, 518; taste threshold, 519
Mirić, D.: hive-temperature control, 265; propolis collection, 268n
"Misdirection": importance, 192, 217–222, 313; experiments, 196–212, 233–239; cause, 213–217

"Misdirection": importance (*Cont.*)
"residual": detour experiment, 176, 182; sense organs, 204; effect on dance, 204–207, 229; fundamental orientation, 221–222; magnetic-field experiment, 460–461
Moody, M. F.: polarized light, 414, 432, 441; eye structure, 433
Morita, H., taste organs, 517
Morse, R. A., olfactory perception, 288
Moth: tail-wagging, 322–323, 327; polarized light, 409; eye structure, 422; pollination, 484; color response, 488, 490; olfactory organs, 498
Mueller, H. C., orientation, 458
Müller, B., odor trails, 317
Müller, Hermann, floral odor, 506
Musca domestica, 421, 422
Mustard, 258, 483
Myrmica ruginodis, 326
Mysidium, 414
Myosotis silvatica, 484–485

Naka, K.: polarized light, 417; eye structure, 423, 434
Narcissus poeticus, 483–484, 488
Nassa, 414
Nasturtium, 509, 510
Nasutitermes cornigera, 317
Nectar: markers, 486–491, 508–509, 521; glands, 511; components, 511–513
Nedel, J. O.: gravity perception, 144, 145; mandibular glands, 289, 507; stingless bees, 311, 312; orientation, 327; taste organs, 517
Neese, V.: eye-bristle perception, 187; "misdirection," 210–212; eye structure, 479
Nettle, 509
Neuhaus, W., olfactometer, 503
New, D. A. T., solar orientation, 162–163, 351
Nicotiana tabacum, 48
Nielsen, E. T., migrations, 451
Nixon, H. L., hive food exchange, 266
Nocturnal dance, 351–353, 378
Nolte, D. J., eye structure, 421, 434
Nonnea pulla, 485
Notonecta, 415; *glauca,* 477
Nuptial flight, mandibular glands, 51, 289

Octopus: orientation, 414, 440–441; eye structure, 432
Ocypode, 447
Oenothera fruticosa, 35
Oettingen-Spielberg, Therese: sensory perception, 247; foraging factors, 252
Olfactory organs: perception of queen, 288, of drones, 289, of danger, 290; localizing, 449–502, location, 495–499, 521–522; measuring apparatus, 503; importance, 507
Omophron limbatum, 445
Oncorhynchus nerka, 456
Onion, 258

Onobrychis sativa, 258
Opfinger, Elisabeth, hive placement, 469–470
Optic sensitivity: solar orientation, 374–377, 380; polarization, 380–381, 385, 392, 405, 409–414, 435–438; color response, 401–404, 436, 467–476; dioptric apparatus, 416–417
Orchestia, 445, 446
Oschmann, D. H., stimulatory feeding, 263
Otto, F.: transport experiment, 117–118; displacement experiment, 169–173, 333

Paederus, 445, 446, 448
Pain, J., nuptial flight, 289
Palitschek von Palmforst, E.: dance measuring, 25, 26; "misdirection," 213n
Pankowa, S. W., tail-wagging dance, 70
Paparia, stimulatory feeding, 257
Papaver: nudicaule, 34; *somniferum,* 258
Papi, F.: polarized light, 429, 431; orientation, 440, 444, 445, 447
Pardi, L., orientation, 440, 441, 444, 445, 447
Park, O. W., gravity perception, 145
Park, W., dances, 5, 266
Parker, R. L., diurnal productivity, 254
Parriss, J. R.: orientation, 414, 433, 441; eye structure, 432
Parthenocissus quinquefolia, 48, 249
Paschke, I., time training, 355
Perdeck, A. C., migrations, 454
Pedestrian bees: detour experiment, 183–186; propolis collection, 268n
Peer, D., flight range, 66
Perch, orientation, 456
Periplaneta americana, 422, 477
Peters, W., taste organs, 517
Petunia, 263
Phaleria, 445, 449
Phaseolus multiflorus, 48
Philanthus triangulum, 331, 470
Phillips, E. F., eye structure, 434
Philpott, D. E., eye structure, 421–423
Phlox, 47, 224, 241
Phormia, 321–322, 479
Pieris brassicae, 450
Pirus malus, 258
Piscitelli, Annemarie: impeded-bee experiments, 115–116; taste threshold, 266
Pliny, observation hive, 4n
Pohl, R., color vision, 472
Polarized light: vibration plane, 381–384, 386, 392, 409, 412–414, 435–436; effect on dance, 383–385, 387–391, 393–394, 397–398, 407, 434–436; lamp effect, 383, 395, 434; effect on flight path, 392–393, 407; vs. solar orientation, 398–400
Pollen: collecting, 34–36, 240, 268; message, 34–35, 49; odor, 35; traps, 62
Poppy, 34, 258, 482, 485
Porsche, O., floral color, 482
Portillo, J. del, compound eye, 162
Potentilla reptans, 485–486, 488, 489

Prestage, J. J., olfactory organs, 498
Pritsch, G., stimulatory feeding, 259, 263
Propolis: collection, 268, 277; use, 268
Pyrameis atalanta, 519

Quadbeck, G., taste organs, 517
Queen: sounds of, 3, 286–287; cells, 288; mandibular glands, 288, 292; nuptial flight, 289; stinging, 291

Rademacher, B., orientation, 145
Ranunculus acer, 48
Rape, 258, 262, 263, 483
Raspberry, 258
Rau, P., solar orientation, 366
Rauschmayer, F., hive color, 469
Réaumur, observation hive, 4n
Renner, O.: bee room, 13–14; feeding device, 18–19; recording device, 20–22; mandibular glands, 51, 289; scent-gland experiment, 54–55; diurnal rhythm, 253, 354, 357; displacement experiment, 356–364
Rensing, L., eye structure, 428
Reptiles, orientation, 457–458
Resin. *See* Propolis
Rhein, W. von, stimulatory feeding, 259, 263
Ribbands, C. R.: odor, 33–34, 50, 54, 495; waggling, 104; consistency, 249; hive food exchange, 266; solar orientation, 366; hive placement, 467; olfactory acuity, 503
Ribes: rubrum, 249; *grossularia,* 512
Richards, G., olfactory organs, 497
Robert, P., orientation, 450, 459
Robertis, E. de, eye structure, 432
Robertson, J. D., eye structure, 432
Robinia viscosa, 31, 53, 238
Robins, orientation, 454
Roccus chrysops, 455
Roeder, K. D., taste organs, 517
Roepke, W., "shaking" dance, 302
Rollwagen, polarized light, 372
Rosa: moschata, 34, 35, 54; *polyantha,* 35
Rösch, G. A.: odor, 54; propolis collection, 268
Round dance: food message, 4, 5, 29–30, 42–43, 46, 55–56, 324; pattern, 29; tempo, 44–45, 55; distance index, 61; transition dance, 61; vs. tail-wagging dance, 149–153; elevation experiment, 168; sound, 238; crossbred colonies, 294–298
Rowell, C. H. Fr., polarized light, 414, 441
Rubus idaeus, 258
Ruttner, F., nuptial flight, 289

Safflower, 258
Sainfoil, 258
Saint Paul, Ursula von, orientation, 451
Sakagami, S. F., taste thresholds, 513
Salamanders, orientation, 456, 457

Salmon, orientation, 456
Sanchez, eye structure, 434
Santschi, F., orientation, 134, 439, 445
Sarcophaga: aldrichi, 406; *bullata,* 422
Sauer, F., migrations, 454
Savely, H. E.: eye structure, 429; polarized light, 429, 431
Scarites terricola, 495
Scarlet runner, 48
Scent organs: location, 50–51; eversion, 50–56; direction index, 51–56, 224, 234, 289–290
Schaller, A., taste organs, 492
Schardt, Helga: floral color, 484; nectar markers, 487, 488
Schick, W., trembling dance, 282
Schifferer, G., weighted bees, 109–110, 113–114, 239–240, 282
Schifferer, Ilse, orientation, 439–440
Schistocerca gregaria, 406, 422
Schlote, F. W., eye structure, 423–427
Schmeidler, F., polarized light, 372–376
Schmid, H., "jostling run," 278
Schmidt, Anneliese, taste organs, 519n
Schneider, D., olfactory organs, 498
Schneider, F., orientation, 450, 459
Schneider, Wilfriede, vibration response, 98, 287
Scholze, E., impeded-bee experiment, 115–116
Schöne, H., orientation, 414, 444
Schöntag, Adele, nectar components, 511, 512
Schreindler, Elfriede, senility, 75
Schricker, B.: foraging factors, 251; ocelli, 406, 407
Schuà, L., noontime sluggishness, 244
Schüz, E., migrations, 453
Schwarz, R., olfactory acuity, 493, 503–504
Schwassmann, H. O., orientation, 455
Schweiger, Elisabeth, dance tempo, 73–75, 149, 212
Schweiger, M., dance tempo, 67
Seitz, A., interplay of stimuli, 245
Sekera, Z., polarized light, 282
Senility: dance tempo, 74, 104; orientation, 231
Sensory perception: orientation, 182, 194, 508; external fragrance, 223, 234, 248–250, 508; direction index, 222–223; gravity perception, 231; internal fragrance, 234, 280; taste, 244; queen scent, 292; olfactory acuity, 492–494, 502–505, 521–522; color vs. odor, 506–508; visual acuity, 479–491, 520–521; humidity response, 495; localization, 499–502; taste threshold, 512–516, 523
Seybold, A., floral colors, 482
"Shaking dance," 280–281, 283
Shaposhnikova, N. G., transport experiment, 117n
Shearer, D. A., geraniol, 50–51
Sickle dance: performers, 279, 283; transition form, 279; "twitching segments," 279; crossbred colonies, 293–298
Silene: inflata, 484; *natans,* 484

"Silent dance," 58
Silk moth, 510
Simpson, J., queen sounds and substance, 287, 288
Sinapis arvensis, 258, 483
Singh, S.: consistency, 249; orientation, 366
Sjöstrand, F. S., eye structure, 432
Sladen, F. W., scent glands, 4, 51
Slifer, Eleanor, olfactory organs, 497, 498
Smith, F. E.: orientation, 407; polarized light, 409, 412
Smith, F. G., colony comparison, 296
Snails, orientation, 414, 463
Solar orientation: effect on dance, 130–131, 141–144, 196–198, 233, 349–353; sun substitutes, 135–136, 368–374; vs. gravity perception, 144, 160, 197–198; displacement experiment, 170–171; wind, 191–195; flight path, 195; horizontal comb, 197, 200, 230–231; vertical comb, 197–198, 200–202, 231; "misdirection," 205–208; measuring apparatus, 348; displaced bees, 349; time index, 360–361; optic sensitivity, 374–377, 380–381; learned art, 327, 379–380, 447
Sols, A., toxic sugars, 515
Sorokin, stimulatory feeding, 257
Spasmodic dance, 279, 283
Speck, U., orientation, 459–460
Speirs, N., hive placement, 467
Spencer-Booth, Yvette, hive placement, 467
Spider: orientation, 185, 414, 440, 444–448; eye structure, 429–432
Spitzner, Ernest: dance, 6; buzzing, 285
Spontaneous dance, 349–350
Spontaneous orientation, 407–409, 436, 439
Spoor, A., eye structure, 433
Sprengel, C. K., floral structure, 481
Spurway, H.: waggling, 104; distance curve, 121
Squid: orientation, 414; eye structure, 432
Ssacharow, stimulatory feeding, 262
Stapel, C., stimulatory feeding, 263
Staudenmayer, T., toxic sugars, 515
Steche, W.: dance measurement, 97, 99; waggling, 104; crossbred colonies, 298–300; toxic sugars, 515
Stein, G., odor trails, 315
Steinhoff, Hildtraut: odor, 49–50; stimulatory feeding, 258
Stimulatory feeding: value, 257–258, 261–264; preparation, 258–261; effect on dance, 259
Stingless bees. *See* Meliponini
Stockhammer, K.: polarized light, 406, 415, 416; eye structure, 419, 420, 433, 434
Stoecker, Marieluise, visual acuity, 481
Stephens, G. C., polarized light, 416
Stepwise experiments: forager control, 87–88; relation to dance, 96–97, 127; importance, 106, 108
Streck, P., visual acuity, 478
Stuart, A. M., odor trail, 317
Stumper, M. R., energy expenditure, 114
Stumpf, Hildegard, polarized light, 416
Sturckow, Brunhild, taste organs, 517

Sudd, J., odor trails, 316, 317
Sugar solution: ingredients, 19, 237; effect on dance, 45, 236–239, 241, 244, 245, 282; use in stepwise experiment, 88–94; directed-search experiment, 150–154; fan-shaped experiment, 157, 203; scent-trail experiment, 223–224; effect on sound, 237; effect on honey stomach, 236–237, 239; vs. water, 267
Sunflower, 258
Swarming: swarm structure, 269; informant dancers, 269–271, 277; "marathon," 350; dance duration, 269; sensory perception, 271; dance tempo, 273, 276; "artificial" swarms, 273; controlled hive sites, 273–275; feeding of scouts, 275–276; swarming time, 276; buzzing run, 276, 277; swarm movement, 276; "preswarm," 386; "afterswarm," 386
Sylvia communis, 454

Tail-wagging dance: compared to round dance, 45n, 149–153; message, 57; pattern, 57, 63; "silent dance," 58; importance of sound, 57–58, 63, 325; direction index, 57, 59, 61–62, 64–65, 97–100, 109, 230–233, 324–325; use of honey stomach, 59; variations, 71–74; measured, 97–99; waggling, see Waggling run; error percentage, 103; age of bee, 75, 104; temperature, 104, 127; response to artificial bee, 104–105; terrain, 109; wind, 109; weighted bee, 109–110; horizontal comb, 131–133, 197, 230–231; solar orientation, 196–197; vertical comb, 197–198; effect of sugar solution, 238; crossbred colonies, 295–298
Takeda, K.: humidity perception, 495; taste organs, 517
Talitrus, 440, 444–446, 449
Talorchestia, 445, 446
Taraxacum officinale, 258
Taricha rivularis, 457
Taste organs, location, 517–519, 523
Temperature factor: hive, 76; body, 76; dance, 76–77; food solution, 78; tail-wagging dance, 104–127
Tenckhoff-Eikmanns, Inge, orientation, 217, 326
Tenebrio molitor, 217
Termites: odor trail, 317, 320; orientation, 459
Terrain, effect on dance tempo, 127
Thaker, C. V., queen cells, 301
Thistle, 48, 258
Thorpe, W. H., dance, 227
Thurm, U., gravity perception, 145
Tilia, 512; *platyphyllos,* 32
Tinbergen, N., orientation, 331, 470
Titow, J., stimulatory feeding, 257
Toads, orientation, 456
Tobacco, 48
Tonapi, K. V., queen cells, 301
Tongiorgi, P.: polarized light, 429, 431; orientation, 447

Transport experiments: flight vs. dance tempo, 116–121; species used, 125–126
Trembling dance: length, 282; purpose, 282, 283; cause, 282, 284; age of bees, 282, 283
Trifolium: pratense, 258; *repens,* 258; *hybridium,* 262
Trigona. See Meliponini
Tritonia cocosmaeflora, 48
Tropaeolum majus, 509
Tschumi, P.: sickle dances, 293; crossbred colonies, 296–298
Turano, A., eye structure, 421
Turnip, 258, 262
Turtles, orientation, 457
Twitty, V., orientation, 457
Tylos, 444, 445, 446

Ucatangeri, 444
Uchida, T., orientation, 465
Ultraviolet light: from sky, 376, 382, 398–400; reflected by flowers, 482–491
Unhock, N., dance, 5–6
Unterholzner, stimulatory feeding, 258

Vaccinium myrtillus, 249, 510, 511
Vanessa, 450, 517
Velia, 218, 428, 444, 446, 449, 453
Verheijen-Voogd, Christine, mandibular glands, 288
Vetch, 258
Vicia: vilossa, 258; *faba,* 512
Visual organs: response to light, 404–406; eye structure, 423–428, 478–479; response to vibration, 436
Viswanathan, H., *A. dorsata,* 301
Vogel, Berta, sorbitol, 237, 515
Vogel, R., antennae, 496, 498
Vogel, St., nectar markers, 489
Vowles, D. M.: polarized light, 415; vibratory perception, 440
Vries, H. de: brightness pattern, 409; polarized light, 418, 433

Waggling run: defined, 98n; measurement, 102–104; distance index, 127, 130, 232; solar orientation, 130–132; sound, 286
Wagner, W., sound, 314
Wahl, O.: foraging factors, 252, 254; temporal memory, 254; diurnal rhythm, 354–355
Wald, G., eye structure, 432
Walther, J. B., color perception, 477
Wanke, L., transport experiment, 117
Washizu, Y., visual acuity, 478
Wasps: communication, 315–316; orientation, 331, 333, 470–471, 520
Water: flights over, 111–113; need, 265; use, 265; dance stimulant, 266; substitutes, 266, 277; sugar vs. pure, 267; age of bee, 267
Water bug, color vision, 477

Water strider: orientation, 218, 446–447; polarized light, 416, 428, 444
Waterman, T. H.: orientation, 414; polarized light, 415, 444
Watson, J. B., orientation, 331
Wax, 50, 54
Weather: wind on dance tempo, 79–81, 127; stepwise experiment, 88–96; flight speed, 109; distortion, 157–159; orientation, 186–195; bee sensitivity, 245
Weaver, N., consistency, 249
Webb, H. M., orientation, 463
Weber-Fechner law, 122
Weighted bees, 239–240, 241, 282
Weis, Ilse, taste organs, 518
Weissweiler, A., floral colors, 482
Wellington, W. G., orientation, 406
Wellmann, P., ultraviolet light, 376
Wells, M. J., orientation, 414, 441
Wendler, Lotte, polarized light, 417
Wenner, A. M.: dance sound, 58, 97, 98; flight duration, 114; distance curve, 124; auditory testing, 286–287
Werner, Grete: temperature, 76; drugged feeding, 83, 354
Weyer, F., senility, 74
Whitethroats, 454
Wiechert, Elsbeth, hive placement, 467
Wiedemann, Ingrid, visual acuity, 478
Williams, C. B., orientation, 450
Wilson, E. O., odor trails, 316
Wiltschko, W., migrations, 454–455
Wisby, W. J., orientation, 455
Wittekindt, W.: photographing dance, 25; tail-wagging dance, 45n; nocturnal dance, 245–246; "marathon" dance, 350
Wojtusiak, R., odor, 54
Wolbarsht, M. L., taste organs, 517
Wolf, E.: transport experiment, 118; orientation, 134, 331, 348, 465; antennae, 144; visual acuity, 478, 480
Wolken, J. J., eye structure, 421–423, 432
Wykes, G. R.: nectar components, 511; taste threshold, 514, 516

Xylocopa, 366

Yasamuzi, G., eye structure, 421

Zander, E.: flight range, 66; gravity perception, 145
Zerrahn, Gertrud, visual acuity, 478, 480
Zeutschel, B., polarized light, 421, 422
Ziegenspeck, H., floral colors, 482
Zimmermann, M., nectar glands, 511
Zmarlicki, C., olfactory perceptions, 288
Zonana, H. V., optic structure, 432
Zootermopsis nevadensis, 317
Zwehl, Vera von: polarized light, 417, 418, 420; color sensitivity, 476, 477